Principles of Measurement Systems

Third edition

Addison Wesley Longman Limited
Edinburgh Gate
Harlow, Essex CM20 2JE
United Kingdom
and Associated Companies throughout the world

First published 1983
Second edition 1988
Third edition 1995
Reprinted 1995, 1996 and 1997

British Library Cataloguing in Publication Data
A catalogue entry for this book is available from the British Library.

ISBN 0-582-23779-3

Library of Congress Cataloging-in-Publication Data
Bentley, John P., 1943-
Principles of measurement systems / John P. Bentley, -- 3rd ed.
p. cm.
Includes bibliographical references and index
1. Physical instruments. 2. Physical measurements.
3. Engineering instruments. 4. Automatic control. I. Title.
QC53. B44 1995
530.8--dc20 94-29098
CIP

Set by 4 in 10/12 pt Compugraphic Times

Produced by Longman Singapore Publishers (Pte) Ltd.
Printed in Singapore

To Pauline, Sarah and Victoria

Contents

Preface to the third edition

Measurement is an essential activity in every branch of technology and science. We will need to know the speed of a car, the temperature of our working environment, the flow rate of liquid in a pipe, the amount of oxygen dissolved in river water. It is important, therefore, that the study of measurement forms part of engineering and science courses in further and higher education. The aim of this book is to provide the fundamental principles of measurement which underly these studies.

The book treats measurement as a coherent and integrated subject by presenting it as the study of measurement systems. A measurement system is an information system which presents an observer with a numerical value corresponding to the variable being measured. A given system may contain four types of elements; sensing, signal conditioning, signal processing and data presentation elements.

The book is divided into three parts. *Part A* (*Chapters 1 to 7*) examines general systems principles. This part begins by discussing the static and dynamic characteristics that individual elements may possess and how they are used to calculate the overall system measurement error, under both steady and unsteady conditions. In later chapters, the principles of loading and two-port networks, the effects of interference and noise on system performance, reliability, maintainability and choice using economic criteria are explained. *Part B* (*Chapters 8 to 11*) examines the principles, characteristics and applications of typical sensing, signal conditioning, signal processing and data presentation elements in wide current use. *Part C* (*Chapters 12 to 18*) examines a number of specialised measurement systems which have important industrial applications. These are flow measurement systems, intrinsically safe systems, heat transfer, optical, ultrasonic, gas chromatography, data acquisition and communications systems.

The third edition has been extensively extended and updated to reflect developments in technology since the second edition was published in 1988. Chapter 5 has been strengthened by including material on **two-port networks**. The basic principles of **reliability and maintainability** in Chapter 7 have been extended and there is new material on **failure rate data** and models. The section on **current transmitters** in Chapter 9 has been updated and extended and now includes material on intelligent or **smart transmitters**. Developments in **microcomputers** and increased use of **high-level languages** are reflected in an expanded Chapter 10. This chapter also gives examples of how **signal processing** elements perform steady-state and dynamic compensation calculations. Chapter 13 now looks at **intrinsically safe systems**, both pneumatic and electronic, which can be safely used in explosive atmospheres. A section on **optical interferometers** has been added to Chapter 15 and the material in Chapter 16 on ultrasonic **pulse reflection systems** updated. Finally new material on communications systems and **protocols**, including Fieldbus, has been added to Chapter 18.

Each chapter in the book is clearly divided into sections, the topics to be covered are introduced at the beginning and reviewed in a conclusion at the end. Basic and important equations are highlighted and a number of references are given at the end of each chapter, these should provide useful supplementary reading. The book contains about 250 line diagrams and tables and about 100 problems. At the end of the book there are answers to all the numerical problems and a comprehensive index.

This book is primarily aimed at students taking modules in measurement and instrumentation as part of degree courses in instrumentation/control, mechanical, manufacturing, electrical, electronic, chemical engineering and applied physics. Much of the material will also be helpful to lecturers and students involved in HNC/HND courses in industrial measurement and control. The book should also be useful to professional engineers and technicians engaged in solving practical measurement problems.

I would like to thank academic colleagues, industrial contacts and countless students for their helpful comments and criticism over many years. Thanks are again especially due to my wife Pauline, for her constant support and patient typing of the manuscript.

John P. Bentley
Guisborough 1994

Acknowledgements

The author gratefully acknowledges permission for the reproduction of illustrations, tables and data from the following bodies:
National Physical Laboratory — Tables 2.3, 2.4 and 2.5 and Fig. 15.22(a), (b)
The Institute of Measurement and Control — Fig. 2.15(a), Tables 5.1 and Table 7.3
Council of the Institution of Mechanical Engineers — Fig. 2.15(b)
The Institution of Chemical Engineers — Table 7.1
Professor F. Lees, Loughborough University — Table 7.2
Mullard Ltd — Fig. 8.2(a)
British Standards Institution — Table 8.2, Fig. 11.1(b), Table 12.1 and Fig. 12.8
Oxford University Press — Fig. 8.14(a) and Table 8.3
The Institute of Physics — Figs 8.1(b) and 16.14 and Table 7.1
Transducer Technology — Figs 8.16(b) and 8.17
John Wiley and Sons Ltd — Fig. 8.18(b)
Kent Industrial Measurements Ltd, E.I.L. Analytical Instruments — Fig. 8.18(c)
E.D.T. Research — Table 8.4
Hodder and Stoughton — Figs 10.11 and 10.12
Intel Corporation — Table 10.5
Dr Martin Healey, U.C.S.W., Cardiff — Fig. 11.3(b)
Kent Process Control Ltd, Flow Products — Figs 12.11(a) and 12.13
McGraw-Hill Book Co. (USA.) — Figs 15.10 and 15.12(b)
The Open University Press — Fig. 11.8(c)
Butterworth and Co. — Table 16.2
Academic Press — Table 16.1
Pergamon Press Ltd — Fig. 17.5
Kent Automation Systems Ltd — Error Detection System in Section 18.5.2

Part A

General principles

1

The General Measurement System

The purpose of a measurement system is to present an observer with a numerical value corresponding to the variable being measured. In general this numerical value or measured value does not equal the true value of the variable. Thus the measured value of the flow rate down a pipe as presented on a pointer-scale indicator may be $11.0\ m^3\ hr^{-1}$, whereas the true flow may be $11.2\ m^3\ hr^{-1}$; the measured speed of an engine as indicated on a digital display may be 3140 r.p.m. whereas the true speed may be 3133 r.p.m. The problems involved in trying to establish the true value of a variable will be discussed in the next chapter. For the present, it is sufficient to realise that the input to the measurement system is the true value of the variable and the output is the measured value (see Fig. 1.1).

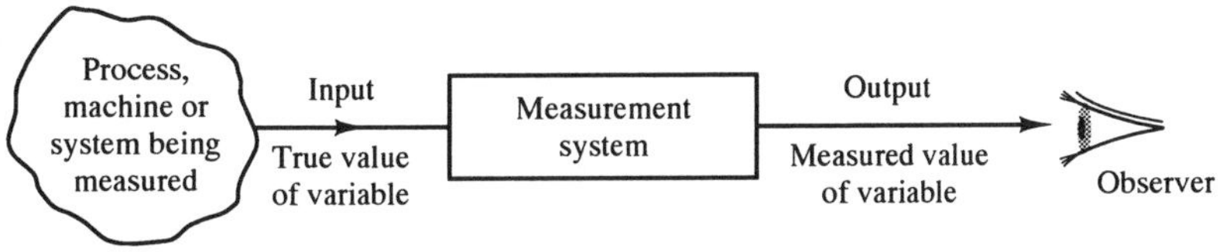

Fig. 1.1 Purpose of measurement system

The measurement system consists of several elements or blocks. It is possible to identify four types of elements, although in a given system one type of element may be missing or may occur more than once. The four types are shown in Fig. 1.2 and can be defined as follows.

Sensing element

This is in contact with the process and gives an output which depends in some way on the variable to be measured. Examples are:

thermocouple where millivolt e.m.f. depends on temperature;
strain gauge where resistance depends on mechanical strain;
orifice plate where pressure drop depends on flow rate.

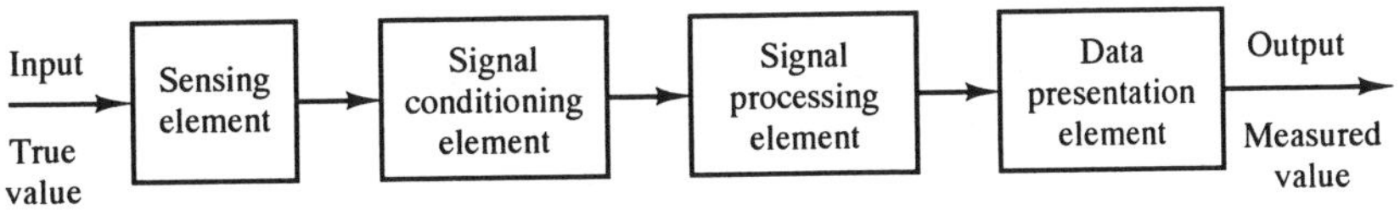

Fig. 1.2 General structure of measurement system

If there is more than one sensing element in a system, the element in contact with the process is termed the primary sensing element, the others secondary sensing elements.

Signal conditioning element

This takes the output of the sensing element and converts it into a form more suitable for further processing, usually a d.c. voltage, d.c. current or frequency signal. Examples are:

deflection bridge which converts an impedance change into a voltage change;
amplifier which amplifies millivolts to volts;
oscillator which converts an impedance change into a variable frequency voltage.

Signal processing element

This takes the output of the conditioning element and converts it into a form more suitable for presentation. Examples are:

analogue-to-digital converter which converts a voltage into a digital form for input to a computer;
a microcomputer which calculates the measured value of the variable from the incoming digital data.

Typical calculations are:

the computation of total mass of product gas from flow rate and density data;
the integration of chromatograph peaks to give the composition of a gas stream;
correction for sensing element non-linearity.

Data presentation element

This presents the measured value in a form which can be easily recognised by the observer. Examples are:

a simple pointer-scale indicator;
chart recorder;
alphanumeric display;
visual display unit.

Figure 1.3 shows a system for weight measurement which incorporates all the elements mentioned above.

The word '**transducer**' is commonly used in connection with measurement and instrumentation. This is a manufactured package which gives an output voltage (usually) corresponding to an input variable such as pressure or acceleration. We see therefore that such a transducer may incorporate both sensing and signal conditioning elements; for example a weight transducer would incorporate the first four elements shown in Fig. 1.3.

It is also important to note that each element in the measurement system may itself be a system made up of simpler components. Chapters 8 to 11 discuss typical examples of each type of element in common use.

Fig. 1.3 Weight measurement system

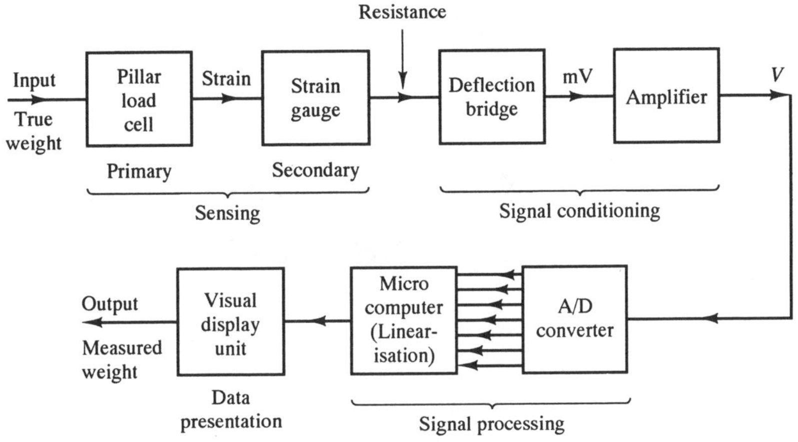

Fig. 1.4 Block diagram symbols

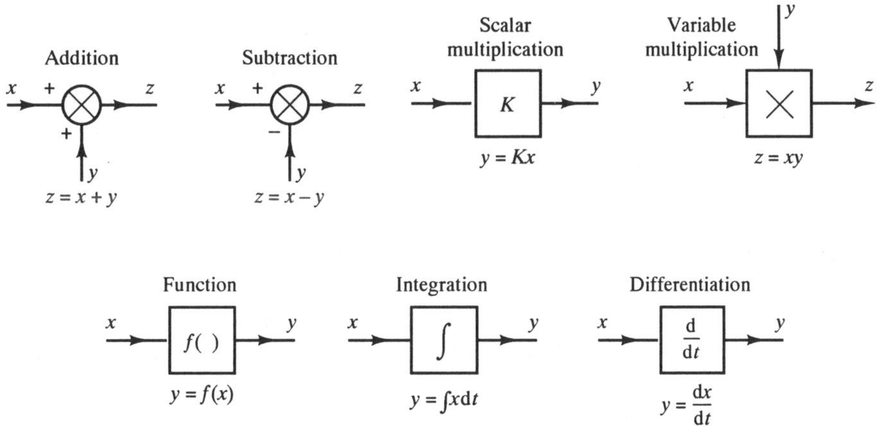

A block diagram approach is very useful in discussing the properties of elements and systems. Figure 1.4 shows the main block diagram symbols used in this book.

Conclusions

This chapter has defined the purpose of a measurement system and explained that a system contains four types of element: **sensing, signal conditioning, signal processing** and **data presentation** in display elements. The function of each type has been defined and examples given.

2

Static Characteristics of Measurement System Elements

In the previous chapter we saw that a measurement system consists of different types of elements. The following chapters discuss the characteristics that typical elements may possess and their effect on the overall performance of the system. This chapter is concerned with static or steady-state characteristics; these are the relationships which may occur between the output O and input I of an element when I is either at a constant value or changing slowly.

2.1 Systematic characteristics

Systematic characteristics are those that can be exactly quantified by mathematical or graphical means. These are distinct from statistical characteristics which cannot be exactly quantified and are discussed in Section 2.3.

Range

The input of an element is specified by the minimum and maximum values of I, i.e. I_{MIN} to I_{MAX}. The output range is specified by the minimum and maximum values of O, i.e. O_{MIN} to O_{MAX}. Thus a pressure transducer may have an input range of 0 to 10^4 Pa and an output range of 4 to 20 mA; a thermocouple may have an input range of 100 to 250 °C and an output range of 4 to 10 mV.

Span

Span is the maximum variation in input or output, i.e. input span is $I_{MAX} - I_{MIN}$, and output span is $O_{MAX} - O_{MIN}$. Thus in the above examples the pressure transducer has an input span of 10^4 Pa and an output span of 16 mA; the thermocouple has an input span of 150 °C and an output span of 6 mV.

Linearity

An element is said to be linear if corresponding values of I and O lie on a straight line. The **ideal straight line** connects the minimum point $A(I_{MIN}, O_{MIN})$ to maximum point $B(I_{MAX}, O_{MAX})$ and therefore has the equation:

$$O - O_{MIN} = \left[\frac{O_{MAX} - O_{MIN}}{I_{MAX} - I_{MIN}}\right](I - I_{MIN}) \qquad [2.1]$$

Ideal straight line equation

$$O_{IDEAL} = KI + a \qquad [2.2]$$

where:

$$K = \text{ideal straight-line slope} = \frac{O_{MAX} - O_{MIN}}{I_{MAX} - I_{MIN}}$$

and

$$a = \text{ideal straight-line intercept} = O_{MIN} - KI_{MIN}$$

Thus the ideal straight line for the above pressure transducer is:

$$O = 1.6 \times 10^{-3}\, I + 4.0$$

Non-linearity

In many cases the straight-line relationship defined by eqn [2.2] is not obeyed and the element is said to be **non-linear**. Non-linearity can be defined (Fig. 2.1) in terms of a function $N(I)$ which is the difference between actual and ideal straight-line behaviour.

i.e.

$$N(I) = O(I) - (KI + a) \qquad [2.3]$$

or

$$O(I) = KI + a + N(I) \qquad [2.4]$$

Non-linearity is often quantified in terms of the maximum non-linearity $\hat{N}$ expressed as a percentage of full-scale deflection (f.s.d.) i.e. as a percentage of span. Thus:

$$\text{Max. non-linearity as a percentage of f.s.d.} = \frac{\hat{N}}{O_{MAX} - O_{MIN}} \times 100\% \qquad [2.5]$$

Fig. 2.1 Definition of non-linearity

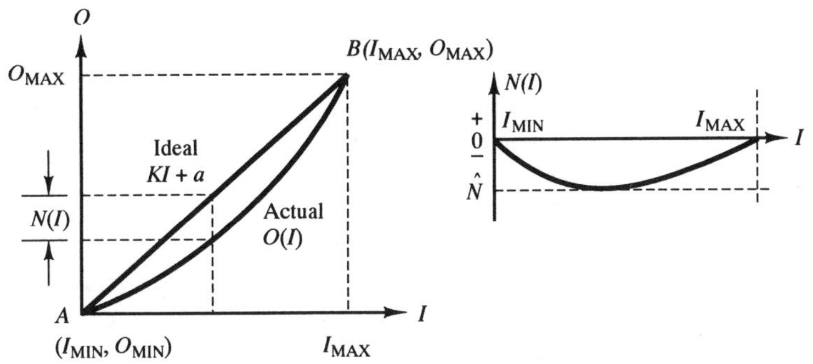

In many cases $O(I)$ and therefore $N(I)$ can be expressed as a polynomial in I i.e.:

$$O(I) = a_0 + a_1 I + a_2 I^2 + \ldots + a_q I^q + \ldots + a_m I^m = \sum_{q=0}^{q=m} a_q I^q \qquad [2.6]$$

An example is the temperature variation of the thermoelectric e.m.f. at the junction of two dissimilar metals. For a copper–constantan (Type T) thermocouple junction, the first four terms in the polynomial relating e.m.f. $E(T)$, expressed in μV, and junction temperature T°C are:

$$E(T) = 38.74T + 3.319 \times 10^{-2}T^2 + 2.071 \times 10^{-4}T^3 - 2.195 \times 10^{-6}T^4 + \text{higher order terms up to } T^8 \qquad [2.7a]$$

for the range 0 to 400 °C.[1] Since $E = 0\,\mu$V at $T = 0$ °C and $E = 20\,869\,\mu$V at $T = 400$ °C, the equation to the ideal straight line is:

$$E_{\text{IDEAL}} = 52.17T \qquad [2.7b]$$

and the non-linear correction function is:

$$\begin{aligned} N(T) &= E(T) - E_{\text{IDEAL}} \\ &= -13.43T + 3.319 \times 10^{-2}T^2 + 2.071 \times 10^{-4}T^3 \\ &\quad -2.195 \times 10^{-6}T^4 + \text{higher order terms} \end{aligned} \qquad [2.7c]$$

In some cases expressions other than polynomials are more appropriate; for example the resistance $R(T)$ ohms of a thermistor at T°C is given by:

$$R(T) = 0.04 \exp\left(\frac{3300}{T + 273}\right)$$

Sensitivity

This is the rate of change of O with respect to I, i.e. $dO/dI = K + dN/dI$. Thus for an ideal element $dO/dI = K$, i.e. for the above pressure transducer $dO/dI = 1.6 \times 10^{-3}$ mA/Pa. For the copper–constantan thermocouple the sensitivity dE/dT at T°C is given by:

$$dE/dT = 38.74 + 6.638 \times 10^{-2}T + 6.213 \times 10^{-4}T^2 - 8.780 \times 10^{-6}T^3 + \text{higher order terms} \qquad [2.8]$$

which has an approximate value of 50 μV °C^{-1} at 200 °C.

Environmental effects

In general, the output O depends not only on the signal input I but on environmental inputs such as ambient temperature, atmospheric pressure, relative humidity, supply voltage, etc. Thus if eqn [2.4] adequately represents the behaviour of the element under 'standard' environmental conditions, e.g. 25 °C ambient temperature, 1000 millibars atmospheric pressure, 50% RH and 10 V supply voltage; then the equation must be modified to take account of deviations in environmental conditions from 'standard'. There are two main types of environmental input. A **modifying** input causes the linear sensitivity of the element to change. Thus if I_M is the **deviation**

Fig. 2.2 Effects of modifying and interfering inputs

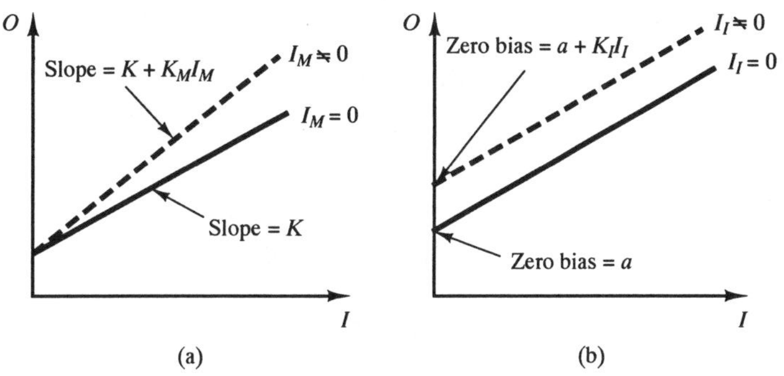

in a modifying environmental input from 'standard' value (I_M is zero at standard conditions), then this produces a change in linear sensitivity from K to $K + K_M I_M$ (Fig. 2.2(a)). An **interfering** input causes the intercept or zero bias of the element to change. Thus if I_I is the **deviation** in an interfering environmental input from 'standard' value (I_I is zero at standard conditions); then this produces a change in zero bias from a to $a + K_I I_I$ (Fig. 2.2(b)). K_M, K_I are referred to as environmental coupling constants or sensitivities. Thus we must now correct eqn [2.4], replacing KI with $(K + K_M I_M)I$ and replacing a with $a + K_I I_I$ to give:

$$O = KI + a + N(I) + K_M I_M I + K_I I_I \qquad [2.9]$$

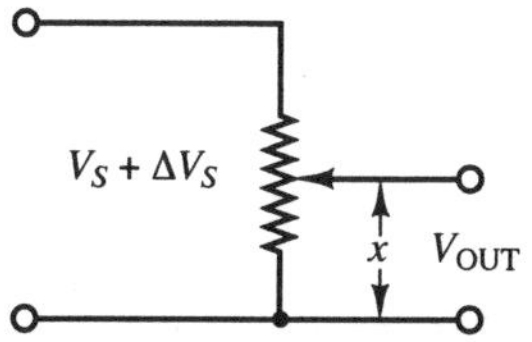

If x is the fractional displacement, then $V_{OUT} = (V_S + \Delta V_S)x$ $= V_S x + \Delta V_S x$

Fig. 2.3

An example of a modifying input is the variation ΔV_s in the supply voltage V_s of the potentiometric displacement sensor shown in Fig. 2.3. An example of an interfering input is provided by variations in the reference junction temperature T_2 of the thermocouple (see following section and Section 8.5).

Hysteresis

For a given value of I, the output O may be different depending on whether I is increasing or decreasing. Hysteresis is the difference between these two values of O (Fig. 2.4) i.e.

$$\text{Hysteresis } H(I) = O(I)_{I\downarrow} - O(I)_{I\uparrow} \qquad [2.10]$$

Again hysteresis is usually quantified in terms of the maximum hysteresis $\hat{H}$ expressed as a percentage of f.s.d., i.e. span. Thus:

$$\text{Maximum hysteresis as a percentage of f.s.d.} = \frac{\hat{H}}{O_{MAX} - O_{MIN}} \times 100\% \qquad [2.11]$$

A simple gear system (Fig. 2.5) for converting linear movement into angular rotation provides a good example of hysteresis. Due to the 'backlash' or 'play' in the gears the angular rotation θ, for a given value of x, is different depending on the direction of the linear movement.

Fig. 2.4 Hysteresis

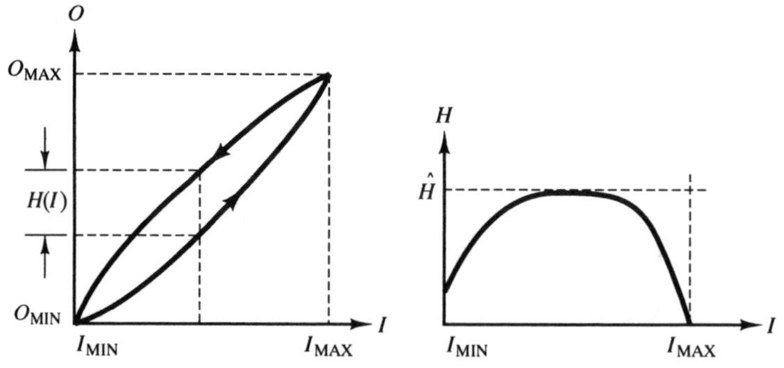

Fig. 2.5 Backlash in gears

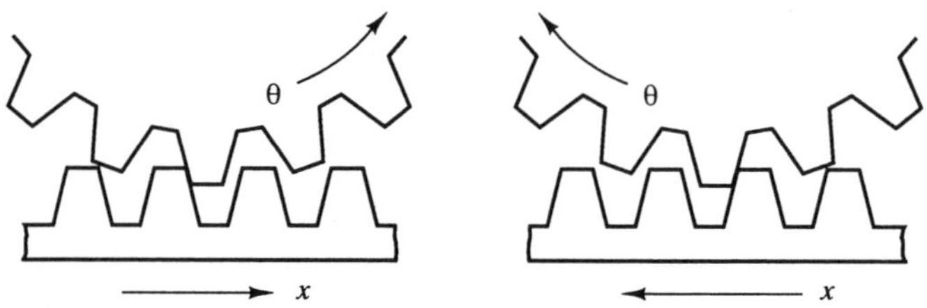

Resolution

Some elements are characterised by the output increasing in a series of discrete steps or jumps in response to a continuous increase in input (Fig. 2.6). Resolution is defined as the largest change in I that can occur without any corresponding change in O. Thus in Fig. 2.6 resolution is defined in terms of the width ΔI_R of the widest step; resolution expressed as a percentage of f.s.d. is thus:

$$\frac{\Delta I_R}{I_{MAX} - I_{MIN}} \times 100\%.$$

A common example is a wire-wound potentiometer (Fig. 2.6); in response to a continuous increase in x the resistance R increases in a series of steps, the size of

Fig. 2.6 Resolution and potentiometer example

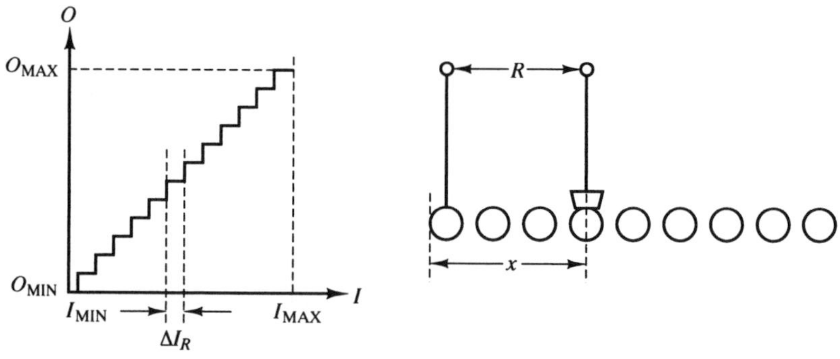

each step being equal to the resistance of a single turn. Thus the resolution of a 100 turn potentiometer is 1 per cent. Another example is an analogue-to-digital converter (Ch. 10); here the output digital signal responds in discrete steps to a continuous increase in input voltage; the resolution is the change in voltage required to cause the output code to change by the least significant bit.

Wear and ageing

These effects can cause the characteristics of an element, e.g. K and a, to change slowly but systematically throughout its life. One example is the stiffness of a spring $k(t)$ decreasing slowly with time due to wear, i.e.

$$k(t) = k_0 - bt \qquad [2.12]$$

where k_0 is the initial stiffness and b is a constant. Another example is the constants a_1, a_2, etc. of a thermocouple, measuring the temperature of gas leaving a cracking furnace, changing systematically with time due to chemical changes in the thermocouple metals.

Error bands

Non-linearity, hysteresis and resolution effects in many modern sensors and transducers are so small that it is difficult and not worthwhile to exactly quantify each individual effect. In these cases the manufacturer defines the performance of the element in terms of error bands (Fig. 2.7). Here the manufacturer states that for any value of I, the output O will be within $\pm h$ of the ideal straight-line value O_{IDEAL}. Here an exact or systematic statement of performance is replaced by a statistical statement in terms of a probability density function $p(O)$. In general a probability density function $p(x)$ is defined so that the integral $\int_{x_1}^{x_2} p(x)\,dx$ (equal to the area under the curve in Fig. 2.8 between x_1 and x_2) is the probability P_{x_1,x_2} of x lying between x_1 and x_2 (section 6.2).
In this case the probability density function is rectangular (Fig. 2.7) i.e.

$$p(O)\begin{cases} = \dfrac{1}{2h} & O_{IDEAL} - h \le O \le O_{IDEAL} + h \\ = 0 & O > O_{IDEAL} + h \\ = 0 & O_{IDEAL} - h > O \end{cases} \qquad [2.13]$$

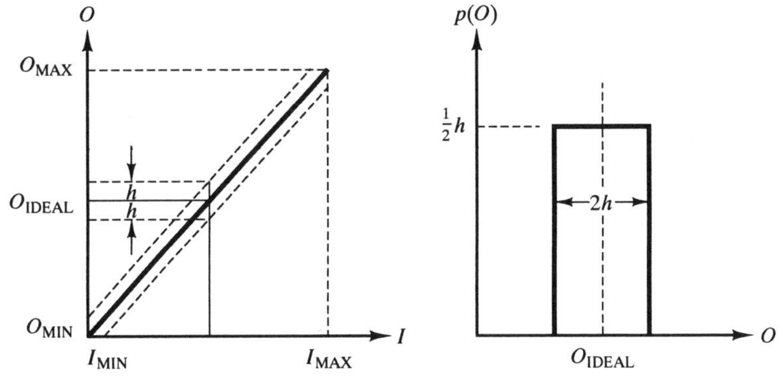

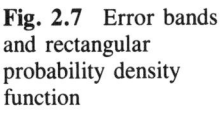

Fig. 2.7 Error bands and rectangular probability density function

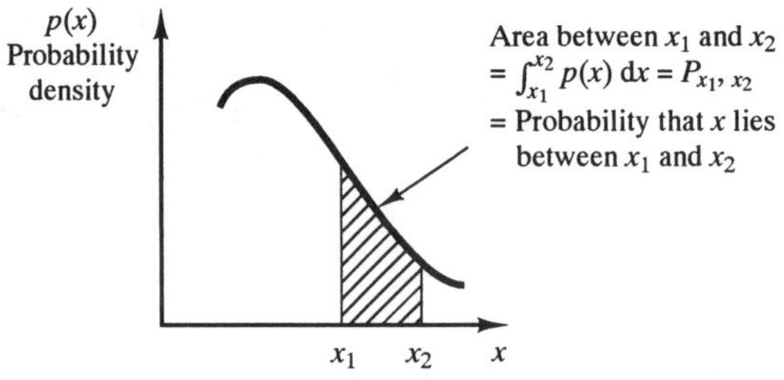

Fig. 2.8 Probability density function

We note that the area of the rectangle is equal to unity: this is the probability of O lying between $O_{IDEAL} - h$ and $O_{IDEAL} + h$.

2.2 Generalised model of a system element

If hysteresis and resolution effects are not present in an element but environmental and non-linear effects are, then the steady-state output O of the element is in general given by eqn [2.9] i.e.:

$$O = KI + a + N(I) + K_M I_M I + K_I I_I \qquad [2.9]$$

Figure 2.9 shows this equation in block diagram form to represent the **static** characteristics of an element. For completeness the diagram also shows the transfer function $G(s)$ which represents the **dynamic** characteristics of the element. The meaning of transfer function will be explained in Chapter 4 where the form of $G(s)$ for different elements will be derived.

Examples of this general model are shown in Fig. 2.10(a), (b), and (c) which summarise the static and dynamic characteristics of a strain gauge, thermocouple and accelerometer respectively.

The strain gauge has an unstrained resistance of 100 Ω and gauge factor (Section 8.1) of 2.0. Non-linearity and dynamic effects can be neglected, but the resistance of the gauge is affected by ambient temperature as well as strain. Here temperature acts as both a modifying and interfering input, i.e. it affects both gauge sensitivity and resistance at zero strain.

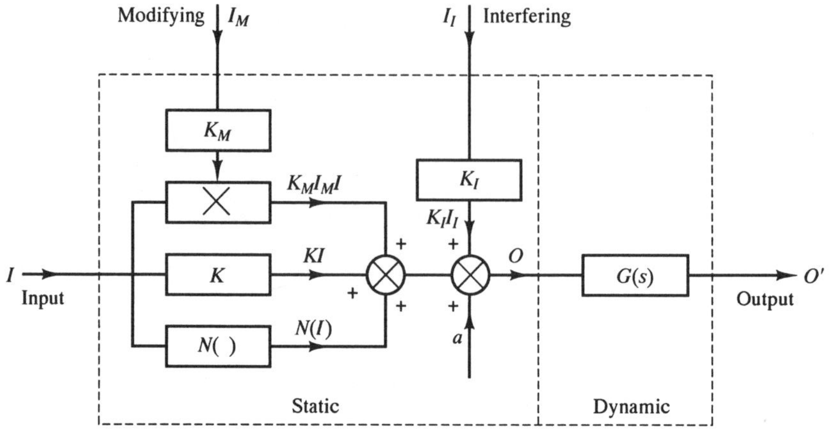

Fig. 2.9 General model of element

Fig. 2.10 Examples of element characteristics
(a) Strain gauge
(b) Copper–constantan thermocouple
(c) Accelerometer

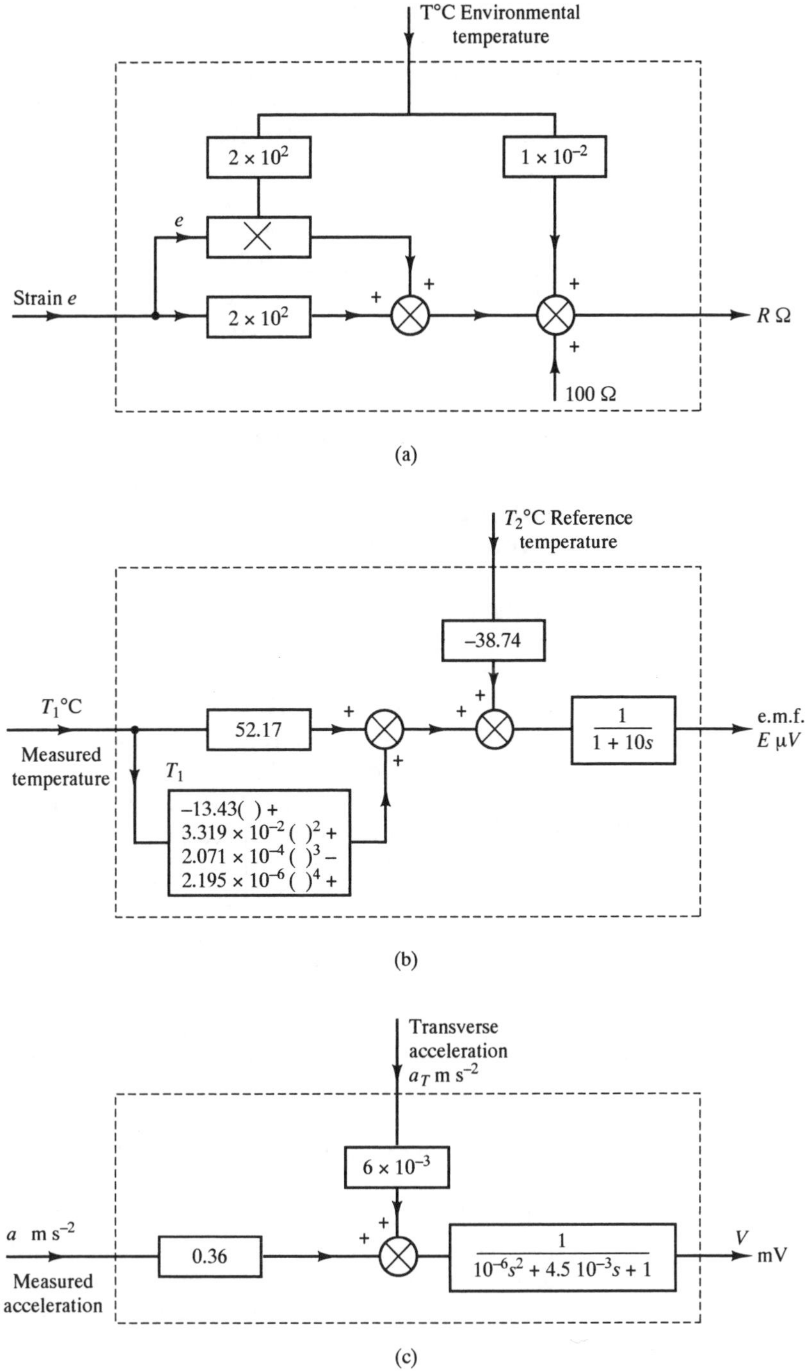

Figure 2.10(b) represents a copper–constantan thermocouple between 0 and 400 °C. The figure is drawn using eqns [2.7] (b) and (c) for ideal straight-line and non-linear correction function; these apply to a single junction. A thermocouple installation consists of two junctions (section 8.5) — a measurement junction at

T_1 °C and a reference junction at T_2 °C. The resultant e.m.f. is the difference of the two junction potentials and thus depends on both T_1 and T_2, i.e. $E(T_1, T_2) = E(T_1) - E(T_2)$; T_2 is thus an interfering input. The model applies to the situation where T_2 is small compared to T_1, so that $E(T_2)$ can be approximated by $38.74\,T_2$, the largest term in eqn [2.7a]. The dynamics are represented by a first-order transfer function of time constant 10 seconds (Chs 4 and 14).

Figure 2.10(c) represents an accelerometer with a linear sensitivity of $0.35\,\text{mV m}^{-1}\,\text{s}^2$ and negligible non-linearity. Any transverse acceleration a_T, i.e. any acceleration perpendicular to that being measured, acts as an interfering input. The dynamics are represented by a second-order transfer function with a natural frequency of 250 Hz and damping coefficient of 0.7 (Chs 4 and 8).

2.3 Statistical characteristics

2.3.1 Statistical variations in the output of a single element with time — repeatability

Suppose that the input I of a single element e.g. a pressure transducer, is held constant, say at 0.5 bar, for several days. If a large number of readings of the output O are taken, then the expected value of 1.0 volt is not obtained on every occasion; a range of values such as 0.99, 1.01, 1.00, 1.02, 0.98, etc., scattered about the expected value, is obtained. This effect is termed a lack of **repeatability** in the element. Repeatability is the ability of an element to give the same output for the same input, when repeatedly applied to it. Lack of repeatability is due to random effects in the element and its environment. An example is the vortex flowmeter (Section 12.2.4): for a fixed flow rate $Q = 1.4 \times 10^{-2}\,\text{m}^3\,\text{s}^{-1}$, we would expect a constant frequency output $f = 209$ Hz. Because the output signal is not a perfect sine wave, but is subject to random fluctuations, the measured frequency varies between 207 and 211 Hz.

The most common cause of lack of repeatability in the output O are random fluctuations with time in the environmental inputs I_M, I_I: if the coupling constants K_M, K_I are non-zero, then there will be corresponding time variations in O. Thus random fluctuations in ambient temperature cause corresponding time variations in the resistance of a strain gauge or the output voltage of an amplifier; random fluctuations in the supply voltage of a deflection bridge affect the bridge output voltage.

By making reasonable assumptions for the probability density functions of the inputs I, I_M, I_I (in a measurement system random variations in the input I to a given element can be caused by random effects in the previous element) the probability density function of the element output O can be found. The most likely probability density function for I, I_M, I_I is the normal or Gaussian distribution function (Fig. 2.11):

Gaussian probability density function

$$p(x) = \frac{1}{\sigma\sqrt{(2\pi)}}\exp\left[-\frac{(x-\bar{x})^2}{2\sigma^2}\right] \qquad [2.14]$$

Fig. 2.11 Gaussian probability density function with $\bar{x} = 0$

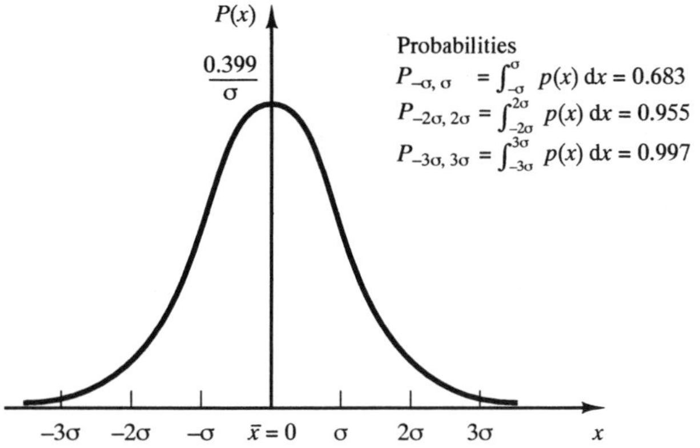

where: $\bar{x}$ = mean or expected value (specifies centre of distribution)
σ = standard deviation (specifies spread of distribution).

Equation [2.9] expresses the independent variable O in terms of the independent variables I, I_M, I_I. Thus if ΔO is a small deviation in O from mean value $\bar{O}$, caused by deviations ΔI, ΔI_M, ΔI_I from respective mean values $\bar{I}$, $\bar{I}_M$, $\bar{I}_I$, then:

$$\Delta O = \left(\frac{\partial O}{\partial I}\right)\Delta I + \left(\frac{\partial O}{\partial I_M}\right)\Delta I_M + \left(\frac{\partial O}{\partial I_I}\right)\Delta I_I \qquad [2.15]$$

Thus ΔO is a linear combination of the variables, ΔI, ΔI_M, ΔI_I; the partial derivatives can be evaluated using equation [2.9]. It can be shown[2] that if a dependent variable y is a linear combination of independent variables x_1, x_2, x_3 i.e.

$$y = a_1x_1 + a_2x_2 + a_3x_3 \qquad [2.16]$$

and if x_1, x_2, x_3 have Gaussian distributions with standard deviations σ_1, σ_2, σ_3 respectively, then the probability distribution of y is also Gaussian with standard deviation σ given by:

$$\sigma = \sqrt{(a_1^2\sigma_1^2 + a_2^2\sigma_2^2 + a_3^2\sigma_3^2)} \qquad [2.17]$$

From [2.15] and [2.17] we see that the standard deviation of ΔO, i.e. of O about mean $\bar{O}$, is given by:

Standard deviation of output for a single element

$$\sigma_0 = \sqrt{\left[\left(\frac{\partial O}{\partial I}\right)^2\sigma_I^2 + \left(\frac{\partial O}{\partial I_M}\right)^2\sigma_{I_M}^2 + \left(\frac{\partial O}{\partial I_I}\right)^2\sigma_{I_I}^2\right]} \qquad [2.18]$$

where σ_I, σ_{I_M}, σ_{I_I} are the standard deviations of the inputs. Thus σ_0 can be calculated using [2.18] if σ_I, σ_{I_M}, σ_{I_I} are known; alternatively if a calibration test (see following section) is being performed on the element then σ_0 can be estimated directly from the experimental results. The corresponding mean or expected value $\bar{O}$ of the element output is given by:

Mean value of output for a single element

$$\bar{O} = K\bar{I} + a + N(\bar{I}) + K_M\bar{I}_M\bar{I} + K_I\bar{I}_I \qquad [2.19]$$

and the corresponding probability density function is:

$$p(O) = \frac{1}{\sigma_O\sqrt{(2\pi)}} \exp\left[\frac{-(O - \bar{O})^2}{2\sigma^2}\right] \qquad [2.20]$$

2.3.2 Statistical variations amongst a batch of similar elements — tolerance

Suppose that a user buys a batch of similar elements, e.g. a batch of one hundred resistance thermometers, from a manufacturer. If he then measures the resistance R_0 of each thermometer at 0 °C he finds that the resistance values are not all equal to the manufacturer's quoted value of 100.0 Ω. A range of values such as 99.8, 100.1, 99.9, 100.0, 100.2 Ω, distributed statistically about the quoted value, is obtained. This effect is due to small random variations in manufacture and is often well represented by the Gaussian probability density function given earlier. In this case we have:

$$p(R_0) = \frac{1}{\sigma_{R_0}\sqrt{(2\pi)}} \exp\left[\frac{-(R_0 - \bar{R}_0)^2}{2\sigma^2_{R_0}}\right] \qquad [2.21]$$

where $\bar{R}_0$ = mean value of distribution = 100 Ω and σ_{R_0} = standard deviation, typically 0.1 Ω. However, a manufacturer may state in his specification that R_0 lies within ±0.15 Ω of 100 Ω for all thermometers, i.e. he is quoting **tolerance** limits of ±0.15 Ω. Thus in order to satisfy these limits he must reject for sale all thermometers with $R_0 < 99.85$ Ω and $R_0 > 100.15$ Ω, so that the probability density function of the thermometers bought by the user now has the form shown in Fig. 2.12.

The user has two choices:

(a) He can design his measurement system using the manufacturer's value of $R_0 = 100.0$ Ω and accept that any individual system, with $R_0 = 100.1$ Ω say, will have a small measurement error. This is the usual practice.
(b) He can perform a calibration test to measure R_0 as accurately as possible for each element in the batch. This theoretically removes the error due to uncertainty in R_0 but is time-consuming and expensive. There is also a small remaining

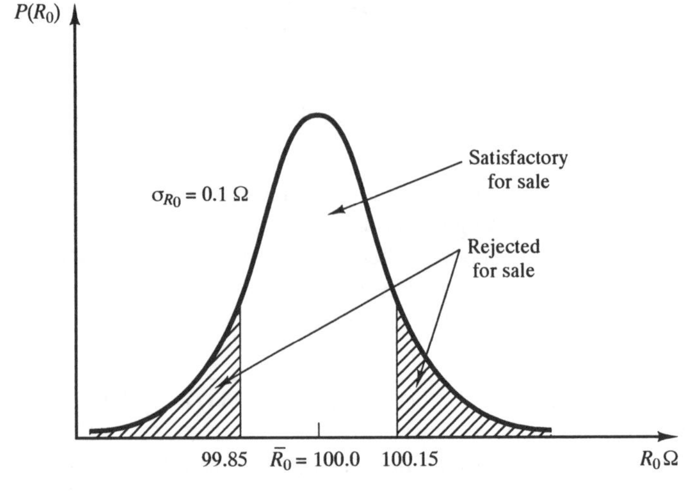

Fig. 2.12 Tolerance limits

uncertainty in the value of R_0 due to the limited accuracy of the calibration equipment.

This effect is found in any batch of 'identical' elements; significant variations are found in batches of thermocouples and thermistors, for example. In the general case we can say that the values of parameters, such as linear gain K and zero bias a, for a batch of elements are distributed statistically about mean values $\bar{K}$, $\bar{a}$.

2.3.3 Summary

In the general case of a batch of several 'identical' elements, where each element is subject to random variations in environmental conditions with time, both inputs I, I_M, I_I and parameters K, a, etc. are subject to statistical variations. If we assume that each statistical variation can be represented by a Gaussian probability density function, then the probability density function of the element output O is also Gaussian i.e.:

$$p(O) = \frac{1}{\sigma_0\sqrt{(2\pi)}}\exp\left[\frac{-(O-\bar{O})^2}{2\sigma_0^2}\right] \qquad [2.20]$$

where the mean value $\bar{O}$ is given by:

Mean value of output for a batch of elements

$$\bar{O} = \bar{K}\bar{I} + \bar{N}(\bar{I}) + \bar{a} + \bar{K}_M\bar{I}_M\bar{I} + \bar{K}_I\bar{I}_I \qquad [2.22]$$

and the standard deviation σ_0 is given by:

Standard deviation of output for a batch of elements

$$\sigma_0 = \sqrt{\left[\left(\frac{\partial O}{\partial I}\right)^2\sigma_I^2 + \left(\frac{\partial O}{\partial I_M}\right)^2\sigma_{I_M}^2 + \left(\frac{\partial O}{\partial I_I}\right)^2\sigma_{I_I}^2 + \left(\frac{\partial O}{\partial K}\right)^2\sigma_K^2 + \left(\frac{\partial O}{\partial a}\right)^2\sigma_a^2 + \ldots\right]} \qquad [2.23]$$

Tables 2.1 and 2.2 summarise the static characteristics of a platinum resistance

Table 2.1 Model for platinum resistance thermometer

Model equation	$R_T = R_0(1 + \alpha T + \beta T^2)$
Individual mean values	$\bar{R}_0 = 100.0\,\Omega$, $\bar{\alpha} = 3.909 \times 10^{-3}$, $\bar{\beta} = -5.897 \times 10^{-7}$ (between 100 and 130 °C)
Individual standard deviations	$\sigma_{R_0} = 4.33 \times 10^{-2}$ Grade I, $\sigma_\alpha = 0.0$ $\sigma_{R_0} = 1.15 \times 10^{-1}$ Grade II, $\sigma_\beta = 0.0$
Partial derivatives	$\left(\frac{\partial R_T}{\partial R_0}\right) = 1.449$ at $T = 117$ °C
Overall mean value	$\bar{R}_T = \bar{R}_0(1 + \bar{\alpha}\bar{T} + \bar{\beta}\bar{T}^2)$
Overall standard deviation	$\sigma_{R_T}^2 = \left(\frac{\partial R_T}{\partial R_0}\right)^2\sigma_{R_0}^2$

Table 2.2 Model for resistance to current converter

Model equation	4 to 20 mA O/P for 138.5 Ω to 149.8 Ω I/P (100 to 130 °C) ΔT_a = deviation of amb. temp from 20 °C $i = KR_T + K_M R_T \Delta T_a + K_I \Delta T_a + a$
Individual mean values	$\bar{K} = 1.4134$, $\bar{K}_M = 1.4134 \times 10^{-4}$, $\bar{K}_I = -1.637 \times 10^{-2}$ $\bar{a} = -191.76$, $\overline{\Delta T}_a = -10$
Individual standard deviations	$\sigma_K = 0.0$, $\sigma_{K_M} = 0.0$, $\sigma_{K_I} = 0.0$ $\sigma_a = 0.24$, $\sigma_{\Delta T_a} = 6.7$
Partial derivatives	$\left(\frac{\partial i}{\partial R_T}\right) = 1.413$, $\left(\frac{\partial i}{\partial \Delta T_a}\right) = 4.11 \times 10^{-3}$, $\left(\frac{\partial i}{\partial a}\right) = 1.00$
Overall mean value	$\bar{i} = \bar{K}\bar{R}_T + \bar{K}_M \bar{R}_T \overline{\Delta T}_a + \bar{K}_I \overline{\Delta T}_a + \bar{a}$
Overall standard deviation	$\sigma_i^2 = \left(\frac{\partial i}{\partial R_T}\right)^2 \sigma_{R_T}^2 + \left(\frac{\partial i}{\partial \Delta T_a}\right)^2 \sigma_{\Delta T_a}^2 + \left(\frac{\partial i}{\partial a}\right)^2 \sigma_a^2$

thermometer and a resistance to current converter. The resistance thermometer is characterised by a small amount of non-linearity and a spread of values of R_0 (resistance at 0 °C) amongst a given batch. The resistance to current is characterised by ambient temperature T_a acting as both a modifying and interfering input. The zero and sensitivity of this element are adjustable. We cannot be certain that the current converter will be set up exactly as required and this is reflected in the non-zero value of σ_a.

2.4 Identification of static characteristics — calibration

2.4.1 Standards

The static characteristics of an element can be found experimentally by measuring corresponding values of the input I, the output O and the environmental inputs I_M, I_I, when I is either at a constant value or changing slowly. This type of experiment is referred to as **calibration**, and the measurement of the variables I, O, I_M, I_I must be accurate if meaningful results are to be obtained. The instruments and techniques used to quantify these variables are referred to as **standards** (Fig. 2.13).

The **accuracy** of a measurement of a variable is the closeness of the measurement

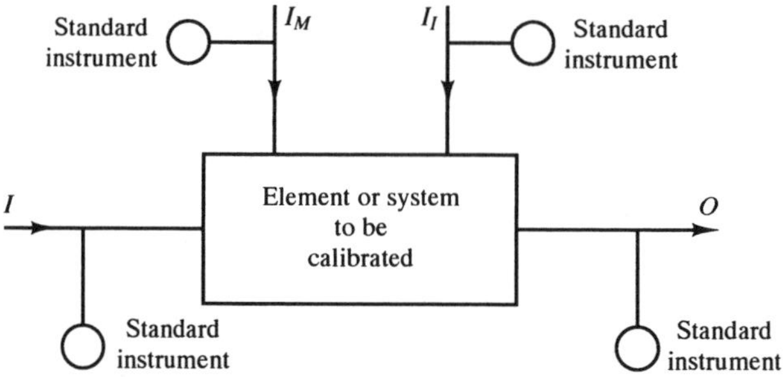

Fig. 2.13 Calibration of an element

Fig. 2.14 Simplified traceability ladder

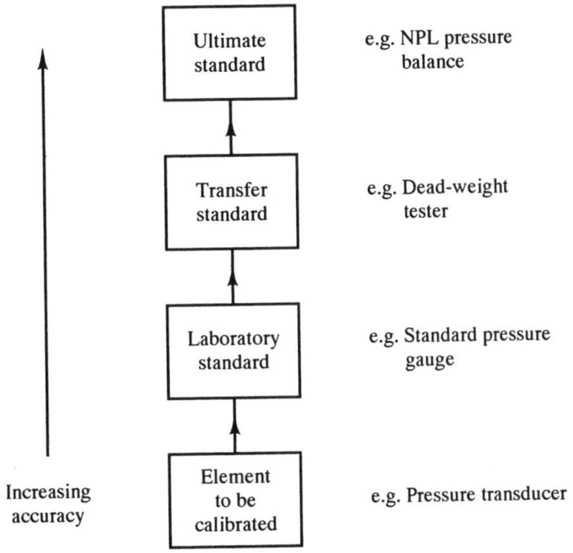

to the true value of the variable. It is quantified in terms of measurement error, i.e. the difference between the measured value and the true value (Ch. 3). Thus the accuracy of a laboratory standard pressure gauge is the closeness of the reading to the true value of pressure. This brings us back to the problem, mentioned in the previous chapter, of how to establish the true value of a variable. We define the true value of a variable as the measured value obtained with a standard of ultimate accuracy. Thus the accuracy of the above pressure gauge is quantified by the difference between the gauge reading, for a given pressure, and the reading given by the ultimate pressure standard. However, the manufacturer of the pressure gauge may not have access to the ultimate standard to measure the accuracy of his products. He can, however, measure the accuracy of his gauges relative to a portable intermediate or transfer standard, e.g. a dead weight pressure tester. The accuracy of the transfer standard must be found by calibration against the ultimate pressure standard. This introduces the concept of a **traceability ladder**, which is shown in simplified form in Fig. 2.14.

The element is calibrated using the laboratory standard, which should itself be calibrated using the transfer standard, and this in turn should be calibrated using the ultimate standard. Each element in the ladder should be significantly more accurate than the one below it.

2.4.2 SI Units

Having introduced the concepts of standards and traceability we can now discuss different types of standards in more detail. The International System of Units (SI) comprises seven base and two supplementary units which are listed and defined in Table 2.3. The units of all physical quantities can be derived from these base and supplementary units. Table 2.4 lists common physical quantities and shows the derivation of their units from the base units. In the United Kingdom the National

Table 2.3 SI base and supplementary units (after National Physical Laboratory 'Units of Measurement' poster 1984[7])

Base units	
Time: second (s)	The second is the duration of 9 192 631 770 periods of the radiation corresponding to the transition between the two hyperfine levels of the ground state of the caesium-133 atom.
Length: metre (m)	The metre is the length of the path travelled by light in vacuum during a time interval of 1/299 792 458 of a second.
Mass: kilogram (kg)	The kilogram is the unit of mass; it is equal to the mass of the international prototype of the kilogram.
Electric current: ampere (A)	The ampere is that constant current which, if maintained in two straight parallel conductors of infinite length, of negligible circular cross-section, and placed 1 metre apart in vacuum, would produce between these conductors a force equal to 2×10^{-7} newton per metre of length.
Thermodynamic temperature: kelvin (K)	The kelvin, unit of thermodynamic temperature, is the fraction 1/273.16 of the thermodynamic temperature of the triple point of water.
Amount of substance: mole (mol)	The mole is the amount of substance of a system which contains as many elementary entities as there are atoms in 0.012 kilogram of carbon 12.
Luminous intensity: candela (cd)	The candela is the luminous intensity, in a given direction, of a source that emits monochromatic radiation of frequency 540×10^{12} hertz and that has a radiant intensity in that direction of (1/683) watt per steradian.
Supplementary units	
Plane angle: radian (rad)	The radian is the plane angle between two radii of a circle which cut off on the circumference an arc equal in length to the radius.
Solid angle: steradian (sr)	The steradian is the solid angle which, having its vertex in the centre of a sphere, cuts off an area of the surface of the sphere equal to that of a square with sides of length equal to the radius of the sphere.

Physical Laboratory (NPL) is responsible for the physical realisation of all of the base units and many of the derived units mentioned above. The NPL is therefore the custodian of ultimate or primary standards in the UK. There are secondary standards held at British Calibration Service (BCS) centres. These have been calibrated against NPL standards and are available to calibrate transfer standards.

At NPL, the **metre** is realised using the wavelength of the 633 nm radiation from an iodine stabilised helium-neon laser. The reproducibility of this primary standard is about 3 parts in 10^{11} and the wavelength of the radiation has been accurately related to the definition of the metre in terms of the velocity of light. The primary standard is used to calibrate secondary laser interferometers which are in turn used to calibrate precision length bars, gauges and tapes. A simplified traceability ladder for length[3] is shown in Fig. 2.15(a).

The international prototype of the **kilogram** is made of platinum-iridium and is kept at the International Bureau of Weights and Measures (BIPM) in Paris. The British national copy is kept at NPL and is used in conjunction with a precision balance to calibrate secondary and transfer kilogram standards. Figure 2.15(b) shows a simplified traceability ladder for mass and weight.[4] The weight of a mass m is the force mg it experiences under the acceleration of gravity g; thus if the local value of g is known accurately, then a force standard can be derived from mass standards. At NPL dead-weight machines covering a range of forces from 450 N to 30 MN are used to calibrate strain-gauge load cells and other weight transducers.

Table 2.4 SI derived units (after National Physical Laboratory Units of Measurement poster, 1984[7])

Examples of SI derived units expressed in terms of base units

Quantity	*SI units* Name	Symbol
area	square metre	m^2
volume	cubic metre	m^3
speed, velocity	metre per second	$m\ s^{-1}$
acceleration	metre per second squared	$m\ s^{-2}$
wave number	1 per metre	m^{-1}
density, mass density	kilogram per cubic metre	$kg\ m^{-3}$
specific volume	cubic metre per kilogram	$m^3\ kg^{-1}$
current density	ampere per square metre	$A\ m^2$
magnetic field strength	ampere per metre	$A\ m^{-2}$
concentration (of amount of substance)	mole per cubic metre	$mol\ m^{-3}$
luminance	candela per square metre	$cd\ m^{-2}$

SI derived units with special names

Quantity	*SI unit* Name	Symbol	Expression in terms of other units	Expression in terms of SI base units
frequency	hertz	Hz		s^{-1}
force	newton	N		$m\ kg\ s^{-2}$
pressure, stress	pascal	Pa	N/m^2	$m^{-1}\ kg\ s^{-2}$
energy, work, quantity of heat	joule	J	Nm	$m^2\ kg\ s^{-2}$
power, radiant flux	watt	W	J/s	$m^2\ kg\ s^{-3}$
electric charge, quantity of electricity	coulomb	C		sA
electric potential, potential difference electromotive force	volt	V	W/A	$m^2\ kg\ s^{-3}\ A^{-1}$
capacitance	farad	F	C/V	$m^{-2}\ kg^{-1}\ s^4\ A^2$
electric resistance	ohm	Ω	V/A	$m^2\ kg\ s^{-3}\ A^{-2}$
electric conductance	siemens	S	A/V	$m^{-2}\ kg^{-1}\ s^3\ A^2$
magnetic flux	weber	Wb	Vs	$m^2\ kg\ s^{-2}\ A^{-1}$
magnetic flux density	tesla	T	Wb/m^2	$kg\ s^{-2}\ A^{-1}$
inductance	henry	H	Wb/A	$m^2 kg s^{-2} A^{-2}$
Celsius temperature	degree Celsius	°C		K
luminous flux	lumen	lm		cd sr
illuminance	lux	lx	lm/m^2	m^{-2} cd sr
activity (of a radionuclide)	becquerel	Bq		s^{-1}
absorbed dose, specific energy imparted, kerma, absorbed dose index	gray	Gy	J/kg	$m^2 s^{-2}$
dose equivalent, dose equivalent index	sievert	Sv	J/kg	$m^2\ s^{-2}$

(continued)

The **ampere** has traditionally been the electrical base unit and has been realised at NPL using the Ayrton-Jones current balance; here the force between two current-carrying coils is balanced by a known weight. The accuracy of this method is limited by the large dead-weight of the coils and formers and the many length measurements necessary. For this reason the two electrical base units are now chosen to be the **farad** and the **volt** (or watt); the other units such as the ampere, ohm, henry, joule are derived from these two base units with time or frequency units, using Ohm's law where necessary. The farad is realised using a calculable capacitor based on the Thompson–Lampard theorem. Using a.c. bridges, capacitance and frequency standards can then be used to calibrate standard resistors. The primary standard for the volt is based on the Josephson effect in superconductivity; this is used to calibrate secondary voltage standards, usually saturated Weston cadmium cells. The ampere can then be realised using a modified current balance. As before, the force due to a current I is balanced by a known weight mg, but also a separate measurement is

Table 2.4 *(continued)*

Examples of SI derived units expressed by means of special names

Quantity	*SI unit* Name	Symbol	Expression in terms of SI base units
dynamic viscosity	pascal second	Pa s	m^{-1} kg s^{-1}
moment of force	newton metre	Nm	m^2 kg s^{-2}
surface tension	newton per metre	N/m	kg s^{-2}
heat flux density, irradiance	watt per square metre	W/m^2	kg s^{-3}
heat capacity, entropy	joule per kelvin	J/K	m^2 kg s^{-2} K^{-1}
specific heat capacity, specific entropy	joule per kilogram kelvin	J/(kg K)	m^2 s^{-2} K^{-1}
specific energy	joule per kilogram	J/kg	m^2 s^{-2}
thermal conductivity	watt per metre kelvin	w/(m K)	m kg s^{-3} K^{-1}
energy density	joule per cubic metre	J/m^3	m^{-1} kg s^{-2}
electric field strength	volt per metre	V/m	m kg s^{-3} A^{-1}
electric charge density	coulomb per cubic metre	C/m^3	m^{-3} s A
electric flux density	coulomb per square metre	C/m^2	m^{-2} s A
permittivity	farad per metre	F/m	m^{-3} kg^{-1} s^4 A^2
permeability	henry per metre	H/m	m kg s^{-2} A^{-2}
molar energy	joule per mole	J/mol	m^2 kg s^{-2} mol^{-1}
molar entropy, molar heat capacity	joule per mole kelvin	J/(mol K)	m^2 kg s^{-2} K^{-1} mol^{-1}
exposure (X and γ rays)	coulomb per kilogram	C/kg	kg^{-1} s A
absorbed dose rate	gray per second	Gy/s	m^2 s^{-3}

Examples of SI derived units formed by using supplementary units

Quantity	*SI unit* Name	Symbol
angular velocity	radian per second	rad s^{-1}
angular acceleration	radian per second squared	rad s^{-2}
radiant intensity	watt per steradian	W sr^{-1}
radiance	watt per square metre steradian	W m^{-2} sr^{-1}

made of the voltage e induced in the coil when moving with velocity u. Equating electrical and mechanical powers gives the simple equation:

$$eI = m\,g\,u$$

Accurate measurements of m, u and e can be made using secondary standards traceable back to the primary standards of the kilogram, metre, second and volt.

Ideally **temperature** should be defined using the thermodynamic scale, i.e. the relationship $PV = R\theta$ between the pressure P and temperature θ of a fixed volume V of an ideal gas. Because of the limited reproducibility of real gas thermometers the International Practical Temperature Scale (IPTS) was devised. This is shown in Table 2.5 and consists of:

(a) several highly reproducible fixed points corresponding to the melting, boiling or triple points of pure substances under specified conditions;
(b) standard instruments with a known output versus temperature relationship obtained by calibration at fixed points.

The instruments interpolate between the fixed points.

The numbers assigned to the fixed points are such that there is exactly 100 K between the freezing point (273.15 K) and boiling point (373.15 K) of water. This

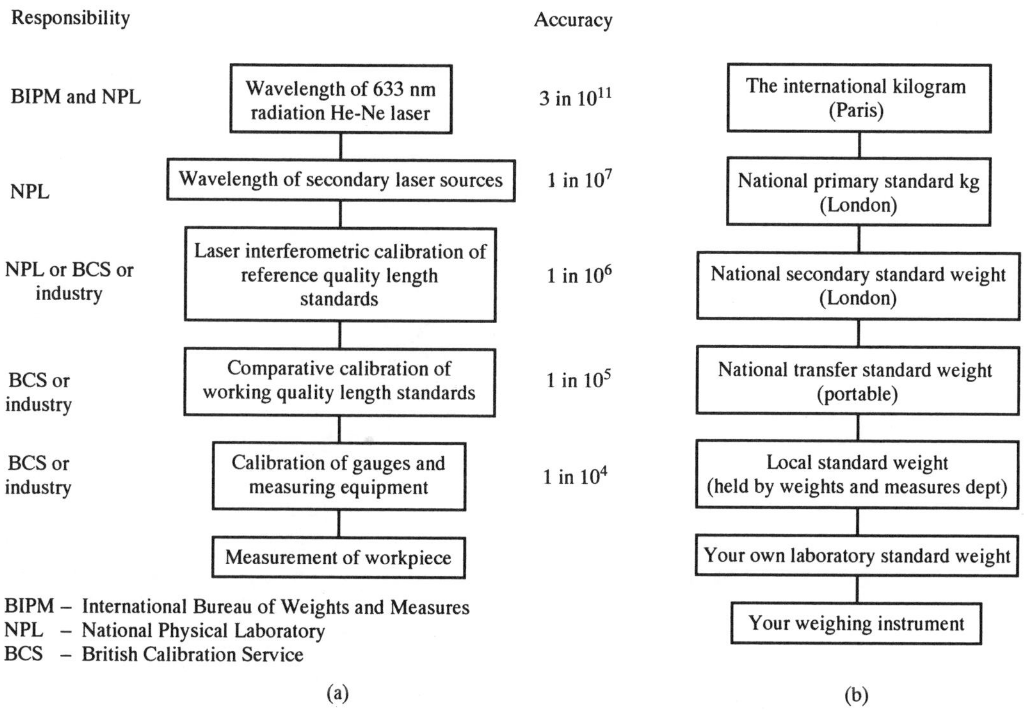

Fig. 2.15 Traceability ladders: (a) Length (adapted from Scarr[3]); (b) Mass (reprinted by permission of the Council of the Institution of Mechanical Engineers from Hayward[4])

means that a change of 1 K is equal to a change of 1 °C on the older Celsius scale. The exact relationship between the two scales is:

$$\theta\,\mathrm{K} = T\,°\mathrm{C} + 273.15$$

The interpolating instruments mentioned in the table are used to calibrate secondary standard instruments; e.g. an interpolating platinum resistance thermometer can be used to calibrate a secondary platinum resistance thermometer.

The standards available for the base quantities, i.e. length, mass, time, current and temperature, enable standards for derived quantities to be realised. This is illustrated in the methods for calibrating liquid flowmeters.[5] The actual flow rate through the meter is found by weighing the amount of water collected in a given time, so that the accuracy of the flow rate standard depends on the accuracies of weight and time standards. Similarly pressure standards can be derived from those for force and area (length).

2.4.3 Experimental measurements and evaluation of results

The calibration experiment is divided into three main parts.

O versus I with $I_M = I_I = 0$

Ideally this test should be held under 'standard' environmental conditions so that

Table 2.5 The International Practical Scale of 1968 (After Physical National Laboratory — booklet on Temperature measurement services[8])

Primary fixed points	Assigned temperature	Interpolating instrument	Method of interpolation
t.p. of e-H_2	13.81 K	PRT	Deviations of the PRT resistance ratio W(T) = R(T)/R(0° C) from a reference function are fitted by simple polynomials in four sub-ranges. The coefficients are obtained by calibration at the fixed points.
b.p. of e-H_2 at 33 330.6 Pa	17.042 K		
b.p. of e-H_2	20.28 K		
b.p. of Ne	27.102 K		
t.p. of O_2	54.361 K		
t.p. of Ar/ b.p. of O_2 (a)	83.798 K 90.188 K		
t.p. of water	273.16 K 0.01 °C	PRT	A quadratic function is specified, with the coefficients determined by calibration at the fixed points.
b.p. of water/ f.p. of Sn (a)	100 °C 231.9681 °C		
f.p. of Zn	419.58 °C		
	630.74 °C		
	(b) 630.74 °C	TC	A quadratic representation of e.m.f. in terms of t_{68}/°C is specified with the constants evaluated by calibration at the two fixed points and at 630.74 °C.
f.p. of Ag	961.93 °C		
f.p. of Au	1064.43 °C		
	1064.43 °C	RP	Defined by Planck's law and realised using a RP calibrated at the Au point

Notes
t.p. = triple point; f.p. = freezing point; b.p. = boiling point, at 101 325 Pa unless otherwise stated; PRT = platinum resistance thermometer; TC = Pt–10%Rh/Pt thermocouple; RP = radiation pyrometer; a = alternatives; b = determined using a PRT; e = equilibrium; Pt = platinum; Rh = rhodium; H_2 = hydrogen; Ne = neon; O_2 = oxygen; Ar = argon; Sn = tin; Zn = zinc; Ag = silver; Au = gold.

$I_M = I_I = 0$; if this is not possible all environmental inputs should be measured. I should be increased slowly from I_{MIN} to I_{MAX} and corresponding values of I and O recorded at intervals of 10 per cent span (i.e. 11 readings), allowing sufficient time for the output to settle out before taking each reading. A further 11 pairs of readings should be taken with I decreasing slowly from I_{MAX} to I_{MIN}. The whole process should be repeated for two further '**ups**' and '**downs**' to yield two sets of data: an '**up**' set $(I_i, O_i)_{I\uparrow}$ and a '**down**' set $(I_j, O_j)_{I\downarrow}$ $i, j = 1, 2, \ldots, n$ $(n = 33)$.

Regression packages are available on most large and medium computers which fit a polynomial i.e. $O(I) = \sum_{q=0}^{q=m} a_q I^q$ to a set of n data points. These packages use a 'least squares' criterion. If d_i is the deviation of the polynomial value $O(I_i)$ from the data value O_i, then $d_i = O(I_i) - O_i$. The program finds a set of coefficients a_0, a_1, a_2, etc. such that the sum of the squares of the deviations i.e. $\Sigma_{i=1}^{i=n} d_i^2$ is a minimum. This involves solving a set of linear equations.[6]

In order to detect any hysteresis, separate regressions should be performed on the two sets of data $(I_i, O_i)_{I\uparrow}$, $(I_j\ O_j)_{I\downarrow}$, to yield two polynomials

$$O(I)_{I\uparrow} = \sum_{q=0}^{q=m} a_q^{\uparrow} I^q \quad \text{and} \quad O(I)_{I\downarrow} = \sum_{q=0}^{q=m} a_q^{\downarrow} I^q$$

If the hysteresis is significant, then the separation of the two curves will be greater than the scatter of data points about each individual curve (Fig. 2.16(a)). Hysteresis $H(I)$ is then given by eqn [2.10], i.e. $H(I) = O(I)_{I\downarrow} - O(I)_{I\uparrow}$. If, however, the scatter of points about each curve is greater than the separation of the curves (Fig. 2.16(b)), then H is not significant and the two sets of data can then be combined to give a single polynomial $O(I)$. The slope K and zero bias a of the ideal straight line joining the minimum and maximum points (I_{MIN}, O_{MIN}) and (I_{MAX}, O_{MAX}) can be found from eqn [2.3]. The non-linear function $N(I)$ can then be found using [2.4]:

$$N(I) = O(I) - (KI + a) \qquad [2.24]$$

Temperature sensors are often calibrated using appropriate fixed points rather than a standard instrument. For example, a thermocouple may be calibrated between 0 and 500 °C by measuring e.m.f.'s at ice, steam and zinc points. If the e.m.f.–temperature relationship is represented by the cubic $E = a_1T + a_2T^2 + a_3T^3$, then the coefficients a_1, a_2, a_3 can be found by solving three simultaneous equations (see Problem 2.1).

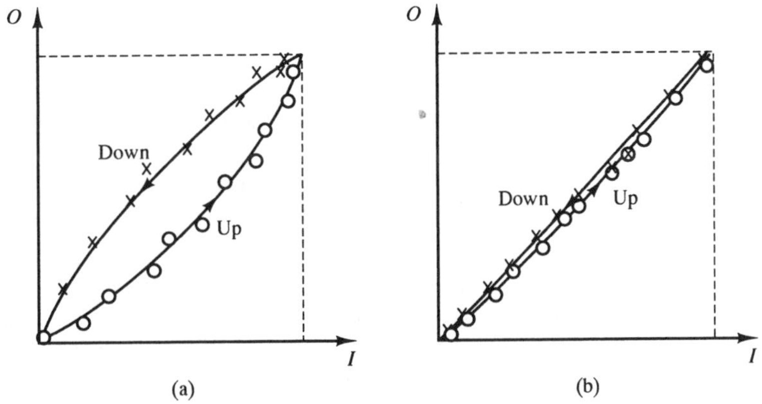

Fig. 2.16
(a) Significant hysteresis
(b) Insignificant hysteresis

O versus I_M, I_I at constant I

We first need to find which environmental inputs are interfering, i.e. which affect the zero bias a. The input I is held constant at $I = I_{MIN}$, and one environmental input is changed by a known amount, the rest being kept at standard values. If there is a resulting change ΔO in O, then the input I_I is interfering and the value of the corresponding coefficient K_I is given by $K_I = \Delta O/\Delta I_I$. If there is no change in O, then the input is not interfering; the process is repeated until all interfering inputs are identified and the corresponding K_I values found.

We now need to identify modifying inputs i.e. those which affect the sensitivity of the element. The input I is held constant at the mid-range value $\frac{1}{2}(I_{MIN} + I_{MAX})$ and each environmental input is varied in turn by a known amount. If a change in input produces a change ΔO in O and is not an interfering input, then it must be a modifying input I_M and the value of the corresponding coefficient K_M is given by:

$$K_M = \frac{1}{I}\frac{\Delta O}{\Delta I_M} = \frac{2}{(I_{MIN} + I_{MAX})}\frac{\Delta O}{\Delta I_M} \quad [2.25]$$

Suppose a change in input produces a change ΔO in O and it has already been identified as an interfering input with a known value of K_I. Then we must calculate a non-zero value of K_M before we can be sure that the input is also modifying. Since

$$\Delta O = K_I \Delta I_{I,M} + K_M \Delta I_{I,M} \frac{(I_{MIN} + I_{MAX})}{2}$$

then

$$K_M = \frac{2}{(I_{MIN} + I_{MAX})}\left[\frac{\Delta O}{\Delta I_{I,M}} - K_I\right] \quad [2.26]$$

Repeatability test

This test should be carried out in the normal working environment of the element e.g. out on the plant, or in a control room, where the environmental inputs I_M, I_I are subject to the random variations usually experienced. The signal input I should be held constant at mid-range value and the output O measured over an extended period, ideally many days, yielding a set of values O_k, $k = 1, 2, \ldots, N$. The mean value of the set can be found using:

$$\bar{O} = \frac{1}{N}\sum_{k=1}^{k=N} O_K \quad [2.27]$$

and the standard deviation (root mean square of deviations from mean) found using:

$$\sigma_0 = \sqrt{\left[\frac{1}{N}\sum_{k=1}^{N}(O_k - \bar{O})^2\right]} \quad (2.28]$$

A histogram of the values O_k should then be plotted in order to estimate the probability density function $p(O)$ and to compare it with the Gaussian form (eqn [2.20]). A repeatability test on a pressure transducer can be used as an example of the construction of a histogram. Suppose that $N = 50$, readings between 0.975 and 1.030 V, are obtained corresponding to an expected value of 1.000 V. The readings are grouped into equal intervals of width 0.005 V and the number in each interval found — e.g. 12, 10, 8, etc. This number is divided by the total number of readings, i.e. 50, to give the probabilities 0.24, 0.20, 0.16, etc. of a reading occurring in a given interval. The probabilities are in turn divided by the interval width 0.005 V to give the probability densities 48, 40, 32 V^{-1} plotted in the histogram (Fig. 2.17). We note that the area of each rectangle represents the probability that a reading lies within the interval, and that the total area of the histogram is equal to unity. The mean and standard deviation are found from [2.27]

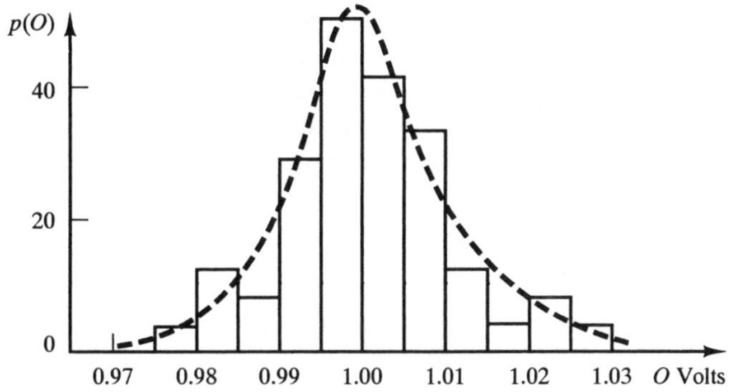

Fig. 2.17 Comparison of histogram with Gaussian probability density function

and [2.28] to be $\bar{O} = 0.999$ V and $\sigma_0 = 0.010$ V respectively. Figure 2.17 also shows a Gaussian probability density function with these values.

Conclusion

The chapter began by discussing the **static** or **steady-state** characteristics of measurement system elements. **Systematic characteristics** such as non-linearity and environmental effects were first explained, this led to the **generalised model** of an element. **Statistical** characteristics i.e. repeatability and tolerance, were then discussed. The last section explained how these characteristics can be measured experimentally i.e. **calibration** and the use and types of **standards**.

References

2.1 B.S. 4937, Part 5 1974 'International Thermocouple Reference Tables', British Standards Institution.

2.2 PARADINE C G, and RIVETT B H P 1966 *Statistical Methods for Technologists*, pp. 64–6, English Universities Press, London.

2.3 SCARR A 1979 'Measurement of length', *Journal Institute Measurement & Control*, 12, July, pp. 265–9.

2.4 HAYWARD A T J 1977 *Repeatability and Accuracy*, p. 34. Mechanical Engineering Publications Ltd, London.

2.5 HAYWARD A T J 1977 Methods of calibrating flowmeters with liquids — a comparative survey, *Transactions Institute Measurement & Control*, pp. 106–16.

2.6 HAWGOOD J 1965 *Numerical Methods in Algol*, pp. 92–103. McGraw-Hill, London.

2.7 NATIONAL PHYSICAL LABORATORY 1984 'Units of Measurement' poster (3rd edn).

2.8 NATIONAL PHYSICAL LABORATORY 1982 Booklet on *Temperature Measurement Services*.

Problems

2.1 The e.m.f. at a thermocouple junction is 645 μV at the steam point, 3375 μV at the zinc point and 9149 μV at the silver point. Given that the e.m.f.–temperature relationship is of the form $E(T) = a_1T + a_2T^2 + a_3T^3$ (T in °C), find a_1, a_2, a_3.

2.2 The resistance $R(\theta)$ of a thermistor at temperature θ K is given by $R(\theta) = \alpha \exp(\beta/\theta)$. Given that the resistance at the ice point ($\theta = 273.15$ K) is 9.00 kΩ and the resistance at the steam point is 0.50 kΩ, find the resistance at 25 °C.

2.3 A displacement sensor has an input range of 0.0 to 3.0 cm and a standard supply voltage $V_s = 0.5$ volts. Using the calibration results given in the table, estimate:

(a) The maximum non-linearity as a percentage of f.s.d.
(b) The constants K_I, K_M associated with supply voltage variations.
(c) The slope K of the ideal straight line.

Displacement x cm	0.0	0.5	1.0	1.5	2.0	2.5	3.0
Output voltage millivolts ($V_s = 0.5$)	0.0	16.5	32.0	44.0	51.5	55.5	58.0
Output voltage millivolts ($V_s = 0.6$)	0.0	21.0	41.5	56.0	65.0	70.5	74.0

2.4 A liquid level sensor has an input range of 0 to 15 cm. Use the calibration results given in the table to estimate the maximum hysteresis as a percentage of f.s.d.

Level h cm	0.0	1.5	3.0	4.5	6.0	7.5	9.0	10.5	12.0	13.5	15.0
O/P volts h increasing	0.00	0.35	1.42	2.40	3.43	4.35	5.61	6.50	7.77	8.85	10.2
O/P volts h decreasing	0.14	1.25	2.32	3.55	4.43	5.70	6.78	7.80	8.87	0.65	10.2

2.5 A repeatability test on a vortex flowmeter yielded the following 35 values of frequency corresponding to a constant flow rate of $1.4 \times 10^{-2}\,\text{m}^3\,\text{s}^{-1}$: 208.6; 208.3 208.7; 208.5; 208.8; 207.6; 208.9; 209.1; 208.2; 208.4; 208.1; 209.2; 209.6; 208.6; 208.5; 207.4; 210.2; 209.2; 208.7; 208.4; 207.7; 208.9; 208.7; 208.0; 209.0; 208.1; 209.3; 208.2; 208.6; 209.4; 207.6; 208.1; 208.8; 209.2; 209.7 Hz.

(a) Using equal intervals of width 0.5 Hz, plot a histogram of probability density values.
(b) Calculate the mean and standard deviation of the data.
(c) Sketch a Gaussian probability density function with the mean and standard deviation calculated in (b) on the histogram drawn in (a).

2.6 A platinum resistance thermometer is used to interpolate between the triple point of water (0 °C), the boiling point of water (100 °C) and the freezing point of zinc (419.6 °C). The corresponding resistance values are 100.0 Ω, 138.5 Ω and 253.7 Ω. The algebraic form of the interpolation equation is:

$$R_T = R_0(1 + \alpha T + \beta T^2)$$

where $R_T\Omega$ = Resistance at T°C
$R_0\Omega$ = Resistance at 0 °C
α, β = constants

Find the numerical form of the interpolation equation.

2.7 The following results were obtained when a pressure transducer was tested in a laboratory under the following conditions:

I Ambient Temperature 20 °C, Supply voltage 10 V (standard)
II Ambient Temperature 20 °C, Supply voltage 12 V
III Ambient Temperature 25 °C, Supply voltage 10 V

Input (barg)	0	2	4	6	8	10
Output (mA)						
I	4	7.2	10.4	13.6	16.8	20
II	4	8.4	12.8	17.2	21.6	28
III	6	9.2	12.4	15.6	18.8	22

(a) Determine the values of K_M, K_I, a, and K associated with the generalised model equation: $O = (K + K_M I_M)I + a + K_I I_I$

(b) Predict an output value when the input is 5 barg, $V_s = 12\,V$ and ambient temperature is 25 °C.

3

The Accuracy of Measurement Systems in the Steady State

In Chapter 1 we saw that the input to a measurement system is the true value of the variable being measured. Also, if the measurement system is complete, the system output is the measured value of the variable. In Chapter 2 we defined accuracy in terms of measurement error i.e. the difference between the measured and true values of a variable. It follows therefore that accuracy is a property of a complete measurement system rather than a single element. Accuracy is quantified using measurement error E where:

$$\begin{aligned} E &= \text{measured value} - \text{true value} \\ &= \text{system output} - \text{system input} \end{aligned} \qquad [3.1]$$

In this chapter we use the static model of a single element, developed previously, to calculate the output and thus the measurement error for a complete system of several elements. The chapter concludes by examining methods of reducing system error.

3.1 Measurement error of a system of ideal elements

Consider the system shown in Fig. 3.1 consisting of n elements in series. Suppose each element is ideal, i.e. perfectly linear and not subject to environmental inputs. If we also assumed the intercept or bias is zero, i.e. $a = 0$, then:

Input/output equation for ideal element with zero intercept

$$O_i = K_i I_i \qquad [3.2]$$

for $i = 1, \ldots, n$, where K_i is the linear sensitivity or slope (eqn [2.3]). It follows that $O_2 = K_2 I_2 = K_2 K_1 I$, $O_3 = K_3 I_3 = K_3 K_2 K_1 I$, and for the whole system:

$$O = O_n = K_1 K_2 K_3 \cdots K_i \cdots K_n I \qquad [3.3]$$

Fig. 3.1

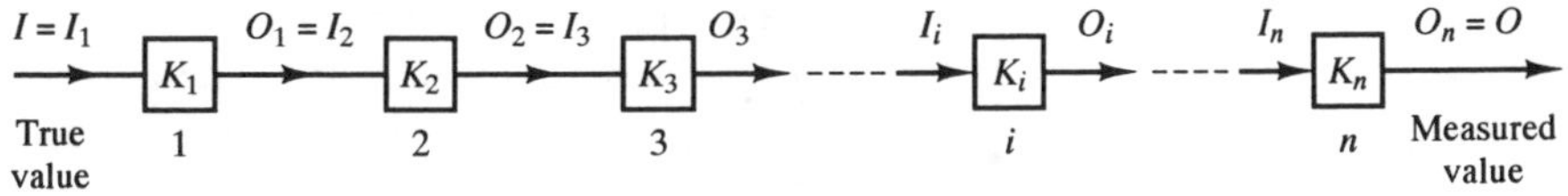

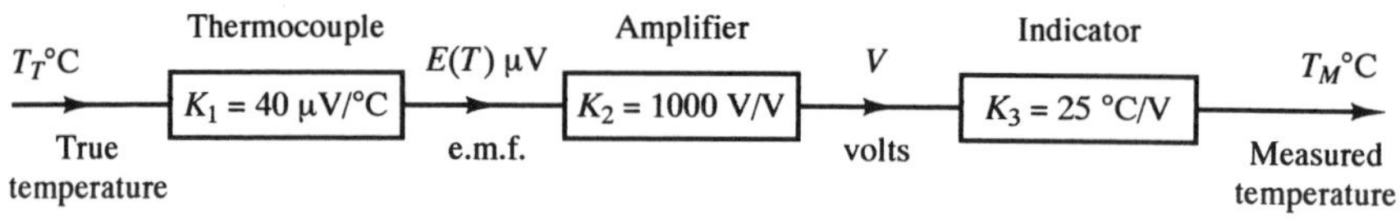

Fig. 3.2 Simple temperature measurement system

If the measurement system is complete, then $E = O - I$, giving:

$$E = (K_1K_2K_3 \cdots K_n - 1)I \tag{3.4}$$

Thus if

$$K_1K_2K_3 \cdots K_n = 1 \tag{3.5}$$

we have $E = 0$ and the system is perfectly accurate. The temperature measurement system shown in Fig. 3.2 appears to satisfy the above condition. The indicator is simply a moving coil voltmeter (see Chapter 11) with a scale marked in degrees Celsius so that an input change of 1 V causes a change in deflection of 25 °C. This system has $K_1K_2K_3 = 40 \times 10^{-6} \times 10^3 \times 25 = 1$ and thus appears to be perfectly accurate. The system is not accurate, however, because none of the three elements present is ideal. The thermocouple is non-linear (Chapter 2), so that as the input temperature changes the sensitivity is no longer $40\,\mu\text{V}\,°\text{C}^{-1}$. Also changes in reference junction temperature (Fig. 2.10(b)) cause the thermocouple e.m.f. to change. The output voltage of the amplifier is also affected by changes in ambient temperature (Chapter 9). The sensitivity K_3 of the indicator depends on the stiffness of the restoring spring in the moving coil assembly. This is affected by changes in environmental temperature and wear, causing K_3 to deviate from $25\,°\text{C}\,\text{V}^{-1}$. Thus the condition $K_1K_2K_3 = 1$ cannot be always satisfied and the system is in error.

In general the error of any measurement system depends on the non-ideal characteristics — e.g. non-linearity, environmental and statistical effects — of every element in the system. Thus in order to quantify this error as precisely as possible we need to use the general model for a single element developed in Sections 2.2 and 2.3.

3.2 The error probability density function of a system of non-ideal elements

In Chapter 2 we saw that the probability density function of the output $p(O)$ of a single element can be represented by a Gaussian distribution (eqn [2.20]). The mean value $\bar{O}$ of the distribution is given by eqn [2.22] which allows for non-linear and environmental effects. The standard deviation σ_0 is given by [2.23], which allows for statistical variations in inputs I, I_M, I_I with time, and statistical variations in parameters K, a, etc. amongst a batch of similar elements. These equations apply to each element in a measurement system of n elements and can be used to calculate the system error probability density function $p(E)$ as shown in Table 3.1.

Equations [3.6] (based on [2.22]) show how to calculate the mean value of the output of each element in turn, starting with $\bar{O}_1$ for the first and finishing with $\bar{O}_n = \bar{O}$ for the nth. The mean value $\bar{E}$ of the system error is simply the difference between the mean value of system output and mean value of system input (eqn [3.7]). Since the probability densities of the outputs of the individual elements are Gaussian,

Table 3.1 General calculation of system $p(E)$

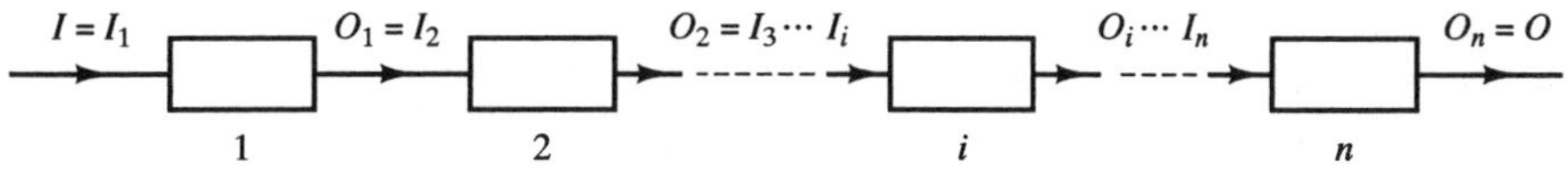

Mean values of element outputs

$$
\begin{aligned}
\bar{I}_1 &= \bar{I} \\
\bar{I}_2 &= \bar{O}_1 = \bar{K}_1\bar{I}_1 + \bar{N}_1(\bar{I}_1) + \bar{a}_1 + \bar{K}_{M_1}\bar{I}_{M_1}\bar{I}_1 + \bar{K}_{I_1}\bar{I}_{I_1} \\
\bar{I}_3 &= \bar{O}_2 = \bar{K}_2\bar{I}_2 + \bar{N}_2(\bar{I}_2) + \bar{a}_2 + \bar{K}_{M_2}\bar{I}_{M_2}\bar{I}_2 + \bar{K}_{I_2}\bar{I}_{I_2} \\
&\vdots \\
\bar{I}_{i+1} &= \bar{O}_i = \bar{K}_i\bar{I}_i + \bar{N}_i(\bar{I}_i) + \bar{a}_i + \bar{K}_{M_i}\bar{I}_{M_i}\bar{I}_i + \bar{K}_{I_i}\bar{I}_{I_i} \\
&\vdots \\
\bar{O} &= \bar{O}_n = \bar{K}_n\bar{I}_n + \bar{N}_n(\bar{I}_n) + \bar{a}_n + \bar{K}_{M_n}\bar{I}_{M_n}\bar{I}_n + \bar{K}_{I_n}\bar{I}_{I_n}
\end{aligned}
\qquad [3.6]
$$

Mean value of system error

$$\bar{E} = \bar{O} - \bar{I} \qquad [3.7]$$

Standard deviations of element outputs

$$
\begin{aligned}
\sigma_{I_1}^2 &= 0 \\
\sigma_{I_2}^2 &= \sigma_{O_1}^2 = \left(\frac{\partial O_1}{\partial I_1}\right)^2 \sigma_{I_1}^2 + \left(\frac{\partial O_1}{\partial I_{M_1}}\right)^2 \sigma_{I_{M_1}}^2 + \left(\frac{\partial O_1}{\partial I_{I_1}}\right)^2 \sigma_{I_{I_1}}^2 + \left(\frac{\partial O_1}{\partial K_1}\right)^2 \sigma_{K_1}^2 + \cdots \\
\sigma_{I_3}^2 &= \sigma_{O_2}^2 = \left(\frac{\partial O_2}{\partial I_2}\right)^2 \sigma_{I_2}^2 + \left(\frac{\partial O_2}{\partial I_{M_2}}\right)^2 \sigma_{I_{M_2}}^2 + \left(\frac{\partial O_2}{\partial I_{I_2}}\right)^2 \sigma_{I_{I_2}}^2 + \left(\frac{\partial O_2}{\partial K_2}\right)^2 \sigma_{K_2}^2 + \cdots \\
&\vdots \\
\sigma_{I_{i+1}}^2 &= \sigma_{O_i}^2 = \left(\frac{\partial O_i}{\partial I_i}\right)^2 \sigma_{I_i}^2 + \left(\frac{\partial O_i}{\partial I_{M_i}}\right)^2 \sigma_{I_{M_i}}^2 + \left(\frac{\partial O_i}{\partial I_{I_i}}\right)^2 \sigma_{I_{I_i}}^2 + \left(\frac{\partial O_i}{\partial K_i}\right)^2 \sigma_{K_i}^2 + \cdots \\
\sigma_O^2 &= \sigma_{O_n}^2 = \left(\frac{\partial O_n}{\partial I_n}\right)^2 \sigma_{I_n}^2 + \left(\frac{\partial O_n}{\partial I_{M_n}}\right)^2 \sigma_{I_{M_n}}^2 + \left(\frac{\partial O_n}{\partial I_{I_n}}\right)^2 \sigma_{I_{I_n}}^2 + \left(\frac{\partial O_n}{\partial K_n}\right)^2 \sigma_{K_n}^2 + \cdots
\end{aligned}
\qquad [3.8]
$$

Standard deviation of system error

$$\sigma_E = \sigma_O \qquad [3.9]$$

Error probability density function

$$p(E) = \frac{1}{\sigma_E\sqrt{(2\pi)}} \exp\left[-\frac{1}{2\sigma_E^2}(E - \bar{E})^2\right] \qquad [3.10]$$

then, using the result outlined in Section 2.3, the probability density function of the system output O and system error E is also Gaussian (eqn [3.10]). Equations [3.8] (based on [2.23]) show how to calculate the standard deviation of the output of each element in turn, starting with σ_{O_1} for the 1st, and finishing with σ_{O_n} for the nth. We note that the standard deviation of the system input is zero and that the standard deviation of the error is equal to that of the system output (eqn [3.9]).

An example of the calculation of $\bar{E}$ *and* σ_E for a temperature measurement system is summarised in Table 3.2. The system (Fig. 3.3) consists of a platinum resistance thermometer, a resistance-to-current converter and a recorder.

Table 3.2 Summary of calculation of $\bar{E}$, σ_E for temperature system

mean $\bar{E}$

$$\bar{T} = 117\,°\text{C} \quad \bar{R}_T = 144.93\,\Omega$$
$$\bar{i} = 13.04\,\text{mA} \quad \bar{T}_M = 116.95\,°\text{C}$$
$$\bar{E} = \bar{T}_M - \bar{T} = -0.005\,°\text{C}$$

Standard deviation σ_E

$$\sigma_T^2 = 0$$
$$\sigma_{R_T}^2 = \left(\frac{\partial R_T}{\partial R_0}\right)^2 \sigma_{R_0}^2 = 39.4 \times 10^{-4} \text{ (Grade I)}$$
$$\sigma_i^2 = \left(\frac{\partial i}{\partial R_T}\right)^2 \sigma_{R_T}^2 + \left(\frac{\partial i}{\partial \Delta T_a}\right)^2 \sigma_{\Delta T_a}^2 + \left(\frac{\partial i}{\partial a}\right)^2 \sigma_a^2$$
$$= 78.7 \times 10^{-4} + 8.18 \times 10^{-4} + 5.76 \times 10^{-2}$$
$$= 6.62 \times 10^{-2}$$
$$\sigma_{T_M}^2 = \left(\frac{\partial T_M}{\partial i}\right)^2 \sigma_i^2 + \left(\frac{\partial T_M}{\partial a}\right)^2 \sigma_a^2 = 24.3 \times 10^{-2}$$
$$\sigma_E = \sigma_{T_M} = 0.49\,°\text{C}$$

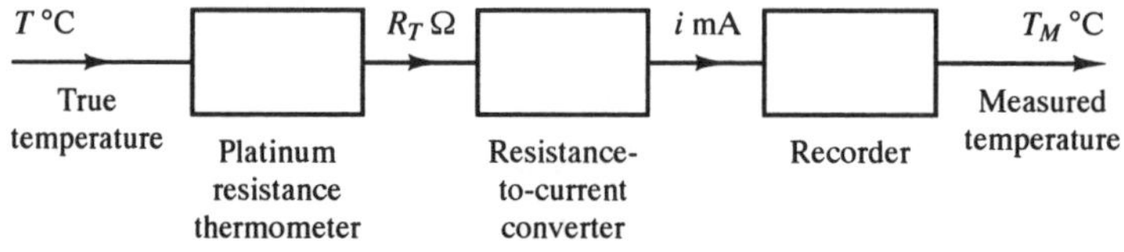

Fig. 3.3 Temperature measurement system

Table 3.3 Model for recorder

Model equation	$T_M = Ki + a$
Individual mean values	$\bar{K} = 1.875$, $\bar{a} = 92.50$ (100 to 130 °C record for 4 to 20 mA input)
Individual standard deviations	$\sigma_k = 0.0$, $\sigma_a = 0.10$
Partial derivatives	$\left(\frac{\partial T_M}{\partial i}\right) = 1.875$, $\left(\frac{\partial T_M}{\partial a}\right) = 1.00$
Overall mean value	$\bar{T}_M = \bar{K}\bar{i} + \bar{a}$
Overall standard deviation	$\sigma_{T_M}^2 = \left(\frac{\partial T_M}{\partial i}\right)^2 \sigma_i^2 + \left(\frac{\partial T_M}{\partial a}\right)^2 \sigma_a^2$

The models for the thermometer, resistance-to-current converter and the recorder are given in Tables 2.1, 2.2 and 3.3 respectively. The mean value $\bar{T}$ of the input temperature is 117 °C; from Table 2.1 the corresponding mean value $\bar{R}_T$ of resistance is 144.93 Ω. Similarly from Table 2.2 we have $\bar{i} = 13.04$ mA, and from Table 3.2 $\bar{T}_M = 116.95$ °C, given $\bar{E} = -0.05$ °C. The standard deviations σ_{R_T}, σ_i, σ_{T_M} are calculated using Tables 2.1, 2.2 and 3.3 to give $\sigma_E = \sigma_{T_M} = 0.49$ °C.

In Section 2.1 we saw that in situations where element non-linearity, hysteresis and environmental effects are small, their overall effect is quantified using error bands. Here a systematic statement of the exact element input/output relationship (e.g. eqn [2.9]) is replaced by a statistical statement. The element output is described by a rectangular probability density function, of width $2h$, centred on the ideal straight line value $O_{IDEAL} = KI + a$. If every element in the system is described in this

Table 3.4 $\bar{E}$ and σ_E for system of elements described by error bands

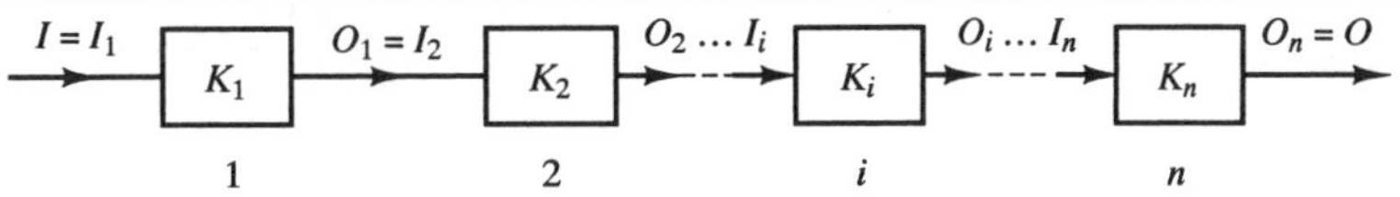

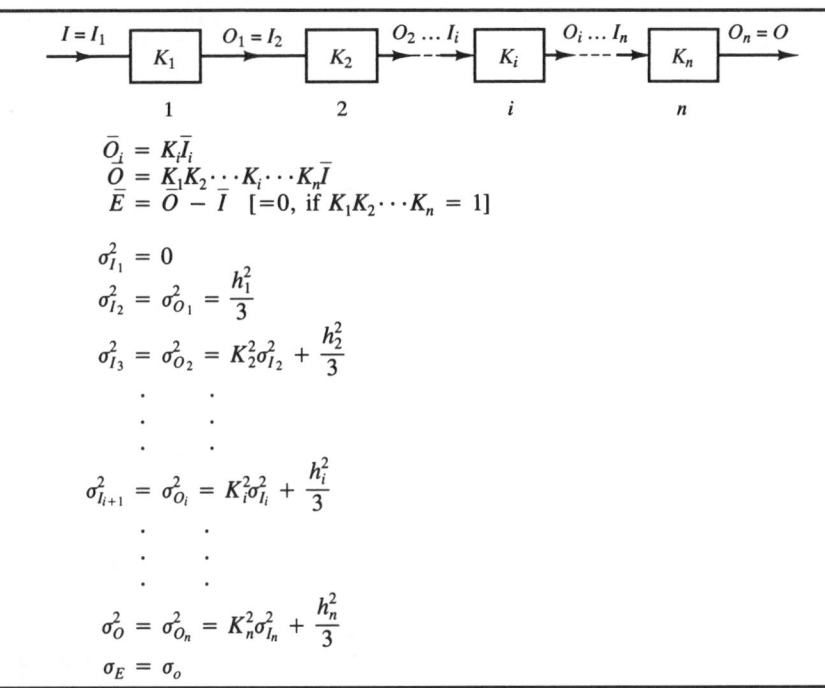

$$\bar{O}_i = K_i\bar{I}_i$$
$$\bar{O} = K_1K_2\cdots K_i\cdots K_n\bar{I}$$
$$\bar{E} = \bar{O} - \bar{I} \quad [=0, \text{ if } K_1K_2\cdots K_n = 1]$$

$$\sigma_{I_1}^2 = 0$$
$$\sigma_{I_2}^2 = \sigma_{O_1}^2 = \frac{h_1^2}{3}$$
$$\sigma_{I_3}^2 = \sigma_{O_2}^2 = K_2^2\sigma_{I_2}^2 + \frac{h_2^2}{3}$$
$$\vdots$$
$$\sigma_{I_{i+1}}^2 = \sigma_{O_i}^2 = K_i^2\sigma_{I_i}^2 + \frac{h_i^2}{3}$$
$$\vdots$$
$$\sigma_O^2 = \sigma_{O_n}^2 = K_n^2\sigma_{I_n}^2 + \frac{h_n^2}{3}$$
$$\sigma_E = \sigma_o$$

way, then the mean output value $\bar{O}_i$ will have the ideal value $\bar{O}_i = K_i\bar{I}_i + a_i$ for each element in the system. In the special case $a_i = 0$ for all i, $\bar{O}_i = K_i\bar{I}_i$ and the mean value $\bar{E}$ of the system error will be zero, provided that $K_1K_2\cdots K_i\cdots K_n = 1$ (Table 3.4). The error probability density function $p(E)$ is the result of combining n rectangular distributions, each of width $2h_i$, $i = 1,2,\ldots,n$. If $n > 3$, then the resultant distribution $p(E)$ approximates to a Gaussian distribution;[1] the larger the value of n the closer the distribution is to Gaussian. The standard deviation σ_E of $p(E)$ can therefore be estimated using the rules for combining several Gaussian distributions given in Section 2.3 (eqns [2.16] and [2.17]). Table 3.4 gives the calculation procedure for $\bar{E}$ and σ_E in this case; note that the standard deviation σ for a rectangular distribution of width $2h$ is $h/\sqrt{3}$. These techniques for combining several rectangular distributions have been used to estimate the accuracy of a gas mass flow computing system.[2]

3.3 Error reduction techniques

In the two previous sections we saw that the error of a measurement system depends on the non-ideal characteristics of every element in the system. Using the calibration techniques of Section 2.4, we can identify which elements in the system have the most dominant non-ideal behaviour. We can then devise compensation strategies for these elements which should produce significant reductions in the overall system error. This section outlines compensation methods for non-linear and environmental effects.

One of the most common methods of correcting a non-linear element is to introduce a **compensating non-linear element** into the system. This method is illustrated in

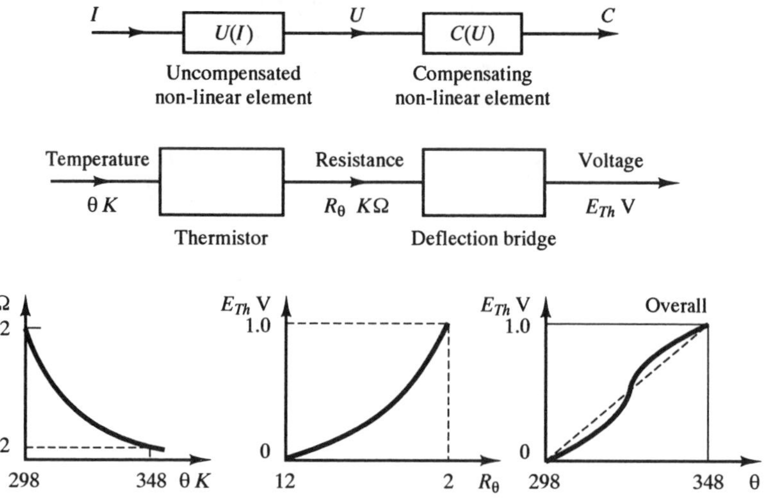

Fig. 3.4 Compensating non-linear element

Fig. 3.4. Given a non-linear element, described by $U(I)$, we need a compensating element $C(U)$, such that the overall characteristics $C[U(I)]$ of the elements together is as close to the ideal straight line as possible. The method is illustrated in Fig. 3.4 by the use of a deflection bridge to compensate for the non-linear characteristics of a thermistor. A detailed procedure for the design of the bridge is given in Section 9.1.

The most obvious method of reducing the effects of environmental inputs is that of **isolation**, i.e. to isolate the transducer from environmental changes so that effectively $I_M = I_I = 0$. Examples of this are the placing of the reference junction of a thermocouple in a temperature-controlled enclosure, and the use of spring mountings to isolate a transducer from the vibrations of the structure to which it is attached. Another obvious method is that of **zero environmental sensitivity**, where the element is completely insensitive to environmental inputs i.e. $K_M = K_I = 0$. An example of this is the use of a metal alloy with zero temperature coefficients of expansion and resistance as a strain gauge element. Such an ideal material is difficult to find, and in practice the resistance of a metal strain gauge is affected slightly by changes in ambient temperature.

A more successful method of coping with environmental inputs is that of **opposing environmental inputs**. Suppose that an element is affected by an environmental input; then a second element, subject to the same environmental input, is deliberately introduced into the system so that the two effects tend to cancel out. This method is illustrated for interfering inputs in Fig. 3.5 and can be easily extended to modifying inputs.

An example is compensation for variations in the temperature T_2 of the reference junction of a thermocouple. For a copper/constantan thermocouple (Fig. 2.10(b)), we have $K_I I_I$ equal to $-38.74 T_2\ \mu V$ so that a compensating element giving an output equal to $+38.74 T_2\ \mu V$ is required. The design of the reference junction compensation element is discussed in Sections 8.5, 9.1 and Problem 9.2.

An example of a **differential system** (Fig. 3.5(b)) is the use of two matched strain gauges, in adjacent arms of a bridge to provide compensation for ambient temperature changes. One gauge is measuring a tensile strain $+e$ and the other an equal

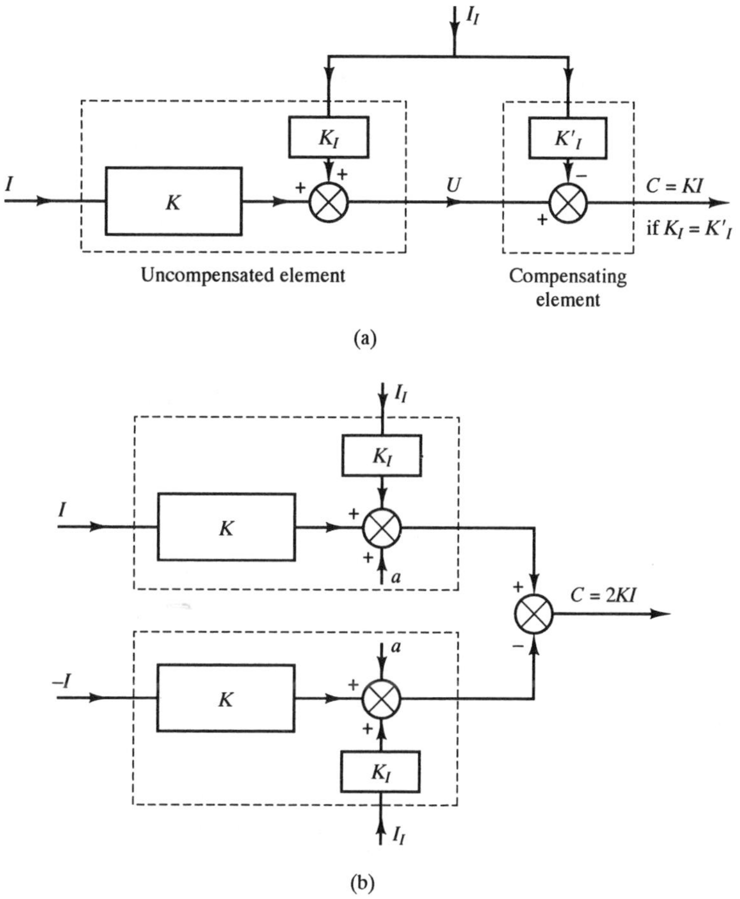

Fig. 3.5 Compensation for interfering inputs
(a) Using opposing environmental inputs
(b) Using a differential system

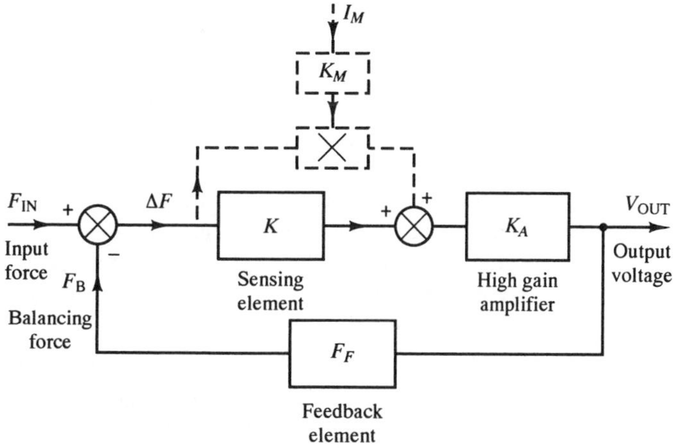

Fig. 3.6 Closed loop force transducer

compressive strain $-e$. The bridge effectively subtracts the two resistances so that the strain effect is doubled and the environmental effects cancel out.

The use of **high-gain negative feedback** is an important method of compensating for modifying inputs and non-linearity. Figure 3.6 illustrates the technique for a force transducer. The voltage output of a force-sensing element, subject to a modifying input, is amplified by a high-gain amplifier. The amplifier output is fed back to an element (e.g. a coil and permanent magnet) which provides a balancing force to oppose the input force.

Ignoring the effects of the modifying input for the moment we have:

$$\Delta F = F_{IN} - F_B \qquad [3.11]$$

$$V_{OUT} = KK_A \Delta F \qquad [3.12]$$

$$F_B = K_F V_{\mathrm{OUT}} \qquad [3.11]$$

i.e.

$$\frac{V_{\mathrm{OUT}}}{KK_A} = F_{\mathrm{IN}} - K_F V_{\mathrm{OUT}}$$

giving

Equation for force transducer with negative feedback

$$V_{\mathrm{OUT}} = \frac{KK_A}{1 + KK_A K_F} F_{\mathrm{IN}} \qquad [3.14]$$

If the amplifier gain K_A is made large such that the condition

$$KK_A K_F \gg 1 \qquad [3.15]$$

is satisfied, then $V_{\mathrm{OUT}} \approx \frac{1}{K_F} F_{\mathrm{IN}}$. This means that the system output depends only on the gain K_F of the feedback element and is independent of the gains K and K_A in the forward path. This means that, providing the above condition is obeyed, changes in K and K_A due to modifying inputs and/or non-linear effects have negligible effect on V_{OUT}. This can be confirmed by repeating the above analysis with K replaced by $K + K_M I_M$, giving

$$V_{\mathrm{OUT}} = \frac{(K + K_M I_M) K_A}{1 + (K + K_M I_M) K_A K_F} F_{\mathrm{IN}} \qquad [3.16]$$

which again reduces to $V_{\mathrm{OUT}} \approx \frac{F_{\mathrm{IN}}}{K_F}$ if $(K + K_M I_M) K_A K_F \gg 1$. We now, of course, have to ensure that the gain K_F of the feedback element does not change due to non-linear or environmental effects. Since the amplifier provides most of the required power, the feedback element can be designed for low power-handling capacity, giving greater linearity and less susceptibility to environmental inputs. A commonly used device which employs this principle is discussed in Section 9.4.

The rapid fall in the cost of digital integrated circuits in recent years has meant that microcomputers are now widely used as signal-processing elements in measurement systems (Ch. 10). This means that the powerful techniques of **computer estimation of measured value** can now be used. For this method, a good model of the elements in the system is required. In Sections 2.1 and 2.2 we saw that the steady state output O of an element is in general given by an equation of the form:

Direct equation

$$O = KI + a + N(I) + K_M I_M I + K_I I_I \qquad [2.9]$$

This is the **direct equation**; here O is the dependent variable which is expressed in terms of the independent variables I, I_M, I_I. In Section 2.4.2 we saw how the direct equation could be derived from sets of data obtained in a calibration experiment. In Section 3.2 eqn [2.9] was used to derive the error probability density function for a complete measurement system.

The steady-state characteristics of an element can also be represented by an alternative equation. This is the **inverse equation**; here the signal input I is the dependent variable and the output O and environmental inputs I_I, I_M the independent variables. The general form of this equation is:

Inverse equation

$$I = K'O + N'(O) + a' + K'_M I_M O + K'_I I \qquad [3.17]$$

where the values of K', $N'(\ )$, a', etc. are quite different from those for the direct equation. For example the direct and inverse equations for a copper/constantan (type T) thermocouple, with reference junction at 0 °C are:

DIRECT

$$E = 3.845 \times 10^{-2}\, T + 4.682 \times 10^{-5}\, T^2 - 3.789 \times 10^{-8}\, T^3 + 1.652 \times 10^{-11}\, T^4 \text{ mV}$$

INVERSE

$$T = 25.55\, E - 0.5973\, E^2 + 2.064 \times 10^{-2}\, E^3 - 3.205 \times 10^{-4}\, E^4 \text{ °C} \qquad [3.18]$$

where E is the thermocouple e.m.f. and T the measured junction temperature between 0 and 400 °C. Both equations were derived using a least squares polynomial fit to BS 4937 data; for the direct equation E is the dependent variable and T the independent variable; for the inverse equation T is the dependent variable and E the independent variable. While the direct equation is more useful for **error estimation**, the inverse equation is more useful for **error reduction**.

The use of the inverse equation in computer estimation of measured value is best implemented in a number of stages; with reference to Fig. 3.7(a), these are:

1. Treat the uncompensated system as a single element. Using the calibration procedure of Section 2.4.2 (or any other method of generating data) the parameters K', a', etc. in the inverse model equation:

   $$I = K'U + N'(U) + a' + K'_M I_M U + K'_I I_I \qquad [3.19]$$

   representing the overall behaviour of the uncompensated system can be found. This procedure will enable major environmental inputs I_M, I_I, to be identified (there may be more than one of each type).
2. The uncompensated system should be connected to the estimator. This consists firstly of a computer which stores the model parameters K', a', $N'(\ )$, etc. If errors due to environmental inputs are considered significant, then environmental sensors to provide the computer with estimates I'_M, I'_I, of these inputs are also necessary. The output U of the uncompensated system is also fed to the computer.
3. The computer then calculates an initial estimate I' of I using the inverse equation:

   $$I' = K'U + N'(U) + a' + K'_M I'_M U + K'_I I'_I \qquad [3.20]$$

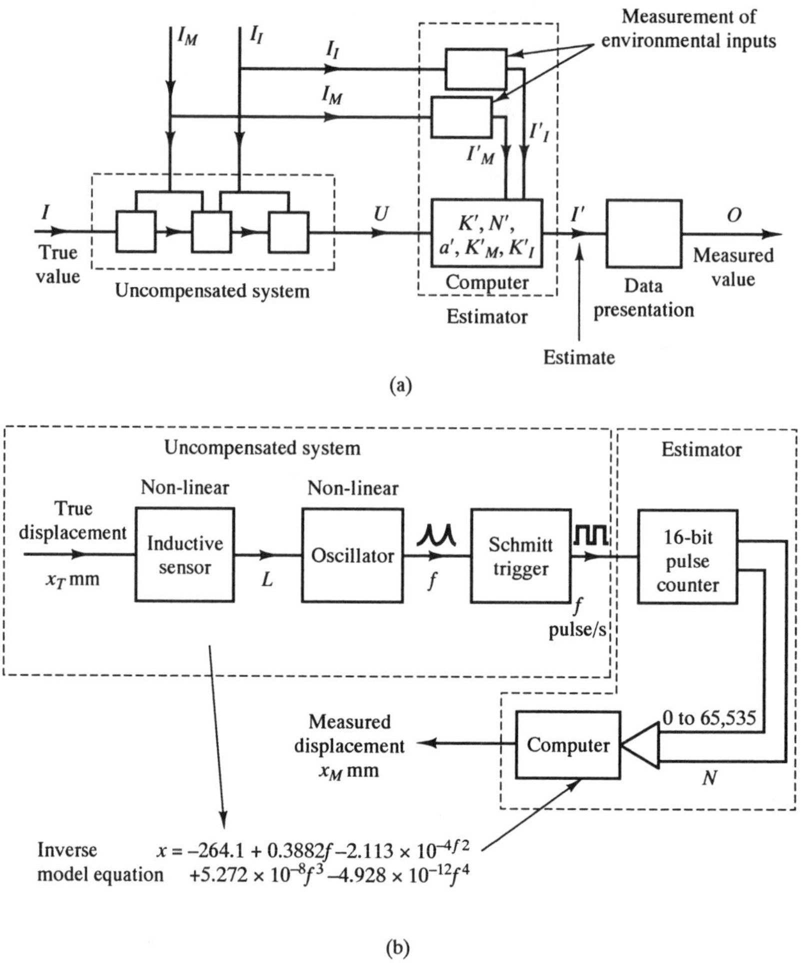

Fig. 3.7 Computer estimation of measured value using inverse model equation
(a) Principles
(b) Example of displacement measurement system

4. The data presentation element then displays the measured value O which should be close to I'. In appications not requiring the highest accuracy the procedure can be terminated at this stage.
5. If high accuracy is required, then it may be possible to further improve the estimator by calibrating the complete system. Values of system output O are measured for a range of known standard inputs I and the corresponding values of system error $E = O - I$ calculated. These error values will be mainly due to random effects but may also contain a small systematic component which can be corrected for.
6. An attempt should now be made to fit the data set (O_i, E_i), $i = 1, 2 \ldots n$, by a least squares straight line of the form:

$$E = kO + b \qquad [3.21]$$

where b is any residual zero error and k specifies any residual scale error. There is little point in attempting a polynomial fit at this stage.
7. The correlation coefficient:

$$r = \frac{\Sigma_{i=1}^{i=n} O_i E_i}{\sqrt{\Sigma_{i=1}^{i=n} O_i^2 \times \Sigma_{i=1}^{i=n} E_i^2}} \qquad [3.22]$$

between E and O data should now be evaluated. If the magnitude of r is greater than 0.5, then there is reasonable correlation between the E and O data; this means the systematic error of eqn [3.7] is present and we can proceed to stage 8 to correct for it. If the magnitude of r is less than 0.5, then there is no correlation between the E and O data; this means that the errors E are purely random and no correction can be made.

8. If appropriate, eqn [3.21] can be used to calculate an improved measured value:

$$O' = O - E = O - (kO + b) \qquad [3.23]$$

The displacement measurement system of Fig. 3.7(b) shows this method. The uncompensated system consists of an inductive displacement sensor, an oscillator (Section 9.5) and a Schmitt trigger (Section 10.1.4). The sensor has a non-linear relation between inductance L and displacement x, the oscillator a non-linear relation between frequency f and inductance L. This means that the inverse model equation, relating displacement x and frequency f of the Schmitt trigger output signal, has the non-linear form shown. The estimator consists of a 16-bit pulse counter and a computer. The computer reads the state of the counter at the beginning and end of a fixed time interval and thus measures the frequency f of the pulse signal. The computer then calculates x from the inverse model equation using model coefficients stored in memory.

Conclusion

This chapter has shown how to find the **error** of a complete measurement system under **steady state** conditions. **Measurement error** was first defined and then the **error probability density function** was derived firstly for a general system of non-ideal elements and then for the typical example of a temperature measurement system. The last section discussed a range of methods for **error reduction**.

References

3.1 PARADINE C G and PIVETT B H P 1966 *Statistical Methods for Technologists*, pp. 165–7, English Universities Press, London

3.2 SARGEANT R A E 1969 'Predicting accuracy in gas mass flow computing systems', *Control*, Jan. 1969.

Problems

3.1 A measurement system consists of a chromel–alumel thermocouple (with cold junction compensation), a millivolt-to-current converter and a recorder. Table Prob. 1 gives the model equations, and parameters for each element. Assuming that all probability distributions are normal, calculate the mean and standard deviation of the error probability distribution, when the input temperature is 117 °C.

Table Prob. 1

	Chromel–alumel thermocouple	e.m.f.-to-current converter	Recorder
Model equation	$E = C_0 + C_1T + C_2T^2$	$i = K_1E + K_MEΔT_a + K_IΔT_a + a_1$	$T_M = K_2i + a_2$
Mean values	$\bar{C}_0 = 0.00$ $\bar{C}_1 = 4.017 \times 10^{-2}$ $\bar{C}_2 = 4.66 \times 10^{-6}$	$\bar{K}_1 = 3.893$, $\overline{\Delta T}_a = -10$ $\bar{a}_1 = -3.864$ $\bar{K}_M = 1.95 \times 10^{-4}$ $\bar{K}_I = 2.00 \times 10^{-3}$	$\bar{K}_2 = 6.25$ $\bar{a}_2 = 25.0$
Standard deviations	$\sigma_{C_0} = 6.93 \times 10^{-2}$ $\sigma_{C_1} = \sigma_{C_2} = 0$	$\sigma_{a_1} = 0.14$, $\sigma_{\Delta T_a} = 10$ $\sigma_{K_1} = \sigma_{K_M} = \sigma_{K_I} = 0$	$\sigma_{a_2} = 0.30$ $\sigma_{K_2} = 0.0$

Table Prob. 2

Element	Linear sensitivity K	Error band width $\pm h$
Pressure sensor	$10^{-4}\ \Omega\ \mathrm{Pa}^{-1}$	$\pm 0.005\ \Omega$
Deflection bridge	$4 \times 10^{-2}\ \mathrm{mV}\ \Omega^{-1}$	$\pm 5 \times 10^{-4}$ mV
Amplifier	$10^3\ \mathrm{mV\ mV}^{-1}$	± 0.5 mV
Recorder	$250\ \mathrm{Pa\ mV}^{-1}$	± 100 Pa

3.2 A pressure measurement system consists of a pressure sensor, deflection bridge, amplifier and recorder. Table Prob. 2 gives the linear sensitivities and error band widths for each element in the system.

(a) Calculate the standard deviation σ_E of the error distribution function.

(b) Given that the recorder is incorrectly adjusted so that its sensitivity is $225\ \mathrm{Pa\ mV}^{-1}$, calculate the mean error $\bar{E}$ for an input pressure of 5×10^3 Pa

3.3 Figure Prob. 3 shows a block diagram of a force transducer using negative feedback. The elastic sensor gives a displacement output for a force input; the displacement sensor gives a voltage output for a displacement input, V_s is the supply voltage for the displacement sensor.

(a) Calculate the output voltage V_0 when
  (i) $V_s = 1.0$ volt, $F = 50$ N;
  (ii) $V_s = 1.5$ volt, $F = 50$ N.

(b) Comment on the practical significance of the variation of the supply voltage V_s.

Fig. Prob. 3

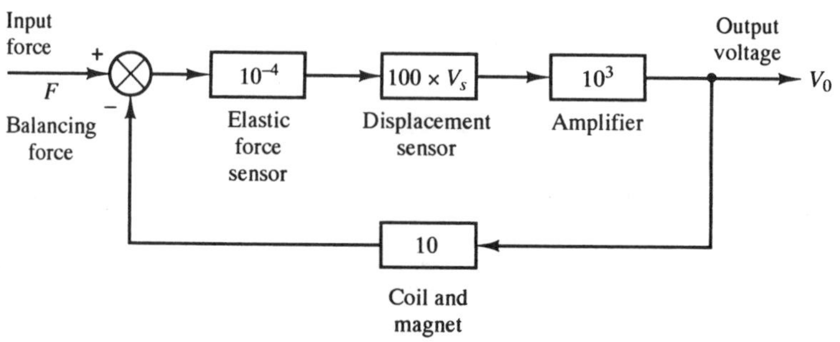

Fig. Prob. 4

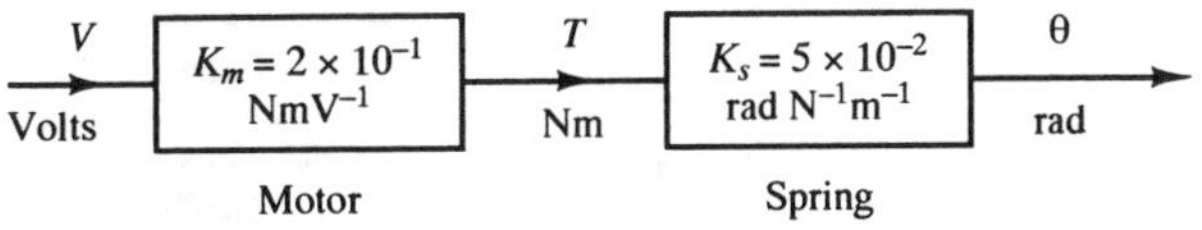

3.4 Figure Prob. 4 is a block diagram of a voltmeter. The motor produces a torque T proportional to voltage V and the output angular displacement of θ is proportional to T. The stiffness K_s of the spring can vary by $\pm 10\%$ about the nominal value of 5×10^{-2} rad $\mathrm{N^{-1}m^{-1}}$. Given that the following are available:

(i) a d.c. voltage amplifier of gain 1000;
(ii) a voltage subtraction unit;
(iii) a stable angular displacement transducer of sensitivity 100 V $\mathrm{rad^{-1}}$;

(a) draw a block diagram of a modified system using these components, which reduces the effect of changes in K_s;
(b) calculate the effect of a 10 per cent increase in K_s on the sensitivity of the **modified** system.

3.5 A temperature measurement system consists of a thermocouple, amplifier, 8-bit analogue-to-digital converter and a microcomputer with display facilities. Table Prob. 5 gives the model equations and parameters for each element in the system. The temperature of the thermocouple measurement junction is T_1 °C and the temperature of the reference junction is T_2 °C. The microcomputer corrects for T_2 having a non-zero mean value.

(a) Estimate the mean and standard deviation of the error probability density function when the input temperature T_1 is 100 °C. Treat rectangular distribution as Gaussian with $\sigma = h/\sqrt{3}$
(b) Explain briefly what modifications should be made to the system to reduce the quantities calculated in (a).

Table Prob. 5

	Thermocouple	Amplifier	Analogue-to-digital converter	Microcomputer with display
Model equations	$E_{T_1,T_2}=a_0 + a_1(T_1-T_2) + a_2(T_1^2-T_2^2)$	$V=K_1E_{T_1,T_2}+b_1$	$n=K_2V+b_2$	$T_m=K_3n+b_3$
Mean values	$\bar{a}_0=0$ $\bar{a}_1=4.3796\times10^{-2}$ $\bar{a}_2=-1.7963\times10^{-5}$ $\bar{T}_2=20$	$\bar{K}_1=255$ $\bar{b}_1=0.0$	$\bar{K}_2=0.1$ $\bar{b}_2=0.0$ n rounded to nearest integer	$\bar{K}_3=1.0$ $\bar{b}_3=20$
Statistical distributions	**Gaussians** with $\sigma_{a_0}=0.05$ $\sigma_{T_2}=2.0$ $\sigma_{a_1}=\sigma_{a_2}=0.0$	**Gaussians** with $\sigma_{b_1}=5.0$ $\sigma_{K_1}=0.0$	b_2 has a **Rectangular distribution** of width ±0.5 $\sigma_{K_2}=0.0$	$\sigma_{K_3}=0.0$ $\sigma_{b_3}=0.0$

3.6 A fluid velocity measurement system consists of a pitot tube, differential pressure transmitter, 8-bit analogue-to-digital converter and a microcomputer with display facilities. Table Prob. 6 gives the model equations and parameters for each element in the system. The microcomputer calculates the measured value of velocity assuming a constant density.

Table Prob. 6

	Pitot tube	Differential pressure transmitter	Analogue-to digital converter	Microcomputer with display
Model equations	$\Delta P = \frac{1}{2}\rho \vartheta_T^2$	$i = K_1 \Delta P + a_1$	$n = K_2 i + a_2$	$\vartheta_M = K_3\sqrt{n - 51}$
Mean values	$\bar{\rho} = 1.2$	$\bar{K}_1 = 0.064$ $\bar{a}_1 = 4.0$	$\bar{K}_2 = 12.80$ $\bar{a}_2 = 0.0$ n rounded off to nearest integer	$\bar{K}_3 = 1.430$
Half-widths of rectangular distribution	$h_\rho = 0.1$	$h_{a_1} = 0.04$	$h_{a_2} = 0.5$	$h_{K_3} = 0.0$

(a) Estimate the mean and standard deviation of the error probability density function assuming the true value of velocity ϑ_T is 14.0 m s^{-1}. Use the procedure of Table 3.1, treating the rectangular distributions as Gaussian with $\sigma = h/\sqrt{3}$.

(b) Explain briefly what modifications could be made to the system to reduce the quantities calculated in (a).

4

Dynamic Characteristics of Measurement Systems

If the input signal I to an element is changed suddenly, from one value to another, then the output signal O will not instantaneously change to its new value. For example, if the temperature input to a thermocouple is suddenly changed from 25 °C to 100 °C, some time will elapse before the e.m.f. output completes the change from 1 mV to 4 mV. The ways in which an element responds to sudden input changes are termed its **dynamic characteristics**, and these are most conveniently summarised using a **transfer function** $G(s)$. The first section of this chapter examines the dynamics of typical elements and derives the corresponding transfer function. The next section examines how standard test signals can be used to identify $G(s)$ for an element. If the input signal to a multi-element measurement system is changing rapidly, then the waveform of the system output signal is in general different from that of the input signal. Section 3 explains how this **dynamic error** can be found and the final section outlines **dynamic compensation** methods that can be used to minimise errors.

4.1 Transfer function $G(s)$ for typical system elements

4.1.1 First-order elements

A good example of a first-order element is provided by a temperature sensor with an electrical output signal, e.g. a thermocouple or thermistor. The bare element (not enclosed in a sheath) is placed inside a fluid (Fig. 4.1). Initially at time $t = 0-$ (just before $t = 0$), the sensor temperature is equal to the fluid temperature i.e. $T(0-) = T_F(0-)$. If the fluid temperature is suddenly raised at $t = 0$, the sensor is no longer in a steady state and its dynamic behaviour is described by the **heat balance equation**:

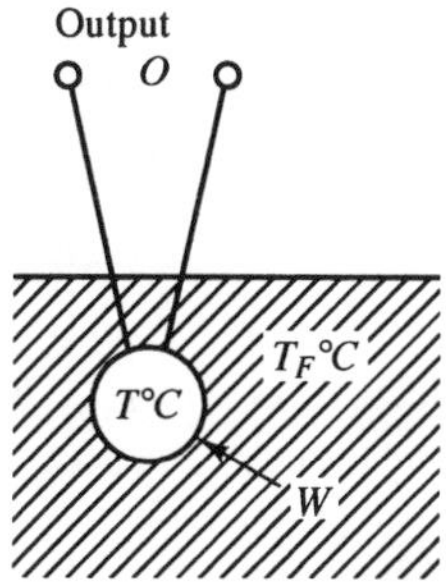

Fig. 4.1 Temperature sensor in fluid

$$\textbf{rate of heat inflow} - \textbf{rate of heat outflow} = \begin{array}{c}\textbf{rate of change of}\\ \textbf{sensor heat content}\end{array} \quad [4.1]$$

Assuming that $T_F > T$, then the rate of heat outflow will be zero, and the rate of heat inflow W will be proportional to the temperature difference $(T_F - T)$. From Chapter 14 we have

$$W = UA(T_F - T) \text{ watts} \quad [4.2]$$

where $U\,\mathrm{W\,m^{-2}\,°C^{-1}}$ is the overall heat transfer coefficient between fluid and sensor and $A\,\mathrm{m^2}$ is the effective heat transfer area. The increase of heat content of the sensor is $MC[T - T(0-)]$ joules, where M kg is the sensor mass and $C\,\mathrm{J\,kg^{-1}\,°C^{-1}}$ is the specific heat of the sensor material. Thus, assuming M and C are constants:

$$\text{rate of increase of sensor heat content} = MC\frac{\mathrm{d}}{\mathrm{d}t}[T - T(0-)] \tag{4.3}$$

Defining $\Delta T = T - T(0-)$ and $\Delta T_F = T_F - T_F(0-)]$ to be the deviations in temperatures from initial steady conditions, the differential equation describing the sensor temperature changes is

$$UA(\Delta T_F - \Delta T) = MC\frac{\mathrm{d}\Delta T}{\mathrm{d}t}$$

i.e.

$$\frac{MC}{UA}\frac{\mathrm{d}\Delta T}{\mathrm{d}t} + \Delta T = \Delta T_F \tag{4.4}$$

This is a **linear differential equation** in which $\mathrm{d}\Delta T/\mathrm{d}t$ and ΔT are multiplied by constant coefficients; the equation is **first order** because $\mathrm{d}\Delta T/\mathrm{d}t$ is the highest derivative present. The quantity MC/UA has the dimensions of time:

$$\left\{\frac{\mathrm{kg \times J \times kg^{-1} \times °C^{-1}}}{\mathrm{W \times m^{-2} \times °C^{-1} \times m^2}} = \frac{\mathrm{J}}{\mathrm{W}} = \mathrm{secs}\right\}$$

and is referred to as the **time constant** τ for the system. The differential equation is now:

Linear first-order differential equation

$$\tau\frac{\mathrm{d}\Delta T}{\mathrm{d}t} + \Delta T = \Delta T_F \tag{4.5}$$

While the above differential equation is a perfectly adequate description of the dynamics of the sensor, it is not the most useful representation. The transfer function based on the Laplace transform of the differential equation provides a convenient framework for studying the dynamics of multi-element systems. The Laplace transform $\bar{f}(s)$ of a time-varying function is defined by:

Definition of Laplace transform

$$\bar{f}(s) = \int_0^\infty e^{-st} f(t)\,\mathrm{d}t \tag{4.6}$$

where s is a complex variable of the form $\sigma = j\omega$ where $j = \sqrt{(-1)}$.

Table 4.1 gives Laplace transforms for some common standard functions $f(t)$. In order to find the transfer function for the sensor we must find the Laplace transform of eqn [4.5]. Using Table 4.1 we have:

$$\tau[s\Delta\bar{T}(s) - \Delta T(0-)] + \Delta\bar{T}(s) = \Delta\bar{T}_F(s) \tag{4.7}$$

where $\Delta T(0-)$ is the temperature deviation at initial conditions prior to $t = 0$. By

Table 4.1 Laplace transforms of common time functions $f(t)$

$$\mathscr{L}[f(t)] = \bar{f}(s) = \int_0^\infty e^{-st} f(t)\,dt$$

Function	Symbol	Graph	Transform
1st Derivative	$\frac{d}{dt}f(t)$		$s\bar{f}(s) - f(0-)$
2nd Derivative	$\frac{d^2}{dt^2}f(t)$		$s^2\bar{f}(s) - sf(0-) - \dot{f}(0-)$
Unit impulse	$\delta(t)$	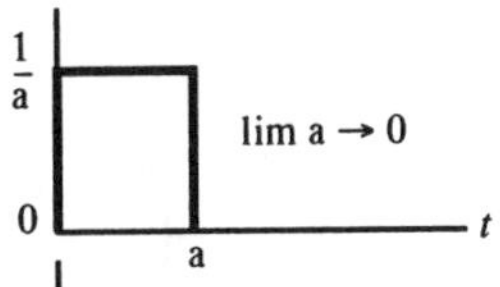	1
Unit step	$\mu(t)$	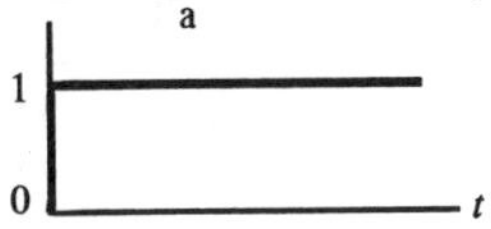	$\frac{1}{s}$
Exponential decay	$\exp(-\alpha t)$		$\frac{1}{s+\alpha}$
Exponential growth	$1 - \exp(-\alpha t)$	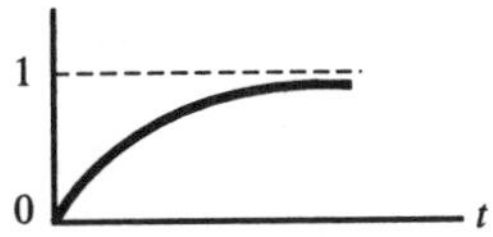	$\frac{\alpha}{s(s+\alpha)}$
Sine wave	$\sin \omega t$	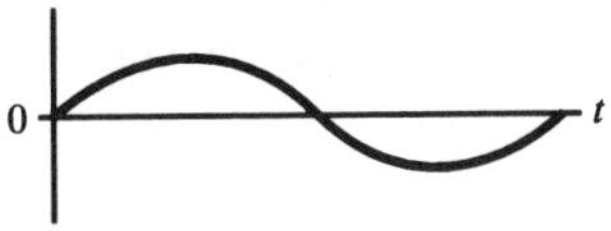	$\frac{\omega}{s^2+\omega^2}$
Phase shifted sine wave	$\sin(\omega t + \phi)$		$\frac{\omega\cos\phi + s\sin\phi}{s^2+\omega^2}$
Exponentially damped sine wave	$\exp(-\alpha t)\sin \omega t$	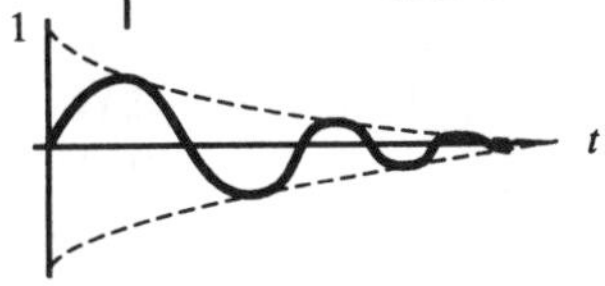	$\frac{\omega}{(s+\alpha)^2+\omega^2}$
Ramp with exponential decay	$t\exp(-\alpha t)$	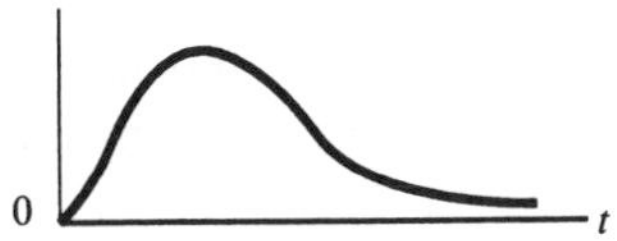	$\frac{1}{(s+\alpha)^2}$

* Initial conditions are at $t = 0-$, just prior to $t = 0$

definition, $\Delta T(0-) = 0$, giving:

$$\tau s \Delta\bar{T}(s) + \Delta\bar{T}(s) = \Delta\bar{T}_F(s)$$

i.e.

$$(\tau s + 1)\Delta\bar{T}(s) = \Delta\bar{T}_F(s) \qquad [4.8]$$

The transfer function $G(s)$ of an element is defined as the ratio of the Laplace transform of the output to the Laplace transform of the input, provided the initial conditions are zero. Thus:

Definition of element transfer function

$$G(s) = \frac{\bar{f}_O(s)}{\bar{f}_i(s)} \quad [4.9]$$

and $\bar{f}_O(s) = G(s)\bar{f}_i(s)$; this means the transfer function of the output signal is simply the product of the element transfer function and the transfer function of the input signal. Because of this simple relationship the transfer function technique lends itself to the study of the dynamics of multielement systems and block diagram representation (Fig. 4.2).

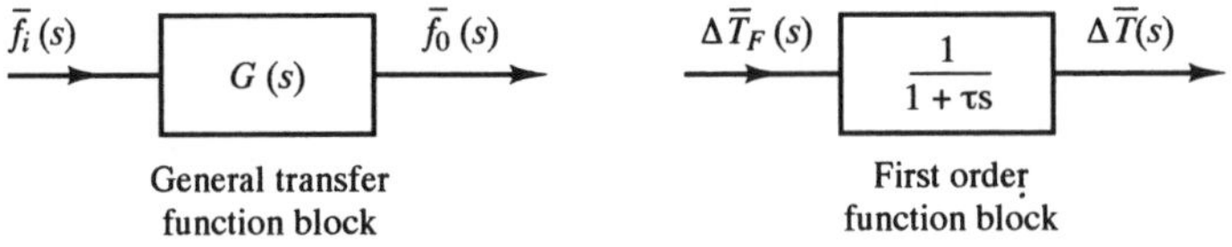

Fig. 4.2 Transfer function representation

From eqns [4.8] and [4.9] the transfer function for a first-order element is:

Transfer function for a first order element

$$G(s) = \frac{\Delta\bar{T}(s)}{\Delta\bar{T}_F(s)} = \frac{1}{1 + \tau s} \quad [4.10]$$

The above transfer function only relates changes in sensor temperature to changes in fluid temperature. The overall relationship between changes in sensor output signal O and fluid temperature is:

$$\frac{\Delta\bar{O}(s)}{\Delta\bar{T}_F(s)} = \frac{\Delta O}{\Delta T}\frac{\Delta\bar{T}(s)}{\Delta\bar{T}_F(s)} \quad [4.11]$$

where $\Delta O/\Delta T$ is the **steady-state sensitivity** of the temperature sensor. For an ideal element $\Delta O/\Delta T$ will be equal to the slope K of the **ideal straight line**. For non-linear elements, subject to small temperature fluctuations, we can take $\Delta O/\Delta T = dO/dT$, the derivative being evaluated at the steady-state temperature $T(0-)$ around which the fluctuations are taking place. Thus for a copper/constantan thermocouple measuring small fluctuations in temperature around 100 °C, $\Delta E/\Delta T$ is found by evaluating dE/dT at 100 °C, using eqn [2.8], to give $\Delta E/\Delta T = 46\ \mu V\ °C^{-1}$. Thus if the time constant of the thermocouple is 10 s the overall dynamic relationship between changes in e.m.f. and fluid temperature is:

$$\frac{\Delta\bar{E}(s)}{\Delta\bar{T}_F(s)} = 46\frac{1}{1 + 10s} \quad [4.12]$$

In the general case of an element with static characteristics given by eqn [2.9], and dynamic characteristics defined by $G(s)$, the effect of small, rapid changes in ΔI is evaluated using Fig. 4.3, in which stead-state sensitivity $(\partial O/\partial I)_{I_0} = K + K_M I_M + (dN/dI)_{I_0}$, and I_0 is the steady-state value of I around which the fluctuations are taking place. Table 4.2 shows analogous fluidic, electrical and mechanical elements, which are also described by a first-order transfer function $G(s) = 1/(1 + \tau s)$. All

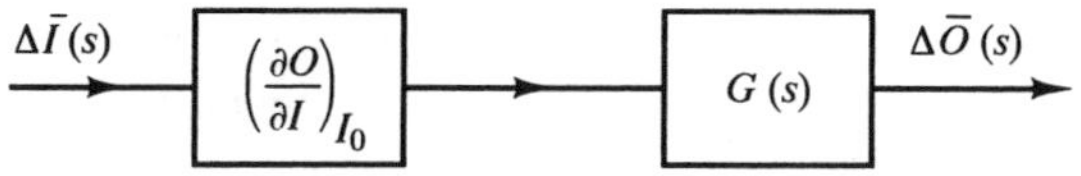

Fig. 4.3 Element model for dynamic calculations

Table 4.2 **Analogous first order elements**

Fluidic

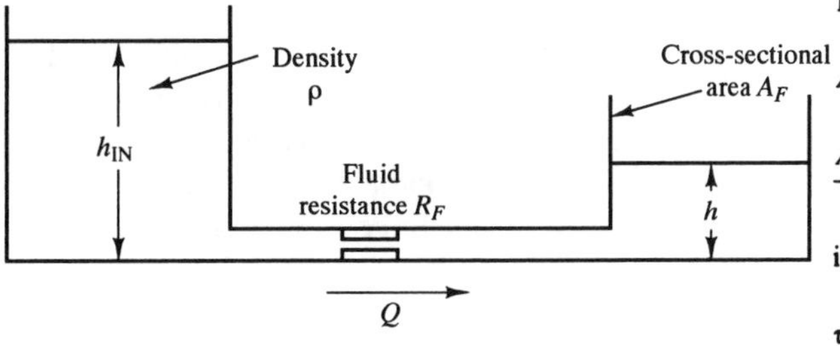

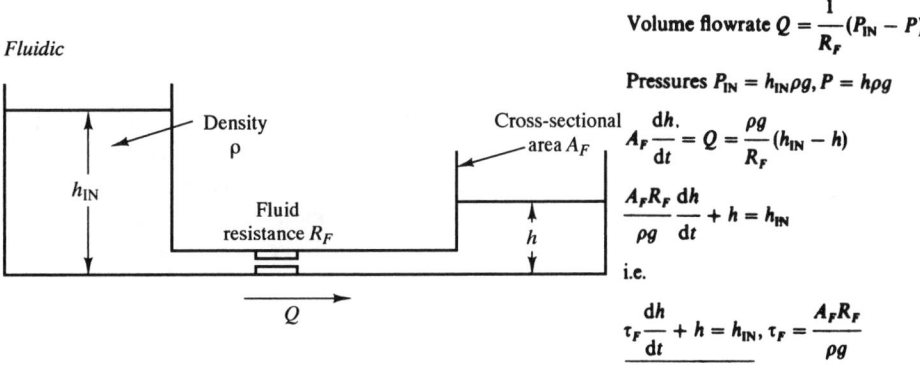

Volume flowrate $Q = \frac{1}{R_F}(P_{IN} - P)$

Pressures $P_{IN} = h_{IN}\rho g, P = h\rho g$

$$A_F \frac{dh}{dt} = Q = \frac{\rho g}{R_F}(h_{IN} - h)$$

$$\frac{A_F R_F}{\rho g}\frac{dh}{dt} + h = h_{IN}$$

i.e.

$$\underline{\tau_F \frac{dh}{dt} + h = h_{IN},\ \tau_F = \frac{A_F R_F}{\rho g}}$$

Electrical

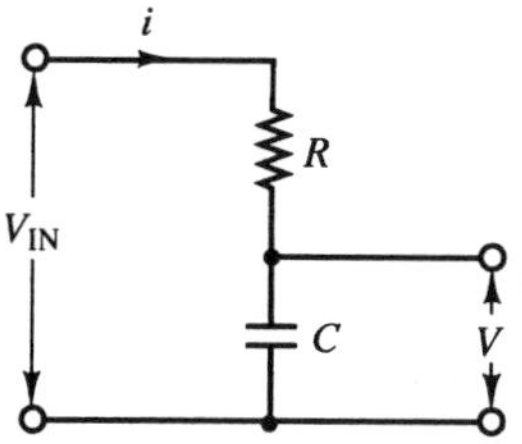

$V_{IN} - V = iR$

Charge $q = CV$, current $i = \frac{dq}{dt} = \frac{C dV}{dt}$

$$RC\frac{dV}{dt} + V = V_{IN}$$

i.e.

$$\underline{\tau_E \frac{dV}{dt} + V = V_{IN},\ \tau_E = RC}$$

Mechanical

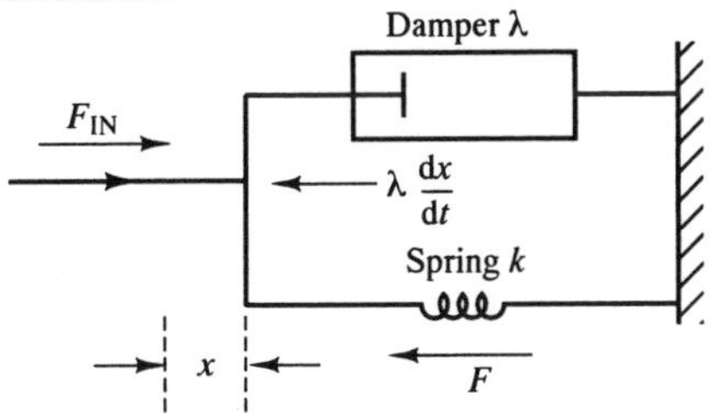

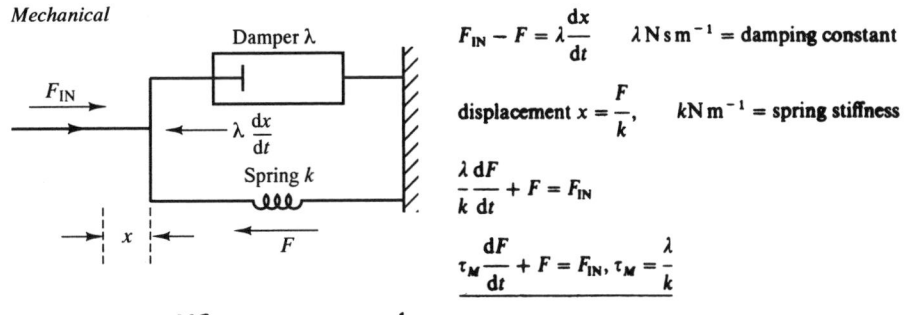

$F_{IN} - F = \lambda \frac{dx}{dt}$ $\lambda\ \mathrm{N\,s\,m^{-1}}$ = damping constant

displacement $x = \frac{F}{k}$, $k\ \mathrm{N\,m^{-1}}$ = spring stiffness

$$\frac{\lambda}{k}\frac{dF}{dt} + F = F_{IN}$$

$$\underline{\tau_M \frac{dF}{dt} + F = F_{IN},\ \tau_M = \frac{\lambda}{k}}$$

Thermal	$\tau_{Th} = \frac{MC}{UA}$	$= R_{Th}C_{Th}$;	$R_{Th} = \frac{1}{UA}$,	$C_{Th} = MC$
Fluidic	$\tau_F = \frac{A_F R_F}{\rho g}$	$= R_F C_F$;	$R_F = R_F$,	$C_F = \frac{A_F}{\rho g}$
Electrical	$\tau_E = RC$	$= R_E C_E$;	$R_E = R$,	$C_E = C$
Mechanical	$\tau_M = \frac{\lambda}{k}$	$= R_M C_M$;	$R_M = \lambda$,	$C_M = \frac{1}{k}$

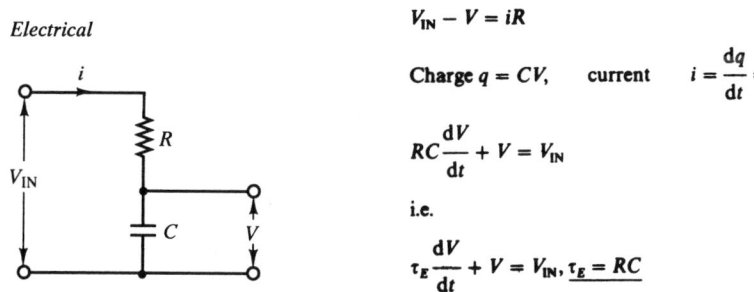

four elements are characterised by 'resistance' and 'capacitance' as illustrated in the table. Temperature, pressure, voltage and force are analogous 'driving' or effort variables; heat flow rate, volume flow rate, current and velocity are analogous 'driven' or flow variables. These analogies are discussed further in Section 5.2.

4.1.2 Second-order elements

The elastic sensor shown in Fig. 4.4, which converts a force input F into a displacement output x, is a good example of a second-order element. The diagram is a conceptual model of the element which incorporates a mass m kg, a spring of stiffness $k\,\text{Nm}^{-1}$, and a damper of constant $\lambda\,\text{N s m}^{-1}$. The system is initially at rest at time $t = 0-$ so that the initial velocity $\dot{x}(0-) = 0$ and the initial acceleration $\ddot{x}(0-) = 0$. The initial input force $F(0-)$ is balanced by the spring force at the initial displacement $x(0-)$ i.e.

$$F(0-) = kx(0-) \tag{4.13}$$

If the input force is suddenly increased at $t = 0$, then the element is no longer in a steady state and its dynamic behaviour is described by Newton's second law, i.e.

resultant force = mass × acceleration

i.e.

$$F - kx - \lambda\dot{x} = m\ddot{x} \tag{4.14}$$

and

$$m\ddot{x} + \lambda\dot{x} + kx = F$$

Defining ΔF and Δx to be the deviations in F and x from initial steady-state conditions:

$$\begin{aligned} \Delta F &= F - F(0-), & \Delta x &= x - x(0-) \\ \Delta\dot{x} &= \dot{x}, & \Delta\ddot{x} &= \ddot{x} \end{aligned} \tag{4.15}$$

The differential equation now becomes:

$$m\Delta\ddot{x} + \lambda\Delta\dot{x} + kx(0-) + k\Delta x = F(0-) + \Delta F$$

which, using [4.13], reduces to:

$$m\Delta\ddot{x} + \lambda\Delta\dot{x} + k\Delta x = \Delta F$$

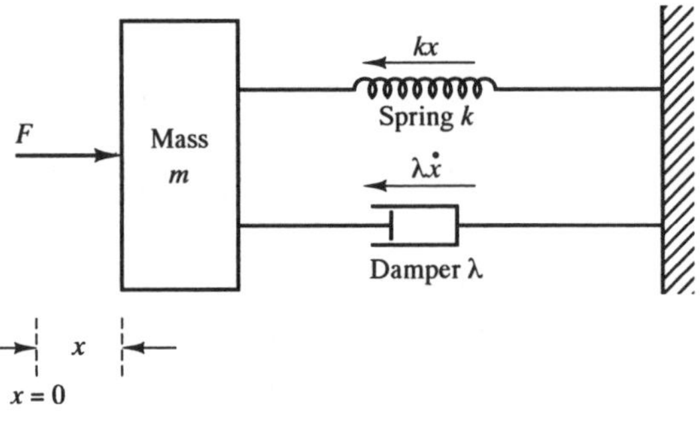

Fig. 4.4 Mass–spring–damper model of elastic force sensor

i.e.

$$\frac{m}{k}\frac{d^2\Delta x}{dt^2} + \frac{\lambda}{k}\frac{d\Delta x}{dt} + \Delta x = \frac{1}{k}\Delta F \qquad [4.16]$$

This is a **second-order linear differential equation** in which Δx and its derivatives are multipled by constant coefficients and the highest derivative present is $d^2\Delta x/dt^2$. If we define

$$\textbf{undamped natural frequency } \omega_n = \sqrt{\frac{k}{m}} \text{rad/sec}$$

and

$$\textbf{damping ratio } \xi = \frac{\lambda}{2\sqrt{(km)}} \qquad [4.17]$$

then $m/k = 1/\omega_n^2$, $\lambda/k = 2\xi/\omega_n$ and [4.16] can be expressed in the standard form:

Linear second-order differential equation

$$\frac{1}{\omega_n^2}\frac{d^2\Delta x}{dt^2} + \frac{2\xi}{\omega_n}\frac{d\Delta x}{dt} + \Delta x = \frac{1}{k}\Delta F \qquad [4.18]$$

In order to find the transfer function for the element we require the Laplace transform of eqn [4.18]. Using Table 4.1 we have:

$$\frac{1}{\omega_n^2}[s^2\Delta\bar{x}(s) - s\Delta x(0-) - \Delta\dot{x}(0-)] + \frac{2\xi}{\omega_n}[s\Delta\bar{x}(s) - \Delta x(0-)] + \Delta\bar{x}(s)$$
$$= \frac{1}{k}\Delta\bar{F}(s) \qquad [4.19]$$

Since $\Delta\dot{x}(0-) = \dot{x}(0-) = 0$ and $\Delta x(0-) = 0$ by definition, [4.19] reduces to:

$$\left[\frac{1}{\omega_n^2}s^2 + \frac{2\xi}{\omega_n}s + 1\right]\Delta\bar{x}(s) = \frac{1}{k}\Delta\bar{F}(s) \qquad [4.20]$$

Thus

$$\frac{\Delta\bar{x}(s)}{\Delta\bar{F}(s)} = \frac{1}{k}G(s).$$

where $1/k$ = steady-state sensitivity K, and

Transfer function for a second-order element

$$G(s) = \frac{1}{\dfrac{1}{\omega_n^2}s^2 + \dfrac{2\xi}{\omega_n}s + 1} \qquad [4.21]$$

Figure 4.5 shows an analogous electrical element, a series L–C–R circuit.

Comparing [4.14] and [4.22] we see that q is analogous to x, V is analogous to F, and L, R and $1/C$ are analogous to m, λ and k respectively (see Table 5.1). The

Fig. 4.5 Series L–C–R circuit

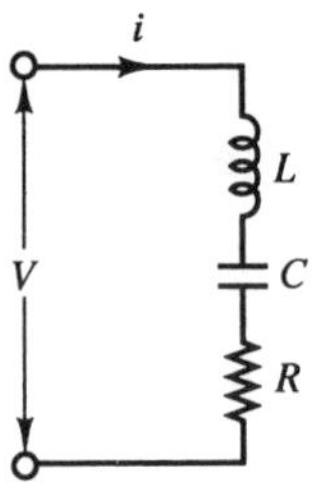

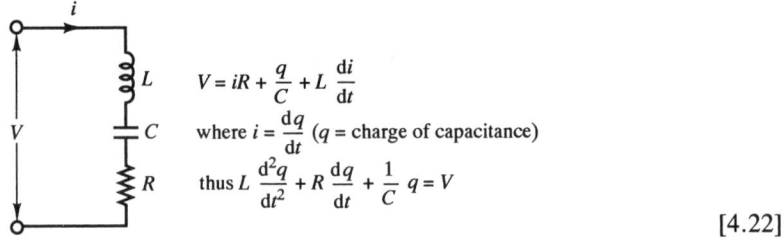

$$V = iR + \frac{q}{C} + L\frac{di}{dt}$$

where $i = \frac{dq}{dt}$ (q = charge of capacitance)

$$\text{thus } L\frac{d^2q}{dt^2} + R\frac{dq}{dt} + \frac{1}{C}q = V \qquad [4.22]$$

electrical circuit is also described by the above second-order transfer function with $\omega_n = 1/\sqrt{(LC)}$ and $\xi = (R/2)\sqrt{(C/L)}$.

4.2 Identification of the dynamics of an element

In order to identify the transfer function $G(s)$ of an element, standard input signals should be used. The two most commonly used standard signals are step and sine wave. This section examines the response of first- and second-order elements to step and sine wave inputs.

4.2.1 Step response of first- and second-order elements

From Table 4.1 we see that the Laplace transform of a step of unit height $u(t)$ is $\bar{f}(s) = 1/s$. Thus if a first-order element with $G(s) = 1/(1 + \tau s)$ is subject to a unit step input signal, the Laplace transform of the element output signal is:

$$\bar{f}_o(s) = G(s)\bar{f}_i(s) = \frac{1}{(1 + \tau s)s} \qquad [4.23]$$

Expressing [4.23] in partial fractions, we have:

$$\bar{f}_o(s) = \frac{1}{(1 + \tau s)s} = \frac{A}{(1 + \tau s)} + \frac{B}{s}$$

Equating coefficients of constants gives $B = 1$, and equating coefficients of s gives $0 = A + B\tau$ i.e. $A = -\tau$. Thus:

$$\bar{f}_o(s) = \frac{1}{s} - \frac{\tau}{(1 + \tau s)} = \frac{1}{s} - \frac{1}{(s + 1/\tau)} \qquad [4.24]$$

Using Table 4.1 in reverse, i.e. finding a time signal $f(t)$ corresponding to a transform $\bar{f}(s)$, we have:

$$f_o(t) = u(t) - \exp\left(\frac{-t}{\tau}\right)$$

and since $u(t) = 1$ for $t > 0$:

Response of first-order element to unit step

$$f_o(t) = 1 - \exp\left(\frac{-t}{\tau}\right) \qquad [4.25]$$

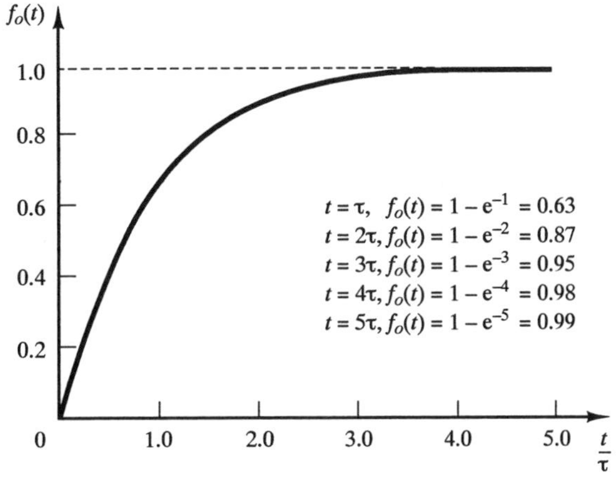

Fig. 4.6 Response of a first-order element to a unit step

The form of the response is shown in Fig. 4.6.

As an example of the use of equation [4.25], consider the temperature sensor of Section 4.1.1. Initially the temperature of the sensor is equal to that of the fluid, i.e. $T(0-) = T_F(0-) = 25\,°\mathrm{C}$, say. If T_F is suddenly raised to 100 °C, then this represents a step change ΔT_F of height 75 °C. The corresponding *change* in sensor temperature is given by $\Delta T = 75(1 - e^{-t/\tau})$ and the actual temperature T of the sensor at time t is given by:

$$T(t) = 25 + 75(1 - e^{-t/\tau}) \qquad [4.26]$$

Thus at time $t = \tau$, $T = 25 + (75 \times 0.63) = 72.3\,°\mathrm{C}$. By measuring the time taken for T to rise to 72.3 °C we can find the time constant τ of the element.

If the second-order element with transfer function

$$G(s) = \frac{1}{\dfrac{1}{\omega_n^2}s^2 + \dfrac{2\xi}{\omega_n}s + 1}$$

is subject to a unit step input signal, then the Laplace transform of the element output signal is:

$$\bar{f}_o(s) = \frac{1}{\left(\dfrac{1}{\omega_n^2}s^2 + \dfrac{2\xi}{\omega_n}s + 1\right)s} \qquad [4.27]$$

Expressing [4.27] in partial fractions we have:

$$\bar{f}_o(s) = \frac{As + B}{\left(\dfrac{1}{\omega_n^2}s^2 + \dfrac{2\xi}{\omega_n}s + 1\right)} + \frac{C}{s} \qquad [4.27]$$

where $A = -1/\omega_n^2$, $B = -\,2\xi/\omega_n$, $C = 1$. This gives:

$$\bar{f}_o(s) = \frac{1}{s} - \frac{(s + 2\xi\omega_n)}{s^2 + 2\xi\omega_n s + \omega_n^2}$$

$$= \frac{1}{s} - \frac{(s + 2\xi\omega_n)}{(s + \xi\omega_n)^2 + \omega_n^2(1 - \xi^2)}$$

$$= \frac{1}{s} - \frac{(s + \xi\omega_n)}{(s + \xi\omega_n)^2 + \omega_n^2(1 - \xi^2)} - \frac{\xi\omega_n}{(s + \xi\omega_n)^2 + \omega_n^2(1 - \xi^2)} \quad [4.28]$$

There are three cases to consider depending on whether ξ is greater than 1, equal to 1, or less than 1. For example if $\xi = 1$ (**critical damping**) then:

$$\bar{f}_o(s) = \frac{1}{s} - \frac{1}{s + \omega_n} - \frac{\omega_n}{(s + \omega_n)^2} \quad [4.29]$$

Using Table 4.1 we have:

Response of second-order element to a unit step, critical damping $\xi = 1$

$$f_o(t) = 1 - e^{-\omega_n t}(1 + \omega_n t) \quad [4.30]$$

Using Standard tables it can be shown,[1] that if $\xi < 1$ (**underdamping**) then:

Second-order step response, underdamping $\xi < 1$

$$f_o(t) = 1 - e^{-\xi\omega_n t}\left[\cos\omega_n\sqrt{(1-\xi^2)}t + \frac{\xi}{\sqrt{(1-\xi^2)}}\sin\omega_n\sqrt{(1-\xi^2)}t\right] \quad [4.31]$$

and if $\xi > 1$ (**overdamping**) then:

Second-order step response, overdamping $\xi > 1$

$$f_o(t) = 1 - e^{-\xi\omega_n t}\left[\cosh\omega_n\sqrt{(\xi^2-1)}t + \frac{\xi}{\sqrt{(\xi^2-1)}}\sinh\omega_n\sqrt{(\xi^2-1)}t\right] \quad [4.32]$$

The form of the responses is shown in Fig. 4.7. As an example consider the step response of a force sensor with stiffness $k = 10^3\ \mathrm{N\,m^{-1}}$, mass $m = 10^{-1}$ kg and damping constant $\lambda = 10\ \mathrm{Ns\,m^{-1}}$. The steady-state sensitivity $K = 1/k = 10^{-3}\ \mathrm{m\,N^{-1}}$, natural frequency $\omega_n = \sqrt{(k/m)} = 10^2\ \mathrm{rad\,s^{-1}}$ and damping coefficient $\xi = \lambda/2\sqrt{(km)} = 0.5$. Initially at time $t = 0-$, a steady force $F(0-) = 10$ N causes a steady displacement of $(1/10^3) \times 10$ metre, i.e. 10 mm. Suppose that at $t = 0$ the force is suddenly increased from 10 to 12 N, i.e. there is a step change ΔF of 2 N. The resulting change $\Delta x(t)$ in displacement is found using

$$\Delta x(t) = \text{steady-state sensitivity} \times \text{step height} \times \text{unit step response } f_o(t) \quad [4.33]$$

i.e.

$$\Delta x(t) = \frac{1}{10^3} \times 2 \times [1 - e^{-50t}(\cos 86.6t + 0.58 \sin 86.6t)] \text{ metre}$$

$$= 2[1 - e^{-50t}(\cos 86.6t + 0.58 \sin 86.6t)] \text{ mm.} \quad [4.34]$$

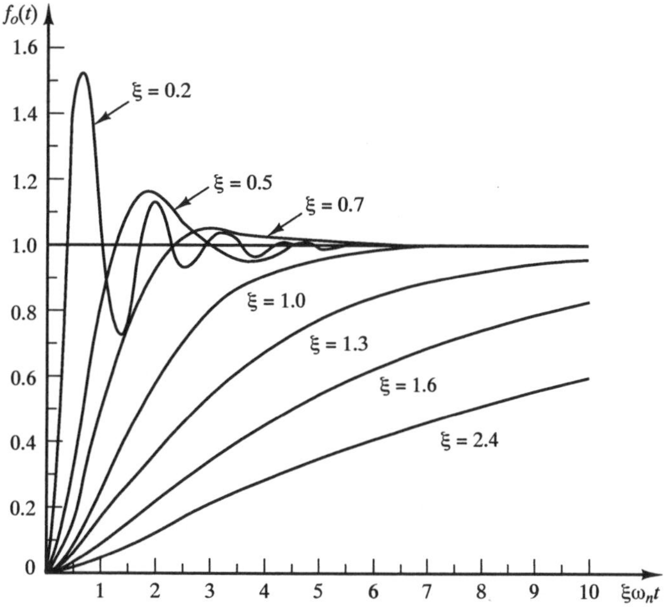

Fig. 4.7 Response of a second-order element to a unit step

Eventually as t becomes large Δx tends to 2 mm, i.e. x settles out at a new steady value of 12 mm. From Fig. 4.7 we see that for $\xi = 0.5, f_o(t)$ has a maximum value $f_o^p = 1.17$ at the peak of the first oscillation, i.e. $\Delta x(t)$ has a maximum value of 2.34 mm. The first oscillation peak occurs at time t_p, where $\xi\omega_n t_p = 1.8$ i.e. $t_p =$ 36 ms. The difference $(f_o^p - 1)$ between maximum and steady-state values of $f_o(t)$ for $\xi < 1$ is termed the maximum overshoot. Maximum overshoot is given by:

$$\exp\left[\frac{-\pi\xi}{\sqrt{(1-\xi^2)}}\right]$$

i.e. it depends only on ξ. For values of $\xi < 1$, an estimate of ξ can be found from measurement of maximum overshoot and then knowing ξ, ω_n can be estimated from measurement of t_p since $\omega_n = \pi/[t_p\sqrt{(1-\xi^2)}]$.

4.2.2 Sinusoidal response of first- and second-order elements

From Table 4.1 we see that the Laplace transform of sine wave $f(t) = \sin \omega t$, with unit amplitude and angular frequency ω, is $\bar{f}(s) = \omega/(s^2 + \omega^2)$. Thus if a sine wave of amplitude $\hat{I}$ is input to a first-order element, then the Laplace transform of the output signal is

$$\bar{f}_o(s) = \frac{1}{(1 + \tau s)} \frac{\hat{I}.\omega}{(s^2 + \omega^2)} \qquad [4.35]$$

Expressing [4.35] in partial fractions we have

$$\bar{f}_o(s) = \frac{A}{(1 + \tau s)} + \frac{Bs + C}{s^2 + \omega^2} \qquad [4.36]$$

where:

$$A = \frac{\omega\tau^2\hat{I}}{(1 + \tau^2\omega^2)}, \quad B = \frac{-\omega\tau\hat{I}}{(1 + \tau^2\omega^2)}, \quad C = \frac{\omega\hat{I}}{(1 + \tau^2\omega^2)}$$

so that:

$$\begin{aligned}\bar{f}_o(s) &= \frac{\omega\tau^2\hat{I}}{(1+\tau^2\omega^2)}\frac{1}{(1+\tau s)} + \frac{\hat{I}}{(1+\tau^2\omega^2)}\left\{\frac{-\omega\tau s+\omega}{s^2+\omega^2}\right\} \\ &= \frac{\omega\tau^2\hat{I}}{(1+\tau^2\omega^2)}\frac{1}{(1+\tau s)} \\ &\quad + \frac{\hat{I}}{\sqrt{(1+\tau^2\omega^2)}}\left\{\frac{\omega\frac{1}{\sqrt{(+\tau^2\omega^2)}}+s\frac{-\omega\tau}{\sqrt{(1+\tau^2\omega^2)}}}{s^2+\omega^2}\right\} \\ &= \frac{\omega\tau^2\hat{I}}{(1+\tau^2\omega^2)}\frac{1}{(1+\tau s)} + \frac{\hat{I}}{\sqrt{(1+\tau^2\omega^2)}}\left\{\frac{\omega\cos\phi+s\sin\phi}{s^2+\omega^2}\right\}\end{aligned} \qquad [4.37]$$

where

$$\cos\phi = \frac{1}{\sqrt{(1+\tau^2\omega^2)}}, \quad \sin\phi = \frac{-\omega\tau}{\sqrt{(1+\tau^2\omega^2)}}$$

Using Table 4.1 we have:

$$f_o(t) = \underbrace{\frac{\omega\tau\hat{I}}{1+\tau^2\omega^2}e^{-t/\tau}}_{\text{Transient term}} + \underbrace{\frac{\hat{I}}{\sqrt{(1+\tau^2\omega^2)}}\sin(\omega t + \phi)}_{\text{Sinusoidal term}} \qquad [4.38]$$

In a sine wave test experiment, we wait until the transient term has decayed to zero and measure the sinusoidal signal:

$$f_o(t) = \frac{\hat{I}}{\sqrt{(1+\tau^2\omega^2)}}\sin(\omega t + \phi) \qquad [4.39]$$

which remains. We see therefore that the output signal is also a sine wave of frequency ω but with amplitude $\hat{I}/\sqrt{(1 + \tau^2\omega^2)}$, and shifted in phase by $\phi = -\tan^{-1}(\omega\tau)$ relative to the input sine wave. These amplitude and phase results can be found directly from the transfer function $G(s) = 1/(1 + \tau s)$ without having to use the table of transforms. If we replace s by $j\omega (j = \sqrt{(-1)})$ in $G(s)$ we form the complex number $G(j\omega) = 1/(1 + j\tau\omega)$. The **magnitude** $|G(j\omega)| = 1/\sqrt{(1 + \tau^2\omega^2)}$ of this complex number is equal to the ratio of output amplitude to input amplitude, and the **angle** or **argument** $\arg G(j\omega) = -\tan^{-1}(\omega\tau)$ is equal to the phase difference ϕ between output and input sine waves. Figure 4.8 shows amplitude ratio versus frequency and phase versus frequency graphs for a first-order element; these are known as the **frequency response** characteristics of the element. From the above equations we see that when $\omega\tau = 1$, i.e. $\omega = 1/\tau$, the amplitude ratio $= 1/\sqrt{2}$ and phase difference $\phi = -45°$. These results enable the value of τ to be found from experimental frequency response data.

The above results can be generalised to an element with steady-state sensitivity

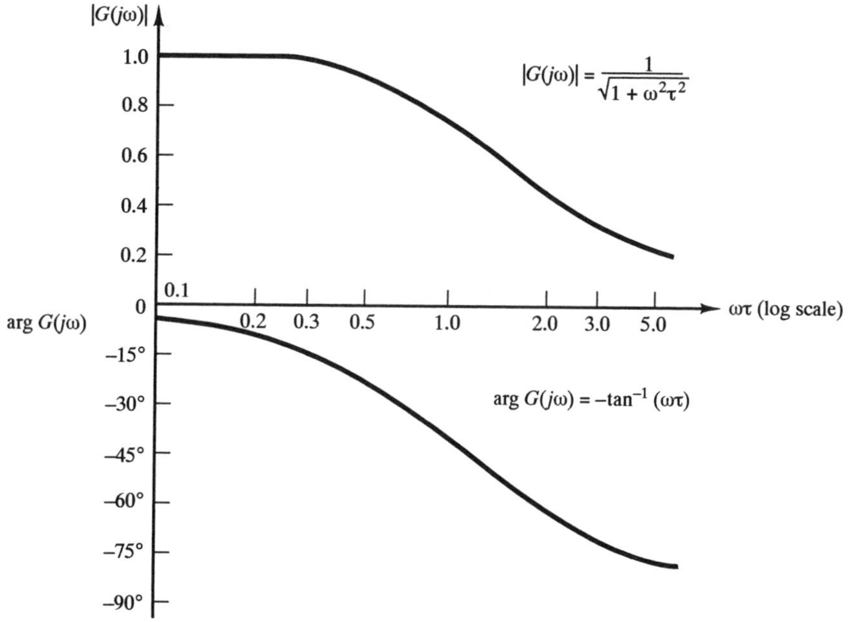

Fig. 4.8 Frequency response characteristics of first-order element with

$$G(s) = \frac{1}{1+\tau s}$$

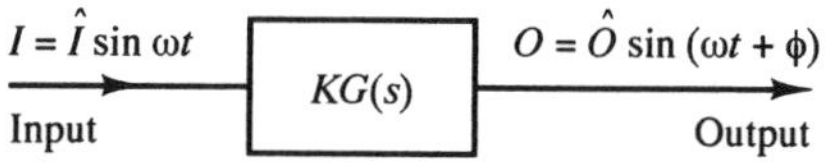

Fig. 4.9 Frequency response of an element with linear dynamics

K(or $\partial O/\partial I$) and transfer function $G(s)$, subject to a sinusoidal input signal $I = \hat{I}$ sin ωt as in Fig. 4.9. In the steady state we can make four statements about the output signal:

(a) O is also a sine wave;
(b) the frequency of O is also ω;
(c) the amplitude of O is $\hat{O} = K|G(j\omega)|\hat{I}$;
(d) the phase difference between O and I is $\phi = \arg G(j\omega)$.

Using the above rules we can quickly find the amplitude ratio and phase relations for a second-order element with:

$$G(s) = \frac{1}{\dfrac{1}{\omega_n^2}s^2 + \dfrac{2\xi}{\omega_n}s + 1}$$

Here we have:

$$G(j\omega) = \frac{1}{\dfrac{1}{\omega_n^2}(j\omega)^2 + \dfrac{2\xi}{\omega_n}(j\omega) + 1}$$

so that

Frequency response characteristics of second-order element

$$\textbf{Amplitude ratio} = |G(j\omega)| = \frac{1}{\sqrt{\left[\left(1 - \dfrac{\omega^2}{\omega_n^2}\right)^2 + 4\xi^2\dfrac{\omega^2}{\omega_n^2}\right]}}$$

$$\textbf{Phase difference} = \arg G(j\omega) = -\tan^{-1}\left[\frac{2\xi\,\omega/\omega_n}{1 - \omega^2/\omega_n^2}\right] \qquad [4.40]$$

These characteristics are shown graphically in Fig. 4.10; both amplitude ratio and phase characteristics are critically dependent on the value of ξ.

We note that for $\xi < 0.7$, $|G(j\omega)|$ has a maximum value which is greater than unity. This maximum value is given by:

$$|G(j\omega)|_{\mathrm{MAX}} = \frac{1}{2\xi\sqrt{(1 - \xi^2)}}$$

and occurs at the **resonant frequency** ω_{R} where:

$$\omega_R = \omega_n\sqrt{(1 - 2\xi^2)} \quad (\xi < 1/\sqrt{2}).$$

Thus by measuring $|G(j\omega|_{\mathrm{MAX}}$ and ω_R, ξ and ω_n can be found. An alternative to plotting $|G(j\omega)|$ versus ω is a graph of the number of decibels N dB versus ω, where $N = 20\log_{10}|G(j\omega)|$. Thus if $|G(j\omega)| = 1$, $N = 0$ dB; $|G(j\omega)| = 10$, $N = +20$ dB; and if $|G(j\omega)| = 0.1$, $N = -20$ dB.

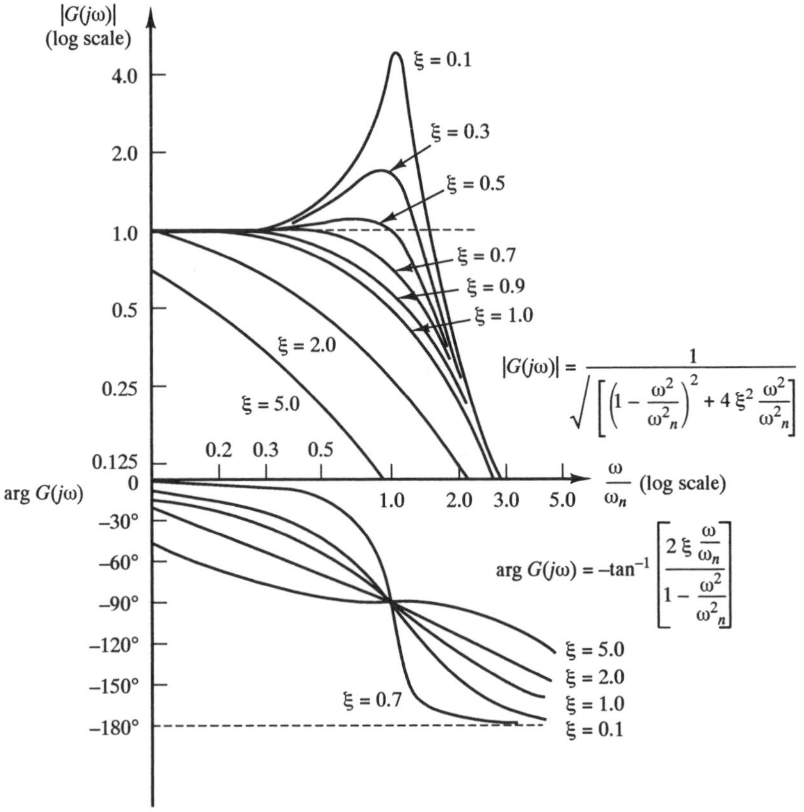

Fig. 4.10 Frequency response characteristics of second-order element with:

$$G(s) = \frac{1}{\dfrac{1}{\omega_n^2}s^2 + \dfrac{2\xi}{\omega_w}s + 1}$$

4.3 Dynamic errors in measurement systems

Figure 4.11 shows a complete measurement system consisting of n elements. Each element i has ideal steady-state and linear dynamic characteristics and can therefore be represented by a constant steady-state sensitivity K_i and a transfer function $G_i(s)$.

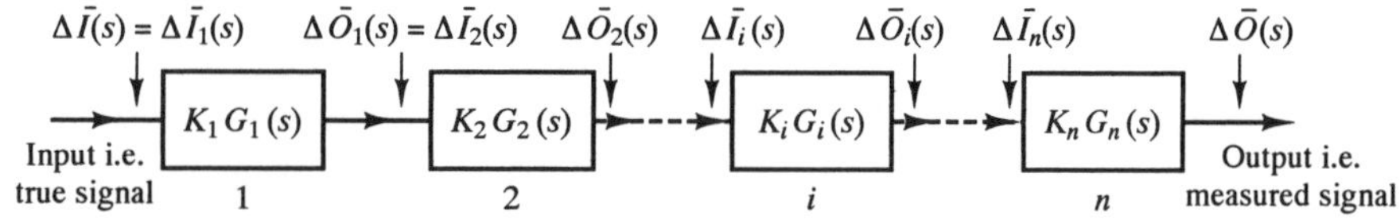

Fig. 4.11 Complete measurement system with dynamics

We begin by assuming that the steady-state sensitivity $K_1 K_2 \cdots K_i \cdots K_n$ for the overall system is equal to 1, i.e. the system has no steady-state error (Section 3.1). The system transfer function $G(s)$ is the product of the individual element transfer functions, i.e.

Transfer function for complete measurement system

$$\frac{\Delta\bar{O}(s)}{\Delta\bar{I}(s)} = G(s) = G_1(s)G_2(s)\cdots G_i(s)\cdots G_n(s) \qquad [4.41]$$

In principle we can use equation [4.41] to find the system output signal $\Delta O(t)$ corresponding to a time varying input signal $\Delta I(t)$. We first find the Laplace transform $\Delta\bar{I}(s)$ of $\Delta I(t)$, then using [4.9] the Laplace transform of the output signal is $\Delta\bar{O}(s) = G(s)\Delta\bar{I}(s)$. By expressing $\Delta\bar{O}(s)$ in partial fractions, and using standard tables of Laplace transforms, we can find the corresponding time signal $\Delta O(t)$. Expressing this mathematically:

$$\Delta O(t) = \mathcal{L}^{-1}[G(s)\Delta\bar{I}(s)] \qquad [4.42]$$

where $\mathcal{L}^{-1}$ denotes the inverse Laplace transform. The dynamic error $E(t)$ of the measurement system is the difference between the measured signal and the true signal, i.e. the difference between $\Delta O(t)$ and $\Delta I(t)$

Dynamic error of a measurement system

$$E(t) = \Delta O(t) - \Delta I(t) \qquad [4.43]$$

Using [4.42] we have:

$$E(t) = \mathcal{L}^{-1}[G(s)\Delta\bar{I}(s)] - \Delta I(t) \qquad [4.44]$$

The simple temperature measurement system (Fig. 4.12), first introduced in Section 3.1, provides a good example of dynamic errors. The thermocouple has a time

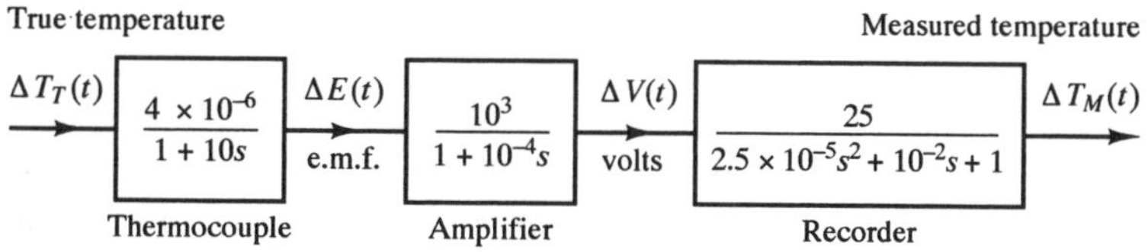

Fig. 4.12 Simple temperature measurement system with dynamics

constant of 10 s, the amplifier a time constant of 10^{-4} seconds (Chapter 9), and the recorder (Chapter 11) is a second-order element with $\omega_n = 200$ rad/s and $\xi = 1.0$. The overall steady-state sensitivity of the system is unity.

We can now calculate the dynamic error of the system for a step input of +20 °C, i.e. $\Delta T_T(t) = 20u(t)$ and $\Delta \bar{T}_T(s) = 20\,(1/s)$. Thus the Laplace transform of the output signal is:

$$\Delta \bar{T}_M(s) = 20\frac{1}{s}\,\frac{1}{(1+10s)}\,\frac{1}{(1+10^{-4}s)}\,\frac{1}{(1+1/200s)^2}$$

$$= 20\left\{\frac{1}{s} - \frac{A}{(s+0.1)} - \frac{B}{(s+10^4)} - \frac{Cs+D}{(s+200)^2}\right\} \qquad [4.45]$$

Using Table 4.1 and equation [4.30],

$$\Delta T_M(t) = 20\{u(t) - Ae^{-0.1t} - Be^{-10^4t} - Ee^{-200t}(1 + 200t)\}$$

and the dynamic error:

$$\begin{aligned} E(t) &= \Delta T_M(t) - \Delta T_T(t) \\ &= -20\{Ae^{-0.1t} + Be^{-10^4t} - Ee^{-200t}(1 + 200t)\} \end{aligned} \qquad [4.46]$$

where the negative sign indicates too low a reading. The Be^{-10^4t} term decays to zero after about 5×10^{-4} s, and the $Ee^{-200t}(1 + 200t)$ term decays to zero after about 25 ms. The $Ae^{-0.1t}$ term, which corresponds to the 10 s time constant of the thermocouple, takes about 50 s to decay to zero and so has the greatest effect on the dynamic error.

We can use the rules developed in Section 4.2.2 to find the dynamic error of a system, with transfer function $G(s)$ subject to a sinusoidal input $\Delta I(t) = \hat{I} \sin \omega t$. From Fig. 4.9 we have:

$$\Delta O(t) = |G(j\omega)|\hat{I} \sin (\omega t + \phi)$$

giving

$$E(t) = \hat{I}\{|G(j\omega)|\sin (\omega t + \phi) - \sin \omega t\} \qquad [4.47]$$

where $\phi = \arg G(j\omega)$.

Suppose that the above temperature measurement system is measuring a sinusoidal temperature variation of amplitude $\hat{T}_T = 20$ °C and period $T = 6$ s, i.e. angular frequency $\omega = 2\pi/T \approx 1.0$ rad s^{-1}. The frequency response function $G(j\omega)$ is:

$$G(j\omega) = \frac{1}{(1+10j\omega)(1+10^{-4}j\omega)(1+10^{-2}j\omega+2.5\times 10^{-5}(j\omega)^2)} \qquad [4.48]$$

so that at $\omega = 1$

$$|G(j\omega)|_{\omega=1} = \frac{1}{\sqrt{\{(1+100)(1+10^{-8})[(1-2.5\times 10^{-5})^2+10^{-4}]\}}} \approx 0.10$$

and [4.49]

$$\arg G(j\omega)_{\omega=1} \approx 0 - \tan^{-1}(10) - \tan^{-1}(10^{-4})) - \tan^{-1}(10^{-2}) \approx -85°$$

We note from the above equations that the values of $|G(j\omega)|$ and $\arg G(j\omega)$ at $\omega = 1$ are determined mainly by the 10 s time constant; the dynamic characteristics of the other elements will only begin to affect the system performance at much higher frequencies. Since $T_T(t) = 20 \sin t$ and $T_M(t) = 0.1 \times 20 \sin(t - 85°)$, the error is:

$$E(t) = 20\{0.1 \sin(t - 85°) - \sin t\} \qquad [4.50]$$

We note that in the case of a sine wave input, the output recording is also a sine wave, i.e. the **waveform** of the signal is unaltered even though there is a reduction in amplitude and a phase shift.

In practice the input signal to a measurement system is more likely to be **periodic** rather than a simple sine wave. A periodic signal is one that repeats itself at equal intervals of time T, i.e. $f(t) = f(t + T) = f(t + 2T)$ etc. where T is the period. One example of a periodic measurement signal is the time variation of the temperature inside a diesel engine; another is the vibration of the casing of a centrifugal compressor. In order to calculate dynamic errors for periodic signals, we need to use **Fourier analysis**. Any periodic signal $f(t)$ with period T, can be expressed as a series of sine and cosine waves; these have frequencies which are **harmonics** of the fundamental frequency $\omega_1 = 2\pi/T \text{ rad s}^{-1}$, i.e.

Fourier series for periodic signal

$$f(t) = a_0 + \sum_{n=1}^{n=\infty} a_n \cos n\omega_1 t = \sum_{n=1}^{n=\infty} b_n \sin n\omega_1 t \qquad [4.51]$$

where

$$a_n = \frac{2}{T}\int_{-T/2}^{+T/2} f(t) \cos n\omega_1 t \, dt$$

$$b_n = \frac{2}{T}\int_{-T/2}^{+T/2} f(t) \sin n\omega_1 t \, dt \qquad [4.52]$$

and

$$a_0 = \frac{1}{T}\int_{-T/2}^{+T/2} f(t) \, dt = \text{average value of } f(t) \text{ over } T$$

If $f(t) = \Delta I(t)$, where $\Delta I(t)$ is the deviation of measurement input signal $I(t)$ from steady-state or d.c. value I_0, then $a_0 = 0$. If we also assume that $f(t)$ is odd, i.e. $f(t) = -f(-t)$, then $a_n = 0$ for all n, i.e. there are only sine terms present in the series. This simplifying assumption does not affect the general conclusions drawn in the following section. The system input signal is thus given by

$$\Delta I(t) = \sum_{n=1}^{n=\infty} \hat{I}_n \sin n\omega_1 t \qquad [4.53]$$

where $\hat{I}_n = b_n$ is the amplitude of the nth harmonic at frequency $n\omega_1$. In order to find $\Delta O(t)$, let us first suppose that only the nth harmonic $\hat{I}_n \sin n\omega_1 t$ is input to the system. From Fig. 4.9 the corresponding output signal is

$$\hat{I}_n |G(jn\omega_1)| \sin(n\omega_1 t + \phi_n)$$

where $\phi_n = \arg G(jn\omega_1)$.

We now require to use the principle of superposition, which is a basic property of linear systems (i.e. systems described by linear differential equations). This can be stated as follows:

> If an input $I_1(t)$ causes an output $O_1(t)$ and an input $I_2(t)$ causes an output $O_2(t)$, then an input $I_1(t) + I_2(t)$ causes an output $O_1(t) + O_2(t)$, provided the system is linear.

This means that if the total input signal is the sum of many sine waves (equation [4.53]), then the total output signal is the sum of the responses to each sine wave, i.e.

$$\Delta O(t) = \sum_{n=1}^{n=\infty} \hat{I}_n |G(jn\omega_1)| \sin(n\omega_1 t + \phi_n) \qquad [4.54]$$

The dynamic error is thus

Dynamic error of system with periodic input signal

$$\Delta E(t) = \sum_{n=1}^{n=\infty} \hat{I}_n \{|G(jn\omega_1)| \sin(n\omega_1 t + \phi_n) - \sin n\omega_1 t\} \qquad [4.55]$$

As an example, suppose that the input to the temperature measurement system is a square wave of amplitude 20 °C and period $T = 6$ s (i.e. $\omega_1 = 2\pi/T \approx 1$ rad s^{-1}) shown in Fig. 4.13. The Fourier Series for the input signal is:

$$\Delta T_T(t) = \frac{80}{\pi}[\sin t + \tfrac{1}{3}\sin 3t + \tfrac{1}{5}\sin 5t + \tfrac{1}{7}\sin 7t + \cdots] \qquad [4.56]$$

Figure 4.13 shows the amplitude–frequency and phase-frequency relationships for the input temperature; these define the frequency spectrum of the signal. The spectrum consists of a number of lines at frequencies ω_1, $3\omega_1$, $5\omega_1$, etc., of decreasing length to represent the smaller amplitudes of the higher harmonics. In practical cases we can terminate or truncate the series at a harmonic where the amplitude is negligible; in this case we choose $n = 7$. In order to find the output signal, i.e. the recorded waveform, we need to evaluate the magnitude and argument of $G(j\omega)$ at $\omega = 1$, 3, 5, 7 rad s^{-1}.

Thus

$$|G(j)| \approx 0.100, \quad |G(3j)| \approx 0.033, \quad |G(5j)| \approx 0.020, \quad |G(7j)| \approx 0.014$$

and

$$\begin{aligned} &\arg G(j) \approx -85°, \quad \arg G(3j) \approx -90°, \\ &\arg G(5j) \approx -92°, \quad \arg G(7j) \approx -93° \end{aligned} \qquad [4.57]$$

Again the above values are determined mainly by the 10 s thermocouple time constant; the highest signal frequency $\omega = 7$ is still well below the natural frequency of the recorder $\omega_n = 200$. The system output signal is:

$$\begin{aligned} \Delta T_M(t) = \frac{80}{\pi}[&0.100 \sin(t - 85°) + 0.011 \sin(3t - 90°) \\ &+ 0.004 \sin(5t - 92°) + 0.002 \sin(7t - 93°)] \end{aligned} \qquad [4.58]$$

Figure 4.13 shows the system frequency response characteristics, output signal frequency spectrum and the output waveform. We note that, in the output signal,

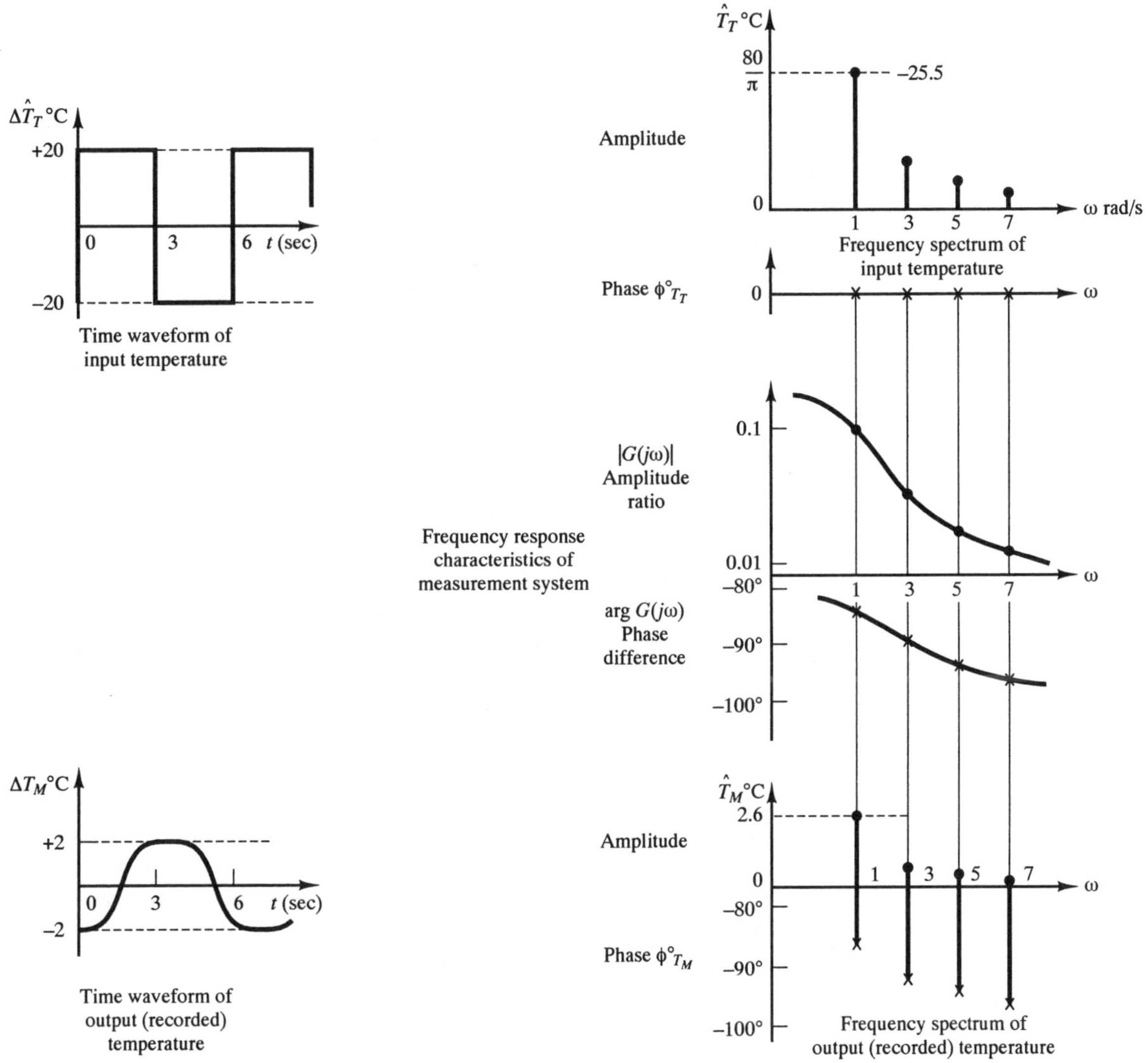

Fig. 4.13 Calculation of dynamic errors with periodic input signal

the amplitudes of the 3rd, 5th and 7th harmonics have been reduced relative to the amplitude of the fundamental. The recorded waveform has therefore a different **shape** from the input signal as well as being reduced in amplitude and changed in phase.

The above ideas can be extended to calculating the dynamic error for **random** input signals. Random signals can be represented by continuous frequency spectra (Chapter 6).

4.4 Techniques for dynamic compensation

From equation [4.55] we see that in order to have $E(t) = 0$ for a periodic signal, the following conditions must be obeyed:

$$|G(j\omega_1)| = |G(j2\omega_1)| = \cdots = |G(jn\omega_1)| = \cdots = |G(jm\omega_1)| = 1$$
$$\arg G(j\omega_1) = \arg G(j2\omega_1) = \cdots = \arg G(jn\omega_1) = \cdots = \arg G(jm\omega_1) = 0 \quad [4.59]$$

where m is the order of the **highest *significant* harmonic**. For a random signal (Chapter 6) with a continuous frequency spectrum containing frequencies between 0 and ω_{max}, we require:

$$|G(j\omega)| = 1 \quad \text{and} \quad \arg G(j\omega) = 0 \quad \text{for} \quad 0 < \omega \leq \omega_{MAX} \quad [4.60]$$

The above conditions represent a theoretical ideal which will be difficult to realise in practice. A more practical criterion is one which limits the variation in $|G(j\omega)|$ to a few per cent for the frequencies present in the signal. For example the condition:

$$0.98 < |G(j\omega)| < 1.02 \quad \text{for} \quad 0 < \omega \leq \omega_{MAX} \quad [4.61]$$

will ensure that the dynamic error is limited to $\approx \pm 2$ per cent for a signal containing frequencies up to $\omega_{MAX}/2\pi$ Hz (Fig. 4.14).

Another commonly-used criterion is that of **bandwidth**. The bandwidth of an element or a system is the range of frequencies for which $|G(j\omega)|$ is greater than $1/\sqrt{2}$. Thus the bandwidth of the system, with frequency response shown in Fig. 4.14, is 0 to ω_B rad s^{-1}. The highest signal frequency ω_{MAX} must be considerably less than ω_B. Since, however, there is a 30% reduction in $|G(j\omega)|$ at ω_B, bandwidth is not a particularly useful criterion for complete measurement systems.

Bandwidth is commonly used in specifying the frequency response of amplifers (Chapter 9); a reduction in $|G(j\omega)|$ from 1 to $1/\sqrt{2}$ is equivalent to a decibel change of $N = 20 \log (1/\sqrt{2}) = -3.0$ dB. A first-order element has a bandwidth between 0 and $1/\tau$ rad s^{-1}.

If a system fails to meet the specified limits on dynamic error $E(t)$; i.e. the system transfer function $G(s)$ does not satisfy a condition such as [4.61], then the first step is to identify which elements in the system dominate the dynamic behaviour. In the temperature measurement system of the previous section, the dynamic error is almost entirely due to the 10 s time constant of the thermocouple.

Fig. 4.14 Percentage limits and bandwidth

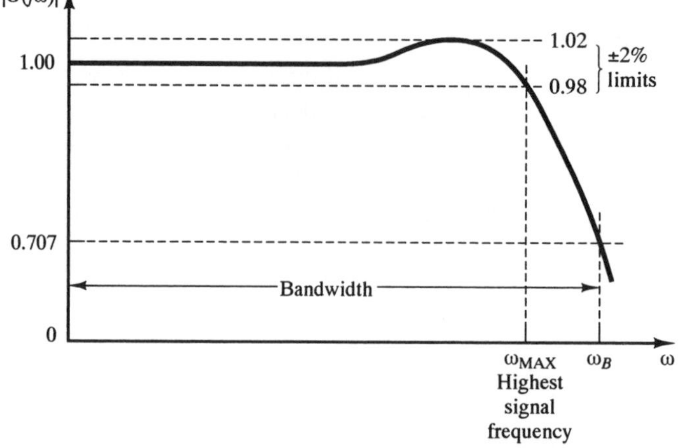

Fig. 4.15 Open-loop dynamic compensation

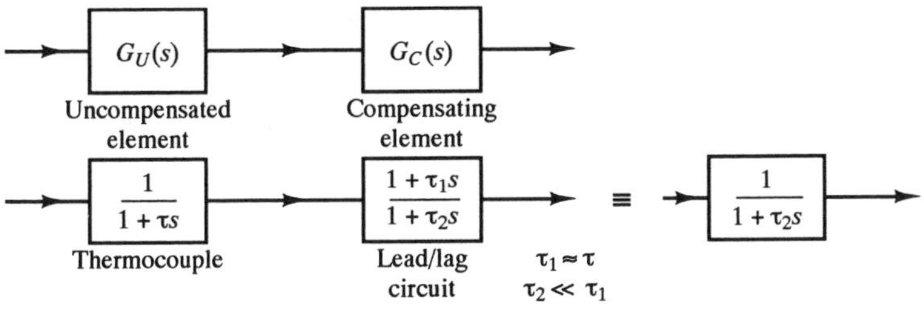

Having identified the dominant elements in the system, the most obvious method of improving dynamic response is that of **inherent design**. In the case of a first-order temperature sensor with $\tau = MC/UA$, τ can be minimised by minimising the mass/area ratio M/A — for example by using a thermistor in the form of a thin flake. In the case of a second-order force sensor with $\omega_n = \sqrt{(k/m)}$, ω_n can be maximised by maximising k/m, i.e. by using high stiffness k and low mass m. Increasing k, however, reduces the steady-state sensitivity $K = 1/k$.

From second-order step and frequency response graphs we see that the optimum value of damping ratio ξ is around 0.7. This value ensures minimum settling time for the step response and $|G(j\omega)|$ closest to unity for the frequency response.

Another possible method is that of **open-loop dynamic compensation** (Fig. 4.15). Given an uncompensated element or system $G_U(s)$, a compensating element $G_C(s)$ is introduced into the system, such that the overall transfer function $G(s) = G_U(s)G_C(s)$ satisfies the required condition (for example eqn [4.61]). Thus if a lead/lag circuit (Fig. 9.7(e)) is used with a thermocouple (Fig. 4.15), the overall time constant is reduced to τ_2 so that $|G(j\omega)|$ is close to unity over a wider range of frequencies. The main problem with this method is that τ can change with heat transfer coefficient U, thus reducing the effectiveness of the compensation (Chapter 14).

Another method is to incorporate the element to be compensated into a closed-loop system with **high gain negative feedback**. An example of this is the constant temperature anemometer system for measuring fluid velocity fluctuations (section 14.3). Another example is the closed-loop accelerometer shown in schematic and block diagram form in Fig. 4.16.

The applied acceleration a produces an inertia force ma on the seismic mass m (Chapter 8). This is balanced by the force of the permanent magnet on the current feedback coil. Any imbalance of forces is detected by the elastic force element to produce a displacement which is detected by a potentiometric displacement sensor (Chapter 5). The potentiometer output voltage is amplified giving a current output which is fed to the feedback coil through a standard resistor to give the output voltage.

Analysis of the block diagram shows that the overall system transfer function is:

$$\frac{\Delta\bar{V}(s)}{\Delta\bar{a}(s)} = \frac{mR}{K_F}\left\{\frac{1}{\dfrac{k}{K_A K_D K_F}\dfrac{1}{\omega_n^2}s^2 + \dfrac{2\xi}{\omega_n}\dfrac{ks}{K_A K_D K_F} + \left(1 + \dfrac{k}{K_A K_D K_F}\right)}\right\} \qquad [4.62]$$

If K_A is made large so that $K_A K_D K_F/k \gg 1$ (Chapter 3), then the system transfer function can be expressed in the form:

Fig. 4.16 Schematic and block diagram of closed-loop accelerometer

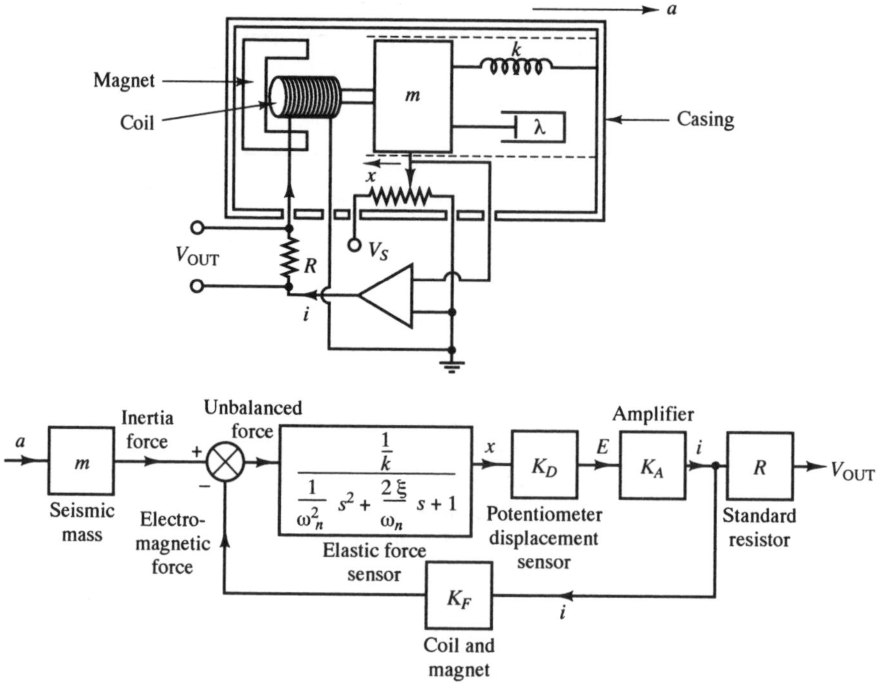

$$\frac{\Delta \bar{V}(s)}{\Delta \bar{a}(s)} = \frac{K_s}{\dfrac{1}{\omega_{ns}^2}s^2 + \dfrac{2\xi_s}{\omega_{ns}}s + 1}$$

where system steady-state sensitivity $K_s = \dfrac{mR}{K_F}$

system natural frequency $\omega_{ns} = \omega_n \sqrt{\left(\dfrac{K_A K_D K_F}{k}\right)}$

system damping ratio $\xi_s = \xi \sqrt{\left(\dfrac{k}{K_A K_D K_F}\right)}$ [4.63]

We see that the system natural frequency ω_{ns} is now much greater than that of the elastic force element itself. The system damping ratio ξ_s is much less than ξ, but by making ξ large a value of $\xi_s \approx 0.7$ can be obtained. Furthermore the system steady-state sensitivity depends only on *m*, K_F and *R* which can be made constant to a high degree.

Conclusion

The **dynamic characteristics** of typical measurement system elements were initially discussed; in particular the **transfer functions** of **first-** and **second-order** elements were derived. The response of both first- and second-order elements to **step** and **sine wave inputs** was then studied. A general description of the **dynamic error** complete measurement system was then developed and applied to a temperature measurement system subject to step, sine wave and **periodic** input signals. Finally methods of **dynamic compensation**, which reduce dynamic error, were explained.

Reference

4.1 HEALEY M 1967 *Principles of Automatic Control*, English Universities Press, London, pp. 308–9.

Problems

4.1 A temperature measurement system consists of linear elements and has an overall steady-state sensitivity of unity. The dynamics of the system are determined by the first-order transfer function of the sensing element. At time $t = 0$, the sensing element is suddenly transferred from air at 20 °C to boiling water. One minute later the element is suddenly transferred back to air. Using the data given below, calculate the system dynamic error at the following times: $t = 10$, 20, 50, 120, and 300 s.

Sensor data
Mass = 5×10^{-2} kg.
Surface area = $10^{-3}\,\text{m}^2$
Specific heat = $0.2\,\text{J kg}^{-1}\,°\text{C}^{-1}$
Heat transfer coefficient for air = $0.2\,\text{W m}^{-2}\,°\text{C}^{-1}$
Heat transfer coefficient for water = $1.0\,\text{W m}^{-2}\,°\text{C}^{-1}$

4.2 A force sensor has a mass of 0.5 kg, stiffness of $2 \times 10^2\,\text{N m}^{-1}$ and a damping constant of $6.0\,\text{N s m}^{-1}$.

(a) Calculate the steady-state sensitivity, natural frequency and damping ratio for the sensor.
(b) Calculate the displacement of the sensor for a steady input force of 2 N
(c) If the input force is suddenly increased from 2 to 3 N, derive an expression for the resulting displacement of the sensor.

4.3 A force measurement system consists of linear elements and has an overall steady-state sensitivity of unity. The dynamics of the system are determined by the second-order transfer function of the sensing element which has a natural frequency $\omega_n = 40\,\text{rad s}^{-1}$ and a damping ratio $\xi = 0.1$. Calculate the system dynamic error corresponding to the periodic input force signal:

$$F(t) = 50\{\sin 10t + \tfrac{1}{3}\sin 30t + \tfrac{1}{5}\sin 50t\}$$

4.4 An uncompensated thermocouple has a time constant of 10 s in a fast moving liquid.

(a) Calculate the bandwidth of the thermocouple frequency response.
(b) Find the range of frequencies for which the amplitude ratio of the uncompensated thermocouple is flat within ±5%.
(c) A lead/lag circuit with transfer function $G(s) = (1 + 10s)/(1 + s)$ is used to compensate for thermocouple dynamics. Calculate the range of frequencies for which the amplitude ratio of the compensated system is flat within ±5%.
(d) The velocity of the liquid is reduced, causing the thermocouple time constant to increase to 20 s. By sketching $|G(j\omega)|$ explain why the effectiveness of the above compensation is reduced.

4.5 An elastic force sensor has an effective seismic mass of 0.1 kg, a spring stiffness of $10\ \mathrm{N\,m^{-1}}$ and a damping constant of $14\ \mathrm{N\,s\,m^{-1}}$.

(a) Calculate the following quantities:
  (i) sensor natural frequency;
  (ii) sensor damping ratio;
  (iii) transfer function relating displacement and force.

(b) The above sensor is incorporated into a closed-loop, force balance accelerometer. The following components are also present:

  Potentiometer displacement sensor: sensitivity $1.0\ \mathrm{V\,m^{-1}}$
  Amplifier: voltage input, current output, sensitivity $40\ \mathrm{A\,V^{-1}}$
  Coil and magnet: current input, force output, sensitivity $25\ \mathrm{N\,A^{-1}}$
  Resistor: $250\ \Omega$.
  (i) Draw a block diagram of the accelerometer.
  (ii) Calculate the overall accelerometer transfer function.
  (iii) Explain why the dynamic performance of the accelerometer is superior to that of the elastic sensor.

4.6 A load cell consists of an elastic cantilever and a displacement transducer. The cantilever has a stiffness of $10^2\ \mathrm{N\,m^{-1}}$, a mass of 0.5 kg and a damping constant of $2\ \mathrm{N\,s\,m^{-1}}$. The displacement transducer has a steady-state sensitivity of $10\ \mathrm{V\,m^{-1}}$.

(a) A package of mass 0.5 kg is suddenly dropped onto the load cell. Use eqn [4.31] to derive a numerical equation describing the corresponding time variation of the output voltage ($g = 9.81\ \mathrm{m\,s^{-2}}$).

(b) The load cell is used to weigh packages moving along a conveyor belt at the rate of 60 per minute. Use the equation derived in (a) to explain why the load cell is unsuitable for this application. Explain what modifications to the load cell are necessary.

4.7 A force measurement system consisting of a piezoelectric crystal, charge amplifier and recorder is shown in Fig. Prob. 7.

(a) Calculate the system dynamic error corresponding to the force input signal:
$F(t) = 50\{\sin 10\,t + \frac{1}{3}\sin 30\,t + \frac{1}{5}\sin 50\,t\}$

(b) Explain briefly the system modifications necessary to reduce the error in (a) (Hint: see Fig. 8.15).

Fig. Prob. 7

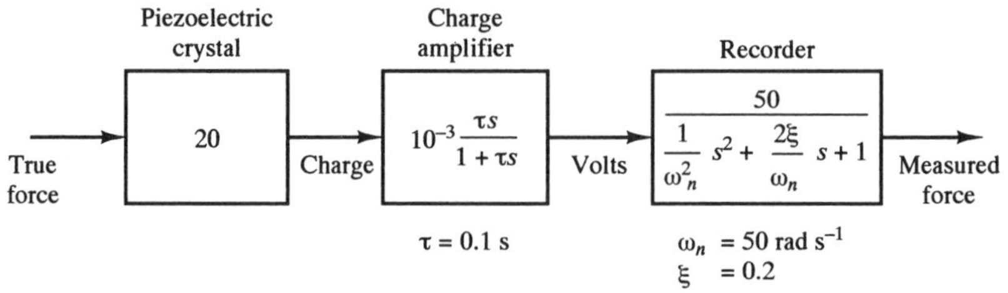

5

Loading effects and two-port networks

In our discussion of measurement systems no consideration has yet been given to the effects of **loading**. One important effect is that of **inter-element loading** where a given element in the system may modify the characteristics of the previous element (for example by drawing current). In turn, the characteristics of this element may be modified by the following element in the system. Inter-element loading is normally an electrical loading effect which is described in the first section of this chapter using **Thévenin** and **Norton equivalent circuits**. The second section begins by discussing the analogies between electrical and non-electrical variables. This means that mechanical and thermal systems can be described by equivalent circuits and sensing elements by two-port networks. Two-port networks are then used to describe **process loading**; here the introduction of the sensing element into the process or system being measured causes the value of the measured variable to change. Finally two-port networks are used to describe **bilateral transducers** which use reversible physical effects.

5.1 Electrical loading

We have so far represented measurement systems as blocks connected by single lines where the transfer of information and energy is in terms of one variable only. Thus in the temperature measurement system of Fig. 3.2 the information transfer between elements is in terms of voltage only. No allowance can therefore by made for the amplifier drawing current from the thermocouple and the indicator drawing current from the amplifier. In order to describe both voltage and current behaviour at the connection of two elements we need to represent each element by equivalent circuits characterised by two terminals. The connection is then shown by two lines.

5.1.1 Thévenin equivalent circuit

Thévenin's theorem states that any network consisting of linear impedances and voltage sources can be replaced by an equivalent circuit consisting of a voltage source E_{Th} and a series impedance Z_{Th} (Fig. 5.1). The source E_{Th} is equal to the open

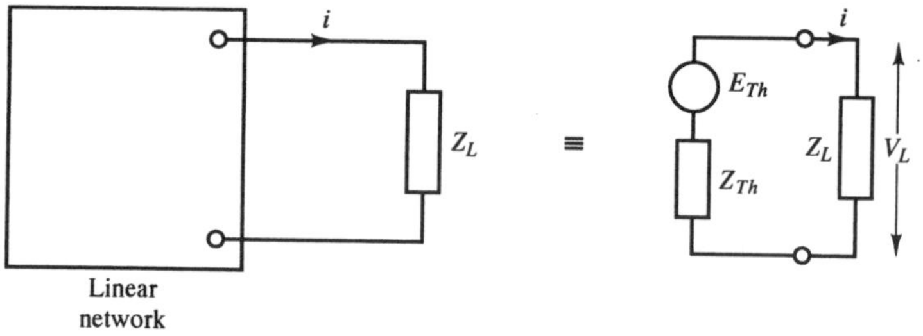

Fig. 5.1 Thévenin equivalent circuit

circuit voltage of the network across the output terminals, and Z_{Th} is the impedance looking back into these terminals with all voltage sources reduced to zero and replaced by their internal impedances. Thus connecting a load Z_L across the output terminals of the network is equivalent to connecting Z_L across the Thévenin circuit. The current i in Z_L is then simply given by:

$$i = \frac{E_{Th}}{Z_{Th} + Z_L} \qquad [5.1]$$

and the voltage V_L across the load by:

Loading of Thévenin equivalent circuit

$$V_L = iZ_L = E_{Th}\frac{Z_L}{Z_{Th} + Z_L} \qquad [5.2]$$

From 5.2 we see that if $Z_L \gg Z_{Th}$, then $V_L \to E_{Th}$; i.e. in order to get **maximum voltage transfer** from the network to the load, the load impedance should be *far greater* than the Thévenin impedance for the network. In order to get **maximum power transfer** from network to load, the load impedance should be equal to the network impedance; i.e. $Z_L = Z_{Th}$.[1] (An example of the calculation of E_{Th} and Z_{Th} for a potentiometer displacement transducer is given in the following section and for a deflection bridge in Section 9.1.)

We can now discuss the Thévenin equivalent circuit for the temperature measurement system of Fig. 3.2. The thermocouple may be represented by $Z_{Th} = 20\,\Omega$ (resistive) and $E_{Th} = 40T\,\mu\text{V}$, where T is the measurement junction temperature, if non-linear and reference junction temperature effects are ignored. The amplifier acts both as a load for the thermocouple and as a voltage source for the indicator. Figure 5.2 shows a general equivalent circuit for an amplifier with two pairs of terminals. Using typical amplifier data (Chapter 9), we have input impedance $Z_{IN} = R_{IN} = 2 \times 10^6\,\Omega$, closed-loop voltage gain $A = 10^3$, output impedance $Z_{OUT} = R_{OUT} = 75\,\Omega$. The indicator is a resistive load of $10^4\,\Omega$. The complete

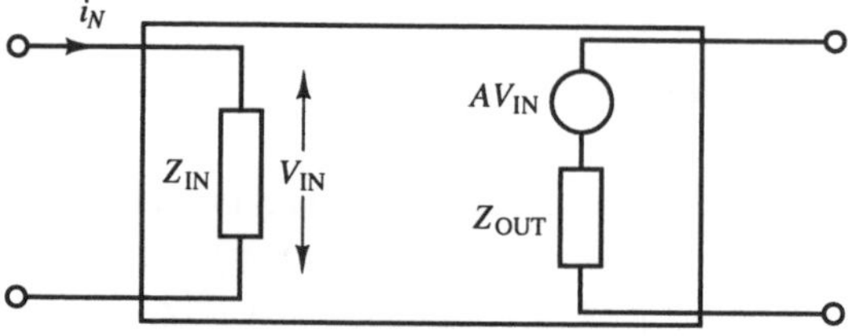

Fig. 5.2 Equivalent circuit for amplifier

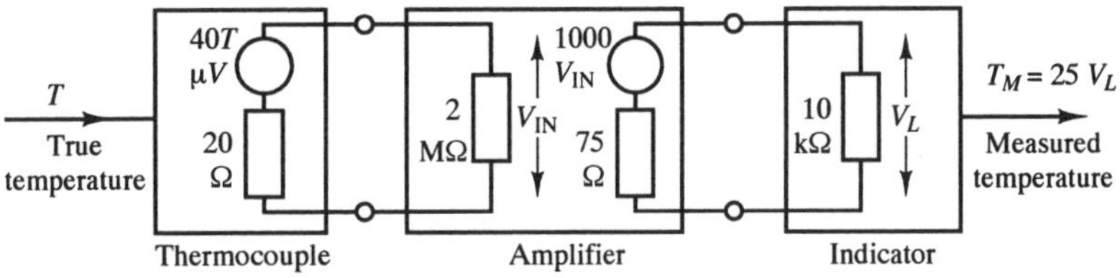

Fig. 5.3 Thévenin equivalent to temperature measurement system

equivalent circuit for the system is shown in Fig. 5.3, and using eqn [5.2] we have:

$$V_{\text{IN}} = 40 \times 10^{-6} T \frac{2 \times 10^6}{2 \times 10^6 + 20} \quad \text{and} \quad V_L = 1000\, V_{\text{IN}} \frac{10^4}{10^4 + 75} \tag{5.3}$$

If the indicator scale is drawn so that a change of 1 V in V_L causes a change in deflection of 25 °C, then the measured temperature $T_M = 25V_L$. This gives:

$$T_M = \left(\frac{2 \times 10^6}{2 \times 10^6 + 20}\right)\left(\frac{10^4}{10^4 + 75}\right) T = 0.9925T \tag{5.4}$$

i.e. we have to introduce the factor $Z_L/(Z_{Th} + Z_L)$ at every interconnection of two elements to allow for loading. The loading error $= -0.0075T$; this is in addition to the steady-state error due to element inperfections calculated in Chapter 3.

The loading error in the above example is small, but if care is not taken it can be very large. Suppose a pH glass electrode (Chapter 8), with sensitivity 59 millivolts per pH i.e. $E_{Th} = 59$ pH mV and $Z_{Th} = R_{Th} = 10^9\,\Omega$, is directly connected to an indicator with $Z_L = R_L = 10^4\,\Omega$ and a scale of sensitivity $\frac{1}{59}$pH/mV. The measured pH is:

$$\text{pH}_M = 59\,\text{pH}\left(\frac{10^4}{10^4 + 10^9}\right)\frac{1}{59} \approx 10^{-5}\,\text{pH} \tag{5.5}$$

i.e. there will be effectively a zero indication for any non-zero value.

This problem is solved by connecting the electrode to the indicator via a **buffer amplifier**. This is characterised by large Z_{IN}, small Z_{OUT} and unity gain $A = 1$. For example, an operational amplifier with a field effect transistor (FET) input stage connected as a voltage follower (Fig. 9.7(c)), would have $Z_{\text{IN}} = 10^{12}\,\Omega$, $Z_{\text{OUT}} = 10\,\Omega$. The indicated pH value for the modified system (Fig. 5.4) is:

$$\text{pH}_M = \frac{10^{12}}{10^{12} + 10^9} \times \frac{10^4}{10^4 + 10}\text{pH} \tag{5.6}$$

and the loading error is now -0.002 pH, which is negligible.

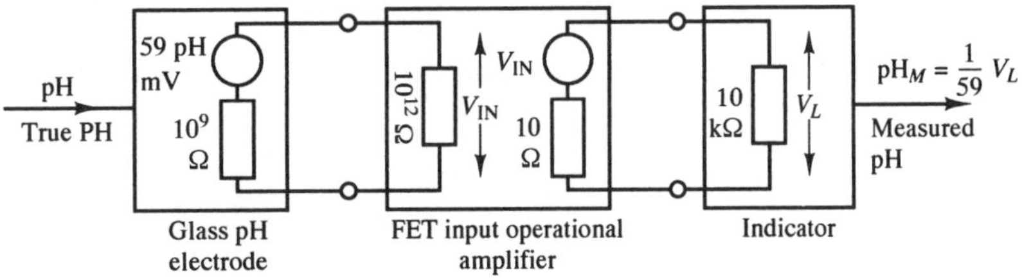

Fig. 5.4 Equivalent circuit for pH measurement system

Fig. 5.5 A.c. loading of tachogenerator

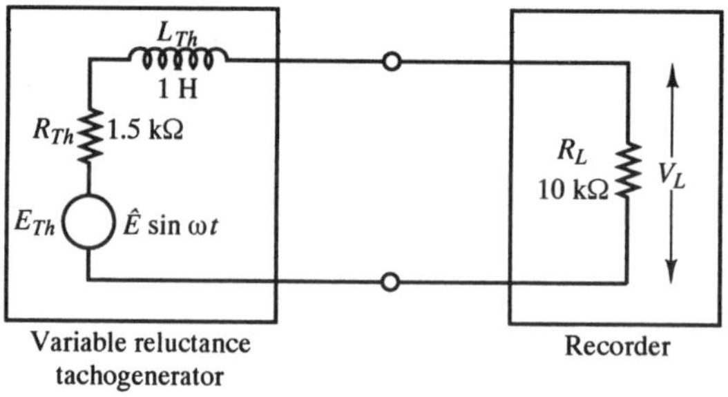

An example of a.c. loading effects is given in Fig. 5.5, which shows the equivalent circuit for a variable reluctance tachogenerator connected to a recorder. The Thévenin voltage E_{Th} for the tachogenerator is a.c. with an amplitude $\hat{E}$ and angular frequency ω, both proportional to the mechanical angular velocity ω_r (section 8.4). In this example, $\hat{E} = (5.0 \times 10^{-3})\ \omega_r$ volts and $\omega = 6\omega_r$ rad s^{-1}. The Thévenin impedance Z_{Th} for the tachogenerator is an inductance and resistance in series (coil surrounding magnet) i.e. $Z_{Th} = R_{Th} + j\omega L_{Th}$. Thus if $\omega_r = 10^3$ rad s^{-1}:

$$\hat{E} = 5\,\text{V},\ \omega = 6 \times 10^3\ \text{rad s}^{-1}$$

and

$$Z_{Th} = 1.5 + 6.0j\text{k}\Omega,$$

so that the amplitude of the recorded voltage is

$$\hat{V}_L = \hat{E}\frac{R_L}{|Z_{Th} + R_L|} = 5\frac{10}{\sqrt{[(11.5)^2 + (6.0)^2]}} = 3.85\ \text{V} \qquad [5.7]$$

If the recorder scale sensitivity is set at $1/(5 \times 10^{-3})$ rad s^{-1} V^{-1}, then the recorded angular velocity is 770 rad s^{-1}. This error can be removed either by increasing the recorder impedance, or by changing the recorder sensitivity to allow for loading effects. A better alternative is to replace the recorder by a counter which measures the frequency rather than the amplitude of the tachogenerator signal (Section 10.3).

5.1.2 Example of Thévenin equivalent circuit calculation: potentiometric displacement sensor

Figure 5.6 shows a schematic diagram of a potentiometric sensor for measuring displacements d. The resistance of the potentiometer varies linearly with displacement.

Fig. 5.6 Potentiometer displacement sensor and Thévenin equivalent circuit

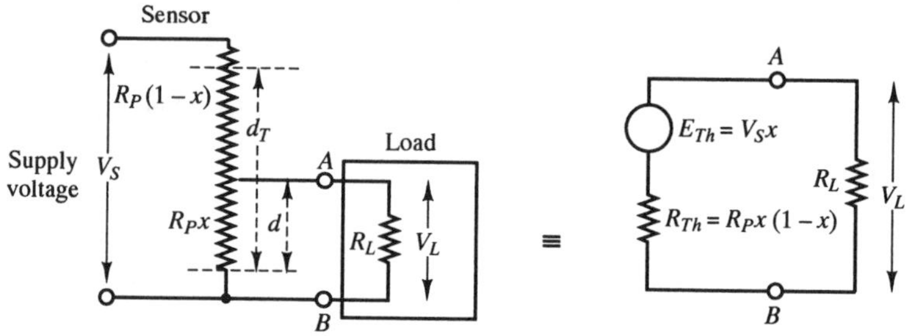

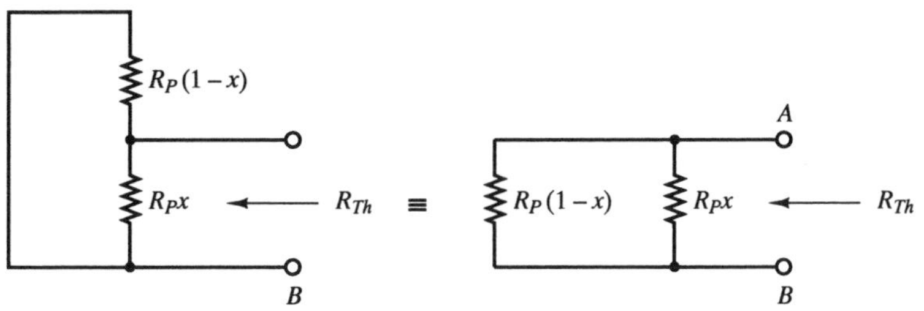

Fig. 5.7 Calculation of R_{Th} for potentiometer

Thus if $x = d/d_T$ is the fractional displacement, the corresponding resistance is R_Px, where R_P Ω is the total resistance of the potentiometer. The Thévinin voltage E_{Th} is the open circuit voltage across the output terminals AB. The ratio between E_{Th} and supply voltage V_s is equal to the ratio of fractional resistance R_Px to total resistance R_P; that is

$$\frac{E_{Th}}{V_s} = \frac{R_Px}{R_P}, \quad \text{giving} \quad E_{Th} = V_s x \qquad [5.8]$$

The Thévenin impedance Z_{Th} is found by setting supply voltage $V_s = 0$, replacing the supply by its internal impedance (assumed to be zero), and calculating the impedance looking back into the terminals AB as shown in Fig. 5.7. Thus:

$$\frac{1}{R_{Th}} = \frac{1}{R_Px} + \frac{1}{R_P(1 - x)}$$

giving

$$R_{Th} = R_Px(1 - x) \qquad [5.9]$$

Thus the effect of connecting a resistive load R_L (recorder or indicator) across the terminals AB is equivalent to connecting R_L across the Thévenin circuit. The load voltage is thus:

$$V_L = E_{Th}\frac{R_L}{R_{Th} + R_L} = V_s x\frac{R_L}{R_Px(1 - x) + R_L}$$

i.e.

Voltage-displacement relationship for a loaded potentiometer

$$V_L = V_s x\frac{1}{(R_P/R_L)x(1 - x) + 1} \qquad [5.10]$$

The relationship between V_L and x is non-linear, the amount of non-linearity depending on the ratio R_P/R_L (Fig. 5.8). Thus the effect of loading a linear potentiometric sensor is to introduce a non-linear error into the system given by:

$$N(x) = E_{Th} - V_L = V_s x\left\{1 - \frac{1}{(R_P/R_L)x(1 - x) + 1}\right\}$$

i.e.

$$N(x) = V_s\left\{\frac{x^2(1 - x)(R_P/R_L)}{1 + (R_P/R_L)x(1 - x)}\right\} \qquad [5.11]$$

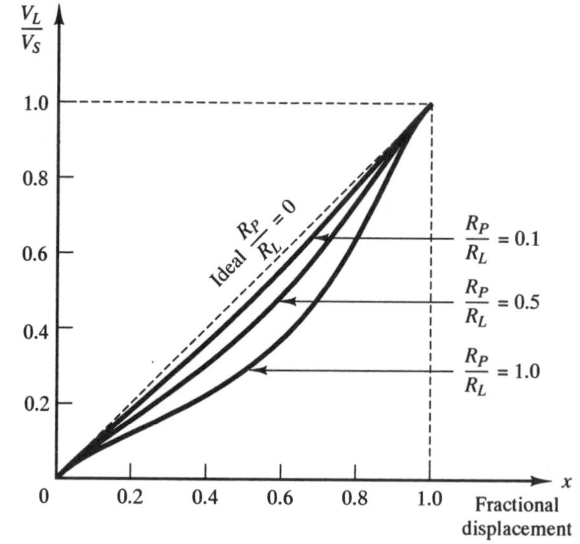

Fig. 5.8 Non-linear characteristics of loaded potentiometer

which reduces to:

$$N(x) \approx V_s(R_P/R_L)(x^2 - x^3)$$

if $R_P/R_L \ll 1$ (the usual situation). $N(x)$ has a maximum value of $\hat{N} = \frac{4}{27}V_s(R_P/R_L)$ when $x = \frac{2}{3}$, corresponding to $dN/dx = 0$ and negative d^2N/dx^2. Expressing $\hat{N}$ as a percentage of full-scale deflection or span V_s volts gives:

$$\hat{N} = \frac{400}{27}\frac{R_P}{R_L} \text{ per cent} \approx 15\frac{R_P}{R_L} \text{ per cent} \qquad [5.12]$$

Non-linearity, sensitivity and maximum power requirements are used to specify the values of R_P and V_s for a given application. Suppose that a 10 cm range potentiometer is to be connected to a 10 kΩ recorder. If the maximum non-linearity must not exceed 2%, then we require 15 $R_P/R_L \le 2$, i.e.$R_P \le \frac{20}{15} \times 10^3\ \Omega$; thus a 1 kΩ potentiometer would be suitable.

Since sensitivity $dV_L/dx \approx V_s$, the greater V_s the higher the sensitivity, but we must satisfy the requirement that power dissipation V_s^2/R_P should not exceed maximum value $\hat{W}$ watts. If $\hat{W} = 0.1$ W we require $V_s \le \sqrt{(0.1 \times 10^3)}$, i.e. $V_s \le 10$ V; if $V_s = 10$ V, then corresponding sensitivity $= 1.0\ \text{V cm}^{-1}$.

5.1.3 Norton equivalent circuit

Norton's theorem states that any network consisting of linear impedances and voltage sources can be replaced by an equivalent circuit consisting of a current source i_N in parallel with an impedance Z_N (Fig. 5.9). Z_N is the impedance looking back into the output terminals with all voltage sources reduced to zero and replaced by their internal impedances, and i_N is the current which flows when the terminals are short circuited. Connecting a load Z_L across the output terminals of the network is equivalent to connecting Z_L across the Norton circuit. The voltage V_L across the load is given by $V_L = i_N Z$, where $1/Z = 1/Z_N + 1/Z_L$, giving:

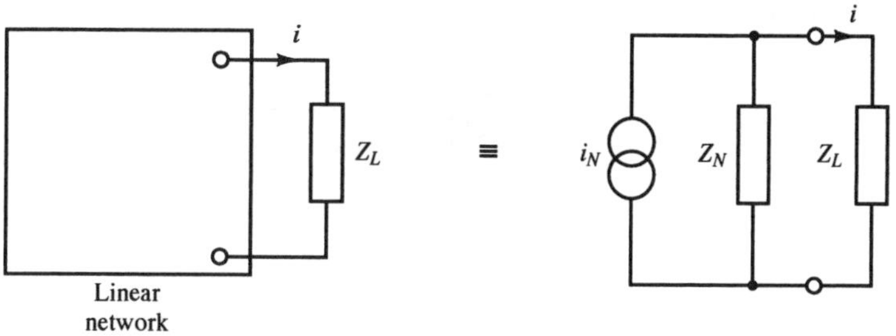

Fig. 5.9 Norton equivalent circuit

Loading of Norton equivalent circuit

$$V_L = i_N \frac{Z_N \cdot Z_L}{Z_N + Z_L} \qquad [5.13]$$

From eqn [5.13] we note that if $Z_L \ll Z_N$, then $V_L \to i_N Z_L$; i.e. in order to develop the maximum current through the load, the load impedance should be *far smaller* than the Norton impedance for the network.

A common example of a current source is an electronic differential pressure transmitter giving an output current signal, range 4 to 20 mA, proportional to an input differential pressure, typical range 0 to 2×10^4 Pa (Section 9.4). Figure 5.10 shows a typical equivalent circuit for the transmitter connected to a recorder via a cable. Using eqn [5.13], the voltage across the total load $R_C + R_R$ of recorder and cable is:

$$V_L = i_N \frac{R_N(R_C + R_R)}{R_N + R_C + R_R} \qquad [5.14]$$

and the ratio $V_R/V_L = R_R/(R_C + R_R)$ giving the recorder voltage:

$$V_R = i_N R_R \frac{R_N}{R_N + R_C + R_R} \qquad [5.15]$$

Using the data given, we have $V_R = 0.9995\, i_N R_R$, so that the recorded voltage deviates from the desired range of 1 to 5 V by only 0.05 per cent.

A second example of a current generator is provided by a piezoelectric crystal acting as a force sensor. If a force F is applied to any crystal, then the atoms of the crystal undergo a small displacement x proportional to F. For a piezoelectric material the crystal acquires a charge q proportional to x i.e. $q = Kx$. The crystal can therefore be regarded as a Norton current source of magnitude $i_N = \mathrm{d}q/\mathrm{d}t =$

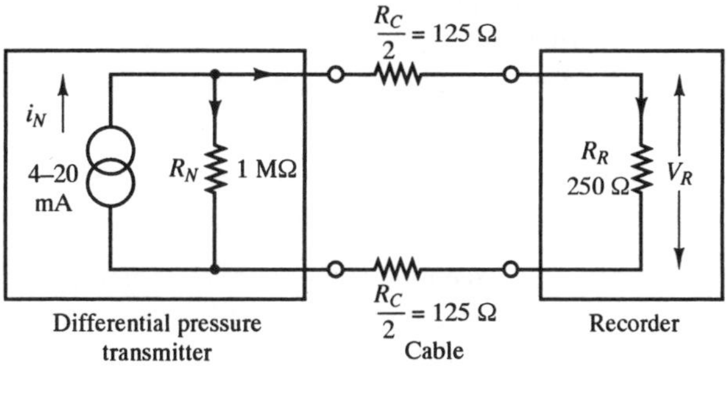

Fig. 5.10 Typical current source and load

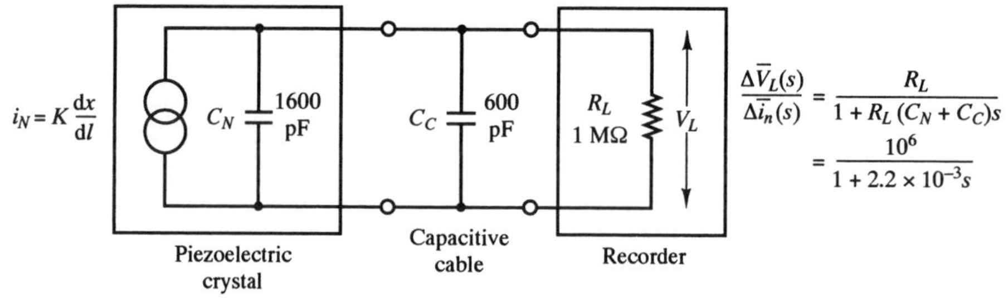

Fig. 5.11 Piezoelectric force measurement system

$K(\mathrm{d}x/\mathrm{d}t)$, where $\mathrm{d}x/\mathrm{d}t$ is the velocity of the atomic deformations. This effect is discussed more fully in Section 8.7, where we see that the crystal acts as a capacitance C_N in parallel with the current source i_N. Figure 5.11 shows the equivalent circuit and typical component values for a crystal connected via a capacitive cable C_C to a recorder acting as a resistive load R_L. The voltage V_L across the load is given by $i_N Z$, where Z is the impedance of C_N, C_C and R_L in parallel. Since

$$\frac{1}{Z} = C_N s + C_C s + \frac{1}{R_L}$$

$$Z = \frac{R_L}{1 + R_L(C_N + C_C)s}$$

where s denotes the Laplace operator. The transfer function relating dynamic changes in source current and recorder voltage is thus:

$$\frac{\Delta \bar{V}_L(s)}{\Delta \bar{i}_N(s)} = \frac{R_L}{1 + R_L(C_N + C_C)s} \qquad [5.16]$$

Thus the effect of electrical loading in this example is to introduce a first order transfer function into the force measurement system; this will affect dynamic accuracy.

5.2 Two-port networks

5.2.1 Generalised effort and flow variables

We have seen in the previous section how electrical loading effects can be described using a pair of variables, voltage and current. Voltage is an example of an **across** or **effort variable** y, current is an example of a **through** or **flow variable** $\dot{x}$. An effort variable drives a flow variable through an impedance. Other examples of effort/flow pairs are force/velocity; torque/angular velocity; pressure difference/volume flow rate; temperature difference/heat flow rate.[2] Each pair y-$\dot{x}$ has the following properties:

(a) the product $y\dot{x}$ represents power in watts;
(b) the ratio $y/\dot{x}$ represents impedance.

The only exception is the thermal variables where the product has the dimensions of watts and temperature. Table 5.1 (adapted from ref. [2]) lists the effort/flow pairs for different forms of energy and for each pair defines the related quantities of

Table 5.1 Flow/effort variables and related quantities (adapted from Finkelstein and Watts, 1971[2])

Variables	$\frac{d}{dt}\dot{x}$	$\int \dot{x} dt$	Flow $\dot{x}$	Effort y	Impedance $\frac{y}{\dot{x}}$	Stiffness $\frac{y}{\int \dot{x} dt}$	Compliance $\frac{\int \dot{x} dt}{y}$	Inertance $\frac{y}{d\dot{x}/dt}$
Mechanical-translation	acceleration	displacement	velocity	force	$\frac{\text{force}}{\text{velocity}} =$ damping constant	$\frac{\text{force}}{\text{displacement}} =$ mechanical stiffness	$\frac{\text{displacement}}{\text{force}} = \frac{1}{\text{mech. stiffness}}$	$\frac{\text{force}}{\text{acceleration}} =$ mass
Mechancial-rotation	angular acceleration	angular displacement	angular velocity	torque	$\frac{\text{torque}}{\text{ang. velocity}} =$ damping constant	$\frac{\text{torque}}{\text{angular disp.}} =$ mechancial stiffness	$\frac{\text{angular disp.}}{\text{torque}} = \frac{1}{\text{mech. stiffness}}$	$\frac{\text{torque}}{\text{angular accn}} =$ moment of inertia
Electrical	$\frac{d}{dt}$(current)	charge	current	voltage	$\frac{\text{voltage}}{\text{current}} =$ electrical resistance	$\frac{\text{voltage}}{\text{charge}} = \frac{1}{\text{elect. capacitance}}$	$\frac{\text{charge}}{\text{voltage}} =$ electrical capacitance	$\frac{\text{voltage}}{\frac{d}{dt}(\text{current})} =$ inductance
Fluidic		volume	volume flow rate	pressure	$\frac{\text{pressure}}{\text{vol. flow rate}} =$ fluidic resistance	$\frac{\text{pressure}}{\text{volume}} = \frac{1}{\text{fluid capacitance}}$	$\frac{\text{volume}}{\text{pressure}} =$ fluidic capacitance	$\frac{\text{pressure}}{\frac{d}{dt}(\text{flow rate})} =$ fluidic inertance
Thermal		heat	heat flow rate	temperature	$\frac{\text{temperature}}{\text{heat flow rate}} =$ thermal resistance	$\frac{\text{temperature}}{\text{heat}} = \frac{1}{\text{therm. capacitance}}$	$\frac{\text{heat}}{\text{temperature}} =$ thermal capacitance	

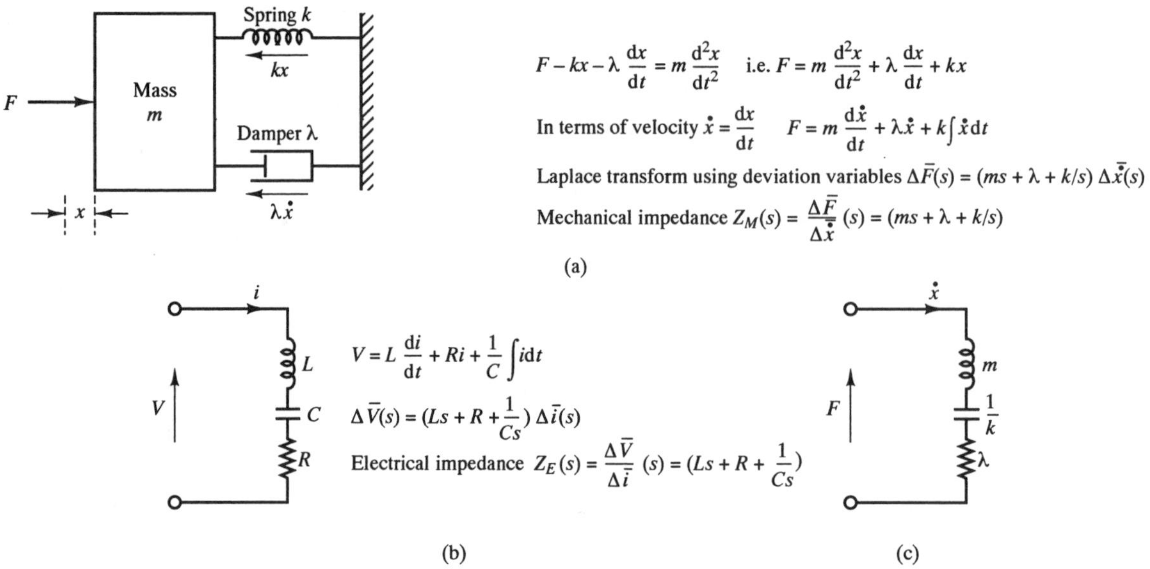

Fig. 5.12 Equivalent circuit for a mechanical system
(a) Parallel mechanical system
(b) Series electrical circuit
(c) Equivalent mechanical circuit

impedance, stiffness, compliance and inertance. Thus we see that the concept of impedance is applicable to mechanical, fluidic and thermal systems as well as electrical. For a mechanical system, mass is analogous to electrical inductance, damping constant is analogous to electrical resistance and 1/stiffness is analogous to electrical capacitance. For a thermal system, thermal resistance is analogous to electrical resistance, thermal capacitance is analogous to electrical capacitance. This means we can generalise the electrical equivalent circuits of Thévenin and Norton to non-electrical systems.

Figure 5.12(a) shows a parallel mechanical system consisting of a mass m, spring stiffness k and damper constant λ. Figure 5.12(b) shows a series electrical circuit consisting of an inductance L, capacitance C and resistance R. Since mechanical impedance is the ratio of force/velocity, the mechanical impedance transfer function is:

$$Z_M(s) = \frac{\Delta\bar{F}}{\Delta\bar{\dot{x}}} = ms + \lambda + \frac{k}{s} \qquad [5.17]$$

The impedance transfer function for the electrical circuit is:

$$Z_E(s) = \frac{\Delta\bar{V}}{\Delta\bar{i}} + Ls + R + \frac{1}{Cs} \qquad [5.18]$$

We see that these have a similar form with m corresponding to L, λ corresponding to R and k corresponding to $1/C$. Thus the parallel mechanical system can be represented by an equivalent circuit consisting of an inductive element m, a resistive element λ and a capacitive element $1/k$ in series (Fig. 5.12(c)).

Figure 5.13(a) shows a thermal system consisting of a body at temperature T immersed in a fluid at temperature T_F. The body has a mass M, specific heat C_H and surface area A; U is the heat transfer coefficient between the body and the fluid. Figure 5.13(b) shows a series electrical circuit with resistance R, capacitance C,

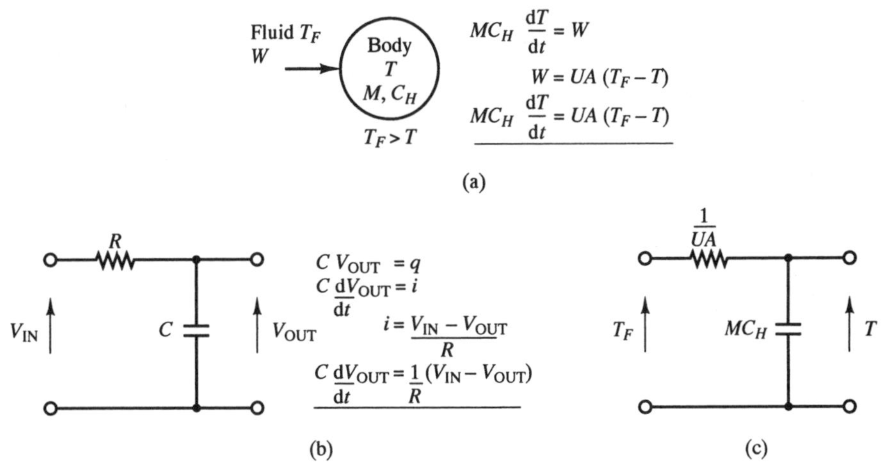

Fig. 5.13 Equivalent circuit for a thermal system
(a) Thermal system
(b) Electrical circuit
(c) Equivalent thermal circuit

input voltage V_{IN} and output voltage V_{OUT}. The differential equation for the thermal system is:

$$MC_H \frac{dT}{dt} = UA(T_F - T) \qquad [5.19]$$

and the differential equation for the electrical circuit is:

$$C \frac{dV_{OUT}}{dt} = \frac{1}{R}(V_{IN} - V_{OUT}) \qquad [5.20]$$

We see that the equations have a similar form and the termperature effort variables T_F, T are analogous to the voltage effort variables V_{IN}, V_{OUT}. Heat flow W is analogous to current flow i, UA is analogous to $1/R$ i.e. the reciprocal of electrical resistance and MC_H is analogous to electrical capacitance C. Thus the thermal system can be represented by an equivalent circuit consisting of a resistive element $1/UA$ in series with a capacitive element MC_H as in Fig. 5.13(c).

5.2.2 Two-port networks

We saw in section 5.1 that the electrical output of a sensing element such as a thermocouple or piezoelectric crystal can be represented by a Thévenin or Norton equivalent circuit. The sensor has therefore two output terminals which allow both voltage and current flow to be specified; this is referred to as an electrical output port. The sensing element will have a mechnical, thermal or fluidic input; we saw in the previous section that mechanical and thermal systems can be represented by equivalent circuits which show the relation between the corresponding effort and flow variables. Thus the input to a mechanical or thermal sensor can be represented by two input terminals which allow both the effort and flow variables to be specified; this is either a mechanical or thermal input port. Thus a sensing element can be represented by a **two-port** or four-terminal network. Figure 5.14(a) shows a two-port representation of a mechanical sensor with input mechanical port and output electrical port; Fig. 5.15(a) shows the two-port representation of a thermal sensor.

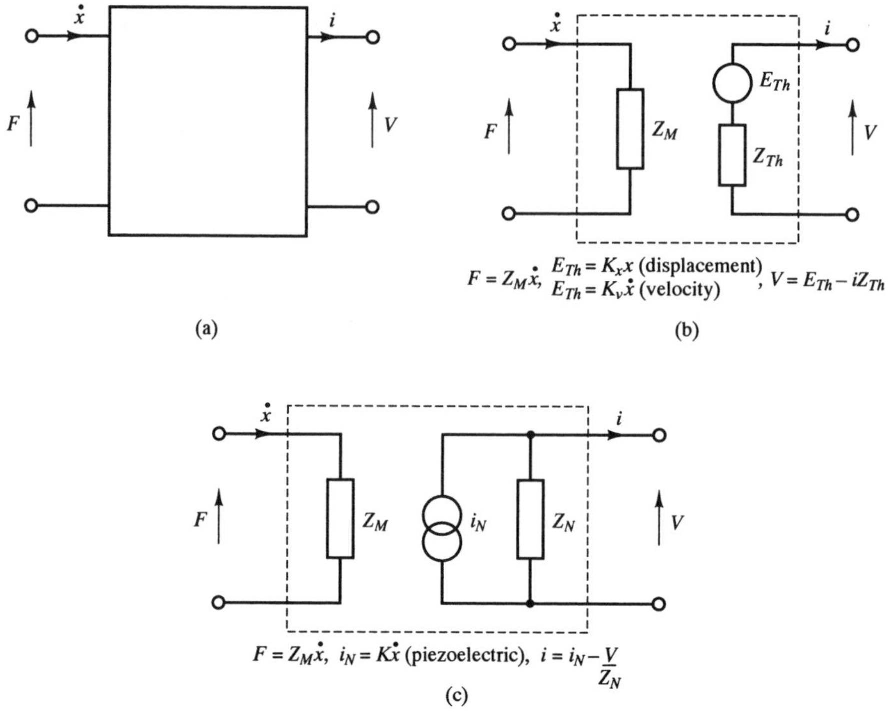

Fig. 5.14 Mechanical sensing elements as two-port networks
(a) Overall two-port representation
(b) Equivalent circuit with Thévenin output
(c) Equivalent circuit with Norton output

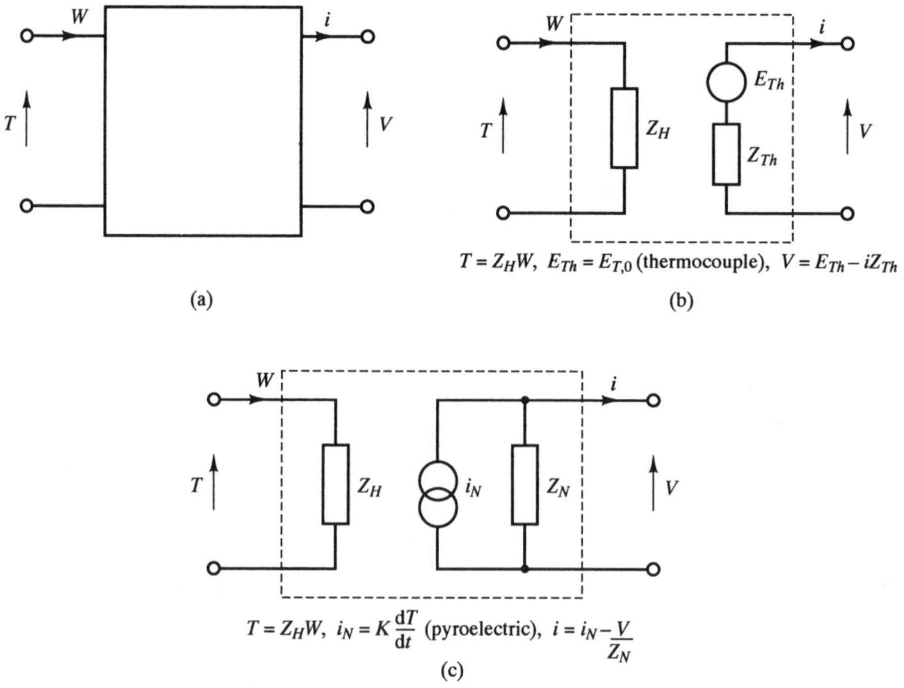

Fig. 5.15 Thermal sensing elements as two-port networks
(a) Overall two-port representation
(b) Equivalent circuit with Thévenin output
(c) Equivalent circuit with Norton output

Figures 5.14(b) and (c) shows the detailed two-port networks for a range of mechanical sensing elements. Figure 5.14(b) shows the equivalent circuit for sensing elements with a Thévenin equivalent circuit at the electrical output port. Z_M is the input mechanical impedance, E_{Th} and Z_{Th} are the Thévenin voltage and impedance. For a displacement sensor E_{Th} is proportional to displacement x i.e.:

$$E_{Th} = K_x x \qquad [5.21)$$

where K_x is the sensitivity and $x = \int \dot{x} dt$

For a potentiometer displacement sensor (Section 5.1.2) $K_x = V_s$ (supply voltage), and for a linear variable differential transformer (LVDT, Section 8.3.2) K_x is the slope of the linear portion of the a.c. voltage versus displacement characteristics. For a velocity sensor, E_{Th} is proportional to velocity $\dot{x}$ i.e.:

$$E_{Th} = K_V \dot{x} \qquad [5.22]$$

For an electromagnetic linear velocity sensor (Section 12.5.1) $K_V = Bl$ where B is the applied magnetic field and l the length of the conductor. In the case of an electromagnetic angular velocity sensor or tachogenerator (Section 8.4) we have $E_{Th} = K_V \omega_r$, where ω_r is the angular velocity and $K_V = dN/d\theta$ the rate of change of flux N with angle θ. Figure 5.14(c) shows the equivalent circuit for sensing elements with a Norton equivalent circuit i_N, Z_N at the electrical output port. For a piezoelectric sensor i_N is proportional to velocity $\dot{x}$ (Section 8.7) i.e.

$$i_N = K\dot{x} \qquad [5.23]$$

where $K = dk$, d = charge sensitivity to force and k is the stiffness of the crystal.

Figures 5.15(b) and (c) show the detailed two-port networks for two examples of thermal sensing elements. Figure 5.15(b) shows the equivalent circuit for a sensing element with a Thévenin equivalent circuit E_{Th}, Z_{Th} at the electrical ouput port, Z_H is the thermal input impedance. For a thermocouple temperature sensor (Section 8.5) with the reference junction at 0 °C we have $E_{Th} = E_{T,0}$ where $E_{T,0}$ is the contact e.m.f at the measured junction at T°C and is given by the power series:

$$E_{T,0} = a_1 T + a_2 T^2 + a_3 T^3 + \ldots \qquad [5.24]$$

Figure 5.15(c) shows the equivalent circuit for a sensing element with a Norton equivalent circuit i_N, Z_N at the electrical output port. For a pyroelectric detector (Section 15.5.1) i_N is proportional to rate of change of temperature dT/dt i.e.

$$i_N = K \frac{dT}{dt} \qquad [5.25]$$

where $K = A\, dP/dT$, and A is the area of the electrodes and dP/dT is the slope of the polarisation temperature characteristics.

5.2.3 Process loading

Having introduced the concepts of equivalent circuits and two-port networks for mechanical and thermal systems, we can now use these concepts to study examples of how a primary sensing element can 'load' the process or element being measured.

Figure 5.16 shows a mechanical system or 'process' represented by a mass, spring

Fig. 5.16 Loading of mechanical system by force sensor

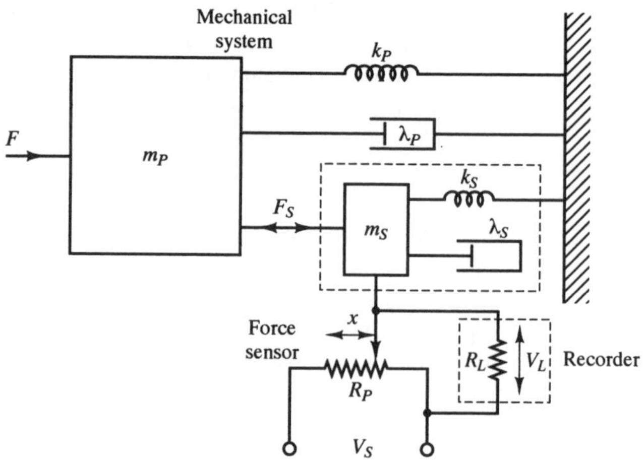

and damper. The force F applied to the 'process' is being measured by a force sensor, consisting of an elastic element in conjunction with a potentiometric displacement sensor. The elastic force sensor can also be represented by a mass spring and damper (Section 4.1.2). Under steady-state conditions when both velocity $\dot{x} = 0$ and acceleration $\ddot{x} = 0$, we have the following force balance equations:

$$\begin{aligned} &\textbf{process} \quad F = k_P x + F_S \\ &\textbf{sensor} \quad F_S = k_S x \end{aligned} \qquad [5.26]$$

showing that the relationship between the measured force F_S and the true force F is:

Steady-state loading of mechanical system

$$F_S = \frac{k_S}{k_S + k_P} F = \frac{1}{1 + k_P/k_S} F \qquad [5.27]$$

We see that in order to minimise the loading error in the steady state the sensor stiffness k_S should be very much greater than the process stiffness k_P.

Under unsteady conditions when $\dot{x}$ and $\ddot{x}$ are non-zero, Newton's second law gives the following differential equations:

$$\begin{aligned} &\textbf{process} \quad F - k_P x - \lambda_P \dot{x} - F_S = m_P \ddot{x} \\ &\textbf{sensor} \quad F_S - k_S x - \lambda_S \dot{x} = m_S \ddot{x} \end{aligned} \qquad [5.28]$$

i.e.

$$\begin{aligned} &m_P \frac{d\dot{x}}{dt} + \lambda_P \dot{x} + k_P \int \dot{x}\, dt = F - F_S \\ &m_S \frac{d\dot{x}}{dt} + \lambda_S \dot{x} + k_S \int \dot{x}\, dt = F_S \end{aligned} \qquad [5.29]$$

Using the analogues given earlier, the sensor can be represented by F_S driving $\dot{x}$ through the mechanical L, C, R circuit m_S, $1/k_S$, λ_S; and the 'process can be represented by $F - F_S$ driving $\dot{x}$ through the mechanical L, C, R circuit m_P, $1/k_P$, λ_P. If Δx, ΔF and ΔF_S are deviations from initial steady conditions, then the Laplace transforms of eqns [5.29] are:

$$\left(m_P s + \lambda_P + \frac{k_P}{s}\right)\Delta\bar{\dot{x}} = \Delta\bar{F} - \Delta\bar{F}_S$$
$$\left(m_S s + \lambda_S + \frac{k_S}{s}\right)\Delta\bar{\dot{x}} = \Delta\bar{F}_S \qquad [5.30]$$

Using Table 5.1 we can define mechanical impedance transfer functions by $Z_M(s) = \Delta\bar{F}(s)/\Delta\bar{\dot{x}}(s)$, so that:

process impedance $$Z_{MP}(s) = m_P s + \lambda_P + \frac{k_P}{s}$$
sensor impedance $$Z_{MS}(s) = m_S s + \lambda_S + \frac{k_S}{s} \qquad [5.31]$$

From [5.30] and [5.31] the relationship between measured and actual dynamic changes in force is:

Dynamic loading of mechanical system

$$\Delta\bar{F}_S(s) = \frac{Z_{MS}}{Z_{MS} + Z_{MP}}\Delta\bar{F}(s) \qquad [5.32]$$

Thus in order to minimise dynamic loading effects, sensor impedance Z_{MS} should be very much greater than process impedance Z_{MP}. Figure 5.17 shows the equivalent circuit for the system: process, force sensor and recorder.

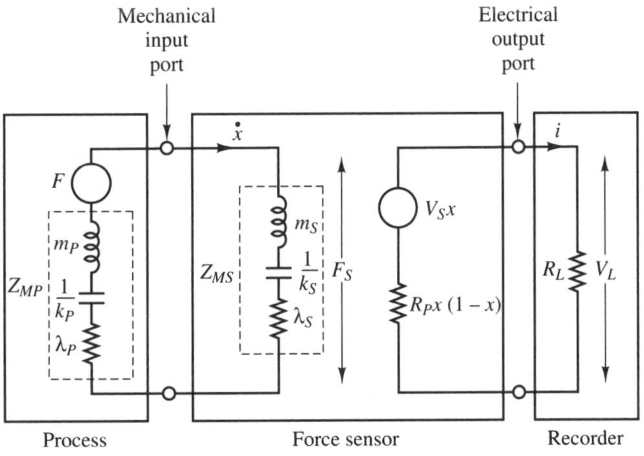

Fig. 5.17 Equivalent circuit for complete system showing a force sensor as a two-port network

Figure 5.18 shows a hot body, i.e. a thermal 'process' whose temperature T_P is being measured by a thermocouple sensor. Under unsteady conditions, heat flow rate considerations give the following differential equations (Section 4.1):

process $$M_P C_P \frac{dT_P}{dt} = W_P - W_S, \quad W_P = U_P A_P (T_F - T_P)$$
sensor $$M_S C_S \frac{dT_S}{dt} = W_S, \quad W_S = U_S A_S (T_P - T_S) \qquad [5.33]$$

where M denotes masses
C denotes specific heats
U denotes heat transfer coefficients
A denotes heat transfer areas.

Fig. 5.18 Loading of thermal 'process' by thermocouple temperature sensor

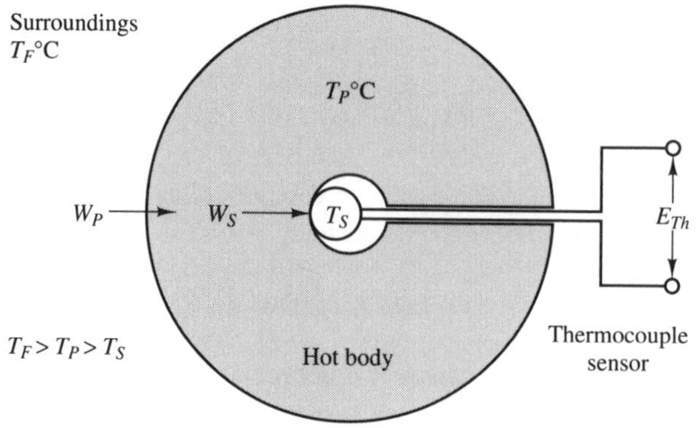

Fig. 5.19 Equivalent circuit for thermal system showing thermocouple as a two-port network

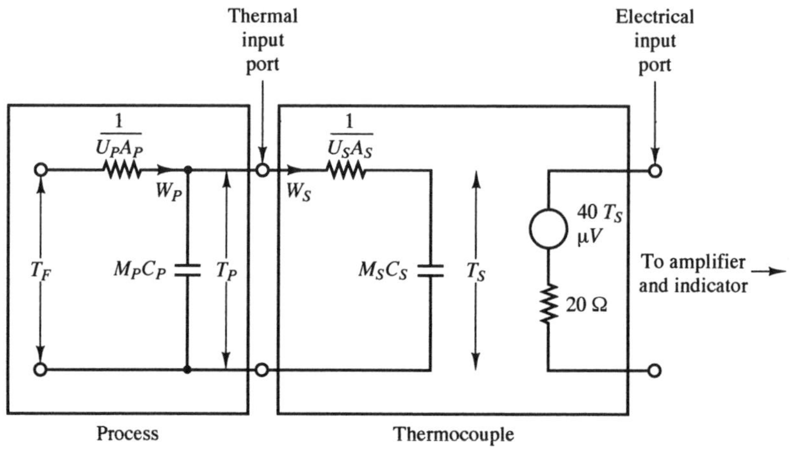

The quantities $M_P C_P$, $M_S C_S$ have the dimensions of heat/temperature and are analogous to electrical capacitance. The quantities $U_P A_P$, $U_S A_S$ have the dimensions of heat flow rate/temperature and are analogous to 1/(electrical resistance). The equivalent circuit for the process and thermocouple is shown in Fig. 5.19. We see that the relationship between T_F and T_P depends on the potential divider $1/U_P A_P$, $M_P C_P$ and the relationship between T_P and T_S depends on the potential divider $1/(U_S A_S)$, $M_S C_S$. Again the thermocouple can be represented as a two-port network with a thermal input port and an electrical output port.

In conclusion we see that the representation of measurement system elements by two-port networks enables both **process** and **inter-element** loading effects to be quantified.

5.2.4 Bilateral transducers

Bilateral transducers are associated with reversible physical effects. In a reversible effect the same device can for example, convert mechanical energy into electrical energy and also convert electrical energy into mechanical energy. When the device converts electrical energy into mechanical energy it acts as a **transmitter** or **sender**.

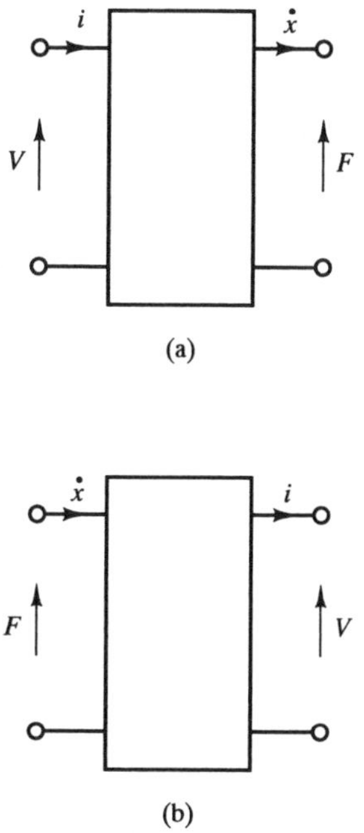

Fig. 5.20 Bilateral transducers
(a) Transmitter/sender
(b) Receiver/sensor

This can be represented by a two-port network with an input electrical port and an output mechanical port as in Fig. 5.20(a). When the device converts mechanical energy into electrical energy it acts as a **receiver** or **sensor**, and can be represented by an input mechanical port and an output electrical port as in Fig. 5.20(b).

The piezoelectric effect is a common example of a reversible effect (Section 8.7). In the **direct effect** a force F applied to the crystal produces a charge q, proportional to F, according to:

$$q = \mathrm{d}F \qquad [5.34]$$

i.e. this is a conversion of mechanical energy to electrical and the device acts as a receiver or sensor. In the **inverse effect** a voltage V applied to the crystal produces a mechanical deformation x, proportional to V, according to:

$$x = \mathrm{d}V \qquad [5.35]$$

This is a conversion of electrical energy to mechanical and the device acts as a transmitter or sender. The detailed equivalent circuits for a piezoelectric transmitter and receiver are given in Section 16.2.1.

Another reversible physical effect is the **electromagnetic effect**. In the direct effect, a conductor of length l moving with velocity $\dot{x}$ perpendicular to a magnetic field B has a voltage:

$$E = Bl\dot{x} \qquad [5.36]$$

induced across the ends of the conductor. This is a conversion of mechanical to electrical energy and the device acts as a receiver or sensor. In the inverse effect a conductor length l carrying a current i in a transverse magnetic field B experiences a force:

$$F = Bli \qquad [5.37]$$

This is a conversion of electrical to mechanical energy and the device acts as a transmitter or sender. Figure 15.21 gives the detailed equivalent circuits for an electromagnetic transmitter and receiver. At the electrical ports the applied voltage drives a current through the electrical impedance L, R; at the mechanical ports the applied force drives a velocity through the mechanical impedance m, $1/k$, λ. The transmitter can be used as a voltage indicator (Section 11.2) and the receiver as a velocity sensor (Sections 5.1.1. and 8.4).

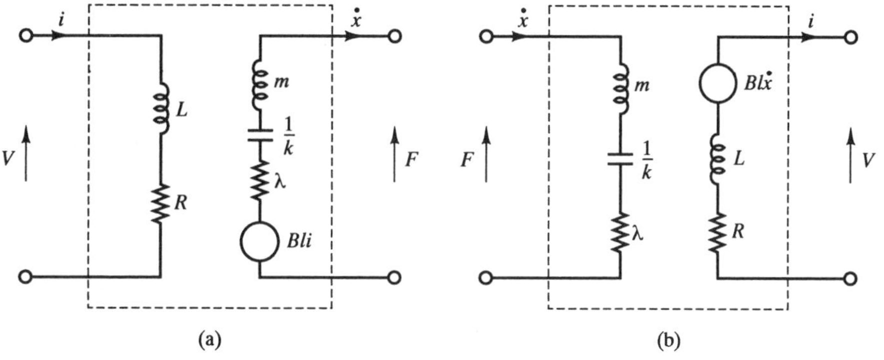

Fig. 5.21 Equivalent circuits for bilateral electromagnetic transducers
(a) Transmitter/sender
(b) Receiver/sensor

Conclusion

We have seen how the use of equivalent circuits and two-port networks has enabled both inter-element and process loading effects to be described.

References

5.1 BELL E C and WHITEHEAD R W 1979 *Basic Electrical Engineering and Instrumentation for Engineers*, Granada, London, pp. 46–7.

5.2 FINKELSTEIN L and WATTS R D 1971 'Systems analysis of instruments', *Jnl Inst. Measmt & Contr.*, vol. 4, Sept., pp. 236–7.

Problems

5.1 A glass pH electrode with a sensitivity of 59 mV pH^{-1} and resistance of $10^9\ \Omega$ is used to measure pH in the range 0 to 15. The electrode is to be connected to a recorder of input range 0 to 100 mV and resistance 100 Ω using a buffer amplifier of unity gain and output resistance 100 Ω

(a) Calculate the input impedance of the amplifier, and the sensitivity of the recorder scale necessary to obtain an accurate recording of pH.

(b) The resistance of the electrode increases to $2 \times 10^9\ \Omega$ due to chemical action. Calculate the resulting measurement error in the above system, as a percentage of full scale, for a true pH of 7.

5.2 The motion of a hydraulic ram is to be recorded using a potentiometer displacement sensor connected to a recorder. The potentiometer is 25 cm long and has linear resistance displacement characteristics. A set of potentiometers with maximum power rating of 5 W and resistance values ranging from 250 to 2500 Ω in 250 Ω steps is available. The recorder has a resistance of 5000 Ω and the non-linear error of the system must not exceed 2% of full scale. Find:

(a) The maximum potentiometer sensitivity that can be obtained.

(b) The required potentiometer resistance and supply voltage in order to achieve maximum sensitivity.

5.3 An electronic differential transmitter gives a current output of 4 to 20 mA linearly related to a differential pressure input of 0 to 10^4 Pa. The Norton impedance of the transmitter is $10^5\ \Omega$. The transmitter is connected to an indicator of impedance of 250 Ω via a cable of total resistance 500 Ω. The indicator gives a reading between 0 and 10^4 Pa for an input voltage between 1 and 5 V. Calculate the system measurement error, due to loading, for an input pressure of 5×10^3 Pa.

5.4 A sensor with mass 0.1 kg, stiffness $10^3\ N\ m^{-1}$ and damping constant $10\ N\ s\ m^{-1}$ is used to measure the force on a mechanical structure of mass 5 kg, stiffness $10^2\ N\ m^{-1}$ and damping constant $20\ N\ s\ m^{-1}$. Find the transfer function relating measured and actual changes in force.

6

Signals and noise in measurement systems

6.1 Introduction

In Chapter 4 we studied the dynamic response of measurement systems to step, sine wave and square wave input signals. These signals are examples of **deterministic signals**: a deterministic signal is one whose value at any future time can be exactly predicted. Thus if we record these signals for an observation period T_O (Fig. 6.1), the future behaviour of the signal, once the observation period is over, is known exactly. The future behaviour of real processes, such as chemical reactors, blast furnaces and aircraft, will depend on unknown factors such as the type of feedstock, reliability of equipment, changes in throughput, atmospheric conditions, and cannot be known in advance. This means that the future value of measured variables, such as reactor temperature, flow in a pipe, aircraft speed, *cannot* be exactly predicted. Thus in real measurement applications the input signal to the measurement system is not deterministic but **random**. If a random signal is recorded for an observation period T_O (Fig. 6.1) the behaviour of the signal, once the observation period is over, is *not* known exactly. However, five statistical quantities: mean, standard deviation, probability density function, power spectral density and autocorrelation function are used to estimate the behaviour of random signals. These are explained and defined in Section 6.2.

Random variations in the measured input variable produce corresponding random variations in the electrical output signals of elements in the system. More precisely the Thévenin voltage and Norton current signals defined in Chapter 5 vary randomly with time and will be referred to as the **measurement signal**. Examples are random fluctuations in the millivolt output of a thermocouple, random fluctuations in the current output of a differential pressure transmitter and random fluctuations in the amplitude and frequency of the a.c. output voltage of a variable reluctance tachogenerator. However, unwanted electrical signals may also be present in the measurement circuit. These may be due to sources inside the measurement circuit or caused by coupling to sources outside the circuit. The magnitude of the unwanted signal may be comparable or larger than that of the measurement signal itself, resulting in a measurement error for the overall system. This is an example of the **interfering inputs** discussed in Chapter 2. The unwanted signal then may be either random, e.g. signals caused by the random motion of electrons; or deterministic, e.g.

Fig. 6.1 Deterministic and random signals

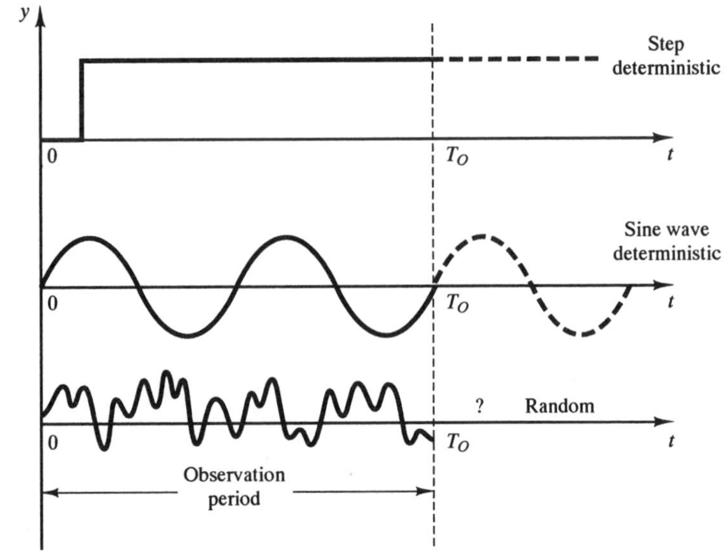

sinusoidal signals at 50 Hz caused by power cables. Unwanted random signals are usually referred to as **noise signals** and unwanted deterministic signals as **interference signals**. Section 6.3 and 6.4 discuss the sources of noise and interference signals and how they affect the measurement circuit. Section 6.5 examines ways of reducing the effects of noise and interference.

6.2 Statistical representation of random signals

Figure 6.2 shows a recording of a section of a random signal obtained during an observation period T_O. Since the signal is random we cannot write down a continuous algebraic equation $y(t)$ for the signal voltage y at time t. We can, however, write down the values y_1 to y_N of N samples taken at equal intervals ΔT during T_O. The first sample y_1 is taken at $t = \Delta T$, the second y_2 is taken at $t = 2\Delta T$, the ith y_1 is taken at $t = i\Delta T$, where $i = 1, \ldots, N$. The sampling interval $\Delta T = T_O/N$ must satisfy the Nyquist sampling theorem which is explained in Section 10.1. We can now use these samples to calculate statistical quantities for the observed section of the signal. These observed statistical quantities will provide a good estimate of

Fig. 6.2 Sampling of a random signal

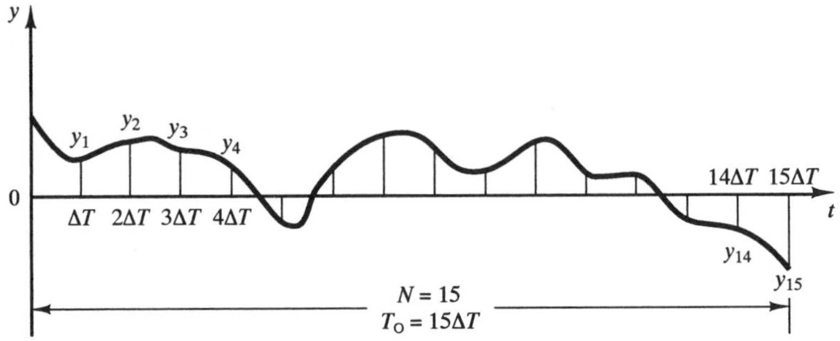

the future behaviour of the signal, once the observation period is over, provided:

(a) T_O is sufficiently long, i.e. N is sufficiently large.
(b) the signal is **stationary**, i.e. long-term statistical quantities do not change with time.

6.2.1 Mean $\bar{y}$

For a signal defined in terms of a continuous function $y(t)$ in the interval 0 to T_O, the mean is given by:

Mean for continuous signal

$$\bar{y} = \frac{1}{T_O}\int_0^{T_O} y(t)\,dt \qquad [6.1]$$

If, however, the signal is represented by the set of sampled values y_i we have

Mean for sampled signal

$$\bar{y} = \frac{1}{N}\sum_{i=1}^{i=N} y_i \qquad [6.2]$$

6.2.2 Standard deviation σ

This is a measure of the average spread or deviation of the signal from the mean value $\bar{y}$. In the continuous case:

Standard deviation for continuous signal

$$\sigma^2 = \frac{1}{T_O}\int_0^{T_O} [y(t) - \bar{y}]^2\,dt \qquad [6.3]$$

and in the sampled case:

Standard deviation for sampled signal

$$\sigma^2 = \frac{1}{N}\sum_{i=1}^{i=N} (y_i - \bar{y})^2 \qquad [6.4]$$

In the special case that $\bar{y} = 0$, the standard deviation σ is equal to the **root mean square** (r.m.s.) value y_{rms},

where:

$$y_{rms} = \sqrt{\frac{1}{T_O}\int_0^{T_O} y^2\,dt} \qquad [6.5]$$

or

$$y_{rms} = \sqrt{\frac{1}{N}\sum_{i=1}^{i=N} y_i^2} \qquad [6.6]$$

6.2.3 Probability density function *p(y)*

This is a function of signal value y and is a measure of the probability that the signal will have a certain range of values. Figure 6.3 shows the set of sample values y_i and the y axis divided into m sections each of width Δy. We can then count the number of samples occurring within each section i.e. n_1 in section 1, n_2 in Section 2, n_j in Section j, etc., where $j = 1, \ldots, m$. the **probability** P_j of the signal occurring in the jth section is thus:

$$P_j = \frac{\text{number of times sample occurs in the } j\text{th section}}{\text{total number of samples}} \qquad [6.7]$$

$$= \frac{n_j}{N}, \quad j = 1, \ldots, m$$

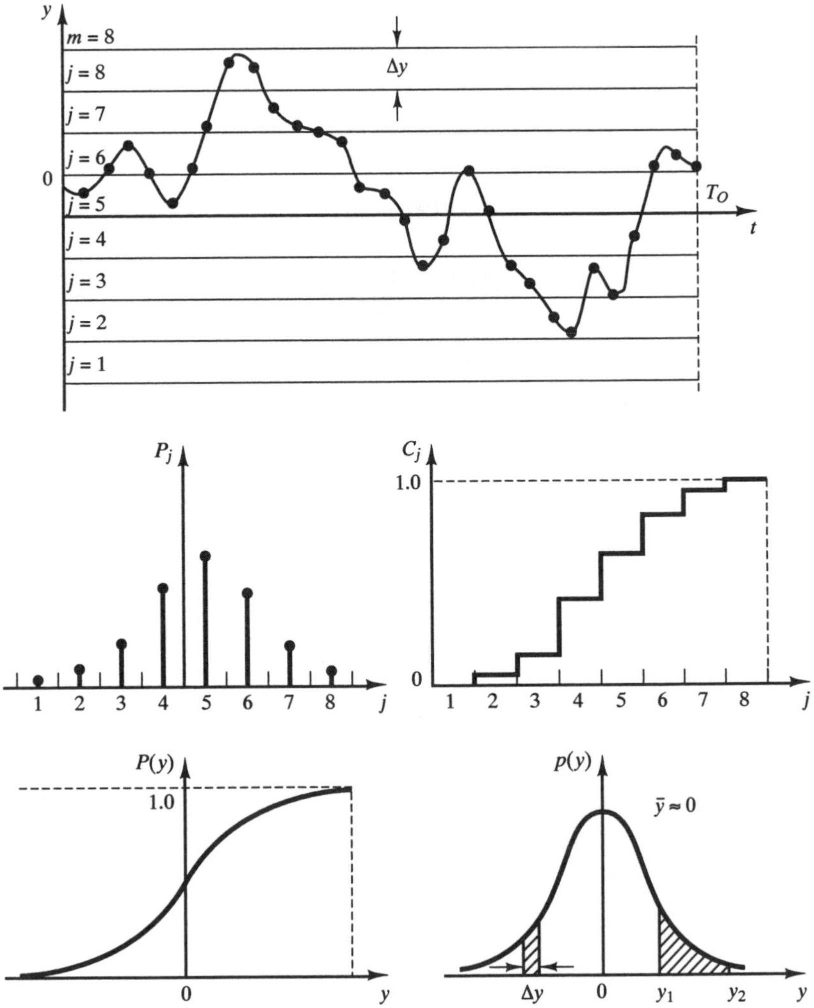

Fig. 6.3 Probability and probability density

The **cumulative probability** C_j is the total probability that the signal will occur in the first j sections and is given by:

Cumulative probability

$$C_j = P_1 + P_2 + \cdots + P_j = \frac{1}{N}(n_1 + n_2 + \cdots + n_j) \qquad [6.8]$$

Figure 6.3 shows the corresponding forms of P_j and C_j. The final value of C_j when $j = m$ is:

$$C_m = \frac{1}{N}(n_1 + n_2 + \cdots + n_m) = \frac{1}{N} \times N = 1 \quad (\text{since } n_1 + n_2 + \cdots + n_m = N) \qquad [6.9]$$

i.e. the total probability of finding the signal in all m sections is unity. In the limit as the interval Δy tends to zero the discrete cumulative probability C_j tends to a continuous function (Fig. 6.3). This is the **cumulative probability distribution function** $P(y)$ (c.d.f.) which is defined by:

Cumulative probability distribution function

$$P(y) = \lim_{\Delta y \to 0} C_j \qquad [6.10]$$

The **probability density function** (p.d.f.) $p(y)$ is more commonly used and is the derivative of $P(y)$, i.e.

Probability density function

$$p(y) = \frac{dP}{dy} \qquad [6.11]$$

Thus the probability $P_{y,y+\Delta y}$ that the signal will lie between y and $y + \Delta y$ is given by:

$$P_{y,y+\Delta y} = \Delta P = p(y)\Delta y \qquad [6.12]$$

i.e. by the area of a strip of height $p(y)$ and width Δy. Similarly the probability $P_{y1,y2}$ that the signal will lie between y_1 and y_2 is given by:

$$P_{y1,y2} = \int_{y1}^{y2} p(y)\, dy \qquad [6.13]$$

i.e. the area under the probability density curve between y_1 and y_2. The total area under the probability density curve is equal to unity corresponding to the total probability of the signal having any value of y. The Gaussian probability density function (Section 2.3).

$$p(y) = \frac{1}{\sigma\sqrt{(2\pi)}} \exp\left[-\frac{(y - \bar{y})^2}{2\sigma^2}\right] \qquad [6.14]$$

usually provides an adequate description of the amplitude distribution of random noise signals.

6.2.4 Power spectral density $\phi(\omega)$

If we record a random signal for several observation periods, each of length T_O (Fig. 6.4), the waveform will be different for each period. However, the average signal power will be approximately the same for each observation period. This means that signal power is a **stationary** quantity which can be used to quantify random signals. In Section 4.3 we saw that a **periodic** signal can be expressed as a **Fourier series**, i.e. a sum of sine and cosine waves with frequencies which are harmonics of the fundamental frequency. The power in a periodic signal is therefore distributed amongst these harmonic frequencies. A random signal is not periodic and cannot be represented by Fourier series but does contain a large number of closely spaced frequencies. Power spectral density is a stationary quantity which is a measure of how the power in a random signal is distributed amongst these different frequencies.

In order to explain the meaning of $\phi(\omega)$ we approximate the random signal by a periodic signal (Fig. 6.4) in which the waveform recorded during the first observation period T_O is exactly repeated during each subsequent observation period. This periodic approximation is valid:

(a) in the limit that the observation period, i.e. signal period T_O becomes large,
(b) providing we use it to calculate the power distribution in the signal.

Fig. 6.4 Power spectrum and power spectral density

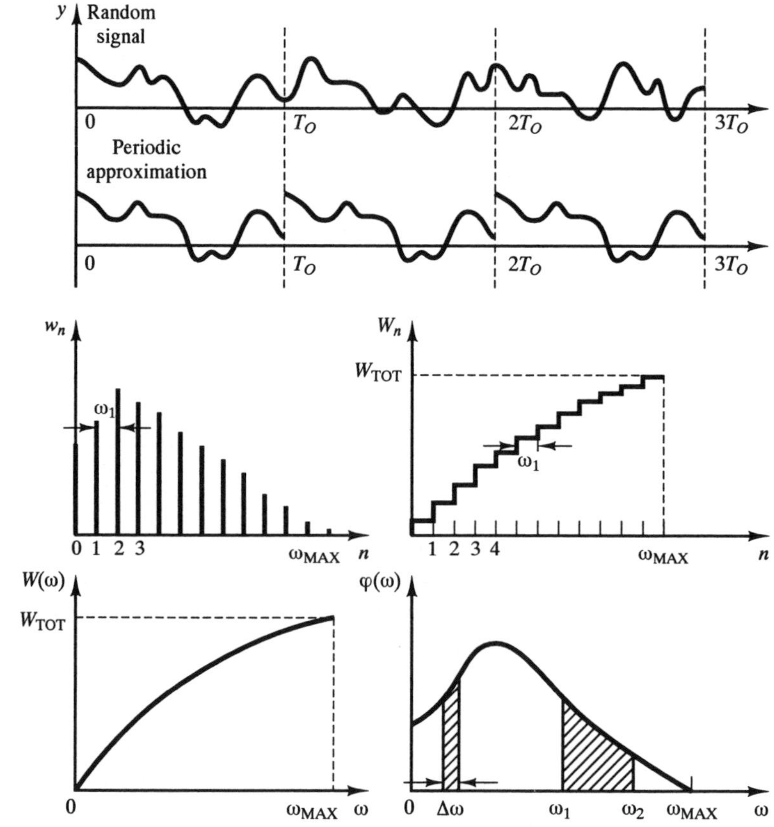

The Fourier series for a voltage signal with period T_O is

$$y(t) = a_0 + \sum_{n=1}^{n=\infty} a_n \cos n\omega_1 t + \sum_{n=1}^{n=\infty} b_n \sin n\omega_1 t \qquad [4.51a]$$

where the fundamental frequency $\omega_1 = 2\pi/T_O$, a_0 = mean $\bar{y}$, and

Fourier series coefficients for sampled signal

$$a_n = \frac{2}{N}\sum_{i=1}^{i=N} y_i \cos (n\omega_1 i\Delta T) = \frac{2}{N}\sum_{i=1}^{i=N} y_i \cos\left(2\pi n\frac{i}{N}\right)$$
$$b_n = \frac{2}{N}\sum_{i=1}^{i=N} y_i \sin (n\omega_1 i\Delta T) = \frac{2}{N}\sum_{i=1}^{i=N} y_i \sin\left(2\pi n\frac{i}{N}\right) \qquad [6.15]$$

Equations [6.15] use the set of sample values y_1 and are equivalent to eqns [4.52) for the Fourier coefficients of a continuous signal. If the nth harmonic $a_n \cos n\omega_1 t$ is applied across a 1 Ω resistor, the instantaneous power in the resistor at time t is $a_n^2 \cos^2 n\omega_1 t$ watts, and the average power over period T_O is:

$$a_n^2 \frac{1}{T_O}\int_0^{T_O} \cos^2 n\omega_1 t \, dt = \frac{a_n^2}{2} \qquad [6.16]$$

Similarly the average power in a 1 Ω resistor due to $b_n \sin n\omega_1 t$ is $b_n^2/2$, so that the power due to the nth harmonic at frequency $n\omega_1$ is:

Power due to nth harmonic

$$w_n = \tfrac{1}{2}(a_n^2 + b_n^2) \qquad [6.17]$$

Figure 6.4 shows the relationship between w_n and ω is a series of lines at the harmonic frequencies $n\omega_1$. This is referred to as the **power spectrum** of the signal and is terminated at ω_{MAX}, the harmonic frequency beyond which w_n becomes negligible. The **cumulative power** W_n is the total power in a 1 Ω resistor due to the first n harmonics and the d.c. component a_0 i.e.

Cumulative power

$$W_n = w_0 + w_1 + w_2 + \cdots + w_n \qquad [6.18]$$

The diagram shows the relation between W_n and ω; we see it is in the form of a staircase, each step having width ω_1. At $\omega = \omega_{MAX}$, $W_n = W_{TOT}$, the total power in the signal. However, in the limit that signal period $T_O \to \infty$, $\omega_1 \to 0$ and W_n becomes a continuous function of ω. This is the **cumulative power function** (c.p.f.) $W(\omega)$, which is defined by:

Cumulative power function

$$W(\omega) = \lim_{\omega_1 \to 0} W_n \qquad [6.19]$$

The **power spectral density** (PSD) $\phi(\omega)$ is more commonly used and is the derivative of $W(\omega)$ (Fig. 6.4), i.e.

Power spectral density

$$\phi(\omega) = \frac{dW}{d\omega} \qquad [6.20]$$

ϕ has the units of watts sec rad^{-1} or W/Hz. Thus the power produced in a 1 Ω resistor due to frequencies between ω and $\omega + \Delta\omega$ is:

$$W_{\omega,\omega+\Delta\omega} = \Delta W = \phi(\omega)\Delta\omega \text{ watts} \quad [6.21)$$

i.e. by the area of a strip height $\phi(\omega$ and width $\Delta\omega$. Similarly the power due to frequencies between ω_1 and ω_2 is:

$$W_{\omega_1,\omega_2} = \int_{\omega_1}^{\omega_2} \phi(\omega)\, d\omega \text{ watts} \quad [6.22]$$

i.e. the area under the power spectral density curve between ω_1 and ω_2. The total area under the power spectral density curve is the total power in the signal, i.e.

$$W_{TOT} = \int_0^{\omega_{MAX}} \phi(\omega)\, d\omega \text{ watts} \quad [6.23]$$

Internal noise sources in electrical circuits can often be regarded as **white noise**, which has a uniform power spectral density over an infinite range of frequencies, i.e.

$$\phi(\omega) = A, \quad 0 \leq \omega \leq \infty \quad [6.24]$$

Another useful representation for both noise and measurement signals is a power spectral density which is constant up to a cut-off frequency ω_C and zero for higher frequencies (**band limited white noise**).

$$\phi(\omega = \begin{matrix} A, & 0 \leq \omega \leq \omega_C \\ 0, & \omega > \omega_C \end{matrix} \quad [6.25]$$

i.e. $W_{TOT} = A\omega_C$.

The maximum frequency ω_{MAX} for a measurement signal depends on the nature of the measured variable. Thus the vibration displacement of part of a machine may contain frequencies up to many kHz, whereas the temperature variations in a chemical reactor may only contain frequencies up to 0.01 Hz. In order to accurately measure random fluctuations in a measured variable, the transfer function $G(s)$ for the measurement system must satisfy the conditions of Section 4.4., i.e. $|G(j\omega)| = 1$ and arg $G(j\omega) = 0$ for all ω up to ω_{MAX}.

6.2.5 Autocorrelation function

Figure 6.5 shows a block diagram of a simple correlator. The input signal $y(t)$ is passed through a variable time delay unit to give a delayed signal $y(t - \beta)$. The signals $y(t)$ and $y(t - \beta)$ are multiplied together to give the product waveform $y(t)y(t - \beta)$. This product waveform is passed through an averager and the average value $\overline{y(t)y(t - \beta)}$ is displayed on a meter: this is the **autocorrelation coefficient** R_{yy}. If the time delay β is altered, the shape of the product waveform changes, causing the meter reading R_{yy} to change. The relationship between R_{yy} and time delay β is the **autocorrelation function** $R_{yy}(\beta)$ of the signal. We note that $R_{yy}(\beta)$ has a maximum value $R_{yy}(0)$ when $\beta = 0$; this is because the corresponding product waveform is $y^2(t)$, which is always positive and has a maximum average value.

If the signal is defined by a continuous function $y(t)$ in the interval 0 to T_O, then $R_{yy}(\beta)$ can be evaluated using:

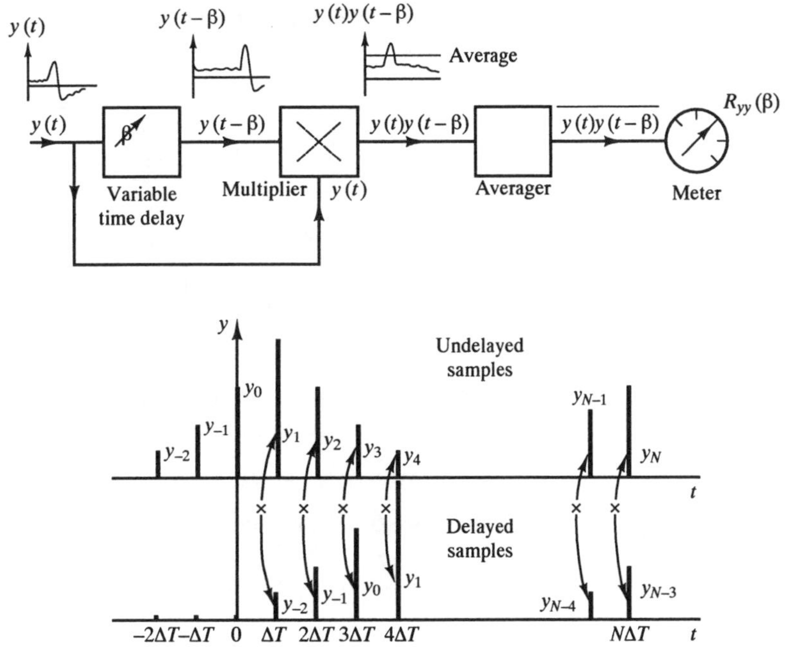

Fig. 6.5 Autocorrelation and evaluation of autocorrelation coefficient R_{yy} (3Δ*T*)

Autocorrelation function of a continuous signal

$$R_{yy}(\beta) = \lim_{T_O\to\infty} \frac{1}{T_O}\int_0^{T_O} y(t)y(t-\beta)\,dt \qquad [6.26]$$

Thus, if $y(t) = b\sin(\omega_1 t + \phi)$,

$$R_{yy}(\beta) = b^2 \lim_{T_O\to\infty} \frac{1}{T_O}\int_0^{T_O} \sin(\omega_1 t + \phi)\sin[\omega_1(t-\beta)+\phi]\,dt$$

$$= b^2 \lim_{T_O\to\infty} \frac{1}{T_O}\int_0^{T_O} \tfrac{1}{2}\{\cos\omega_1\beta - \cos[\omega_1(2t-\beta)+2\phi]\}\,dt$$

$$= \frac{b^2}{2}\cos\omega_1\beta \lim_{T_O\to\infty}\frac{1}{T_O}\int_0^{T_O} dt - \frac{b^2}{2}\lim_{T_O\to\infty}\frac{1}{T_O}\int_0^{T_O}\cos[\omega_1(2t-\beta)+2\phi]\,dt$$

$$= \frac{b^2}{2}\cos\omega_1\beta \qquad [6.27]$$

since the average value of $\cos[\omega_1(2t-\beta)+2\phi]$ is zero. Thus the autocorrelation function of a sinusoidal signal is a cosine function of the same frequency, but the phase information ϕ in the sine wave is lost. The autocorrelation function of any periodic signal has the same period as the signal itself (see Problem 6.2).

A random signal is usually characterised by a set of N sample values y_i. Since information is only available at discrete time intervals, the time delay β is normally an integer multiple of the sampling interval ΔT, i.e.

$$\beta = m\Delta T, \quad m = 0, 1, 2\ldots \qquad [6.28]$$

In this case the autocorrelation function of the signal is found by evaluating the set of autocorrelation coefficients $R_{yy}(m\Delta T)$ which are given by:

Autocorrelation coefficient for sampled signal

$$R_{yy}(m\Delta T) = \lim_{N\to\infty} \frac{1}{N} \sum_{i=1}^{i=N} y_i y_{i-m} \qquad [6.29]$$

where y_i is the sample value at time $i\Delta T$ and y_{i-m} the value at time $(i - m)\Delta T$ (m sampling intervals earlier). Figure 6.4 shows the evaluation of $R_{yy}(3\Delta T)$ for a sampled waveform; this involves calculating the products $y_1y_{-2}, y_2y_{-1}, y_3 y_0, \ldots y_Ny_{N-3}$, summing, and dividing by N.

The autocorrelation function of a random signal can also be found from the power spectral density $\phi(\omega)$. To illustrate this we first consider a periodic signal which is the sum of three harmonics:

$$y(t) = b_1 \sin \omega_1 t + b_2 \sin 2\omega_1 t + b_3 \sin 3\omega_1 t$$

Using eqn [6.27] the autocorrelation function is:

$$R_{yy}(\beta) = \frac{b_1^2}{2} \cos \omega_1\beta + \frac{b_2^2}{2} \cos 2\omega_1\beta + \frac{b_3^2}{2} \cos 3\omega_1\beta$$

which has period $2\pi/\omega_1$ (Fig. 6.6). Using eqn [6.17], the power spectrum of the signal consists of three lines at frequencies ω_1, $2\omega_1$, $3\omega_1$ with heights $b_1^2/2$, $b_2^2/2$, $b_3^2/2$ respectively. Thus the power spectrum can be obtained from the autocorrelation function by Fourier analysis. Similarly the autocorrelation function can be obtained from the power spectrum by adding the harmonics together as in Fourier synthesis. For random signals, $R_{yy}(\beta)$ and $\phi(\omega)$ are related by the Fourier transform or Wiener–Khinchin relations:

$$\begin{aligned} R_{yy}(\beta) &= \int_0^\infty \phi(\omega) \cos \omega\beta \, d\omega \\ \phi(\omega) &= \frac{2}{\pi}\int_0^\infty R_{yy}(\beta) \cos \omega\beta \, d\beta \end{aligned} \qquad [6.30]$$

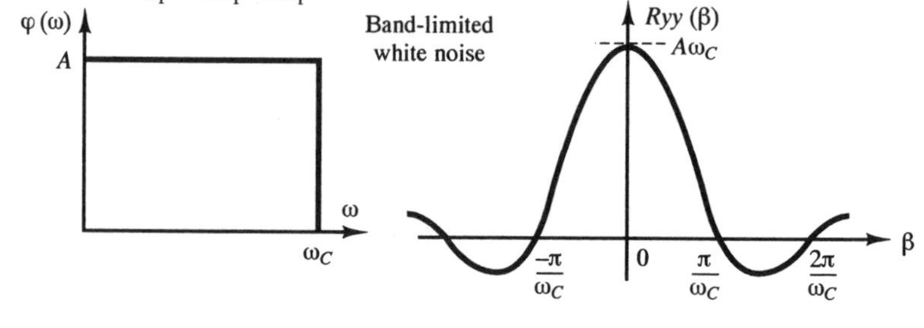

Fig. 6.6 Relationships between power spectrum and autocorrelation function for periodic and random signals

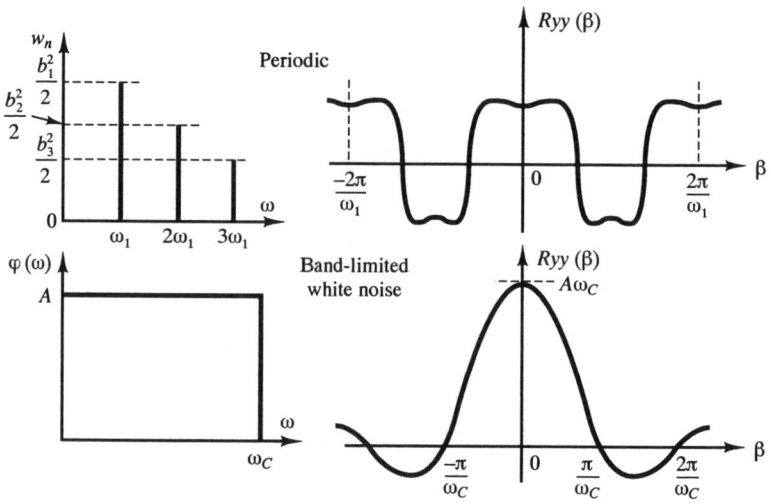

Thus for a signal with $\phi(\omega)$ constant up to ω_C and zero for higher frequencies we have:

$$R_{yy}(\beta) = \int_0^{\omega_C} A \cos \omega\beta \, d\omega = A \left[\frac{\sin \omega\beta}{\beta} \right]_0^{\omega_C}$$

$$= A \frac{\sin \omega_C \beta}{\beta} \qquad [6.31]$$

The form of both functions is shown in Fig. 6.6; we see that $R_{yy}(\beta)$ has its first zero crossings at $\beta = \pm \pi/\omega_C$, i.e. the width of the central 'spike' is $2\pi/\omega_C$. Thus a rapidly varying random signal has a high value of ω_C, i.e. a broad power spectrum but a narrow autocorrelation function, that falls off sharply as β is increased. A slowly varying random signal, however, has a low value of ω_C, i.e. a narrow power spectrum but a broad autocorrelation function that falls slowly as β is increased.

6.2.6 Summary

In order to specify a random signal we need to know:

either **probability density function** / or **mean and standard deviation** — to specify amplitude behaviour

and

either **power spectral density** / or **autocorrelation function** — to specify frequency/time behaviour

Important relations between these different quantities can be derived by considering $R_{yy}(0)$ the autocorrelation coefficient at zero time delay. From eqns [6.26] and [6.30] we have

$$R_{yy}(0) = \lim_{T_O \to \infty} \frac{1}{T_O} \int_0^{T_O} y^2 \, dt = \int_0^{\infty} \phi(\omega) \, d\omega \qquad [6.32]$$

From eqn [6.5], the first expression is equal to y^2_{RMS} in the limit of infinitely long observation time T_O. From equation [6.23] the second expression is equal to W_{TOT}, the total power produced by the signal in a 1 Ω resistor. Thus:

$$R_{yy}(0) = y^2_{\text{RMS}} = W_{\text{TOT}} \qquad [6.33]$$

6.3 Effects of noise and interference on measurement circuits

In Section 5.1 we saw that the interconnection of two measurement system elements; e.g. a thermocouple and amplifier or differential pressure transmitter and recorder, could be represented by an equivalent circuit in which either a Thévenin voltage source or a Norton current source is connected to a load. In industrial installations, source and load may be typically 100 metres apart and noise and/or interference voltages may also be present.

Figure 6.7(a) shows a **voltage transmission system** subject to **series mode**

Fig. 6.7 Effects of interference on measurement circuit
(a) Voltage transmission — series mode interference
(b) Current transmission — series mode interference
(c) Voltage transmission — common mode interference

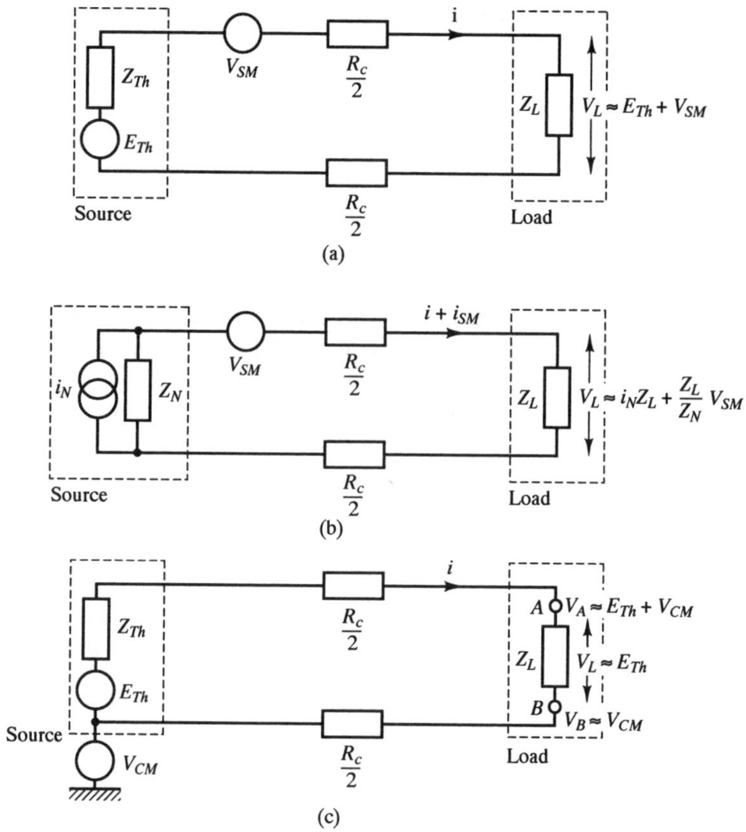

interference; here a noise or interference voltage V_{SM} is in series with the measurement signal voltage E_{Th}. The current i through the load is:

$$i = \frac{E_{Th} + V_{SM}}{Z_{Th} + R_c + Z_L}$$

and the corresponding voltage across the load is:

$$V_L = \frac{Z_L}{Z_{Th} + R_c + Z_L}(E_{Th} + V_{SM}) \qquad [6.34]$$

Normally we make $Z_l \gg R_c + Z_{Th}$ to obtain maximum voltage transfer to the load (Section 5.1.1); under these conditions eqn [6.34] becomes:

$$V_L \approx E_{Th} + V_{SM} \qquad [6.35]$$

This means that with a voltage transmission system all of V_{SM} is across the load, this affects the next element in the system and possibly results in a system measurement error. We define **signal-to-noise** or **signal to interference ratio** *S/N* in decibels by:

Signal-to-noise ratio

$$\frac{S}{N} = 20 \log_{10}\left(\frac{E_{Th}}{V_{SM}}\right) = 10 \log_{10}\left(\frac{W_S}{W_N}\right)\text{dB} \qquad [6.36]$$

where E_{Th} and V_{SM} are the r.m.s. values of the voltages, W_S and W_N are the corresponding total signal and noise powers. Thus if $E_{Th} = 1$ V, $V_{SM} = 0.1$ V, $S/N = +20$ dB.

Figure 6.7(b) shows a **current transmission system** subject to the same series mode interference voltage V_{SM}. The Norton source current i_N divides into two parts, one part through the source impedance Z_N, the other part through Z_L. Using the current divider rule, the current through the load due to the source is:

$$i = i_N \frac{Z_N}{Z_N + R_c + Z_L}$$

In addition there is an interference current

$$i_{SM} = \frac{V_{SM}}{Z_N + R_c + Z_L}$$

through the load due to the interference voltage. The total voltage across the load is therefore:

$$V_L = iZ_L + i_{SM}Z_L$$

$$= i_N Z_L \cdot \frac{Z_N}{Z_N + R_c + Z_L} + V_{SM} \cdot \frac{Z_L}{Z_N + R_c + Z_L} \qquad [6.37]$$

Normally we make $R_c + Z_L \ll Z_N$ to obtain maximum current transfer to the load (Section 5.1.3); under these conditions eqn [6.37] becomes:

$$V_L \approx i_N Z_L + \frac{Z_L}{Z_N} V_{SM} \qquad [6.38]$$

Since $Z_L/Z_N \ll 1$, this means that with a current transmission system only a small fraction of V_{SM} is across the load. Thus a current transmission system has far greater inherent immunity to series mode interference than a voltage transmission system. In a thermocouple temperature measurement system, therefore, it may be better to convert the thermocouple millivolt e.m.f. into a current signal (Section 9.4.1) prior to transmission, rather than transmit the e.m.f. directly.

Figure 6.7(c) shows a voltage transmission system subject to **common mode interference** in which the potentials of both sides of the signal circuit are raised by V_{CM} relative to a common earth plane. If as above $Z_L \gg R_c + Z_{Th}$, then current $i \to 0$ so that the potential drops $iR_c/2$ etc. can be neglected. Under these conditions:

$$\text{Potential at } B = V_{CM}$$
$$\text{Potential at } A = V_{CM} + E_{Th}$$

and $\quad V_L = V_B - V_A = E_{Th}$

This means that the voltage across the load is unaffected by V_{CM}; there is, however, the possibility of conversion of a common mode voltage to series mode.

6.4 Noise sources and coupling mechanisms

6.4.1 Internal noise sources

The random, temperature-induced motion of electrons and other charge carriers in resistors and semiconductors gives rise to a corresponding random voltage which is called thermal or Johnson noise. This has a power spectral density which is uniform over an infinite range of frequencies (white noise) but proportional to the absolute temperature θK of the conductor, i.e.

$$\phi = 4Rk\theta \text{ watts/Hz} \qquad [6.39]$$

where $R\Omega$ is the resistance of the conductor and k is the Boltzmann constant $= 1.4 \times 10^{-23}\,\text{J K}^{-1}$. From eqn [6.22] the total thermal noise power between frequencies f_1 and f_2 is:

$$W = \int_{f_1}^{f_2} 4Rk\theta \, df = 4Rk\theta(f_2 - f_1) \text{ watts} \qquad [6.40]$$

and from [6.33] the corresponding r.m.s. voltage is:

$$V_{RMS} = \sqrt{W} = \sqrt{[4Rk\theta(f_2 - f_1)]} \text{ volts} \qquad [6.41]$$

Thus if $R = 10^6\,\Omega$, $f_2 - f_1 = 10^6$ Hz and $\theta = 300$K, $V_{RMS} = 130\,\mu$V and is therefore comparable with low level measurement signals such as the output from a strain gauge bridge.

A similar type of noise is called **shot noise**; this occurs in transistors and is due to random fluctuations in the rate at which carriers diffuse across a junction. This is again characterised by a uniform power spectral density over a wide range of frequencies.

6.4.2 External noise and interference sources[1]

The most common sources of external interference are nearby **a.c. power circuits** which usually operate at 240 V, 50 Hz. These can produce corresponding sinusoidal interference signals in the measurement circuit, referred to as **mains pick-up** or **hum**. Power distribution lines and heavy rotating machines such as turbines and generators, can cause serious interference.

D.C. power circuits are less likely to cause interference because d.c. voltages are not coupled capacitively and inductively to the measurement circuit.

However, **switching** often occurs in both a.c. and d.c. power circuits when equipment such as motors and turbines are being taken off line or brought back on line. This causes sudden large changes in power, i.e. steps and pulses, which can produce corresponding **transients** in the measurement circuit.

The air in the vicinity of high voltage power circuits can become ionised and a **corona discharge** results. Corona discharge from d.c. circuits can result in random noise in the measurement circuit and that from a.c. circuits results in sinusoidal interference at the power frequency or its second harmonic.

Fluorescent lighting is another common interference source; arcing occurs twice per cycle so that most of the interference is at twice the power frequency.

Radio-frequency transmitters, welding equipment and electric arc furnaces can produce **r.f. interference** at frequencies of several MHz.

6.4.3 Coupling mechanisms to external sources

Inductive coupling[2]

Figure 6.8(a) shows inductive or **electromagnetic coupling** between the measurement circuit and a nearby power circuit. If the circuits are sufficiently close together, then there may be a significant mutual inductance M between them. This means that an alternating current i in the power circuit induces a series mode interference voltage in the measurement circuit of magnitude:

$$V_{SM} = M \frac{di}{dt} \qquad [6.42]$$

Thus if $M \approx 1\ \mu H$ and $di/dt \approx 10^3\ A\ s^{-1}$ (possible in a 1 horsepower motor) then $V_{SM} \approx 1\ mV$, which can be comparable with the measurement signal. The mutual inductance M depends on the geometry of the two circuits, namely on the overlapping length and separation, but is distributed over the entire length of the circuits rather than the 'lumped' equivalent value shown in Fig. 6.8. Inductive coupling will occur even if the measurement circuit is completely isolated from earth.

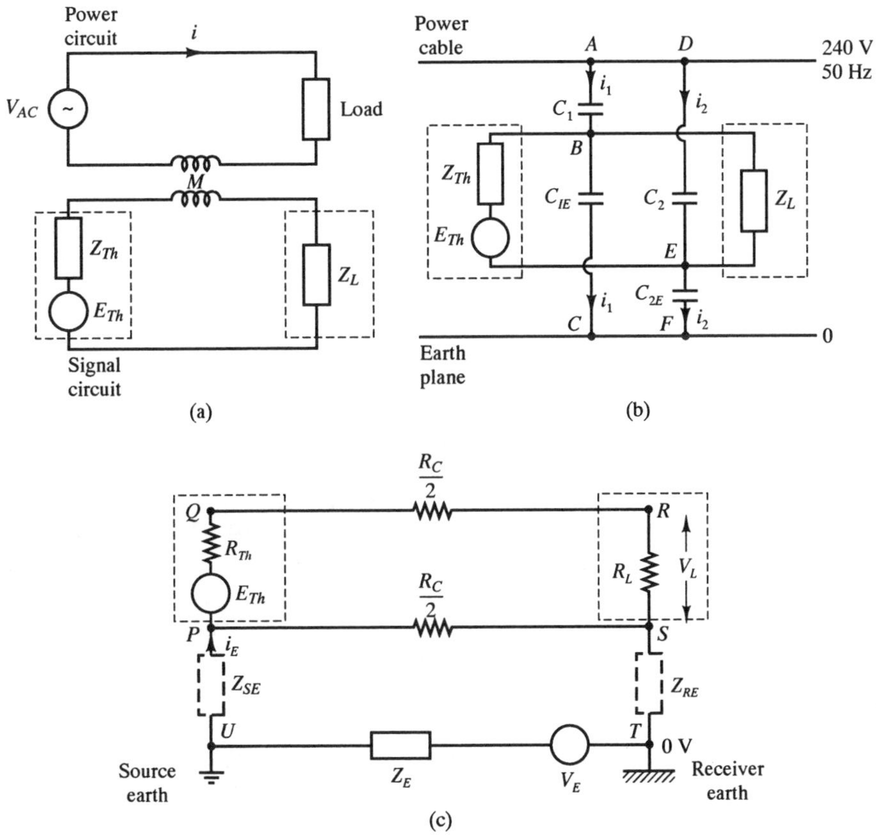

Fig. 6.8 Coupling mechanisms to external sources
(a) Electromagnetic coupling
(b) Electrostatic coupling
(c) Multiple earths

Capacitive coupling[1,2]

Another important coupling mechanism is capacitive or **electrostatic coupling**, which is illustrated in Fig. 6.8(b). The diagram shows the measurement circuit close to an a.c. power cable which is at a potential of 240 V (r.m.s.) relative to the earth plane. The power cable, earth plane and signal leads are all conductors, so that there may be some capacitance between the power cable and the signal leads and between the signal leads and the earth plane. These capacitances will be distributed over the entire length of the measurement circuit, but are represented by 'lumped' equivalents. C_1 and C_2 are the capacitances between the power cable and signal leads, and C_{1E} and C_{2E} the capacitances between the signal lead and the earth plane; all four capacitances will be proportional to the length of the measurement circuit, which could be tens of metres in an industrial installation. Ignoring the measurement signal voltage E_{Th} for the moment, the potentials at B and E are determined by the potential dividers ABC and DEF:

$$\begin{aligned} V_B &= 240\left[\frac{1/(j\omega C_{1E})}{1/(j\omega C_{1E}) + 1/(j\omega C_1)}\right] = 240\,\frac{C_1}{C_1 + C_{1E}} \\ V_E &= 240\left[\frac{1/(j\omega C_{2E})}{1/(j\omega C_{2E}) + 1/(j\omega C_2)}\right] = 240\,\frac{C_2}{C_2 + C_{2E}} \end{aligned} \qquad [6.43]$$

Thus we have a common mode interference voltage $V_{CM} = V_E$ and a series mode interference voltage:

$$V_{SM} = V_B - V_E = 240\left\{\frac{C_1}{C_1 + C_{1E}} - \frac{C_2}{C_2 + C_{2E}}\right\} \qquad [6.44]$$

Thus series mode interference is zero only if there is perfect balance between the coupling capacitances i.e. $C_1 = C_2$ and $C_{1E} = C_{2e}$; in practice small imbalances are present due to slightly different distances between each signal lead and the power cable/earth plane.

Multiple earths[1]

The above explanation assumes an earth plane having a potential of 0 volts at every point on its surface. Heavy electrical equipment can, however, cause currents to flow through the earth causing different potentials at different points. If the measurement circuit is completely isolated from the earth plane there is no problem. In practice, however, there may be a leakage path connecting the signal source to one earth point and another leakage path connecting the recorder or indicator to a different earth point, some distance away. If the two earth points are at different potentials, then common and series mode interference voltages are produced in the measurement circuit.

Figure 6.8(c) illustrates the general problem of **multiple earths**. The measurement signal source E_{Th} is connected via a resistive cable to a receiver represented by a resistive load R_L. Provided $R_L \gg R_c + R_{Th}$, the current flow in $PQRS$ is negligible and $V_L \approx E_{th}$, providing the circuit is completely isolated from earth. However, leakage paths Z_{SE} and Z_{RE} exist between source/source earth and receiver/receiver earth. If V_E is the difference in potential between source earth and receiver earth, then a current i_E flows in the circuit *UPST*, given by:

$$i_E = \frac{V_E}{Z_E + Z_{SE} + (R_C/2) + Z_{RE}} \qquad [6.45]$$

Thus

$$\begin{aligned}
\text{potential at } P &= V_E - i_E(Z_E + Z_{SE}) \\
\text{potential at } Q, R &= V_E - i_E(Z_E + Z_{SE}) + E_{Th} \\
\text{potential at } S &= i_E Z_{RE}
\end{aligned}$$

Thus there is a common mode interference voltage:

$$V_{CM} = V_S = V_E \frac{Z_{RE}}{Z_E + Z_{SE} + (R_C/2) + Z_{RE}} \qquad [6.46]$$

To find the series mode interference voltage we need to calculate the voltage across R_L:

$$\begin{aligned}
V_L = V_R - V_S &= V_E - i_E(Z_E + Z_{SE} + Z_{RE}) + E_{Th} \\
&= E_{Th} + i_E R_C/2
\end{aligned}$$

i.e. there is a series mode interference voltage:

$$V_{SM} = V_E \frac{R_C/2}{Z_E + Z_{SE} + (R_C/2) + Z_{RE}} \qquad [6.47]$$

Ideally we require both Z_{SE} and Z_{RE} to be as large as possible in order to minimise i_E and V_{SM}; this, however, is not always possible in an industrial application. A common example is a thermocouple installation where in order to achieve as good a speed of response as possible the tip of the thermocouple touches the thermowell or sheath (Section 14.2). The thermowell is itself bolted to a metal vessel or pipe which is in turn connected to one point in the earth plane. Thus Z_{SE} will be very small, say $Z_{SE} = 10\,\Omega$ (resistive), so that the receiver must be isolated from earth to minimise V_{SM}. Taking $Z_E = 1\,\Omega$, $R_C/2 = 10\,\Omega$, $V_E = 1$ V and $Z_{RE} = 10^6\,\Omega$, we have:

$$V_{SM} = 1 \times \frac{10}{1 + 10 + 10 + 10^6}\,\text{V} \approx 10\,\mu\text{V}$$

The worst case is when the receiver is also directly connected to earth i.e. $Z_{RE} \approx 0$, giving $V_{SM} = 0.48$ V. Thus if the measurement circuit must be connected to earth, the connection must be made at one point only.

6.5 Methods of reducing effects of noise and interference

6.5.1 Physical separation

Since mutual inductances and coupling capacitances between measurement and power circuits are inversely proportional to the distance between them, this distance should be as large as possible.

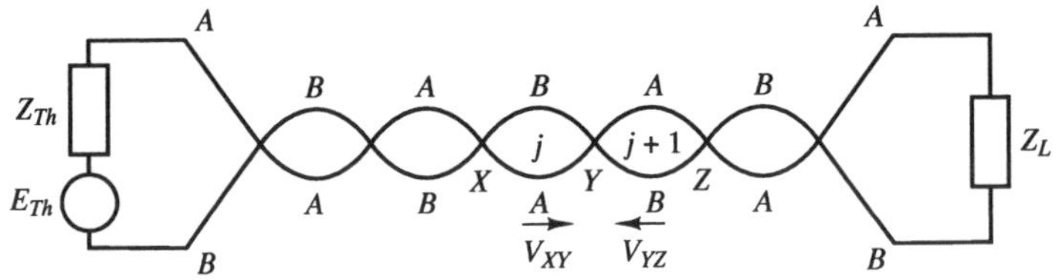

Fig. 6.9 Reduction of electromagnetic coupling by twisted pairs

6.5.2 Electromagnetic shielding

The simplest way of reducing the effects of inductive coupling to an external interference source is shown in Fig. 6.9. The two conductors A and B of the measurement circuit are twisted into loops of approximately equal area. This arrangement is commonly known as **twisted pairs** and is explained in ref. [2]. The magnitude of the interference voltage induced in a given loop is proportional to the area of the loop and the rate of change of the external magnetic field. The sign of the induced voltage depends on the orientation of conductors A and B. Thus if a voltage V_{XY} is inducted in the jth loop between points X and Y, then an opposing voltage V_{YZ} is induced in the $(j + 1)$th loop between Y and Z. In the ideal case of both loops having the same area and experiencing the same magnetic fields, $|V_{XY}| = |V_{YZ}|$, i.e. there is a zero resultant induced voltage between X and Z. This process is repeated for the whole length of the twisted pair, giving a reduced overall interference voltage.

6.5.3 Electrostatic screening and shielding

The best method of avoiding the problem of capacitive coupling to a power circuit (Section 6.4.3) is to enclose the entire measurement circuit in an earthed metal screen or shield. Figure 6.10(a) shows the ideal arrangement; the screen is connected directly to earth at a single point, either at the source or the receiver. There is no direct connection between the screen and the measurement circuit, only high impedance leakage paths via the small (screen/measurement circuit) capacitances C_{SM}. The screen provides a low impedance path to earth for the interfering currents i; the currents through C_{SM} and C_E are small thus reducing series and common mode interference.

The above ideal of the measurement circuit completely insulated from the screen and the screen earthed at one point only may be difficult to achieve in practice for the following reasons:

(a) The signal source may be directly connected to a local earth point via the structure on which it is mounted; an example is the thermocouple installation mentioned in Section 6.4.3.
(b) The receiver may be directly connected to a local earth, an example is in a computer-based system where the receiver must be directly connected to the computer earth.
(c) There may be indirect connections via leakage impedances.

Figure 6.10(b) illustrates the general problem. The measurement circuit PQRS is

Fig. 6.10 Reduction of electrostatic coupling using screening
(a) Ideal arrangement
(b) Practical equivalent circuit

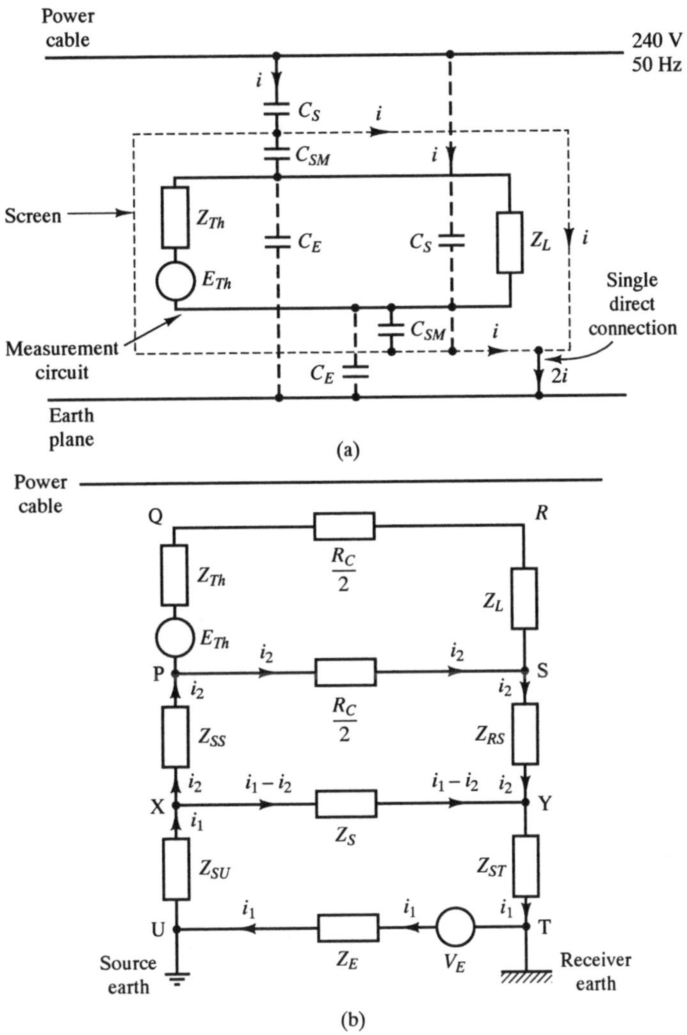

connected to the screen (impedance Z_S) via source/screen impedance Z_{SS} and receiver/screen impedance Z_{RS}. The screen is connected to earth point U at the source end via Z_{SU} and to earth point T at the receiver end via Z_{ST}. The measurement circuit can be affected by interference voltages from both V_E (potential difference between U and T) and nearby power circuits.

Analysis of circuits UXYT and XPSY using Kirchoff's laws gives:

$$\text{UXYT} \quad V_E = i_1 Z_E + i_1 Z_{SU} + (i_1 - i_2) Z_S + i_1 Z_{ST} \qquad [6.48]$$

$$\text{XPSY} \quad 0 = -(i_1 - i_2) Z_S + i_2 Z_{SS} + i_2 R_C/2 + i_2 Z_{RS} \qquad [6.49]$$

Solution of these equations gives:

$$i_1 = \frac{(Z_S + Z_{SS} + R_C/2 + Z_{RS}) V_E}{[(Z_E + Z_{SU} + Z_S + Z_{ST})(Z_S + Z_{SS} + R_C/2 + Z_{RS}) - Z_S^2]} \qquad [6.50]$$

$$i_2 = \frac{Z_S V_E}{[(Z_E + Z_{SU} + Z_S + Z_{ST})(Z_S + Z_{SS} + R_C/2 + Z_{RS}) - Z_S^2]} \qquad [6.51]$$

The series mode interference voltage in the measurement circuit PQRS is the voltage drop across PS i.e.

$$V_{SM} = i_2 R_C/2 = \frac{Z_S R_C/2 V_E}{[(Z_E + Z_{SU} + Z_S + Z_{ST})(Z_S + Z_{SS} + R_C/2 + Z_{RS}) - Z_S^2]} \qquad [6.52]$$

The common mode interference voltage is the voltage drop aross ST i.e.

$$V_{CM} = V_{ST} = V_{SY} + V_{YT} = i_2 Z_{RS} + i_1 Z_{ST} \qquad [6.53]$$

$$= \frac{[Z_{RS}Z_s + Z_{ST}(Z_S + Z_{SS} + R_C/2 + Z_{RS})]V_E}{[(Z_E + Z_{SU} + Z_S + Z_{ST})(Z_S + Z_{SS} + R_C/2 + Z_{RS}) - Z_S^2]}$$

To minimise V_{SM} and V_{CM}, we want the product term:

$$(Z_E + Z_{SU} + Z_S + Z_{ST})(Z_S + Z_{SS} + R_C/2 + Z_{RS})$$

to be large. Since in practice the earth impedance Z_E, screen impedance Z_S and cable resistance $R_C/2$ are all small, the above condition reduces to:

$$(Z_{SU} + Z_{ST})(Z_{SS} + Z_{RS}) \text{ to be large} \qquad [6.54]$$

However, we cannot have *both* Z_{SU} *and* Z_{ST} large; either Z_{ST} or Z_{SU} must be small otherwise the screen will not be earthed and there is therefore no low impedance path to earth for the capacitively coupled interference currents. Condition (6.54) therefore reduces to the two conditions:

$$Z_{SU}(Z_{SS} + Z_{RS}) = \text{HIGH}; \; Z_{ST} = \text{LOW}$$

OR [6.55]

$$Z_{ST}(Z_{SS} + Z_{RS}) = \text{HIGH}; \; Z_{SU} = \text{LOW}$$

which are satisfied if:

$$Z_{SU} = \text{HIGH AND } (Z_{SS} \text{ OR } Z_{RS} \text{ OR both} = \text{HIGH}), \; Z_{ST} = \text{LOW}$$

OR

$$Z_{ST} = \text{HIGH AND } (Z_{SS} \text{ OR } Z_{RS} \text{ OR both} = \text{HIGH}), \; Z_{SU} = \text{LOW} \qquad [6.56]$$

As mentioned above, in many practical situations it may be impossible to have the measurement circuit completely isolated from the screen i.e. *both* Z_{SS} *and* Z_{RS} high. In this situation, possible confusion is avoided if Z_{SU} *and* Z_{SS} are *both* high, *or* Z_{ST} *and* Z_{RS} are *both* high.

In this situation the conditions become:

$$Z_{SU} = \text{HIGH AND } Z_{SS} = \text{HIGH}, \; Z_{RS} = \text{LOW AND } Z_{ST} = \text{LOW}$$

(Isolated source)

OR

$$Z_{ST} = \text{HIGH AND } Z_{RS} = \text{HIGH}, \; Z_{SS} = \text{LOW AND } Z_{SU} = \text{LOW}$$

(Isolated receiver)

[6.57]

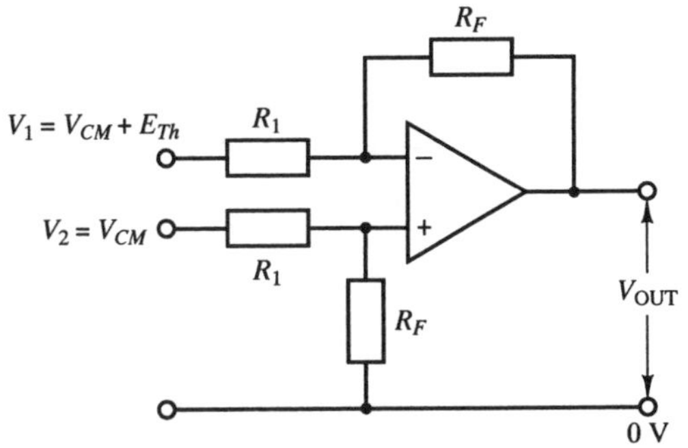

Fig. 6.11 Use of differential amplifier

6.5.4 Use of differential amplifiers

Common mode interference voltages can be successfully rejected by the use of a differential amplifier (Fig. 6.11 and Section 9.2). An ideal differential amplifier has an output:

$$V_{OUT} = \frac{R_F}{R_1}(V_2 - V_1) = -\frac{R_F}{R_1}E_{Th} \qquad [6.58]$$

i.e. only the sensor voltage E_{Th} is amplified. The output of a practical amplifier (Section 9.2.2) contains a contribution proportional to V_{CM}; from eqn [9.33] we have:

$$V_{OUT} = -\frac{R_F}{R_1}E_{Th} + \left(1 + \frac{R_F}{R_1}\right)\frac{V_{CM}}{\text{CMRR}} \qquad [6.59]$$

The common mode rejection ratio (CMRR) of the amplifier is the ratio of differential voltage gain to common mode voltage gain and should be as large as possible to minimise this effect. Thus if we have $E_{Th} = 1$ mV, $R_1 = 1\,\text{k}\Omega$, $R_F = 1\,\text{M}\Omega$, $V_{CM} = 1$V and CMRR $= 10^5$(100 dB) then:

$$V_{OUT} \approx -1.0 + 0.01\,\text{V}$$

i.e. the resultant series mode interference is only 1%.

6.5.5 Filtering

A filter is an element which transmits a certain range (or ranges) of frequencies and rejects all other frequencies. An **analogue filter** is an electrical network, consisting usually of resistors, capacitors and operational amplifiers, which conditions continuous signals. A **digital filter** is usually a digital computer programmed to process sampled values of a signal (Chapter 10). Provided that the power spectrum of the measurement signal occupies a *different* frequency range from that of the noise or interference signal, then filtering improves the signal-to-noise ratio.

Figures 6.12(a)–(d) show the use of **low pass**, **high pass**, **band pass** and **band stop** filters in rejecting noise. In all cases the filter transmits the measurement signal but rejects the noise signal which occupies a different frequency range. The diagrams show the amplitude ratio $|G(j\omega)|$ for each filter and the power spectral densities $\phi(\omega)$ for signal and noise. In order to transmit the measurement signal without distortion the transfer function $G(s)$ of the filter must, ideally, satisfy the conditions

Fig. 6.12 Use of filtering to reject noise

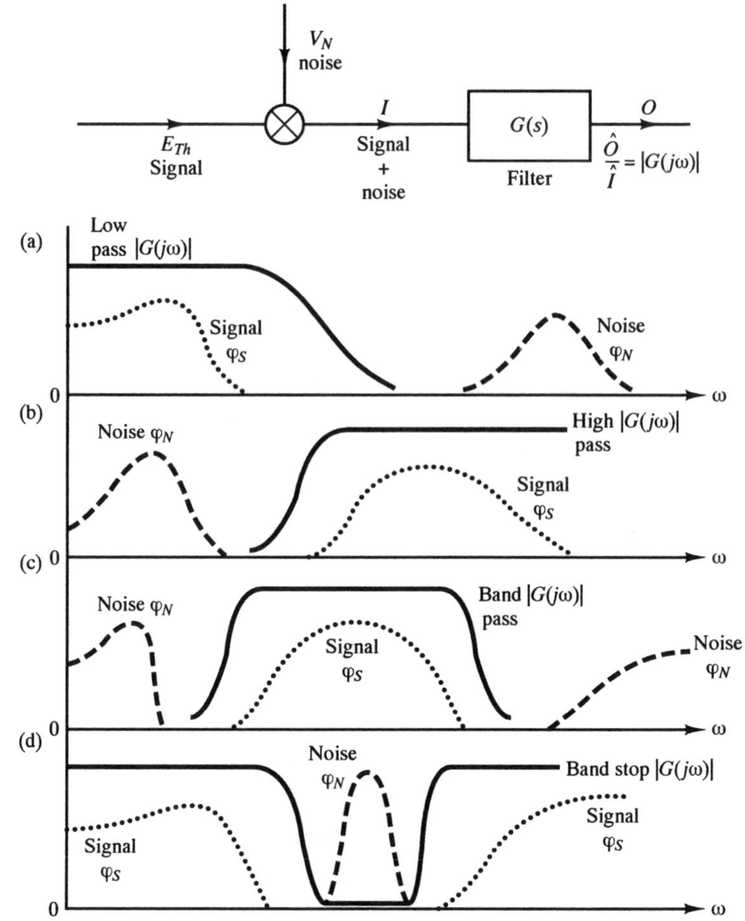

of Section 4.4, i.e. that $|G(j\omega)| = 1$ and $\arg G(j\omega) = 0$ for all the frequencies present in the measurement signal spectrum. Analogue filtering can be implemented at the signal conditioning stage; the a.c. amplifier of Fig. 9.7(e) is an example of a band pass filter. Digital filtering can be implemented at the signal processing stage (Chapter 10).

If, however, measurement signal and noise spectra overlap, filtering has limited value. Figure 6.13(a) shows a measurement signal affected by wide band noise; a low-pass filter with bandwidth matched to the signal spectrum removes as much of the noise as possible, but the noise inside the filter bandwidth still remains. Figure 6.13(b) shows the measurement signal affected by 50 Hz interference; a narrow band-stop filter centred on 50 Hz rejects the interference but also rejects measurement signal frequencies around 50 Hz and causes amplitude and phase distortion over a wider range of frequencies.

6.5.6 Modulation

The problem of Fig. 6.13(b) can be solved by modulating the measurement signal onto a higher frequency carrier signal, e.g. a 5 kHz sine wave as shown in Fig.

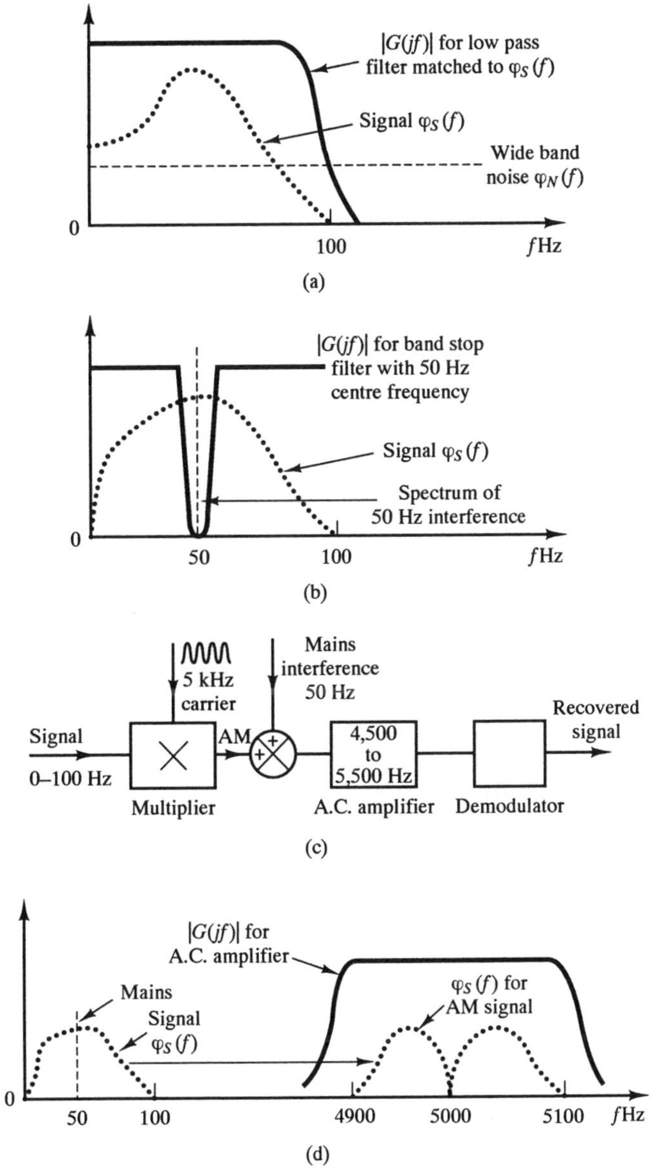

Fig. 6.13 Limitations of filtering and use of modulation

6.13(c). The simplest form of modulation is **amplitude modulation**; this involves the multiplication of measurement and carrier signals and is discussed in detail in Section 9.3.

Modulation causes the spectrum of the measurement signal to be shifted to around 5 kHz (Fig. 6.13(d)). If the 50 Hz interference is added **after** modulation i.e. during transmission from sensor/modulator to a remote amplifier/demodulator, the interference spectrum is not shifted. The interference can then be easily rejected by an a.c. amplifier, i.e. a band pass filter with bandwidth matched to the spectrum of the amplitude modulated signal. Modulation, however, does not help the problem of Fig. 6.14(a); the noise has a uniform power spectral density over a wide band

Fig. 6.14 Signal averaging

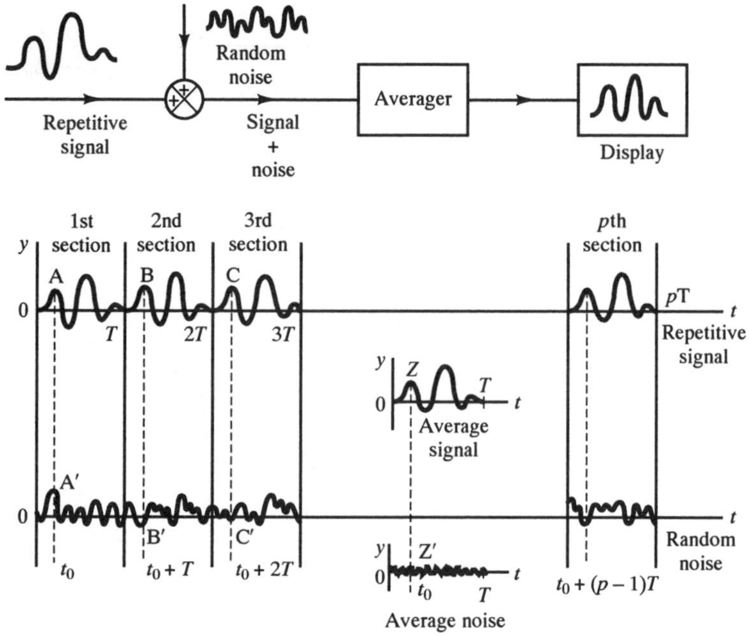

of frequencies, so that moving the measurement signal to a different frequency range does not improve the signal-to-noise ratio.

6.5.7 Averaging

Signal averaging can be used to recover a **repetitive** measurement signal affected by random noise, even if the signal r.m.s. value is much less than that of the noise.[3] The process is shown in Fig. 6.14.

Suppose that T is the time for each complete cycle of the repetitive signal; p sections of the noise affected signal, each of duration T are fed into the averager. N samples from each section are taken and stored, giving pN samples in total, typically we may have $p = 50$, $N = 100$. The sampling is exactly synchronised; i.e. if the ith sample of the 1st section is taken at time t_0, the ith sample of the 2nd section is taken at $t_0 + T$, the ith sample of the 3rd at $t_0 + 2T$, and so on. Corresponding samples from each section are then averaged; e.g. the first sampled values from each of the p sections are added together and divided by p. Thus the average value of the ith sample is:

$$y_i^{AV} = \frac{1}{p}(y_{i1} + y_{i2} + \cdots + y_{ip}), \quad i = 1, \ldots, N \tag{6.60}$$

Each of these N average sample values are then displayed at the appropriate time to give the averaged signal. Corresponding sample values $A, B, C, \ldots$ of the signal component are approximately equal, so that the average value Z has a similar magnitude. Corresponding sample values $A', B', C', \ldots$ of the noise component are very different, some positive and some negative, so that the average value Z' is reduced in magnitude. Averaging therefore maintains the r.m.s. value of the measurement signal while reducing the r.m.s. value of the random noise.

This improvement in signal-to-noise ratio can be readily calculated for random noise with a Gaussian probability density function. Suppose we have p Gaussian signals $y_1(t)$ to $y_p(t)$, with standard deviations σ_1 to σ_p respectively; then from Section 2.3 the average signal

$$y_{AV}(t) = \frac{1}{p}[y_1(t) + y_2(t) + \cdots + y_p(t)] \qquad [6.61]$$

is also Gaussian with standard deviation:

$$\sigma_{AV} = \sqrt{\left(\frac{\sigma_1^2}{p^2} + \frac{\sigma_2^2}{p^2} + \cdots + \frac{\sigma_p^2}{p^2}\right)} \qquad [6.62]$$

If $\sigma_1 = \sigma_2 = \sigma_p = \sigma$, then

Reduction in noise standard deviation due to averaging

$$\sigma_{AV} = \sqrt{\left(\frac{p\sigma^2}{p^2}\right)} = \frac{\sigma}{\sqrt{p}} \qquad [6.63]$$

Thus if we average 50 sections $\sigma_{AV} = \sigma/\sqrt{50} \approx \sigma/7$, i.e. the noise r.m.s. value is reduced by a factor of 7; giving an increase in signal-to-noise ratio of 17 dB.

6.5.8 Autocorrelation

Autocorrelation can be used to **detect** the presence of a sinusoidal or any periodic signal buried in random noise. The actual waveform of the measurement signal cannot be recovered because phase information is lost in correlation, but we can measure the amplitude and period of the signal from the autocorrelation function (ACF) of the noise-affected signal. The ACF for the (signal + noise) is the sum of the signal ACF and noise ACF, i.e.

$$R_{yy}^{S+N}(\beta) = R_{yy}^{S}(\beta) + R_{yy}^{N}(\beta) \qquad [6.64]$$

Thus using eqns [6.27] and [6.31] the ACF for a sinusoidal signal affected by band limited white noise is:

$$R_{yy}^{S+N}(\beta) = \frac{b^2}{2}\cos\omega_1\beta + A\frac{\sin\omega_C\beta}{\beta} \qquad [6.65]$$

The form of $R_{yy}^{S+N}(\beta)$ is shown in Fig. 6.15; at large values of β the $A(\sin\omega_c\beta)/\beta$ term due to the noise decays to zero leaving the $(b^2/2)\cos\omega_1\beta$ term due to the signal. Thus the amplitude b and period $2\pi/\omega_1$ of the original signal can be found from the amplitude and period of the autocorrelation function at large values of time delay. This method can be used in the vortex flowmeter (Chapter 12) to measure the vortex frequency in the presence of random flow turbulence.

Conclusion

The chapter began by defining **random** and **deterministic signals** and explained that in many practical situations the wanted signal may be random. Unwanted signals

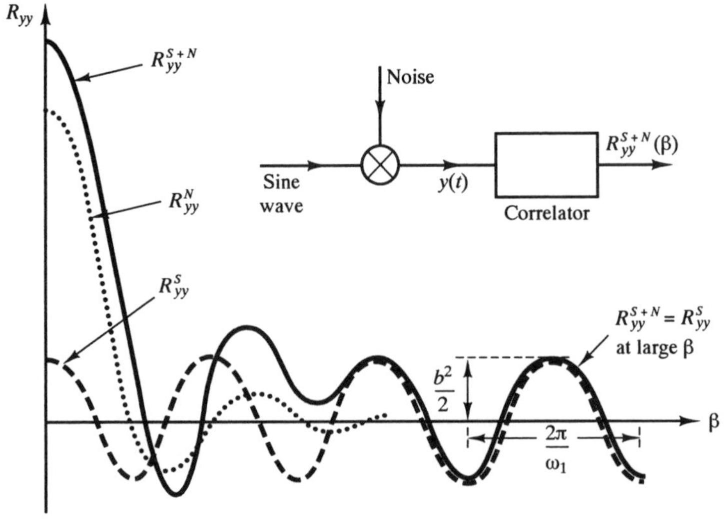

Fig. 6.15 Auto-correlation detection of periodic signal buried in noise

may also be present in the measurement circuit, these can be classified as either **interference** (deterministic) or **noise** (random).

The chapter then explains how random signals can be quantified using the following statistical functions: **mean, standard deviation, probability density function, power spectral density** and **autocorrelation function.**

The effects of noise and interference voltage on measurement circuits using both voltage and current transmission are then discussed. In the following section internal noise and external interference sources are discussed and the mechanisms whereby external sources are coupled to the measurement circuit are explained. The chapter concludes by explaining methods of reducing the effects of noise and interference, these include **electromagnetic shielding, electrostatic screening, filtering, modulation, averaging** and **autocorrelation.**

References

6.1 OLIVER F J 1972 *Practical Instrumentation Transducers.* Pitman, London, pp. 290–333.

6.2 COOK B E 1979 'Electronic noise and instrumentation', *Journal Instrumentation Measurement and Control*, vol. 12, no. 8.

6.3 Hewlett Packard, Technical Literature on Model 3721A Correlator.

Problems

6.1 Table Prob. 1 gives 50 sample values of a random signal:

(a) Estimate the mean and standard deviation of the signal

(b) Using an interval $\Delta y = 0.5$ V, draw the P_j and C_j discrete probability distributions for the signal.

Table Prob. 1

−0.59	1.02	−0.25	−0.34	0.95	1.24	−0.30	0.21	−0.89
−1.00	1.36	0.03	0.04	−0.13	−0.71	−1.23	0.03	−1.00
0.65	0.11	0.99	0.17	0.39	2.61	−0.08	−0.33	0.99
2.15	0.91	0.89	1.43	−1.69	−0.25	2.47	−1.97	−2.26
0.42	0.05	0.26	0.33	−0.42	0.79	−0.07	−0.32	−0.66
−0.63	−0.06	−0.61	0.77	1.90				

6.2 Two complete periods of a square wave can be represented by the following 20 sample values:

$$+1+1+1+1+1 \quad -1-1-1-1-1 \quad +1+1+1+1+1 \quad -1-1-1-1-1$$

Find the autocorrelation function of the signal over one complete period by evaluating the coefficients $R_{yy}(m\Delta T)$ for $m = 0, 1, 2, \ldots, 10$.

6.3 A sinusoidal signal of amplitude 1.4 mV and frequency 5 kHz is 'buried' in Gaussian noise with zero mean value. The noise has a uniform power spectral density of 100 pW Hz^{-1} up to a cut-off frequency of 1 MHz.

(a) Find the total power, r.m.s. value and standard deviation for the noise signal.
(b) What is the signal-to-noise ratio in dB?
(c) Sketch the autocorrelation function for the combined signal and noise.
(d) The combined signal is passed through a band-pass filter with centre frequency 5 kHz and bandwidth 1 kHz. What improvement in signal-to-noise ratio is obtained?
(e) The filtered signal is then passed through a signal averager which averages corresponding samples of 100 sections of signal. What further improvement in signal-to-noise ratio is obtained?

6.4 A thermocouple giving a 10 mV d.c. output voltage is connected to a high impedance digital voltmeter some distance away. A difference in potential exists between earth at the thermocouple and earth at the voltmeter. Using the equivalent circuit given in Fig. Prob. 4.

(a) Calculate the r.m.s. values of the series mode and common mode interference voltages at the voltmeter input.
(b) If given that the d.v.m. has a common mode rejection ratio of 100 dB, find the minimum and maximum possible measured voltages.

Fig. Prob. 4

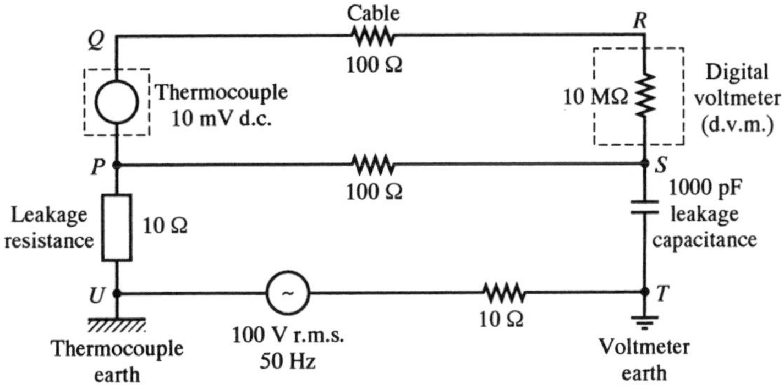

6.5 A sinusoidal signal is transmitted over a noisy transmission link to a remote correlator acting as a receiver. Fig. Prob. 5 shows a typical auto-correlation function. Use the figure to estimate the following quantities:

Fig. Prob. 5

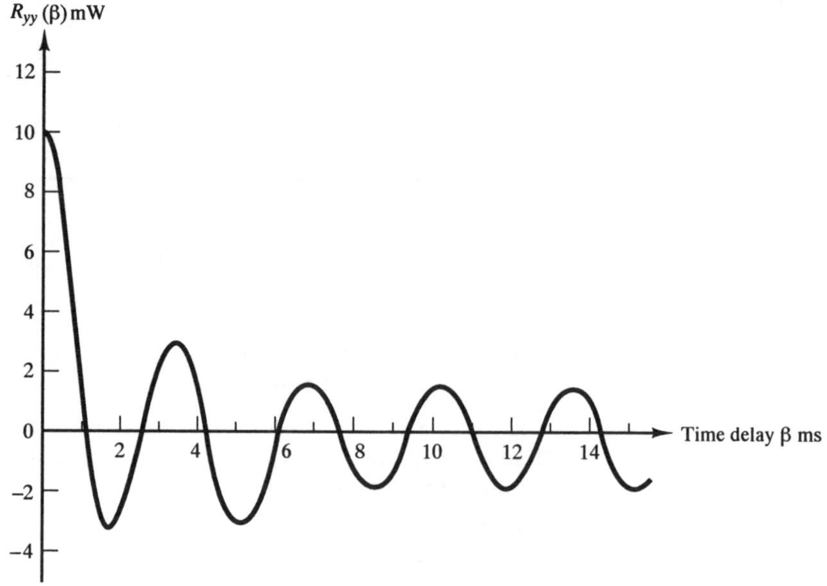

(a) Signal power
(b) Noise power
(c) Signal-to-noise ratio in decibels
(d) Signal amplitude
(e) Signal frequency
(f) Noise standard deviation (assume zero mean)

Hint: use eqns [6.27] and [6.33].

7

Reliability, choice and economics of measurement systems

In Chapters 3 and 4 we defined the accuracy of a measurement system and explained how measurement error can be calculated, under both steady-state and dynamic conditions. **Reliability** is another important characteristic of a measurement system; it is no good having an accurate measurement system which is constantly failing and requiring repair. The first section of this chapter deals with the reliability of measurement systems; first explaining the fundamental principles of reliability and the reliability of practical systems, then failure rate data and finally examining ways of improving reliability. The following section examines the problems of how to choose the most appropriate measurement system, for a given application, from several competing possibilities. Initially a specification for the required application can be drawn up; this will be a list of important parameters such as accuracy, reliability, etc., each with a desired value. This can then be compared with the manufacturer's specification for each of the competing measurement systems and the system with the closest specification is chosen. Even if all the required information is available this procedure is far from satisfactory because it takes no account of the relative importance of each parameter. A better method, explained in the final section, is to choose the system with minimum total lifetime operating cost.

7.1 Reliability of measurement systems

7.1.1 Fundamental principles of reliability

Probability P

If a large number of random, independent trials are made, then the probability of a particular event occurring is given by the ratio:

$$P = \frac{\text{number of occurrences of the event}}{\text{total number of trials}} \qquad [7.1]$$

in the limit that the total number of trials tends to infinity. Thus the probability of a tossed coin showing heads tends to the theoretical value of $\frac{1}{2}$ over a larger number of trials.

Reliability R(t)

The reliability of a measurement element or system can be defined as: 'the probability that the element or system will operate to an agreed level of performance, for a specified period, subject to specified environmental conditions'. In the case of a measurement system 'agreed level of performance' could mean an accuracy of $\pm 1.5\%$. If the system is giving a measurement error outside these limits, then it is considered to have failed, even though it is otherwise working normally. The importance of environmental conditions on the reliability of measurement systems will be discussed more fully later. Reliability decreases with time; a measurement system that has just been checked and calibrated should have a reliability of 1 when first placed in service. Six months later, the reliability may be only 0.5 as the probability of survival decreases.

Unreliability F(t)

This is 'the probability that the element or system will *fail* to operate to an agreed level of performance, for a specified period, subject to specified environmental conditions'. Since the equipment has either failed or not failed the sum of reliability and unreliability must be unity i.e.

$$R(t) + F(t) = 1 \qquad [7.2]$$

Unreliability also depends on time; a system that has just been checked and calibrated should have an unreliability of zero, when first placed in service, increasing to, say, 0.5 after six months.

7.1.2 Practical reliability definitions

Since $R(t)$ and $F(t)$ are dependent on time; it is useful to have measures of reliability which are independent of time. We will consider two cases, in the first the items are non-repairable and in the second the items are repairable.

Non-repairable items

Suppose that N individual items of a given non-repairable component are placed in service and the times at which failures occur are recorded during a test interval T. We further assume that all the N items fail during T and that the ith failure occurs at time T_i, i.e. T_i is the survival time or **up time** for the ith failure. The total up time for N failures is therefore $\Sigma_{i=1}^{i=N} T_i$ and the **mean time to failure** is given by

$$\text{Mean time to fail} = \frac{\text{Total up time}}{\text{Number of failures}}$$

i.e.

$$\text{MTTF} = \frac{1}{N}\sum_{i=1}^{i=N} T_i \qquad [7.3]$$

The **mean failure rate** $\bar{\lambda}$ is correspondingly given by:

$$\text{Mean failure rate} = \frac{\text{Number of failures}}{\text{Total up time}}$$

i.e.

$$\bar{\lambda} = \frac{N}{\sum_{i=1}^{i=N} T_i} \qquad [7.4]$$

i.e. mean failure rate is the reciprocal of MTTF.

There are N survivors at time $t = 0$, $N - i$ at time $t = T_i$, decreasing to zero at time $t = T$; Fig. 7.1(a) shows how the probability of survival, i.e. reliability, $R_i = (N - i)/N$ decreases from $R_i = 1$ at $t = 0$, to $R_i = 0$ at $t = T$. The ith rectangle has height $1/N$ and length T_i and area T_i/N. Therefore from eqn 7.3 we have:

$$\text{MTTF} = \text{Total area under the graph}$$

In the limit that $N \to \infty$, the discrete reliability function R_i becomes the continuous function $R(t)$. The area under $R(t)$ is $\int_0^T R(t)\,dt$ so that we have in general:

$$\text{MTTF} = \int_0^\infty R(t)\,dt \qquad [7.5]$$

The upper limit of $t = \infty$ corresponds to N being infinite.

Repairable items

Figure 7.1(b) shows the failure pattern for N items of a repairable element observed over a test interval T. The **down time** T_{Dj} associated with the jth failure is the total time that elapses between the occurrence of the failure and the repaired item being put back into normal operation. The total down time for N_F failures is therefore $\sum_{j=1}^{j=N_F} T_{Dj}$ and the **mean down time** is given by

$$\text{Mean down time} = \frac{\text{Total down time}}{\text{Number of failures}}$$

i.e.

$$\text{MDT} = \frac{1}{N_F} \sum_{j=1}^{j=N_F} T_{Dj} \qquad [7.6]$$

The **total up time** can be found by subtracting the total down time from NT, i.e.:

$$\text{Total up time} = NT - \sum_{j=1}^{j=N_F} T_{Dj}$$

$$= NT - N_F\,\text{MDT}$$

The mean up time or the **mean time between failures** (MTBF) is therefore given by:

$$\text{Mean time between failures} = \frac{\text{Total up time}}{\text{Number of failures}}$$

Fig. 7.1 Failure patterns:
(a) non-repairable items
(b) repairable items

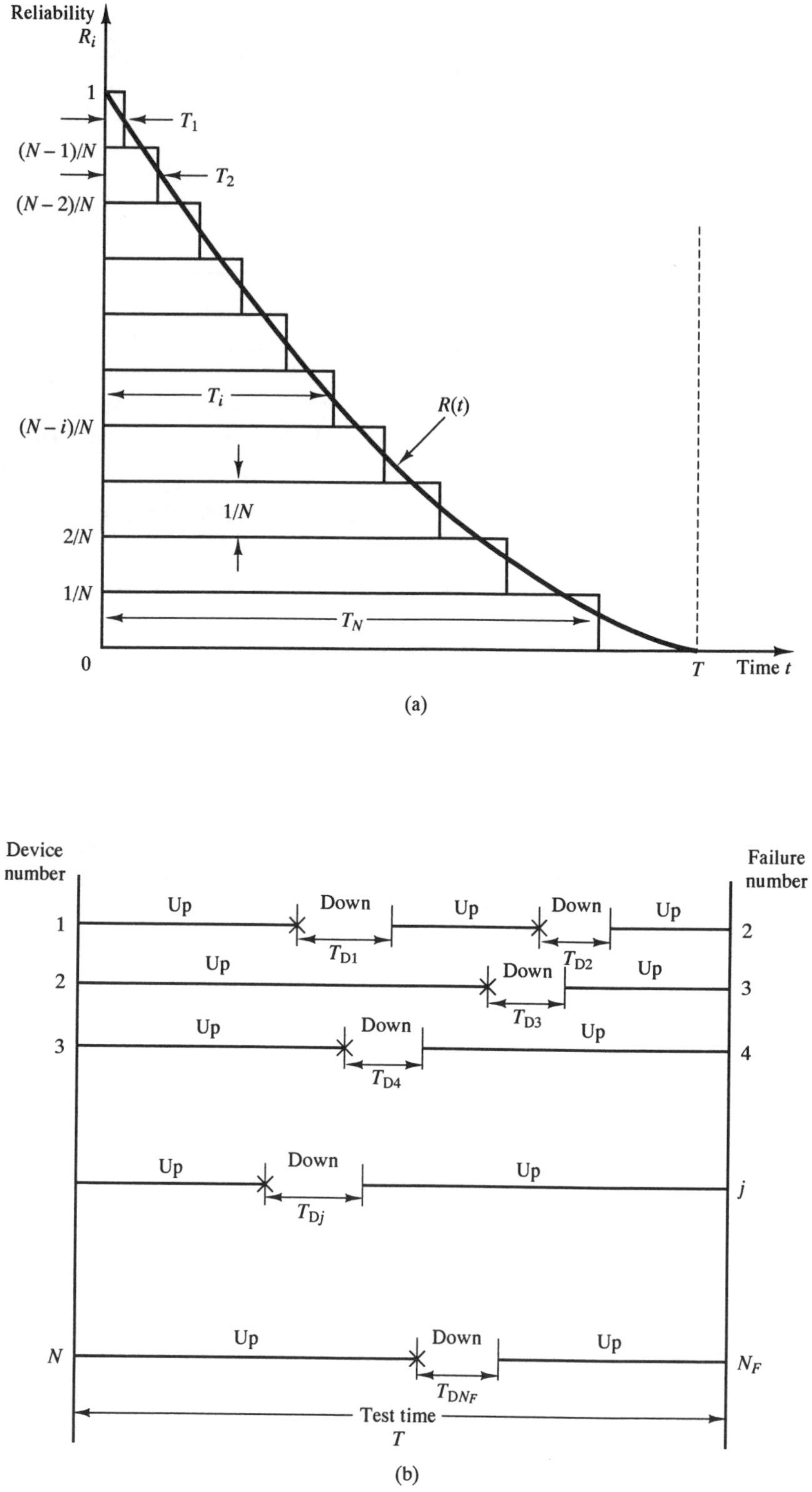

i.e.

$$\text{MTBF} = \frac{NT - N_F \text{ MDT}}{N_F} \qquad [7.7]$$

The **mean failure rate** $\bar{\lambda}$ is correspondingly given by:

$$\text{Mean failure rate} = \frac{\text{Number of failures}}{\text{Total up time}}$$

i.e.

$$\bar{\lambda} = \frac{N_F}{NT - N_F \text{ MDT}} \qquad [7.8]$$

Again mean failure rate is the reciprocal of MTBF.

Thus if 150 faults are recorded for 200 transducers over 1.5 years with a mean down time of 0.002 years, then the observed MTBF is 1.998 years and the mean failure rate 0.5005 yr^{-1}.

The **availability** of the element is the fraction of the total test interval that it is performing within specification, i.e. up; thus we have:

$$\text{Availability} = \frac{\text{Total up time}}{\text{Test interval}}$$

$$= \frac{\text{Total up time}}{\text{Total up time} + \text{Total down time}}$$

$$= \frac{N_F \times \text{MTBF}}{N_F \times \text{MTBF} + N_F \times \text{MDT}}$$

i.e.

$$A = \frac{\text{MTBF}}{\text{MTBF} + \text{MDT}} \qquad [7.9]$$

Using the above data of MTBF = 1.998 years, MDT = 0.002 years, gives A = 0.999.

Unavailability U is similarly defined as the fraction of the total test interval that it is not performing to specification, i.e. failed or down, thus we have:

$$\text{Unavailability} = \frac{\text{Total down time}}{\text{Test interval}}$$

giving:

$$U = \frac{\text{MDT}}{\text{MTBF} + \text{MDT}} \qquad [7.10]$$

It follows from eqns [7.9] and [7.10] that:

$$A + U = 1 \qquad [7.11]$$

7.1.3 Instantaneous failure rate and its relation to reliability

We assume to begin with that n items of an element survive up to time $t = \xi$ and that Δn items fail during the small time interval $\Delta\xi$ between ξ and $\xi + \Delta\xi$. The probability of failure during interval $\Delta\xi$ (given survival to time ξ) is therefore equal to $\Delta n/n$. Assuming no repair during $\Delta\xi$ the corresponding **instantaneous failure rate** or **hazard rate** at time ξ is, from eqn [7.8] given by:

$$\lambda(\xi) = \frac{\Delta n}{n\Delta\xi} = \frac{\text{Failure probability}}{\Delta\xi} \qquad [7.12]$$

The unconditional probability ΔF that an item fails during the iternal $\Delta\xi$ is:

ΔF = Probability that item survives up to time ξ

and

Probability that item fails between ξ and $\xi + \Delta\xi$ (given survival to ξ).

The first probability is given by $R(\xi)$ and from eqn [7.12] the second probability is $\lambda(\xi)\Delta\xi$. The combined probability ΔF is the product of these probabilities:

$$\Delta F = R(\xi)\lambda(\xi)\Delta\xi$$

i.e.

$$\frac{\Delta F}{\Delta\xi} = R(\xi)\lambda(\xi)$$

Thus in the limit that $\Delta\xi \to 0$, we have:

$$\frac{dF}{d\xi} = R(\xi)\lambda(\xi) \qquad [7.13]$$

also since $F(\xi) = 1 - R(\xi)$, $dF/d\xi = -(dR/d\xi)$, giving:

$$-\frac{dR}{d\xi} = R(\xi)\lambda(\xi)$$

i.e.

$$\int_{R(0)}^{R(t)} \frac{dR}{R} = -\int_0^t \lambda(\xi)\, d\xi \qquad [7.14]$$

In eqn [7.14], the left-hand integral is with respect to R and the right-hand integral with respect to ξ. Since at $t = 0$, $R(0) = 1$, we have:

$$[\log_e R]_1^{R(t)} = -\int_0^t \lambda(\xi)\, d\xi$$

$$\log_e R(t) = -\int_0^t \lambda(\xi)\, d\xi$$

i.e.

Relation between reliability and instantaneous failure rate

$$R(t) = \exp\left[-\int_0^t \lambda(\xi)\, d\xi\right] \qquad [7.15]$$

7.1.4 Typical forms of failure rate function

In the previous section instantaneous failure rate or hazard rate $\lambda(t)$ was defined. Figure 7.2 shows the most general form of $\lambda(t)$ throughout the lifetime of an element. This is the so-called **bathtub curve** and consists of three distinct phases: early failure, useful life and wear-out failure. The **early failure region** is characterised by $\lambda(t)$ decreasing with time. When items are new, especially if the element is a new design, early failures can occur due to design faults, poor quality components, manufacturing faults, installation errors, operator and maintenance errors, the latter may be due to unfamiliarity with the product. The hazard rate falls as design faults are rectified, weak components are removed and the user becomes familiar with installing, operating and maintaining the element. The **useful life region** is characterised by a low, constant failure rate. Here all weak components have been removed: design, manufacture, installation, operating and maintenance errors rectified so that failure is due to a variety of unpredictable lower level causes. The **wear-out region** is characterised by $\lambda(t)$ increasing with time as individual items approach the end of the design life for the product; long-life components which make up the element are now wearing out.

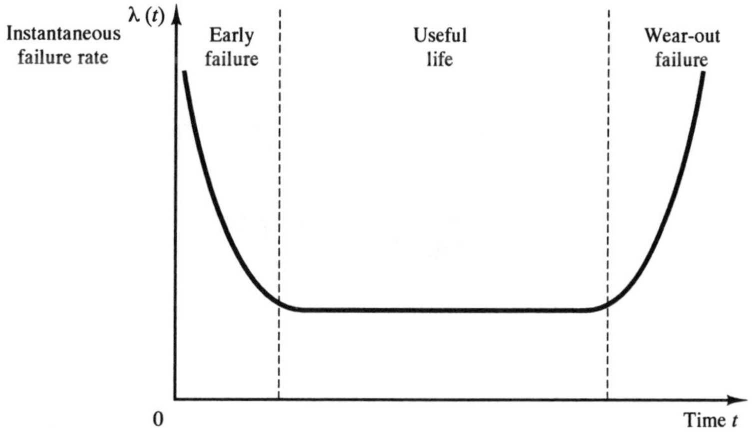

Fig. 7.2 Typical variation in instantaneous failure rate (hazard rate) during the lifetime of element — 'bathtub curve'

Many measurement elements have a useful life region lasting many years, so that a **constant failure rate** model is often a good approximation. Here we have:

$$\lambda(t) = \lambda(\xi) = \lambda = \text{constant} \qquad [7.16]$$

so that:

$$R(t) = \exp\left[-\lambda \int_0^t \xi\right] = \exp(-\lambda t) \qquad [7.17]$$

and:

$$F(t) = 1 - \exp(-\lambda t)$$

Thus a constant failure or hazard rate gives rise to an **exponential reliability** time variation or distribution shown in Fig. 7.3.

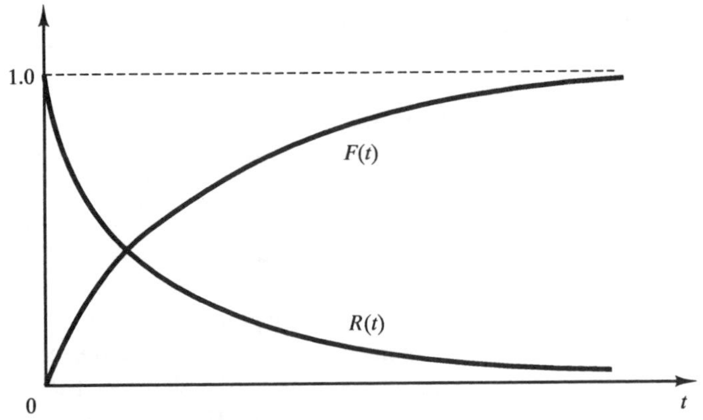

Fig. 7.3 Reliability and unreliability with constant failure rate model

7.1.5 Reliability of systems

Series systems

We saw in Chapter 1 that a complete measurement system consists of several elements usually in series or cascade. Figure 7.4 shows a series system of m elements with individual reliabilities $R_1, R_2, \ldots, R_i, \ldots, R_m$ respectively. The system will only survive if every element survives; if one element fails then the system fails. Assuming that the reliability of each element is independent of the reliability of the other elements, then the probability that the system survives is the probability that element 1 survives *and* the probability that 2 survives *and* the probability that 3 survives, etc. The system reliability R_{SYST} is therefore the product of the individual element reliabilities i.e.

Reliability of series system

$$R_{SYST} = R_1 R_2 \ldots R_i \ldots R_m \qquad [7.18]$$

If we further assume that each of the elements can be described by a constant failure rate λ (Section 7.1.4), and if λ_i is the failure rate of the ith element, then R_i is given by the exponential relation (7.17):

$$R_i = e^{-\lambda_i t} \qquad [7.19]$$

Thus

$$R_{SYST} = e^{-\lambda_1 t}\, e^{-\lambda_2 t} \cdots e^{-\lambda_i t} \cdots e^{-\lambda_m t} \qquad [7.20]$$

so that if λ_{SYST} is the overall system failure rate:

$$R_{SYST} = e^{-\lambda_{SYST} t} = e^{-(\lambda_1 + \lambda_2 + \cdots + \lambda_i + \cdots + \lambda_m)t} \qquad [7.21]$$

and

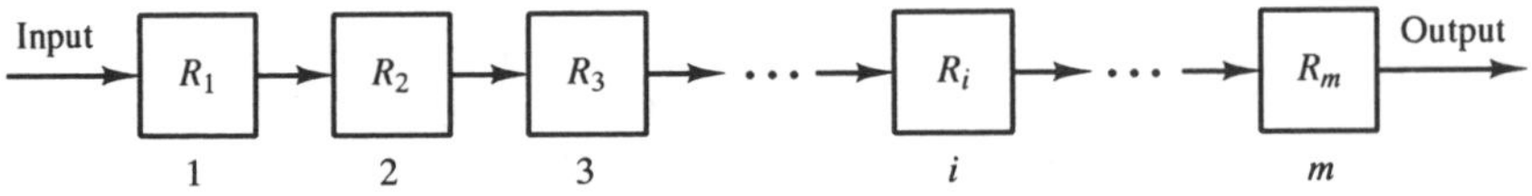

Fig. 7.4 Reliability of series system

Failure rate of system of m elements in series

$$\lambda_{SYST} = \lambda_1 + \lambda_2 + \cdots + \lambda_i + \cdots + \lambda_m \quad [7.22]$$

This means that the overall failure rate for a series system is the sum of the individual element or component failure rates. Equations 7.18 and 7.22 show the importance of keeping the number of elements in a series system to a minimum, if this is done the system failure rate will be minimum and the reliability maximum. A measurement system consisting of a thermocouple ($\lambda_1 = 1.1$), a millivolt to current converter ($\lambda_2 = 0.1$) and a recorder ($\lambda_3 = 0.1$) in series will therefore have a failure rate $\lambda_{SYST} = 1.3$.

Parallel systems

Figure 7.5 shows an overall system consisting of n individual elements or systems in parallel with individual unreliabilities $F_1, F_2, \ldots, F_j, \ldots, F_n$ respectively. Only one individual element or system is necessary to meet the functional requirements placed on the overall system. The remaining elements or systems increase the reliability of the overall system, this is termed **redundancy**. The overall system will only fail if every element/system fails, if one element/system survives the overall system survives. Assuming that the reliability of each element/system is independent of the reliability of the other elements, then the probability that the overall system fails is the probability that element/system 1 fails **and** the probability that 2 fails **and** the probability that 3 fails, etc. The overall system unreliability F_{SYST} is therefore the **product** of the individual element system unreliabilities i.e.

Unreliability of parallel system

$$F_{SYST} = F_1 F_2 \cdots F_j \cdots F_n \quad [7.23]$$

Fig. 7.5 Reliability of parallel system

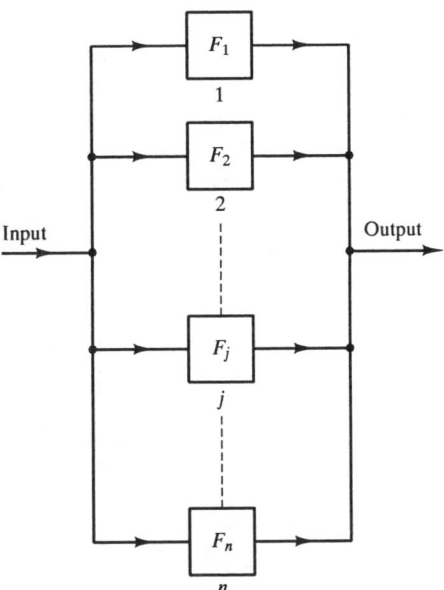

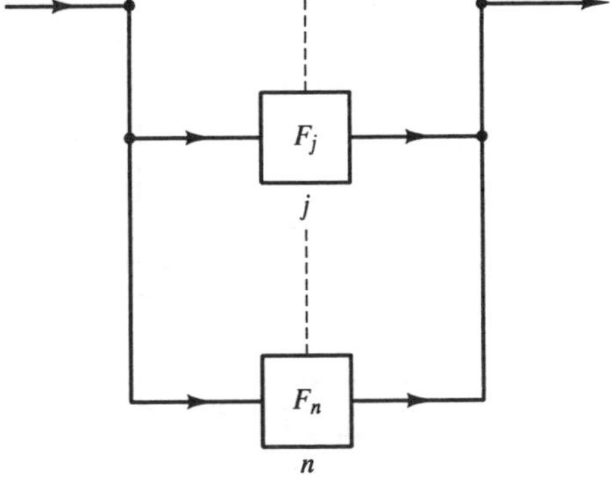

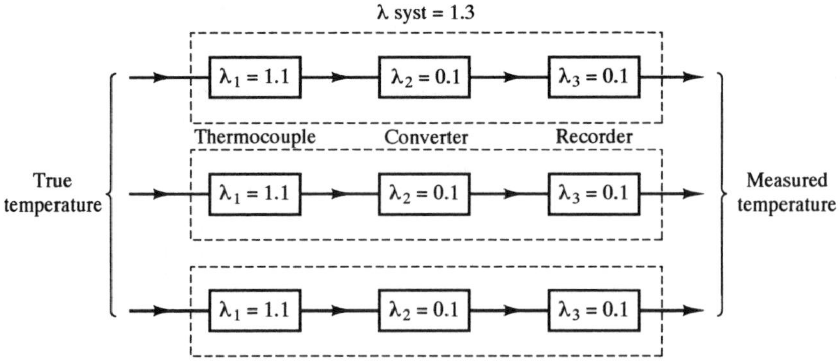

Fig. 7.6 Reliability of three-thermocouple temperature measurement systems in parallel

Comparing eqns [7.18] and [7.23] we see that for series systems system reliability is the product of element reliabilities, whereas for parallel systems, system unreliability is the product of element unreliabilities. Often the individual elements/systems are identical, so that $F_1 = F_2 = \cdots = F_i = \cdots = F_n = F$, that gives:

$$F_{\text{SYST}} = F^n \qquad [7.24]$$

The temperature measurement system of the previous section has a failure rate $\lambda_{\text{SYST}} = 1.3\,\text{yr}^{-1}$; the corresponding unreliability F is given by

$$F = 1 - e^{-\lambda_{\text{SYST}} t}$$

Thus if $t = 0.5$ year then $F = 0.478$. Figure 7.6 shows a redundant system consisting of three single temperature measurement systems in parallel. The overall system unreliability is therefore:

$$F_{\text{OVERALL}} = F^3 = 0.109$$

so that the probability of a failure with the overall system is less than a quarter of that of a single system.

The above parallel system while reliable, is expensive. Since the thermocouple failure rate is 11 times greater than converter and recorder failure rates, a more cost effective redundant system would have three thermocouples in parallel and only one converter and recorder. One possible system is shown in Fig. 7.7, the three thermocouple e.m.f.'s E_1, E_2, E_3 are input into a middle value selector element. The selector output signal is that input e.m.f. which is neither the lowest nor highest; thus if $E_1 = 5.0\,\text{mV}$, $E_2 = 5.2\,\text{mV}$, $E_3 = 5.1\,\text{mV}$, the output signal is E_3. If, however, thermocouple 3 fails so that $E_3 = 0\,\text{mV}$, the selector output signal is E_1.

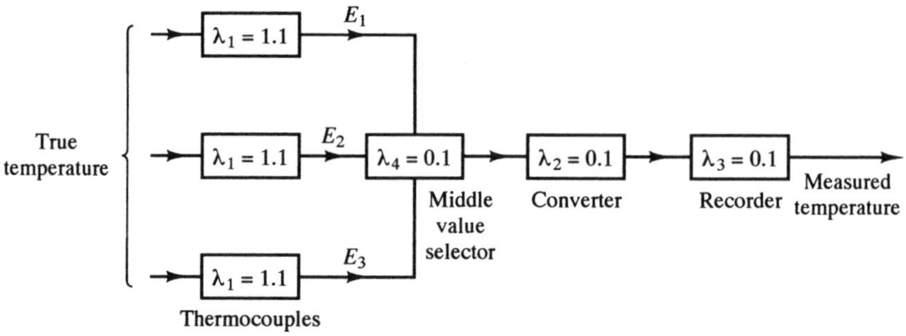

Fig. 7.7 Reliability of system with three thermocouples and middle value selector

The reliability of this sytem can be analysed by replacing the three thermocouples by a single element of unreliability:

$$F_1 = (1 - e^{-\lambda_1 t})^3 = (1 - e^{-0.55})^3 = 0.076$$

or reliability $R_1 = 1 - 0.076 = 0.924$. The reliability of the other elements with $\lambda = 0.1$ is:

$$R_2 = R_3 = R_4 = e^{-\lambda t} = e^{-0.05} = 0.951$$

Using [7.18], the overall system reliability is:

$$R_{\text{OVERALL}} = R_1 R_2 R_3 R_4 = 0.924(0.951)^3 = 0.795$$

i.e. $F_{\text{OVERALL}} = 0.205$. This is almost twice the unreliability of the parallel system but less than half that of a single system.

7.1.6 Failure rate data and models

A distinction must now be made between **components** and **elements**. A component is defined as a 'non-repairable' device, i.e. when it fails it is removed and thrown away. Examples are a resistor or an integrated circuit. An element, however, is a repairable part of a system which is usually made up of several components. Examples are a pressure sensor, temperature transmitter, recorder.

Failure rate data for both components and elements can be found experimentally by direct measurements of the frequency of failure of a number of items of a given type. Equation [7.4] can be used to find the observed failure rate of non-repairable components and eqn [7.8] to find $\bar{\lambda}$ for repairable elements.

Table 7.1 gives observed average failure rates for typical instruments. These data have been taken from the UK data bank operated by the Systems Reliability Service

Table 7.1 Observed average failure rates for instruments (after Wright[11])

Instrument	Environment	Experience (item-years)	No. of failures	Failure rate (failures/y)
Chemical analyser, Oxygen	Poor, chemical/ship	4.34	30	6.92
pH meter	Poor, chemical/ship	28.08	302	10.75
Conductivity indicator	Average, industrial	7.53	18	2.39
Fire detector head	Average, industrial	1470	128	0.09
Flow transmitter, pneumatic	Average, industrial	125	126	1.00
Level indicator, pneumatic	Average, industrial	898	201	0.22
Pressure controller	Average, industrial	40	63	1.58
Pressure indicator, dial, mechanical	Average, industrial	575	178	0.31
Pressure sensor, differential, electronic	Poor, chemical/ship	225	419	1.86
Pressure transmitter	Average, industrial	85045	806	0.01
Recorder, pen	Average, industrial	26.02	7	0.27
Temperature indicator and alarm	Fair, laboratory	47.2	101	2.14
Temperature indicator, resistance thermomenter	Fair, laboratory	212.3	68	0.32
Temperature indicator, bimetal	Average, industrial	165	215	1.30
Temperature trip unit	Average, industrial	120	70	0.58
Thermocouple	Poor, chemical/ship	317	127	0.40
Valve, gate	Poor, chemcial/ship	11564	841	0.07
non-return	Poor, chemical/ship	1530	101	0.07
solenoid	Poor, chemical/ship	1804	66	0.04

Table 7.2 Observed failure rates for instruments in different chemical plant environments (after Lees[2])

Instrument (p = pneumatic)	Observed failure rate, faults/year	Environmental correction factor	Base failure rate, faults/year
Control value (p)			
—	0.25	1	0.25
Works A	0.57	2	0.29
Works B	2.27	4	0.57
Works C	0.127	2	0.064
differential pressure transmitter (p)			
—	0.76	1	0.76
Works A (flow)	1.86	3	0.62
Works A (level)	1.71	4	0.43
Works B (flow)	2.66	4	0.67
Works C (flow)	1.22	2	0.61
Variable area flowmeter transmitter (p)			
—	0.68	1	0.68
Works A	1.01	3	0.34
Thermocouple			
Works A	0.40	3	0.13
Works B	1.34	4	0.34
Works C	1.00	4	0.25
Controller			
—	0.38	1	0.38
Works A	0.26	1	0.26
Works B	1.80	1	1.80
Works C	0.32	1	0.32
Pressure switch			
—	0.14	1	0.14
Works A	0.30	2	0.15
Works B	1.00	4	0.25

(SRS).[1] The table specifies the environment in which each type of instrument is located. For an element located in the process fluid, the environment is determined by the nature of the fluid, e.g. temperature, corrosion properties, presence of dirt or solid particles. For an element located in the atmosphere, the environment is determined by atmospheric conditions, e.g. temperature, humidity, salinity, presence of dust. The failure rate of a given type of element will depend on the environment in which it is located, an iron–copper thermocouple will have a higher failure rate in a corrosive acid than in water.

Table 7.2 shows the observed failure rates for given types of elements at three works A, B, C which process different materials and fluids and have different background environments.[2] The observed failure rates can be regarded as the produce of a **base failure rate** λ_B and an **environmental correction factor** π_E

Element failure rate model

$$\lambda_{OBS} = \pi_E \times \lambda_B \qquad [7.25]$$

Here λ_B corresponds to the best environmental conditions and π_E has values 1, 2, 3 or 4, the highest figure corresponding to the worst environment.

The failure rate of elements can alternatively be calculated from the failure rate data/models for the basic components which make up the element.

Table 7.3 Calculation of overall failure rate for electronic square root module (after Hellyer[3])

Component	Failure rates per 10^{10} hours									
	F1	F2	F3	K1	K2	K3	K4	Rate	Qty	Value
RESISTORS										
Carbon film										
$0 < R \leq 100$ K	100	0	0	1	1	1	1	100	17	1700
100 K $< R \leq 1$ M	100	0	0	1	1.1	1	1	110	2	220
1 M $< R \leq 10$ M	100	0	0	1	1.6	1	1	160	3	480
Metal film										
$0 < R \leq 100$ K	150	0	0	1	1	1	1	150	17	2550
100 K $< R \leq 1$ M	150	0	0	1	1.1	1	1	165	3	495
POTENTIOMETERS										
$0 < R \leq 50$ K	700	0	0	1	1	1	1	700	8	5600
50 K $< A \leq 100$ K	700	0	0	1	1	1.1	1	770	1	770
CAPACITORS										
Metal film										
$0 < C \leq 33$ nF	200	0	0	1	1	1	1	200	3	600
33 nF $< C \leq 1$ μF	200	0	0	1	1	1.3	1	260	2	520
1 μF $< C \leq 10$ μF	200	0	0	1	1	1.5	1	300	1	300
Ceramic										
$0 < C \leq 3.3$ nF	150	0	0	1	1	1	1	150	1	150
Electrolytic										
$3.2 < C \leq 62$ μF	500	0	0	0.29	1	0.7	1	102	1	102
DIODES										
Silicon LP	200	0	0	0.55	1	1	1	110	2	220
Zener	1000	0	0	1.3	1	1	1	1300	1	1300
TRANSISTORS										
NPN LP	400	0	0	1.4	1	1	1	560	2	1120
INTEGRATED CIRCUITS										
OP AMP	160	50	600	1	1	1	1	810	9	7290
Quad switch	38	320	560	1	1	1	1	918	1	918
OTHERS										
Edge connectors	300	0	0	1	1	1	1	300	8	2400
Soldered joints	20	0	0	1	1	1	1	20	167	3340
PCB	60	0	0	1	1	1	1	60	1	60
	Total rate:	3.01×10^{-6} per hour								
	MTBF:	3.32×10^{5} hours								

Notes
(a) Data sources:
1: *Electronic Reliability Data*, National Centre of Systems Reliability, Application Code 2, 25C
2: *Reliability Prediction Manual for Guided Weapon Systems*, MOD
3: Component supplier's information
(b) Rate = (F1 + F2 + F3) × K1 × K2 × K3 × K4

Table 7.3 shows the calculation of the overall failure rate, from basic component data, for an electronic square root extractor module.[3] The module gives an output voltage signal in the range 1–5 V, proportional to the square root of the input signal with range 4–20 mA; this type of module is commonly used in fluid flow rate control systems. The module is made up from basic electronic components of various types, all connected in series. Several components of each type are present. From eqn [7.22] the failure rate λ of a module containing m different component types in series with failure rates $\lambda_1, \lambda_2, \ldots, \lambda_i, \ldots, \lambda_m$, and one of each type present is:

$$\lambda = \lambda_1 + \lambda_2 + \cdots + \lambda_i + \cdots + \lambda_m \qquad [7.26]$$

If there are multiple components of each type, all connected in series, then the module failure rate is given by:

Module failure rate — multiple components of each type

$$\lambda = N_1\lambda_1 + N_2\lambda_2 + \cdots + N_i\lambda_i + \cdots + N_m\lambda_m \quad [7.27]$$

where $N_1, N_2, \ldots, N_i, \ldots, N_m$ are the quantities of each component type. The failure rate of each component type is calculated using the model equation:

$$\lambda_i = (F_{1i} + F_{2i} + F_{3i}) \times K_{1i} \times K_{2i} \times K_{3i} \times K_{4i} \quad [7.28]$$

Table 7.3 gives values of F_1, F_2, F_3, K_1, K_2, K_3, K_4 and failure rate λ_i for each component. Each failure rate is then multiplied by the appropriate quantity N_i and the $N_i\lambda_i$ values added together to give a total module failure rate of 3.01×10^{-6} per hour.

7.1.7 Design and maintenance for reliability

Design for reliability

The following general principles should be observed.

Element selection. Only elements with well-established failure rate data/models should be used. Furthermore some technologies are inherently more reliable than others. Thus an inductive LVDT displacement sensor (Chapter 8) is inherently more reliable than a resistive potentiometer; the latter involves a contact sliding over a wire track which will eventually become worn. A vortex flowmeter (Chapter 12) involves no moving parts and is therefore likely to be more reliable than a turbine flowmeter which incorporates a rotor assembly.

Environment. The environment in which the element is to be located should first be defined and the element should consist of components and elements which are capable of withstanding that environment. Thus the diaphragm of a differential pressure transmitter on a sulphuric acid duty should be made from a special alloy, e.g. Hastelloy C, which is resistant to corrosion.

Minimum complexity. We saw above that for a series system, the system failure rate is the sum of the individual component/element failure rates. Thus the number of components/elements in the system should be the minimum required for the system to perform its function.

Redundancy. We also saw that the use of several identical elements/systems connected in parallel increases the reliability of the overall system. Redundancy should be considered in situations where either the complete system or certain elements of the system have too high a failure rate.

Diversity. In practice faults can occur, which either cause more than one element in a given system, or a given element in each of several identical systems, to fail simultaneously. These are referred to as **common mode failures** and can be caused by incorrect design, defective materials and components, faults in the manufacturing process, or incorrect installation. One common example is an electronic system where several of the constituent circuits share a common electrical power supply; failure of the power supply causes all of the circuits to fail. This problem can be solved using **diversity**; here a given function is carried out by two systems in parallel, but each system is made up of different elements with different operating principles. One example is a temperature measurement system made up of two subsystems in parallel; one electronic and one pneumatic.

Maintenance

The **Mean Down Time** MDT for a number of items of a repairable element has been defined as the mean time between the occurrence of the failure and the repaired element being put back into normal operation. It is important that MDT is as small as possible in order to minimise the financial loss caused by the element being out of action.

There are two main types of maintenance strategy used with measurement system elements. **Breakdown maintenance** simply involves repairing or replacing the element when it fails. Here MDT or **mean repair time**, T_R is the sum of the times taken for a number of different activities. These include **realisation** that a fault has occurred, **access** to the equipment, **fault diagnosis**, **assembly** of repair equipment, components and personnel, **active repair/replacement** and finally **checkout**. **Preventive maintenance** is the servicing of equipment and/or replacement of components at regular fixed intervals; the corresponding **maintenance frequency** is m times per year. Here MDT or **mean maintenance time**, T_M is the sum of times for access, **service/replacement** and checkout activities and therefore should be significantly less than mean repair time with breakdown maintenance.

7.2 Choice of measurement systems

The methods to be used and problems involved in choosing the most appropriate measurement system for a given application can be illustrated by a specific example. The example used will be the choice of the best system to measure the volume flow rate of a clean liquid hydrocarbon, range 0 to $100\,m^3\,hr^{-1}$, in a 0.15 m (6 inch) diameter pipe. The measured value of flow rate must be presented to the observer in the form of a continuous trend on a chart recorder. The first step is to draw up a specification for the required flow measurement system. This will be a list of all important parameters for the complete system such as measurement error, reliability, cost; each with a desired value or range of values. The first two columns of Table 7.4 are an example of such a 'job specification'. As explained in Chapter 3, system measurement error in the steady state can be quantified in terms of the mean $\bar{E}$ and standard deviation σ_E of the error probability distribution $p(E)$. These quantities depend on the imperfections e.g. non-linearity, repeatability — of every element in the system. System failure rate λ and repair time T_R were defined in Section 7.1. Initial cost C_I is the cost of purchase, delivery, installation and commissioning of the complete system. C_R is the average cost of materials for each repair.

In this example the choice could be between four competing systems based on the orifice plate, vortex, turbine and electromagnetic primary sensing elements. The

Table 7.4 Comparison table for selection of flow measurement system

Parameter		Job specification	System (1) Orifice plate	System (2) Vortex	System (3) Turbine	System (4) Electromagnetic
Measurement error (at $50\ m^3\ h^{-1}$) $m^3\ h^{-1}$	$\bar{E}$ σ_E	≤ 0.25 ≤ 0.8	0.2 0.7	0.1 0.3	0.03 0.1	Not technically feasible
Initial cost	C_I £	$\leq 4{,}000$	3,500	3,000	4,200	
Annual failure rate	λ failures yr^{-1}	≤ 2.0	1.8	1.0	2.0	
Average repair time	T_R hours	≤ 8	6	5	7	
Material repair cost	C_R £	≤ 200	100	100	300	

principles and characteristics of all four elements are discussed in Chapter 12. The configuration of the four systems could be:

1. orifice plate — electronic D/P transmitter — square rooter — recorder
2. vortex element — frequency to voltage converter — recorder
3. turbine element — frequency to voltage converter — recorder
4. electromagnetic element — self-balancing potentiometer/recorder.

The next step is to decide whether all the competing systems are technically feasible. The electromagnetic device will not work with electrically non-conducting fluids such as hydrocarbons so that system (4) is technically unsuitable. Systems (1), (2) and (3) are feasible and the specification for each competing system must then be written down to see whether it satisfies the job specification. Table 7.4 gives possible specifications for the orifice plate, vortex and turbine systems. This data is entirely fictitious and is given only to illustrate the problems of choice. In practice a prospective user may not have all the information in Table 7.4 at his disposal. The manufacturer will be able to give estimates of initial cost C_I and the limits of measurement error, e.g. $\pm 2\%$ of full scale for the orifice plate system at $50\ m^3\ hr^{-1}$. He will not, however, be able to give values of mean error $\bar{E}$, failure rates and repair times. The last two quantities will depend on the environment of the user's plants and the maintenance strategy used. This information may be available within the user's company if adequate maintenance records have been kept. From Table 7.4 we see that the turbine flowmeter system (3) does not satisfy the job specification; both initial cost and material repair cost are outside the limits set. This would appear to rule out system (3), leaving (1) and (2). Both these systems satisfy the job specification but the vortex system (2) is cheaper, more accurate and more reliable than the orifice plate system (1). Thus, based on a straightforward comparison of job and system specification, the vortex measurement system (2) would appear to be the best choice for this application.

Under certain circumstances, however, the above conclusion could be entirely wrong. The turbine system is more expensive and less reliable than the vortex system, but is three times more accurate. We must now ask how much this increased accuracy is worth? Suppose the market value of the hydrocarbon is £100 m^{-3}. A measurement error of one standard deviation in the turbine system, where $\sigma_E = 0.1\,m^3\,hr^{-1}$, represents a potential cash loss of £10 per hour or approximately £80 000 per annum. A corresponding error in the vortex system, where $\sigma_E = 0.3\,m^3\,hr^{-1}$, represents an approximate potential loss of £240 000 per annum. The difference between these two figures far outweighs the extra initial and maintenance costs, so that the turbine system (3) is the best choice in this case. We can conclude, therefore, that in order to choose the correct system for a given application, the financial value of each parameter in the job specification must be taken into account. In a costing application of this type, a digital printout of flow rate is more suitable than an analogue trend record.

7.3 Total lifetime operating cost

The total lifetime operating costing (TLOC) of a measurement system is the total cost penalty, incurred by the user, during the lifetime of the system. The TLOC is given by

TLOC = initial cost of system (purchase, delivery, installation and commissioning)
+ cost of failures and maintenance over lifetime of the system
+ cost of measurement error over lifetime of the system. [7.29]

and therefore takes account of the financial value of each parameter in the job specification. The best system for a given appication is then the one with minimum TLOC. This method also enables the user to decide whether a measurement system is necessary at all. If no system is installed, TLOC may still be very large because no measurement implies a large measurement error. A measurement system should be purchased if it produces a significant reduction in TLOC.

Using eqn [7.29], we can derive an algebraic expression for TLOC using the parameters listed in Table 7.4. The initial cost of the system is £C_I. If the system lifetime is T years and average failure rate λ faults yr^{-1}, then the total number of faults is λT. Since the average repair time is T_R hours then the total 'downtime' due to repair is $\lambda T T_R$ hours. The total lifetime cost of failures is the sum of the repair cost (materials and labour) and the process cost, i.e. the cost of lost production and efficiency while the measurement system is withdrawn for repair. If we define

£C_R = average materials cost per repair
£C_L = repair labour cost per hour
£C_P = process cost per hour,

then the total repair cost is $(C_R\lambda T + C_L T_R\lambda T)$ and the total process cost is $C_P T_R \lambda T$, giving:

$$\text{Total lifetime cost of failures} = [C_R + (C_L + C_P)T_R]\lambda T \qquad [7.30]$$

The above costs only apply to breakdown maintenance; many users also practise preventive maintenance in order to reduce failure rates. Suppose preventive maintenance is carried out on a measurement system m times yr^{-1}, the average maintenance time is T_M hours and the materials cost per service is £C_M. The total number of services is mT and the total time taken for preventive maintenance is mTT_M hours. Usually preventive maintenance of measurement systems is carried out at a time when the process of plant itself is shut down for repair and maintenance. This means that no process costs are incurred during preventive maintenance, giving:

$$\text{total lifetime maintenance cost} = (C_M + C_L T_M)mT \qquad [7.31]$$

The last term in eqn [7.29] involves the total lifetime cost of measurement error. In order to evaluate this we first need to evaluate the cost penalty function $C(E)$ £yr^{-1} associated with a given steady-state measurement system error E. The form of $C(E)$ depends on the economics of the process, on which the measurement is being made. A good example is temperature measurement in a chemical reactor where a degradation reaction is taking place.[4] Here two reactions occur simultaneously, the feedstock A is converted into a desired product B but B is also degraded to an undesired product C. The rates of both reactions increase sharply with temperature, the rate of B to C being more temperature sensitive than the rate of A to B.

There is an optimum temperature at which the yield of B is maximum. If the reactor is operated at either above or below this optimum temperature, then the yield of B is sharply reduced and a cost penalty is incurred. This situation will occur if there is a measurement system error E between measured and true values of reactor temperature: the system tells the operator that the reactor is at optimum temperature when it is not. Figure 7.8(a) shows the form of the cost penalty function $C(E)$ in this case. We see that when $E = 0$, $C(E) = 0$ corresponding to optimum temperature; but $C(E)$ increases rapidly with positive and negative values of E as the yield of B decreases.

In Figure 7.8(b) $C(E)$ represents the cost penalty, due to imperfect flow measurement, incurred by a customer receiving a fluid by pipeline from a manufacturer. If E is positive, the customer is charged for more fluid than he actually receives and so is penalised, i.e. $C(E)$ is positive. If E is negative, the customer is charged for less fluid than actually received and $C(E)$ is negative. Figure 7.8(c) refers to a non-critical liquid level measurement in a vessel. The vessel should be about half full, but plant problems will occur if the vessel is emptied or completely

Fig. 7.8 Error cost penalty function $C(E)$ for different processes

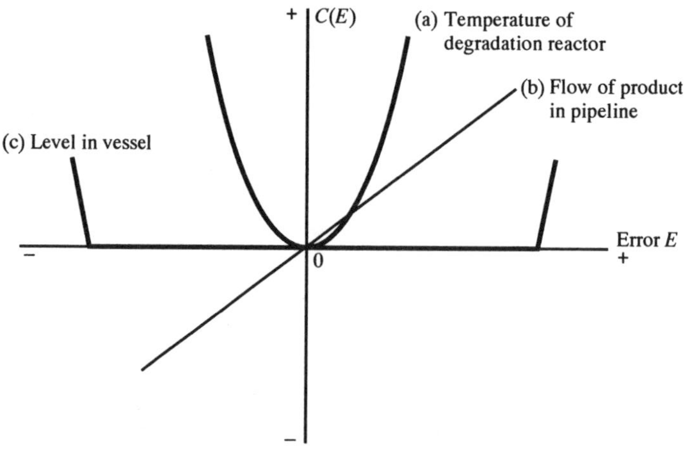

filled. Thus a cost penalty is incurred only if there is gross measurement error, i.e. the measurement system shows the vessel to be half full when almost empty.

We saw in Chapter 3 that the exact value of measurement error E, for a given measurement system, at a given time, cannot be found. We can, however, find the probability that the system will have a certain error. This is quantified using a Gaussian probability density function $p(E)$, with mean $\bar{E}$ and standard deviation σ_E (Table 3.1). The probability of getting a measurement error between E and $E + \mathrm{d}E$ is $p(E)\mathrm{d}E$, the corresponding cost penalty is $C(E)p(E)\mathrm{d}E$ per year, or $TC(E)p(E)\mathrm{d}E$ throughout the system lifetime. The total lifetime cost of measurement error is then found by integrating the above expression over all possible values of E, i.e.

$$\text{total lifetime cost of measurement error} = T\int_{-\infty}^{\infty} C(E)p(E)\mathrm{d}E \qquad [7.32]$$

The integral has a finite value, because the value of $p(E)$ becomes negligible for $|E|$ greater than 3 or 4 standard deviations; it can be evaluated numerically[4] using values of the normalised Gaussian distribution.[5]

From eqn [7.29]–[7.32] we have

$$\text{TLOC} = C_I + [C_R + (C_L + C_P)T_R]\lambda T + (C_M + C_M T_M)mT + T\int_{-\infty}^{\infty} C(E)p(E)\mathrm{d}E \qquad [7.33]$$

The relative importance of the terms in the above equation will depend on the application. In the chemical reactor $C(E)$ and C_P will be the major factors so that accuracy and reliability will be far more important than initial cost. In the tank level application, measurement error is unimportant and minimum TLOC will be obtained by the best trade-off between initial cost, reliability, and maintainability.

Conclusion

The first section of the chapter has discussed the reliability of measurement systems. The fundamental principles and practical definitions of reliability were first explained and the relationship between reliability and instantaneous failure rate derived. The typical variation in instantaneous failure rate throughout the lifetime of an element was then discussed and the reliability of series and parallel systems examined. The section concluded by looking at failure rate data and models and general strategies in design and maintenance for reliability.

The second section dealt with the problem of choice of measurement systems approached by comparing the job specification with those of the competing systems. This method does not take account of the financial value of each item in the specification. A better method, discussed in the final section, is to choose the systems with minimum lifetime operating cost.

References

7.1 WRIGHT R I (SRD Warrington) 1984 'Instrument reliability', *Instrument Science and Technology*, Vol 1, Bristol, Institute of Physics, pp. 82–92.

7.2 LEES F P 1976 'The reliability of Instrumentation', *Chemistry and Industry*, March, pp. 195–205.

7.3 HELLYER F G (Protech Instruments and Systems) 1985 'The application of reliability engineering to high integrity plant control systems' *Measurement and Control*, 18 June, pp. 172–6.

7.4 BENTLEY J P 1979 'Errors in industrial temperature measurement systems and their effect on the yield of a chemical degradation reaction' *8th IMEKO Congress of the International Measurement Confederation*, Moscow, May.

7.5 WHITE J, YEATS A and SKIPWORTH G 1974 *Tables for Statisticians*, Stanley Thorne, London, pp. 18–19.

Problems

7.1 A batch of one hundred identical thermocouples were tested over a twelve-week period. Twenty failures were recorded and the corresponding down times in hours were as follows:

5, 6, 7, 8, 4, 7, 8, 10, 5, 4, 8, 5, 4, 5, 6, 5, 4, 9, 8, 6.

Calculate:

(a) Mean down time
(b) Mean time between failures
(c) Mean failure rate
(d) Availability

7.2 A flow measurement system consists of an orifice plate (λ = 0.75), differential pressure transmitter (λ = 1.0), square root extractor (λ = 0.1) and recorder (λ = 0.1). Calculate the probability of losing the flow measurement after 0.5 year for the following:

(a) a single flow measurement system;
(b) three identical flow measurement systems in parallel;
(c) a system with 3 orifice plates, 3 differential pressure transmitters and a middle value selector relay (λ = 0.1). The selected transmitter output is passed to a single square root extractor and recorder.

Annual failure rate data are given in brackets; assume that all systems were initially checked and found to be working correctly.

7.3 Use the data given in Table Prob. 3 to decide which level measurement system should be purchased. Assume a breakdown maintenance only strategy is practised, each system has the same measurement error and a 10-year total lifetime.

Table Prob. 3

	Parameter	System (1)	System (2)
Initial cost	(£)	1000	2000
Materials cost per repair	(£)	20	15
Labour cost per hour	(£)	10	10
Process cost per hour	(£)	100	100
Repair time	(h)	8	12
Annual failure rate	(yr^{-1})	2.0	1.0

Part B

Typical measurement system elements

8

Sensing elements

In Chapter 1 we saw that, in general, a measurement system consists of four types of elements: sensing, signal conditioning, signal processing and data presentation elements. The sensing element is the first element in the measurement system; it is in contact with, and draws energy from, the process or system being measured. The input to this element is the true value of the measured variable; the output of the element depends on this value. The purpose of this chapter is to discuss the principles of sensing elements in wide current use; more specialised elements are discussed in later chapters. Table 8.1 lists the sensing elements described in this book according to the physical principle involved, e.g. inductive or thermoelectric. The elements are classified according to whether the output signal is electrical or mechanical. Elements with an electrical output are further divided into passive and active. Passive devices such as resistive, capacitive and inductive elements require an external power supply in order to give a voltage or current output signal; active devices, e.g. electromagnetic and thermoelectric elements, need no external power supply.

The table also denotes the input measured variables, which a given element is able to sense, by giving the relevant section number in the book. Thus resistive elements can sense temperature, heat flux, flow velocity, displacement, strain and gas composition. Elastic elements can sense force, pressure torque, level and density. Sensors with a mechanical output are commonly used as the primary sensing element in measurement systems for mechanical variables such as force or flow rate. In order to obtain an electrical signal, this primary element is followed by a secondary sensing element with an electrical output signal. Examples are a resistive strain gauge sensing the strain in an elastic cantilever in a force measurement system, and an electromagnetic tachogenerator sensing the angular velocity of a turbine in a flow measurement system.

8.1 Resistive sensing elements

8.1.1 Potentiometers for linear displacement measurement

The potentiometric displacement sensor has already been introduced in Section 5.1.2.

Table 8.1 Sensing elements and measured variables

	Input measured variable / Physical principle	Temperature	Heat/light flux	Pressure	Force	Torque	Level	Density	Flow rate	Flow velocity	Displacement/ strain	Velocity	Acceleration	Gas composition	Ionic concentration	Humidity
Electrical output passive	Resistive	8.1									5.1					
		15.5	15.5							14.4	8.1			14.5		
	Capacitive			8.2			8.2				8.2					8.2
	Inductive										8.3					
	Piezoresistive			8.8												
	Photovoltaic		15.5													
Electrical output active	Electromagnetic								12.4			8.4				
	Thermoelectric	8.5														
		15.5	15.5													
	Piezoelectric			16.2	8.7								8.7			
	Electrochemical													8.9	8.9	
	Pyroelectric	15.5	15.5													
Mechanical output	Elastic			8.6	4.1											
				9.4	8.6	8.6	9.4									
				9.5	9.5		8.6	9.5					8.6			
	Differential pressure								12.3	12.2						
	Turbine								12.3							
	Vortex								12.3							
	Pneumatic			13.1	13.1		13.1				13.1					
	Coriolis								12.4							

It consists of a cylindrical stator with either a wire-wound track or a film of conductive plastic deposited on it. The resistance per unit length is constant so that the ratio of output voltage to supply voltage is proportional to the fractional displacement x of the slider. The resolution error of a wire-wound potentiometer (Section 2.1) is $100/n$ per cent, where n is the number of turns, and is thus determined by the diameter of the wire. A typical family of wire-wound potentiometers covers displacement spans from 0.5 to 100 inches, with non-linearity from ±0.2%, resolution from 0.008% and resistance values of 1 kΩ inch^{-1}. Conducting plastic film elements have zero resolution error but have higher temperature coefficients of resistance. A family of conductive plastic potentiometers covers displacement spans from 25 to 250 mm, with non-linearity up to ±0.04% and resistance values from 500 Ω to 80 kΩ.

The most modern development is the hybrid track potentiometer which is manufactured by depositing a conductive plastic coating on a precision wire-wound resistance track and incorporates the best features of wire-wound and film types. The resistance of any load (recorder or indicator) used with the potentiometer must be several times greater than the potentiometer resistance otherwise non-linear effects occur (Section 5.1.2).

8.1.2 Resistance thermometers and thermistors for temperature measurement

The resistance of most metals increases reasonably linearly with temperature in the range −100 to +800 °C. The general relationship between the resistance $R_T\,\Omega$ of a metal element and temperature T °C is a power series of the form:

$$R_T = R_0(1 + \alpha T + \beta T^2 + \gamma T^3 + \cdots) \qquad [8.1]$$

where $R_0\,\Omega$ is the resistance at 0 °C and α, β, γ are temperature coefficients of resistance. The magnitude of the non-linear terms is usually small. Figure 8.1(a) shows the variation in the ratio R_T/R_0 with temperature for the metals platinum, copper and nickel. Although relatively expensive, platinum is usually chosen for industrial resistance thermometers; cheaper metals, notably nickel and copper, are used for less demanding applications. Platinum is preferred because it is chemically inert, has linear and repeatable resistance/temperature characteristics, can be used over a wide temperature range (−200 to +800 °C) and in many types of environments. It can be refined to a high degree of purity which ensures that statistical variations in resistance, between similar elements at the same temperature (Section 2.3.2), are small. A typical platinum element has $R_0 = 100.0\,\Omega$, $R_{100} = 138.50\,\Omega$, $R_{200} = 175.83\,\Omega$, $\alpha = 3.91 \times 10^{-3}\,°\text{C}^{-1}$ and $\beta = -5.85 \times 10^{-7}\,°\text{C}^{-2}$. The change in resistance between the ice point and the steam point, i.e. $R_{100} - R_0$, is called the **fundamental interval**; in the above element this is 38.5 Ω. The maximum non-linearity as a percentage of f.s.d. (eqn 2.5), between 0 and 200 °C is +0.76%.

The British Standard BS 1904[1] lays down tolerance limits on the maximum variation in resistance between platinum elements at a given temperature. For Grade 1 elements the tolerance limits are ±0.075 Ω at 0 °C, ±0.13 Ω at 200 °C; and for Grade 2 elements the tolerance limits are ±0.1 Ω at 0 °C, ±0.35 Ω at 200 °C. The amount of electrical power dissipated in the element should be limited in order to

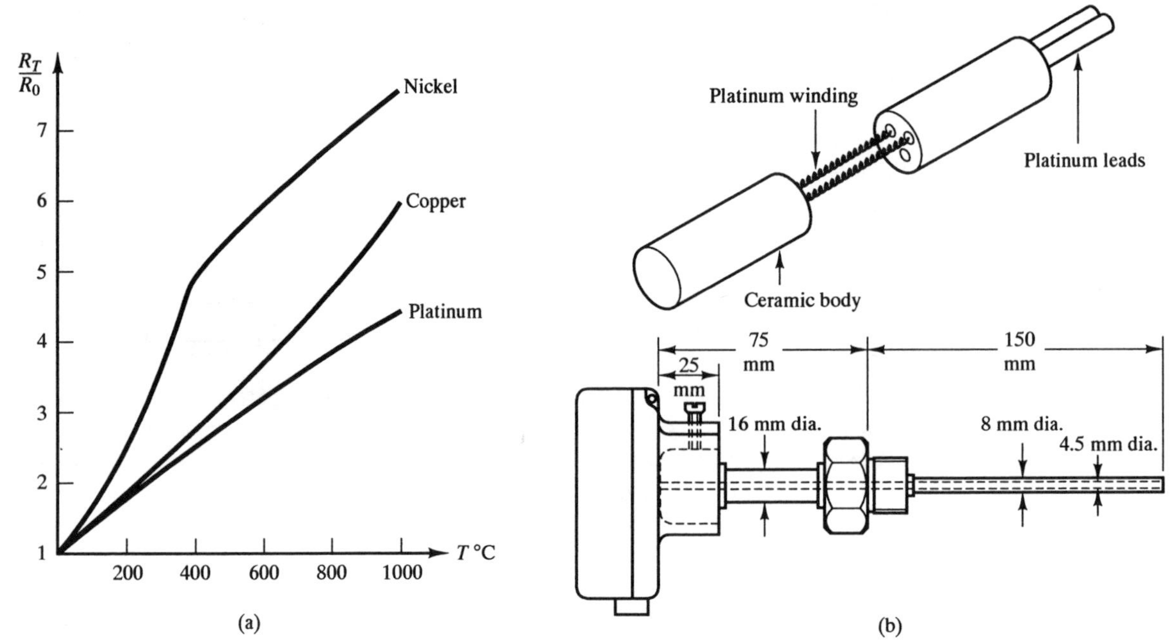

Fig. 8.1 Metal resistance thermometers (a) Resistance/temperature characteristics of community-used metals (b) Typical construction of platinum element probe[3]

avoid self-heating effects (Chapter 14); in a typical element 10 mW of power causes a temperature rise of 0.3 °C.

One type of element is constructed using the partially supported arrangement shown in Fig. 8.1(b).[3] Here fine platinum wire is wound into a very small spiral and is inserted into axial holes in a high purity alumina insulator. A small quantity of glass adhesive is introduced into the holes and the unit is fired, thus securely fixing a part of each turn onto the alumina, the remainder of the wire is free to move. The diagram also shows the element housed in a stainless steel protective sheath.

Resistive temperature elements made from semiconductor materials are known as thermistors. The most commonly used type is prepared from oxides of the iron group of transition metal elements such as chromium, manganese, iron, cobalt and nickel. The resistance of these elements decreases with temperature — in other words there is a negative temperature coefficient (NTC) — in a highly non-linear way. Figure 8.2 shows typical thermistor resistance/temperature characteristics which can be described by the relationship:

$$R_\theta = K \exp\left(\frac{\beta}{\theta}\right) \qquad [8.2]$$

where R_θ is the resistance at temperature θ Kelvin, and K, β are constants for the thermistor. A commonly used alternative equation is:

$$R_\theta = R_{\theta_1} \exp \beta \left[\frac{1}{\theta} - \frac{1}{\theta_1}\right] \qquad [8.3]$$

where R_{θ_1} Ω is the resistance at reference temperature θ_1 K, usually $\theta_1 = 25\,°\mathrm{C} = 298$ K. Thermistors are usually in the form of either beads, rods or discs (Fig. 8.2); bead thermistors are enclosed in glass envelopes. A typical NTC thermistor has a

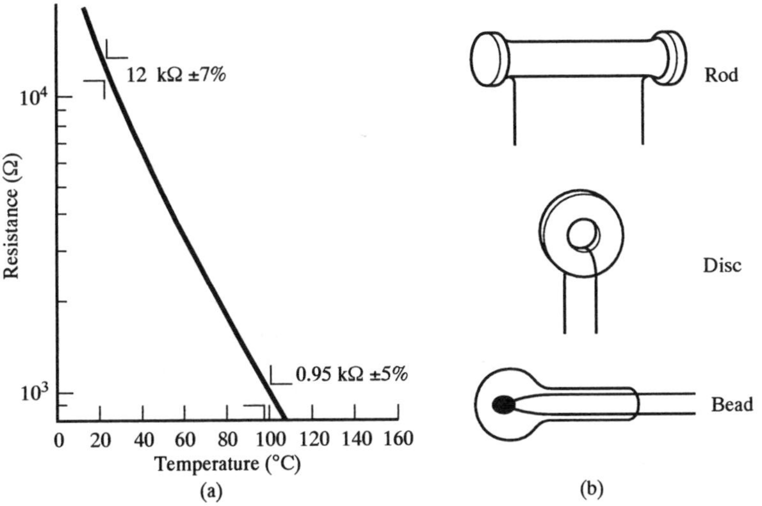

Fig. 8.2 Thermistor resistance-temperature characteristics and types (a) (after Mullard Ltd[2])

resistance of 12 kΩ at 25 °C (298 K), falling to 0.95 kΩ at 100 °C (373 K), and β = 3750 K.[2] The manufacturer's tolerance limits on the above figures are ±7%, i.e. ±840 Ω, at 25 °C and ±5%, i.e. ±47.5 Ω, at 100 °C; which is far wider than for metal elements. The element time constant is 19 sec in air, 3 sec in oil and the self-heating effect is 1 °C rise for every 7 mW of electrical power. Thermistors with positive temperature coefficients (P.T.C.) are also available; the resistance of a typical element increases from 100 Ω at −55 °C to 10 kΩ at 120 °C.

8.1.3 Metal and semiconductor resistance strain gauges

Before discussing strain gauges we must first briefly explain the concepts of stress, strain, elastic modulus and Poisson's ratio.

Stress is defined by force/area, so that in Fig. 8.3(a) the stress experienced by the body is $+F/A$, the positive sign indicating a **tensile stress** which tends to increase the length of the body. In Fig. 8.3(b) the stress is $-F/A$, the negative sign indicating a **compressive stress** which tends to reduce the length of the body. The effect of the applied stress is to produce a **strain** in the body which is defined by (change in length)/(original unstressed length). Thus in Fig. 8.3(a) the strain is $e = +\Delta l/l$ (tensile), and in 8.3(b) the strain is $e = -\Delta l/l$ (compressive); in both cases the strain is **longitudinal** i.e. along the direction of the applied stress. The relationship between

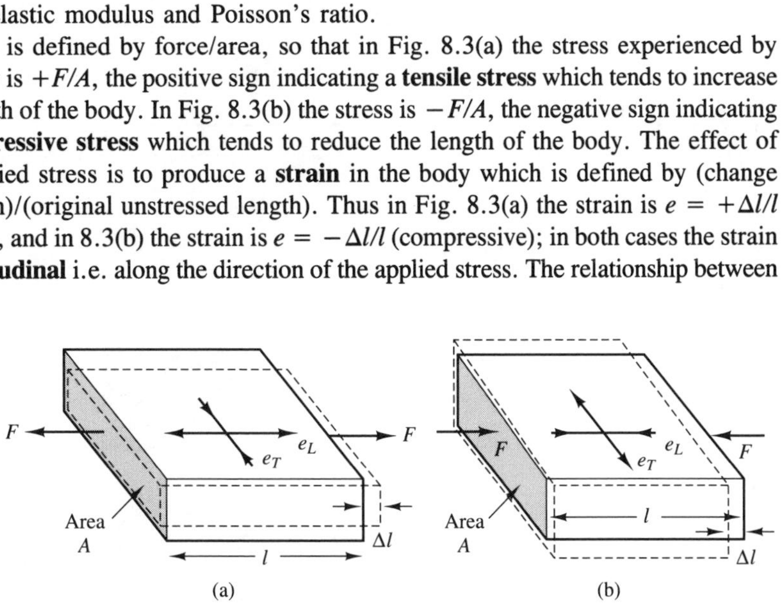

Fig. 8.3 Stress and strain
(a) Effect of tensile stress
(b) Effect of compressive stress

strain and stress is linear for a given body over a certain range of values; the slope of the straight line is termed the **elastic modulus** of the body:

Elastic modulus

$$\text{Elastic modulus} = \frac{\text{stress}}{\text{strain}} \qquad [8.4]$$

For linear tensile or compressive stress the elastic modulus is called **Young's modulus** E; for shear stress the relevant elastic modulus is **shear modulus S**. Returning to Fig. 8.3(a) we note that the increase in length of the body is accompanied by a decrease in cross sectional area i.e. a reduction in width and thickness. Thus in Fig. 8.3(a) the longitudinal tensile strain is accompanied by a transverse compressive strain, and in Fig. 8.3(b) the longitudinal compressive strain is accompanied by a transverse tensile strain. The relation between longitudinal strain e_L and accompanying transverse strain e_T is:

$$e_T = -\nu e_L \qquad [8.5]$$

where ν is Poisson's ratio which has a value between 0.25 and 0.4 for most materials.

A **strain gauge** is a metal or semiconductor element whose resistance changes when under strain. We can derive the relationship between changes in resistance and strain by considering the factors which influence the resistance of the element. The resistance of an element of length l, cross sectional area A and resistivity ρ (Fig. 8.4) is given by:

$$R = \frac{\rho l}{A} \qquad [8.6]$$

In general with strain gauges ρ, l and A can change if the element is strained so that the change in resistance ΔR is given by:

$$\Delta R = \left(\frac{\partial R}{\partial l}\right)\Delta l + \left(\frac{\partial R}{\partial A}\right)\Delta A + \left(\frac{\partial R}{\partial \rho}\right)\Delta \rho \qquad [8.7]$$

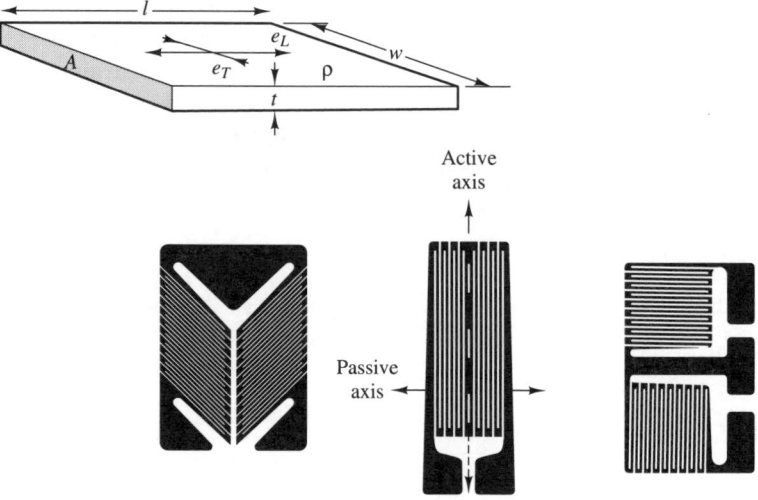

Fig. 8.4 Strain gauges

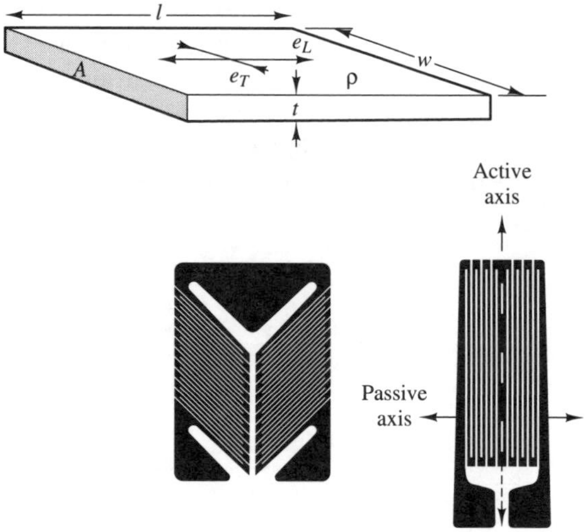

i.e.

$$\Delta R = \frac{\rho}{A}\Delta l - \frac{\rho l}{A^2}\Delta A + \frac{l}{A}\Delta\rho$$

Dividing throughout by $R = \rho l/A$ yields

$$\frac{\Delta R}{R} = \frac{\Delta l}{l} - \frac{\Delta A}{A} + \frac{\Delta\rho}{\rho} \qquad [8.8]$$

The ratio $\Delta l/l$ is the longitudinal strain e_L in the element. Since cross sectional area $A = wt$ (Fig. 8.4),

$$\frac{\Delta A}{A} = \frac{\Delta w}{w} + \frac{\Delta t}{t} = 2e_T$$

where e_T is the transverse strain in the element. From [8.5] and [8.8] we have:

$$\begin{aligned}\frac{\Delta R}{R} &= e_L - 2(-\nu e_L) + \frac{\Delta\rho}{\rho} \\ &= (1 + 2\nu)e_L + \frac{\Delta\rho}{\rho}\end{aligned} \qquad [8.9]$$

We now define the **gauge factor** G of a strain gauge by the ratio (fractional change in resistance)/(strain)

i.e. $$G = \frac{\Delta R/R_0}{e}$$

Hence

Resistance/strain relationship for a strain gauge

$$\frac{\Delta R}{R_0} = Ge \qquad [8.10]$$

where R_0 is the unstrained resistance of the gauge. From [8.9] the gauge factor is given by:

Gauge factor of a strain gauge

$$G = 1 + 2\nu + \frac{1}{e}\frac{\Delta\rho}{\rho} \qquad [8.11]$$

For most metals $\nu \approx 0.3$, and the term $(1/e)(\Delta\rho/\rho)$ representing strain-induced changes in resistivity (piezoresistive effect) is small (around 0.4), so that the overall gauge factor G is around 2.0. A popular metal for strain gauges is the alloy 'Advance'; this is 54% copper, 44% nickel and 1% manganese. This alloy has a low **temperature coefficient of resistance** ($2 \times 10^{-5}\,°C^{-1}$) and a low **temperature coefficient of linear expansion**. Temperature is both an interfering and a modifying input (Section 2.2) and the above properties ensure that temperature effects on zero values and sensitivity are small.

The most common strain gauges are of the bonded type, where the gauge consists of metal foil, cut into a grid structure by a photoetching process, and mounted on a resin film base. The film backing is then attached to the structure to be measured with a suitable adhesive. The gauge should be positioned so that its active axis is along the direction of the measured strain: the change in resistance, due a given strain, along the passive axis is very small compared with that produced by the same strain along the active axis. A typical gauge has:

a gauge factor of 2.0 to 2.2;
unstrained resistance of 120 ± 1 Ω;
linearity within ±0.3%;
maximum tensile strain $+2 \times 10^{-2}$;
maximum compressive strain -1×10^{-2};
maximum operating temperature 150 °C.

The change in resistance at maximum tensile strain is $\Delta R = +4.8\ \Omega$, and $\Delta R = -2.4\ \Omega$ at maximum compressive strain. A maximum gauge current between 15 mA and 100 mA, depending on area, is specified in order to avoid self-heating effects. Unbonded strain gauges consisting of fine metal wire stretched over pillars are used in some applications.

In semiconductor gauges the piezoresistive term $(1/e)(\Delta\rho/\rho)$ can be large, giving large gauge factors. The most common material is silicon doped with small amounts of 'P' type or 'N' type material. Gauge factors of between +100 and +175 are common for P-type silicon, and between −100 and −140 for N-type silicon. A negative gauge factor means a decrease in resistance for a tensile strain. Thus semiconductor gauges have the advantage of greater sensitivity to strain than metal ones, but have the disadvantage of greater sensitivity to temperature changes. Typically a rise in ambient temperature from 0 to 40 °C causes a fall in gauge factor from 135 to 120. Also the temperature coefficient of resistance is larger, so that the resistance of a typical unstrained gauge will increase from 120 Ω at 20 °C to 125 Ω at 60 °C. Strain gauge elements are incorporated in deflection bridge circuits (Section 9.1).

8.2 Capacitive sensing elements

The simplest capacitor or condenser consists of two parallel metal plates separated by a dielectric or insulating material (Fig. 8.5). The capacitance of this parallel plate capacitor is given by:

$$C = \frac{\epsilon_0 \epsilon A}{d} \qquad [8.12]$$

where ϵ_0 is the permittivity of free space (vacuum) of magnitude 8.85 pF m^{-1}, ϵ is the relative permittivity or dielectric constant of the insulating material, A m^2 is the area of overlap of the plates and d m their separation. From [8.12] we see that C can be changed by changing either d, A or ϵ; Fig. 8.5 shows **capacitive displacement sensors** using each of these methods. If the displacement x causes the plate separation to increase to $d + x$ the capacitance of the sensor is:

Variable separation displacement sensor

$$C = \frac{\epsilon_0 \epsilon A}{d + x} \qquad [8.13]$$

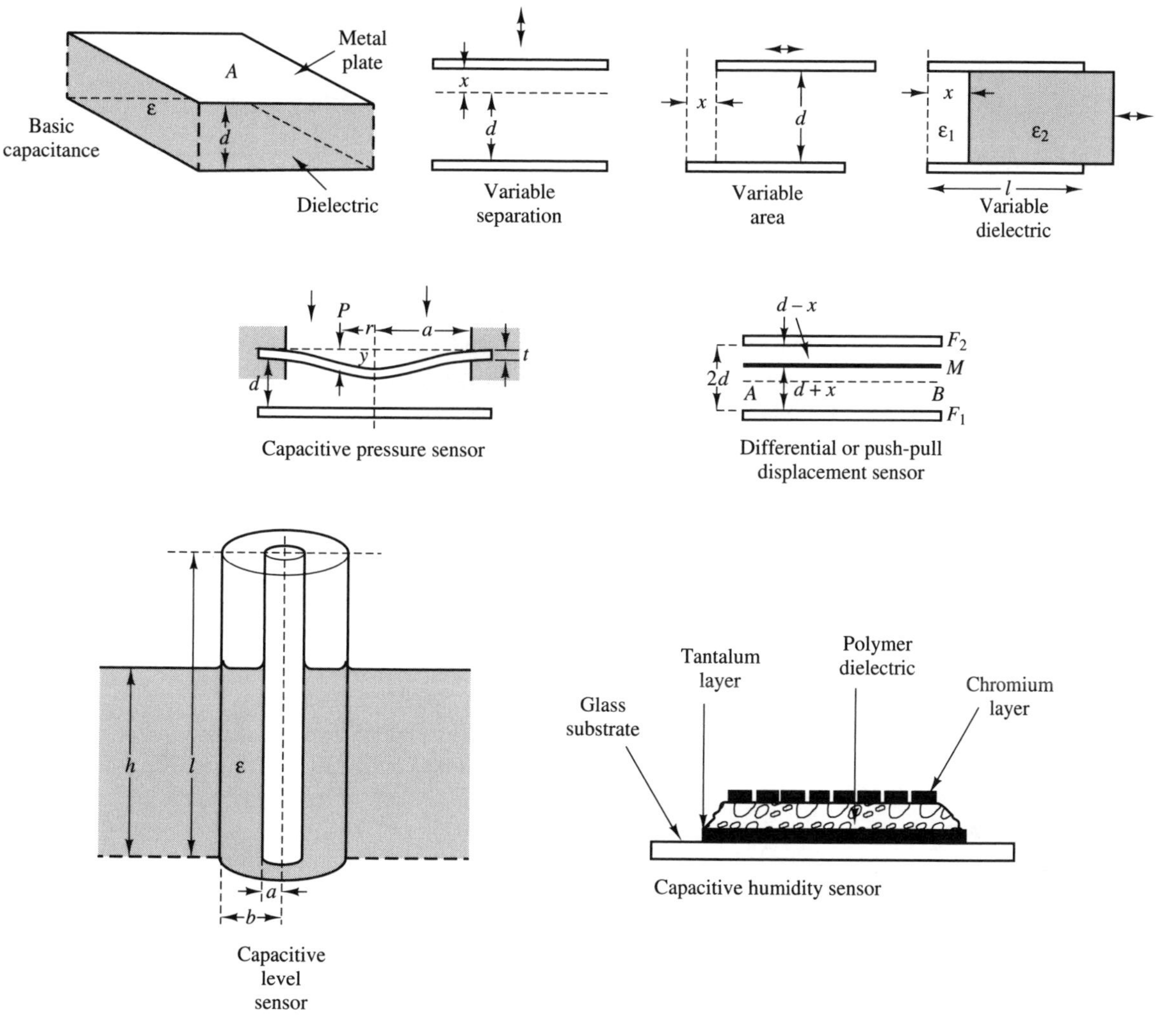

Fig. 8.5 Capacitive sensing elements

i.e. there is a non-linear relation between C and x. In the variable area type, the displacement x causes the overlap area to decrease by $\Delta A = wx$, where w is the width of the plates, giving:

Variable area displacement sensor

$$C = \frac{\epsilon_0 \epsilon}{d}(A - wx) \qquad [8.14]$$

In the variable dielectric type, the displacement x changes the amount of dielectric material $\epsilon_2 (\epsilon_2 > \epsilon_1)$ inserted between the plates. The total capacitance of the sensor is the sum of two capacitances, one with area A_1, dielectric constant ϵ_1, and one with area A_2, dielectric constant ϵ_2, i.e.

$$C = \frac{\epsilon_0 \epsilon_1 A_1}{d} + \frac{\epsilon_0 \epsilon_2 A_2}{d}$$

Since $A_1 = wx$, $A_2 = w(l - x)$, when w is the width of the plates,

Variable dielectric displacement sensor

$$C = \frac{\epsilon_0 w}{d} [\epsilon_2 l - (\epsilon_2 - \epsilon_1)x] \qquad [8.15]$$

A commonly used capacitive pressure sensor is shown in Fig. 8.5. Here one plate is a fixed metal disc, the other is a flexible flat circular diaphragm, clamped around its circumference; the dielectric material is air [$\epsilon \approx 1$). The diaphragm is an elastic sensing element (Section 8.6) which is bent into a curve by the applied pressure P. The deflection y at any radius r is given by:

$$y = \frac{3}{16} \frac{(1 - \nu^2)}{Et^3} (a^2 - r^2)^2 P \qquad [8.16]$$

where

a = radius of diaphragm
t = thickness of diaphragm
E = Young's modulus
ν = Poisson's ratio.

The deformation of the diaphragm means that the average separation of the plates is reduced, the resulting increase in capacitance ΔC is given by[4]

Capacitive pressure sensor

$$\frac{\Delta C}{C} = \frac{(1 - \nu^2)a^4}{16 E d t^3} P \qquad [8.17]$$

where d is the initial separation of the plates and $C = \epsilon_0 \pi a^2 / d$, capacitance at zero pressure.

The variable separation displacement sensor has the disadvantage of being non-linear (eqn [8.13]). This problem is overcome by using the three-plate differential or push-pull displacement sensor shown in Fig. 8.5. This consists of a plate M moving between two fixed plates F_1 and F_2; if x is the displacement of M from the centre line AB, then the capacitances C_1 and C_2 formed by MF_1 and MF_2 respectively, are:

Differential capacitive displacement sensor

$$C_1 = \frac{\epsilon \epsilon_0 A}{d + x}, \quad C_2 = \frac{\epsilon \epsilon_0 A}{d - x} \qquad [8.18]$$

The relations between C_1, C_2 and x are still non-linear, but when C_1 and C_2 are incorporated into the a.c. deflection bridge described in Section 9.1.3, the overall relationship between bridge output voltage and x is linear (eqn [9.23]).

The next sensing element shown in Fig. 8.5 is a level sensor consisting of two concentric metal cylinders. The space between the cylinders contains liquid to the height h of the liquid in the vessel. If the liquid is non-conducting (electrical conductivity less than 0.1 μmho cm^{-3}), it forms a suitable dielectric and the total capacitance of the sensor is the sum of liquid and air capacitances. The capacitance/unit length of two coaxial cylinders, radii b and $a (b > a)$, separated by a dielectric ϵ is $2\pi\epsilon_0\epsilon/\log_e(b/a)$. Assuming the dielectric constant of air is unity, the capacitance of the level sensor is given by:

$$C_h = \frac{2\pi\epsilon_0\epsilon h}{\log_e\left(\frac{b}{a}\right)} + \frac{2\pi\epsilon_0(l - h)}{\log_e\left(\frac{b}{a}\right)}$$

Capacitive level sensor

$$C_h = \frac{2\pi\epsilon_0}{\log_e\left(\frac{b}{a}\right)}[l + (\epsilon - 1)h] \quad [8.19]$$

The sensor can be incorporated into the a.c. deflection bridge of Fig. 9.5(a).

The final sensor shown in Fig. 8.5 is a **thin-film capacitive humidity sensor**.[7] The dielectric is a polymer which has the ability to absorb water molecules; the resulting change in dielectric constant and therefore capacitance is proportional to the percentage relative humidity of the surrounding atmosphere. One capacitor plate consists of a layer of tantalum deposited on a glass substrate, the layer of polymer dielectric is then added, followed by the second plate which is a thin layer of chromium. The chromium layer is under high tensile stress so that it cracks into a fine mosaic which allows water molecules to pass into the dieletric. The stress in the chromium also causes the polymer to crack into a mosaic structure. A sensor of this type has a input range of 0 to 100% RH, a capacitance of 375 pF at 0% RH and a linear sensitivity of 1.7 pF/% RH. The capacitance/humidity relation is therefore the linear equation:

$$C = 375 + 1.7 \text{ RH pF}$$

The maximum departure from this line is 2% due to non-linearity and 1% due to hysteresis.

Capacitive sensing elements are incorporated in either a.c. deflection bridge circuits (Section 9.1) or oscillator circuits (Section 9.5.1). Capacitive sensors are not pure capacitances but have an associated resistance R in parallel, to represent losses in the dielectric; this has an important influence on the design of circuits, particularly oscillator circuits. For example, a typical capacitive humidity sensor would have an approximate parallel dielectric loss resistance R of 100 kΩ at 100 kHz. The quality of a dielectric is often expressed in terms of its 'loss tangent' $\tan\delta$ where:

$$\tan\delta = \frac{1}{\omega CR}$$

In this example, if $C = 500$ pF then $\tan\delta \approx 0.03$. Figure 9.19(b) shows the capacitive sensor incorporated with a pure inductance into an oscillator circuit. The 'quality factor' Q of the circuit is given by

$$Q = R\sqrt{\frac{C}{L}} = \omega_n CR$$

and thus depends crucially on R. If the natural frequency f_n of the circuit with the above humidity sensor is $f_n = 10^5$ Hz, then the circuit Q factor is approximately

30. Precautions should also be taken to minimise the effect of cable and stray capacitances on these circuits.

8.3 Inductive sensing elements

8.3.1 Variable inductance (variable reluctance) displacement sensors

In order to discuss the principles of these elements we must first introduce the concept of a **magnetic circuit**. In an electrical circuit an electromotive force (e.m.f.) drives a current through an electrical resistance and the magnitude of the current is given by

$$\text{e.m.f.} = \text{current} \times \text{resistance} \qquad [8.20]$$

A simple magnetic circuit is shown in Fig. 8.6(a): it consists of a loop or core of ferromagnetic material on which is wound a coil of n turns carrying a current i. By analogy we can regard the coil as a source of magnetomotive force (m.m.f.) which drives a flux ϕ through the magnetic circuit. The equation corresponding to [8.20] for a magnetic circuit is:

$$\text{m.m.f.} = \text{flux} \times \text{reluctance} = \phi \times \mathcal{R} \qquad [8.21]$$

so that reluctance $\mathcal{R}$ limits the flux in a magnetic circuit just as resistance limits the current in an electrical circuit. In this example m.m.f. $= ni$, so that the flux in the magnetic circuit is:

$$\phi = \frac{ni}{\mathcal{R}} \text{ webers} \qquad [8.22]$$

This is the flux linked by a single turn of the coil; the total flux N linked by the entire coil of n turns is:

$$N = n\phi = \frac{n^2 i}{\mathcal{R}} \qquad [8.23]$$

by definition the self-inductance L of the coil is the total flux per unit current i.e.

Self inductance of a coil

$$L = \frac{N}{i} = \frac{n^2}{\mathcal{R}} \qquad [8.24]$$

The above equation enables us to calculate the inductance of a sensing element given the reluctance of the magnetic circuit. The reluctance $\mathcal{R}$ of a magnetic circuit is given by:

$$\mathcal{R} = \frac{l}{\mu\mu_0 A} \qquad [8.25]$$

where l is the total length of the flux path, μ is the relative permeability of the circuit material, μ_0 is the permeability of free space $= 4\pi \times 10^{-7}\,\text{H m}^{-1}$ and A is the cross-sectional area of the flux path. Figure 8.6(b) shows the core separated into two parts by an air gap of variable width. The total reluctance of the circuit is now the reluctance of both parts of the core together with the reluctance of the air gap.

Fig. 8.6 Variable reluctance elements
(a) Basic principle of reluctance sensing elements
(b) Reluctance calculation for typical element
(c) Differential or push/pull reluctance displacement sensor

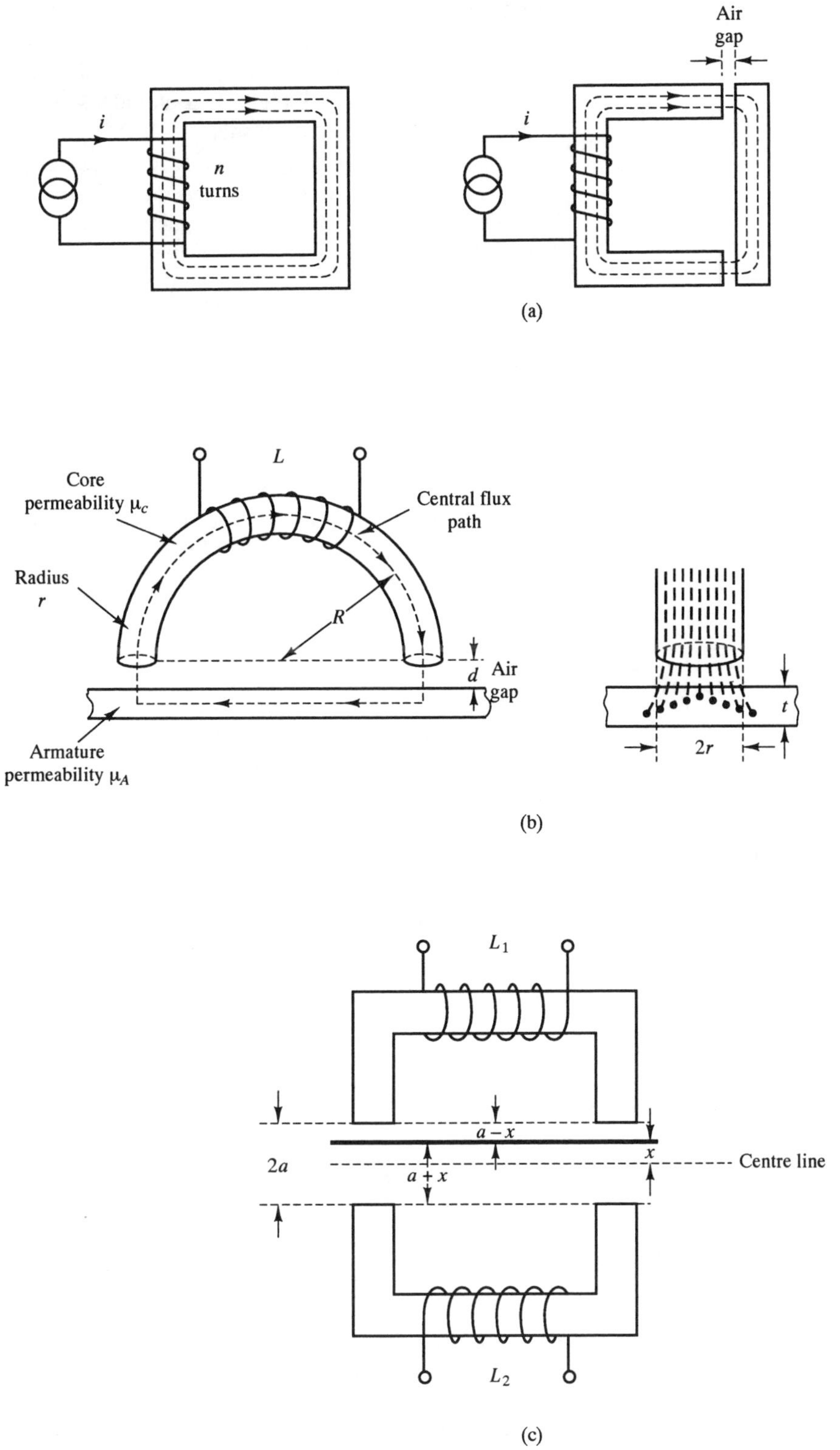

Since the relative permeability of air is close to unity and that of the core material many thousands, the presence of the air gap causes a large increase in circuit reluctance and a corresponding decrease in flux and inductance. Thus a small variation in air gap causes a measurable change in inductance so that we have the basis of an **inductive displacement sensor**.

Figure 8.6(c) shows a typical variable reluctance displacement sensor, consisting of three elements: a ferromagnetic core in the shape of a semitoroid (semicircular ring), a variable air gap and a ferromagnetic plate or armature. The total reluctance of the magnetic circuit is the sum of the individual reluctances, i.e.

$$\mathcal{R}_{\text{TOTAL}} = \mathcal{R}_{\text{CORE}} + \mathcal{R}_{\text{GAP}} + \mathcal{R}_{\text{ARMATURE}} \qquad [8.26]$$

The length of an average, i.e. central path, through the core is πR and the cross-sectional area is πr^2, giving:

$$\mathcal{R}_{\text{CORE}} = \frac{\pi R}{\mu_0 \mu_C \pi r^2} = \frac{R}{\mu_0 \mu_C r^2} \qquad [8.27]$$

The total length of the flux path in air is twice the air gap, i.e. $2d$; also if there is little bending or fringing of the lines of flux in the air gap, then the cross-sectional area of the flux path in air will be close to that of the core. Assuming the relative permeability of air is unity,

$$\mathcal{R}_{\text{GAP}} = \frac{2d}{\mu_0 \pi r^2} \qquad [8.28]$$

The length of an average central flux path in the armature is $2R$; the calculation of the appropriate cross-sectional area is more difficult. A typical flux distribution is shown in Fig. 8.6(c) and for simplicity we assume that most of the flux is concentrated with an area $2rt$, giving

$$\mathcal{R}_{\text{ARMATURE}} = \frac{2R}{\mu_0 \mu_A 2rt} = \frac{R}{\mu_0 \mu_A rt} \qquad [8.29]$$

Thus

$$\mathcal{R}_{\text{TOTAL}} = \frac{R}{\mu_0 \mu_C r^2} + \frac{2d}{\mu_0 \pi r^2} + \frac{R}{\mu_0 \mu_A rt}$$

i.e.

$$\mathcal{R}_{\text{TOTAL}} = \mathcal{R}_0 + kd \qquad [8.30]$$

where

$$\mathcal{R}_0 = \frac{R}{\mu_0 r}\left[\frac{1}{\mu_C r} + \frac{1}{\mu_A t}\right] = \text{reluctance at zero air gap}$$

$$k = \frac{2}{\mu_0 \pi r^2}$$

A typical element with $n = 500$ turns, $R = 2$ cm, $r = 0.5$ cm, $t = 0.5$ cm, $\mu_C = \mu_A = 100$, has $\mathcal{R}_0 = 1.3 \times 10^7\,\text{H}^{-1}$, $k = 2 \times 10^{10}\,\text{H}^{-1}\,\text{m}^{-1}$. This gives $L = 19$ mH at $d = 0$ (zero air gap) and $L = 7.6$ mH at $d = 1$ mm. From [8.24] and [8.30] we have

Inductance of reluctance displacement sensor

$$L = \frac{n^2}{\mathcal{R}_0 + kd} = \frac{L_0}{1 + \alpha d} \qquad [8.31]$$

where $L_0 = n^2/\mathcal{R}_0$ = inductance at zero gap and $\alpha = k/\mathcal{R}_0$. Equation [8.31] is applicable to any variable reluctance displacement sensor; the values of L_0 and α depend on core geometry, permeability, etc. We see that the relationship between L and d is non-linear, but this problem is often overcome by using the push-pull or differential displacement sensor shown in 8.6(d). This consists of an armature moving between two identical cores, separated by a fixed distance of $2a$. From equation [8.31] and Fig. 8.6(d) we have:

Differential reluctance displacement sensor

$$L_1 = \frac{L_0}{1 + \alpha(a - x)}, \quad L_2 = \frac{L_0}{1 + \alpha(a + x)} \qquad [8.32]$$

The relationship between L_1, L_2 and displacement x is still non-linear, but if the sensor is incorporated into the a.c. deflection bridge of Fig. 9.5(b), then the overall relationship between bridge out of balance voltage and x is linear (eqn [9.25]). A typical sensor of this type would have an input span of $\frac{1}{2}$ inch, a coil inductance (L_0) of 25 mH, coil resistance of 70 Ω and maximum non-linearity of 0.5%. Thus, inductive sensors are not pure inductances but have an associated resistance R in series; this has an important influence on the design of oscillator circuits (Fig. 9.18(a)).

8.3.2 Linear Variable Differential Transformer (LVDT) displacement sensor

This sensor is a transformer with a single primary winding and two identical secondary windings wound on a tubular ferromagnetic former (Fig. 8.7). The primary winding is energised by an a.c. voltage of amplitude $\hat{V}_S$, frequency f_S Hz; the two secondaries are connected in series opposition so that the output voltage $\hat{V}_{OUT} \sin(2\pi f_S t + \phi)$ is the difference $V_1 - V_2$ of the voltages induced in the secondaries. A ferromagnetic core or plunger moves inside the former; this alters the mutual inductance between the primary and secondaries. With the core removed the secondary voltages are ideally equal so that $\hat{V}_{OUT} = 0$. With the core in the former, V_1 and V_2 change with core position x, causing amplitude $\hat{V}_{OUT}$ and phase ϕ to change.

The relationships between $\hat{V}_{OUT}$, ϕ and x are shown in Fig. 8.7. We see that there is a null point C at the centre of the sensor ($x = \frac{1}{2}l$) where $\hat{V}_{OUT} = 0$ (ideally); here there is equal coupling between the primary and secondaries so that $V_1 = V_2$. At the points A and B, equally spaced either side of the null point, $\hat{V}_{OUT}$ has the same value V_0. However, at A the output voltage is 180° out of phase with the primary voltage, i.e. $\phi = -180°$ ($V_2 > V_1$), whereas at B the output voltage is in phase with the primary voltage i.e. $\phi = 0°$ ($V_1 > V_2$). Non-linear effects occur at either end (D and E) as the core moves to the edge of the former.

The a.c. output signal is converted into d.c. in a way which distinguishes between

Fig. 8.7 Linear variable differential transformer (LVDT) and characteristics

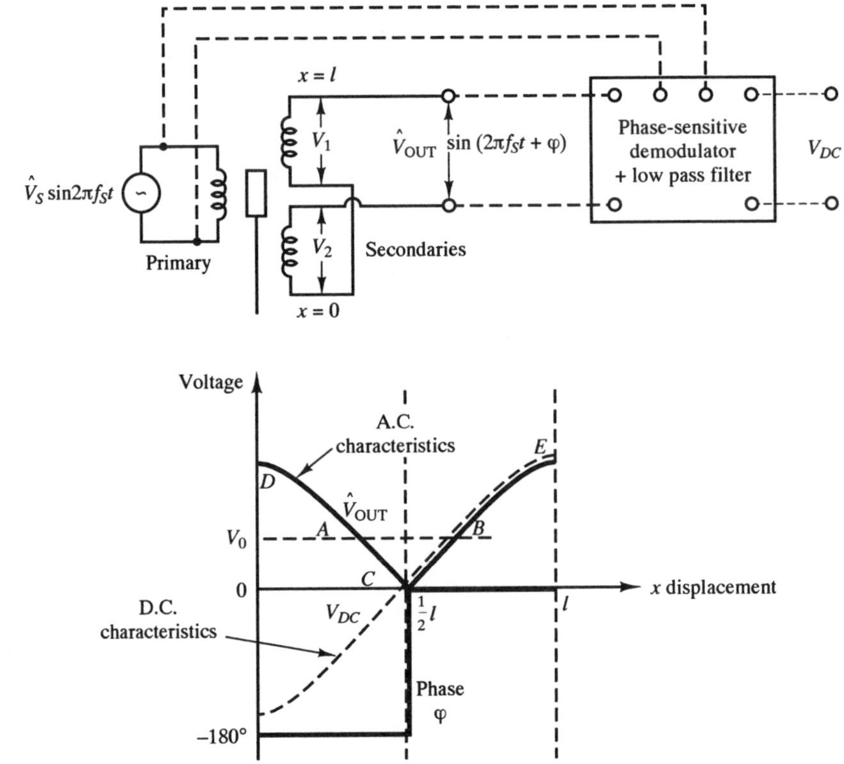

the situations at *A* and *B*, where the amplitude is the same but there is a phase difference of 180°. A **phase sensitive demodulator** is used (Section 9.3) which senses this phase difference and gives a negative voltage at position *A* and a positive voltage at position *B*. The corresponding d.c. characteristics are shown in Fig. 8.7; these show that the non-linearity will increase as the displacement range is increased.

LVDT displacement sensors are available to cover ranges from ±0.25 mm to ± 25 cm. For a typical sensor of range ±2.5 cm, the recommended $\hat{V}_S$ is 4 to 6 V, recommended f_S is 5 kHz (400 Hz minimum, 50 kHz maximum), and maximum non-linearity is 1% f.s.d. over the above range.

8.4 Electromagnetic sensing elements

These elements are used for the measurement of linear and angular velocity and are based on Faraday's law of electromagnetic induction. This states that if the flux *N* linked by a conductor is changing with time, then a back e.m.f. is induced in the conductor with magnitude equal to the rate of change of flux, i.e.

$$E = -\frac{dN}{dt} \qquad [8.33]$$

In an electromagnetic element the change in flux is produced by the motion being investigated; this means that the induced e.m.f. depends on the linear or angular

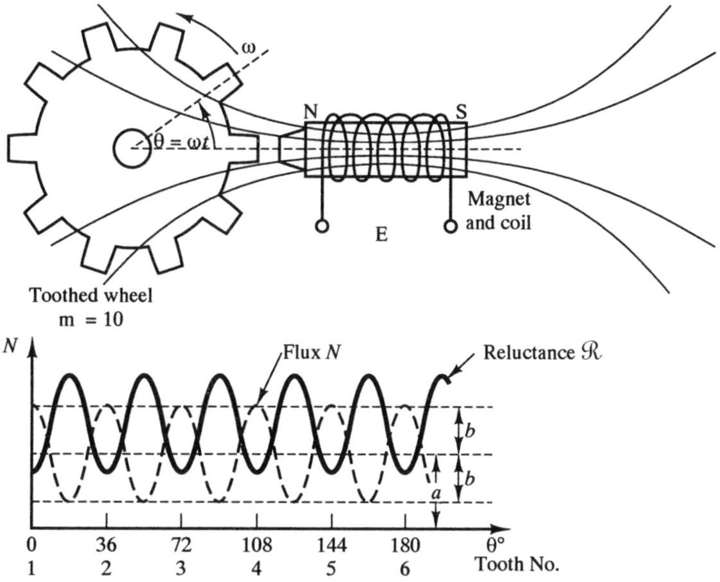

Fig. 8.8 Variable reluctance tachogenerator, angular variations in reluctance and flux

velocity of the motion. A common example of an electromagnetic sensor is the variable reluctance tachogenerator for measuring angular velocity (Fig. 8.8). It consists of a toothed wheel of ferromagnetic material (attached to the rotating shaft) and a coil wound onto a permanent magnet, extended by a soft iron pole piece. The wheel moves in close proximity to the pole piece, causing the flux linked by the coil to change with time, thereby inducing an e.m.f. in the coil.

The magnitude of the e.m.f. can be calculated by considering the magnetic circuit formed by the permanent magnet, air gap and wheel. The m.m.f. is constant with time and depends on the field strength of the permanent magnet. The reluctance of the circuit will depend on the width of the air gap between the wheel and pole piece. When a tooth is close to the pole piece the reluctance is minimum but will increase as the tooth moves away. The reluctance is maximum when a 'gap' is adjacent to the pole piece but falls again as the next tooth approaches the pole piece. Figure 8.8 shows the resulting cyclic variation in reluctance $\mathcal{R}$ with angular rotation θ. From eqn [8.21] the flux in the circuit is given by $\phi = \text{m.m.f.}/\mathcal{R}$, and the total flux N linked by a coil of n turns is:

$$N = n\phi = \frac{n \times \text{m.m.f.}}{\mathcal{R}} \tag{8.34}$$

so that $N \propto 1/\mathcal{R}$. The corresponding variation of flux N with θ is also shown in Fig. 8.8. We see that a reluctance minimum corresponds to a flux maximum and vice versa. This relation may be approximated by

$$N(\theta) \approx a + b \cos m\theta \tag{8.35}$$

where a is the mean flux, b is the amplitude of the flux variation and m is the number of teeth. Using eqn [8.33] the induced e.m.f. is given by:

$$E = -\frac{dN}{dt} = -\frac{dN}{d\theta} \times \frac{d\theta}{dt} \tag{8.36}$$

where

$$\frac{dN}{d\theta} = -\ bm \sin m\theta$$

$$\frac{d\theta}{dt} = \omega_r \text{ (angular velocity of wheel)}$$

and

$$\theta = \omega_r t \text{ (assuming } \theta = 0 \text{ at } t = 0)$$

Thus

Output signal for variable reluctance tachogenerator

$$E = bm\omega_r \sin m\omega_r t \qquad [8.37]$$

This is a sinusoidal signal of amplitude $\hat{E} = bm\omega_r$, and frequency $f = m\omega_r/(2\pi)$; i.e. both amplitude and frequency are proportional to the angular velocity of the wheel.

In principle ω_r can be found from either the amplitude or the frequency of the signal. In practice the amplitude measured by a distant recorder or indicator may be affected by loading effects (Section 5.1) and electrical interference (Section 6.3). A frequency system is therefore preferred (Chapter 10), where the number of cycles in a given time is counted to give a digital signal corresponding to the angular velocity. A variable reluctance tachogenerator is incorporated in the turbine flowmeter (Section 12.2) to give an accurate measurement of volume flow rate or total volume of fluid.

8.5 Thermoelectric sensing elements

Thermoelectric or thermocouple sensing elements are commonly used for measuring temperature. If two different metals A and B are joined together, there is a difference in electrical potential across the junction called the contact potential. This contact potential depends on the metals A and B and the temperature T°C of the junction, and is given by a power series of the form:

$$E_T^{AB} = a_1T + a_2T^2 + a_3T^3 + a_4T^4 + \cdots \qquad [8.38]$$

The values of constants a_1, a_2 etc. depend on the metals A and B. For example, the first four terms in the power series for the e.m.f. of an iron v. constantan (Type J) junction are[5] as follows, expressed in μV.

$$E_T^J = 5.037 \times 10^{+1}T + 3.043 \times 10^{-2}T^2 - 8.567 \times 10^{-5}T^3 + 1.335 \times 10^{-7}T^4 + \text{higher order terms up to } T^7 \qquad [8.39]$$

A thermocouple is a closed circuit consisting of two junctions (Fig. 8.9), at different temperatures T_1 and T_2°C. If a high impedance voltmeter is introduced into the circuit, so that current flow is negligible, then the measured e.m.f. is, to a close approximation, the difference of the contact potentials, i.e.

$$E_{T_1,T_2}^{AB} = E_{T_1}^{AB} - E_{T_2}^{AB} = a_1(T_1 - T_2) + a_2(T_1^2 - T_2^2) + a_3(T_1^3 - T_2^3) + \cdots \qquad [8.40]$$

Fig. 8.9 Thermocouple principles

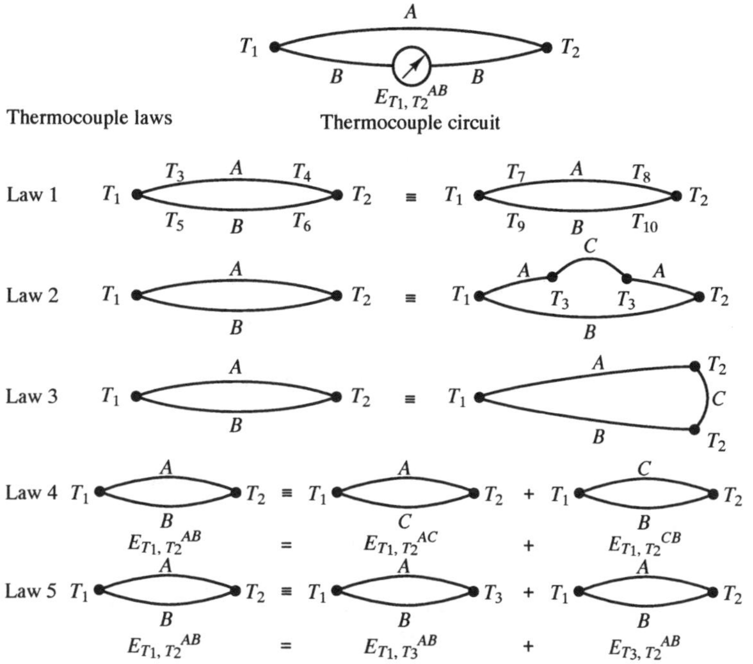

Thus the measured e.m.f. depends on the temperatures T_1, T_2 of both junctions. In the following discussion T_1 will be the temperature to be measured, i.e. the temperature of the measurement junction, and T_2 will be the temperature of the reference junction. In order to accurately infer T_1 from the measured e.m.f., the reference junction temperature T_2 must be known.

Figure 8.9 summarises five 'laws' of thermocouple behaviour which are vital in temperature measurement.[6] Law 1 states that the e.m.f. of a given thermocouple depends only on the temperatures of the junctions and is independent of the temperatures of the wires connecting the junctions. This is important in industrial installations, where the leads connecting measurement and reference junctions may be exposed to large changes in ambient temperature.

Law 2 states that if a third metal C is introduced into A (or B) then, provided the two new junctions are at the same temperature (T_3), the e.m.f. is unchanged. This means that a voltmeter can be introduced into the circuit without affecting the voltage produced.

If a third metal C is inserted between A and B at either junction, then Law 3 states that, provided the two new junctions AC and CB are both at the same temperature (T_1 or T_2), then the e.m.f. is unchanged. This means that at the measurement junction, wires A and B can be soldered or brazed together with a third metal without affecting the e.m.f. A voltage measuring device can be introduced at the reference junction again without affecting the measurement.

Law 4 (**law of intermediate metals**) can be used, for example, to deduce the e.m.f. of a copper–iron (AB) thermocouple, given the e.m.f.s of copper–constantan (AC) and constantan–iron (CB) thermocouples.

The fifth law (**law of intermediate temperatures**) is used in interpreting e.m.f. measurements. For a given pair of metals we have:

$$E_{T_1,T_2} = E_{T_1,T_3} + E_{T_3,T_2} \quad [8.41]$$

where T_3 is the intermediate temperature. If $T_2 = 0\,°C$, then

$$E_{T_1,0} = E_{T_1,T_3} + E_{T_3,0} \quad [8.42]$$

Suppose that we wish to measure the temperature T_1 °C of a liquid inside a vessel with a chromel v. alumel thermocouple. The measurement junction is inserted in the vessel and the reference junction is outside the vessel, where the temperature is measured to be 20 °C, i.e. $T_3 = 20\,°C$. The measured e.m.f. is 5.3 mV using a voltmeter inserted at the reference junction, i.e. $E_{T_1,T_3} = E_{T_1,20} = 5.3$ mV. The value of $E_{T_3,0} = E_{20,0}$ is found to be 0.8 mV using thermcouple tables.[5] These tables give the e.m.f. $E_{T,0}$ for a particular thermocouple, with measured junction at T°C and reference junction at 0 °C. From [8.42], we have $E_{T_1,0} = 5.3 + 0.8 = 6.1$ mV; T_1 is then found to be 149 °C by looking up the temperature corresponding to 6.1 mV.

The importance of correct installation of thermocouples can be illustrated by the problem shown in Fig. 8.10. Here we wish to measure the temperature of high pressure steam in a pipe, at around 200 °C, with a chromel v. alumel thermocouple, for which $E_{200,0} = 8.1$ mV.

Installation (a) with the meter located just outside the pipe is completely useless. The reference junction temperature T_2 can vary widely from sub-zero temperatures

Fig. 8.10 Thermocouple installations

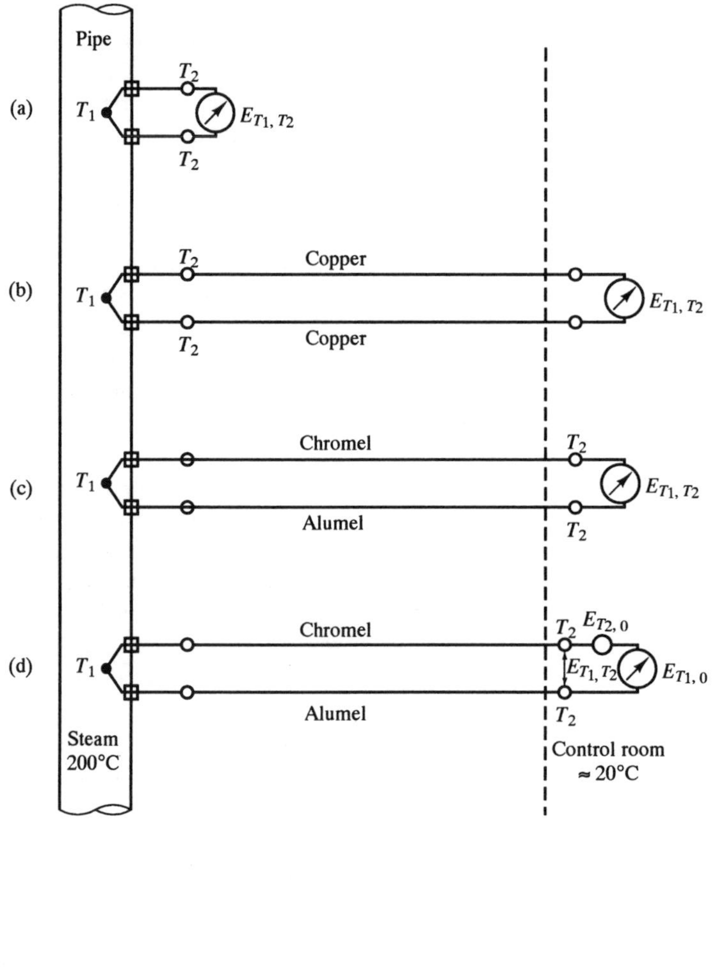

in cold weather to possibly +50 °C if a steam leak occurs; the measured e.m.f. is therefore meaningless. Installation (b) with the meter located in the control room and connected to the thermocouple with copper leads is equally useless — the reference junction is still located outside the pipe.

In installation (c), the thermocouple is extended to the control room using extension or compensation leads made of chromel and alumel. This is an improvement because the reference junction is now in the control room where the variation in ambient temperature is smaller, possibly 10 °C at most. However, this is still not satisfactory for most applications, and one obvious solution is to place the reference in a temperature-controlled environment, e.g. a refrigerator at 0 °C.

An alternative solution which utilises the law of intermediate temperatures is shown in installation (d). The thermocouple e.m.f. is E_{T_1,T_2} for a measured junction temperature of T_1 °C and a reference junction temperature of T_2 °C (T_2 around 20 °C). If we introduce a second source of e.m.f. of magnitude $E_{T_2,0}$ into the circuit in series with E_{T_1,T_2}, then the voltmeter measures $E_{T_1,T_2} + E_{T_2,0}$ which is equal to $E_{T_1,0}$. Thus the voltmeter measures an e.m.f. relative to an apparent fixed reference temperature of 0 °C, even though the actual reference temperature is varying about a mean of 20 °C.

The e.m.f. source producing $E_{T_2,0}$ is known as an automatic reference junction compensation circuit. From eqn [8.40] we have:

$$E_{T_2,0} = a_1T_2 + a_2T_2^2 + a_3T_2^3 + \cdots \qquad [8.43]$$

but since T_2 is small, we can approximate by $E_{T_2,0} \approx a_1T_2$. Thus we require a circuit giving a millivolt output signal proportional to reference temperature T_2. This can be obtained with a metal resistance thermometer incorporated into a deflection bridge circuit, with a large value of R_3/R_2 (Section 9.1). The output voltage of the bridge must be equal to $E_{T_2,0}$, so that using eqn [9.15] we require:

$$E_{T_2,0} = a_1T_2 = V_S \frac{R_2}{R_3} \alpha T_2$$

i.e.

Reference junction compensation bridge

$$V_S \frac{R_2}{R_3} = \frac{a_1}{\alpha} \qquad [8.44]$$

Thus any change in T_2 which causes the thermocouple e.m.f. to alter is sensed by the metal resistance thermometer, producing a compensating change in bridge output voltage. The thermocouple signal $E_{T_1,0}$ is at a low level, typically a few millivolts, and often requires amplification prior to processing and presentation. The open-loop temperature transmitter described in Section 9.4.2, is often used to convert a thermocouple e.m.f. to a current signal in a standard range, e.g. 4 to 20 mA. From eqn [8.40]

$$E_{T_1,0} = a_1T_1 + a_2T_1^2 + a_3T_1^3 + a_4T_1^4 + \cdots \qquad [8.45]$$

so in order to find an accurate estimate of T_1 from $E_{T_1,0}$, the above non-linear equation must be solved. This is an obvious application for a microcomputer (Section 3.3 and Chapter 10).

Table 8.2 Thermocouple data and characteristics (e.m.f. values are after British Standards Institution B.S. 4937, 1974[5])

Type	Temperature ranges (°C)	E.M.F. values (μV)	Tolerances	Extension leads	Characteristics
Iron v. constantan Type J	16 s.w.g. 0–500	$E_{100,0}$ = 5 268 $E_{200,0}$ = 10 777	0 to 300°C ±3°C	as for thermocouple	Oxidising and reducing atmospheres have little effect
	10 s.w.g 0–600	$E_{300,0}$ = 16 325 $E_{500,0}$ = 27 388	above 300°C ±1%		Should be protected from moisture, oxygen and sulphur bearing gases
Copper v. constantan Type T	21 s.w.g. −100 to +400	$E_{-100,0}$ = −3 378 $E_{+100,0}$ = 4 277	0 to 100°C ±1°C	as for thermocouple	Recommended for low and sub-zero temperatures
		$E_{200,0}$ = 9 286 $E_{400,0}$ = 20 869	above 100°C ±1%		Resists oxidising and reducing atmospheres up to approx. 350°C Requires protection from acid fumes
Nickel–Chromium v. Nickel–Aluminium Chromel v. Alumel	16 s.w.g 0–950	$E_{100,0}$ = 4 095 $E_{250,0}$ = 10 151	0 to 400°C ±3°C	as for thermocouple	Recommended for oxidising and neutral conditions
Type K	10 s.w.g. 0–1150	$E_{500,0}$ = 20 640 $E_{1000,0}$ = 41 269	above 400°C ±0.75%		Rapidly contaminated in sulphurous atmospheres Not suitable for reducing atmospheres
Platinum v. Platinum–13% rhodium Type R	0–1400 (depending on sheath)	$E_{300,0}$ = 2 400 $E_{600,0}$ = 5 582	0 to 1100°C ±1°C	Copper Copper–Nickel	Rapidly poisoned by reducing atmospheres.
		$E_{900,0}$ = 9 203 $E_{1200,0}$ = 13 224	1100 to 1400°C ±2°C		Particularly susceptible to many metal vapours, therefore important that non-metal sheaths are used

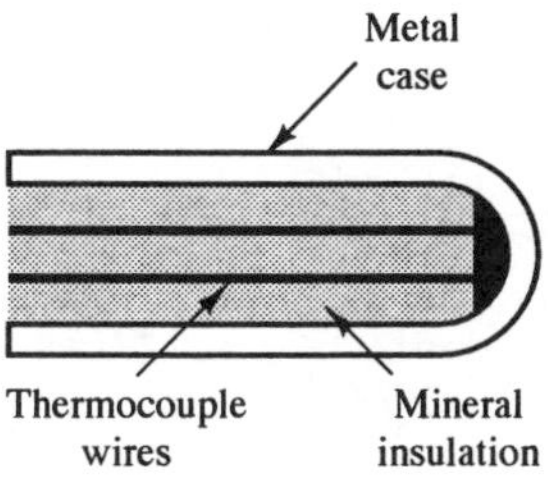

Fig. 8.11 Mineral insulated thermocouple

Table 8.2 summarises the measurement range, e.m.f. values, tolerances and characteristics of four thermocouples in common industrial use. The table can be used to quantify non-linearity: for example a copper v. constantan thermocouple, used between 0 and 400 °C, has an e.m.f. of 9286 μV at 200 °C compared with an ideal straight line value of 10 435 μV. Thus the non-linearity at 200 °C is -1149 μV or -5.5% of f.s.d. Typical tolerances are of the order of $\pm 1\%$, i.e. about 10 times greater than for platinum resistance thermometers. For base metal thermocouples (types J, T, K), the extension or compensation leads are made of the same metals as the thermocouple itself. For rare metal thermocouples (e.g. type R), the metals in the extension leads are copper and copper–nickel which have similar thermoelectric properties to platinum and platinum–rhodium but are far cheaper. In order to give mechanical and chemical protection, the thermocouple is often enclosed in a thermowell or sheath (Fig. 14.1). An alternative is the mineral insulated thermocouple (Fig. 8.11). This is a complete package, where the space between the thermocouple wires and metal case if filled with material which is both a good heat conductor and an electrical insulator. The dynamic characteristics of both types of installation are discussed in Section 14.3.

8.6 Elastic sensing elements

If a force is applied to a spring, then the amount of extension or compression of the spring is approximately proportional to the applied force. This is the principle of elastic sensing elements which convert an input force into an output displacement. Elastic elements are also commonly used for measuring torque, pressure and acceleration which are related to force by the equations:

$$\text{torque} = \text{force} \times \text{distance}$$

$$\text{pressure} = \frac{\text{force}}{\text{area}} \qquad [8.46]$$

$$\text{acceleration} = \frac{\text{force}}{\text{mass}}$$

In a measurement system an elastic element will be followed by a suitable secondary displacement sensor, e.g. potentiometer, strain gauge, LVDT, which converts displacement into an electrical signal. The displacement may be translational or rotational.

Elastic sensing elements have associated mass (inertance) and damping (resistance) as well as spring characteristics. In Section 4.1.2 the dynamics of a mass–spring–damper force sensor were analysed, and shown to be represented by the second-order transfer function:

$$G(s) = \frac{1}{\dfrac{1}{\omega_n^2}s^2 + \dfrac{2\xi}{\omega_n}s + 1}$$

Figure 8.12 shows dynamic models of elastic elements for measuring linear acceleration, torque, pressure and angular acceleration. The dynamics of all four elements are represented by second-order differential equations and transfer functions.

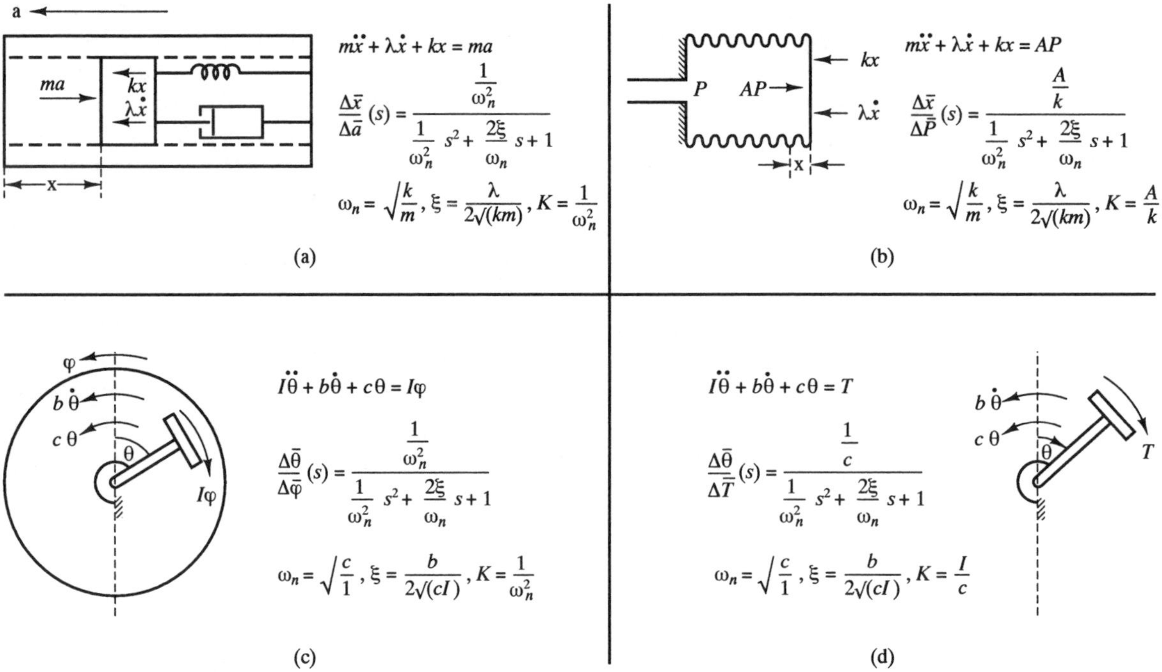

Fig. 8.12 Dynamic models of elastic elements
(a) Linear accelerometer
(b) Pressure sensor
(c) Angular accelerometer
(d) Torque sensor

The figure gives the values of steady-state gain K, natural frequency ω_n and damping ratio ξ for each type of sensor. The differential equation for a torque sensor is an analogue of equation [4.14] for a force sensor; the torque equation involves rotational motion whereas the force equation involves translational motion. Here an input torque T is opposed by the spring torque $c\theta$ and damping torque $\beta\dot{\theta}$, the resultant unbalanced torque is equal to the product of moment of inertia I and angular acceleration $\ddot{\theta}$. In the pressure sensor, the input pressure P produces a force AP over the area A of the bellows which is opposed by the bellows spring force kx and damping force $\lambda\dot{x}$.

The conceptual model of an accelerometer, shown in Fig. 8.12, is a casing containing a 'seismic' mass which moves on frictionless guide rails against a spring. If the casing is given an acceleration a, then the mass m experiences an inertia force ma in an opposite direction. This is the same force a car driver experiences when he is thrown back against his seat as the car accelerates, or thrown forwards during sudden braking. This force is opposed by the spring force kx, where x is the displacement of the mass relative to the casing. Under steady conditions at $t = 0-$ these two forces are in equilibrium, thus:

$$ma_{0-} = kx_{0-} \qquad [8.47]$$

i.e. steady-state sensitivity

$$K = \frac{x_{0-}}{a_{0-}} = \frac{m}{k} = \frac{1}{\omega_n^2}$$

If the input acceleration is suddenly increased to a at time $t = 0-$, then the forces acting on the mass are no longer in equilibrium and the mass moves relative to the casing i.e. velocity $\dot{x}$ and acceleration $\ddot{x}$ are non zero. The resultant unbalanced force is $ma - kx - \lambda\dot{x}$, giving:

$$ma - kx - \lambda\dot{x} = m\ddot{x}$$

i.e. $$m\ddot{x} + \lambda\dot{x} + kx = ma \qquad [8.48]$$

where we note that the acceleration $\ddot{x}$ of the mass relative to the casing is completely different from the acceleration a of the casing itself. By defining deviation variables Δx, Δa, we can use eqns [8.47] and [8.48] to derive the element transfer function (see Section 4.1). The differential equation for an angular acceleration sensor is the rotational analogue of [8.48]; here the inertia torque $I\phi$, due to the input acceleration ϕ, is opposed by spring torque $c\theta$ and damping torque $b\dot{\theta}$.

In Chapter 4 we saw that the optimum value of damping ratio ξ is 0.7; this ensures minimum settling time for step response and $|G(j\omega)|$ closest to unity for the frequency response. Many practical force sensors and accelerometers incorporate liquid or electromagnetic damping in order to achieve this value. However, elements such as diaphragm or bellows pressure sensors with only air damping have a very low damping ratio, typically $\xi = 0.1$ or less. In this case, the natural frequency ω_n of the sensor must be several times greater than the highest signal frequency ω_{MAX}, in order to limit the variation in $|G(j\omega)|$ to a few per cent.

Suppose we wish to measure pressure fluctuations containing frequencies up to 10 Hz, keeping the dynamic error within ±2%, using an elastic pressure element with $\xi = 0.1$. We thus require:

$$0.98 \le |G(j\omega)| \le 1.02 \quad \text{for} \quad \omega \le 62.8 \qquad [8.49]$$

Since ξ is small and $\omega < \omega_n$, eqn [4.40] for $|G(j\omega)|$ can be approximated to:

$$|G(j\omega)| \approx \frac{1}{1 - \omega^2/\omega_n^2} \quad (\xi \text{ small}, \ \omega/\omega_n \ll 1) \qquad [8.50]$$

In this situation $|G(j\omega)|$ cannot be less than unity, so that the above condition simplifies to:

$$\frac{1}{1 - \omega^2/\omega_n^2} \le 1.02 \quad \text{for} \quad \omega \le 62.8 \qquad [8.51]$$

thus

$$\frac{\omega}{\omega_n} \le 0.14 \quad \text{for} \quad \omega \le 62.8, \quad \text{i.e.} \quad \omega_n \ge 450$$

Since $\omega_n = \sqrt{(k/m)}$, ω_n can be increased by increasing the ratio k/m, giving a high stiffness, low mass sensor. However, increasing k reduces the steady-state sensitivity $K = A/k$ of the pressure element, so that a more sensitive secondary displacement sensor is required. The design of any elastic sensing element is therefore a compromise between the conflicting requirements of high steady-state sensitivity and high natural frequency.

Figure 8.13 shows four practical elastic sensing elements. All four elements are fairly stiff, i.e. k and ω_n are high, but steady-state sensitivity K and displacement x are small, so that strain gauges are used as secondary displacement sensors. In

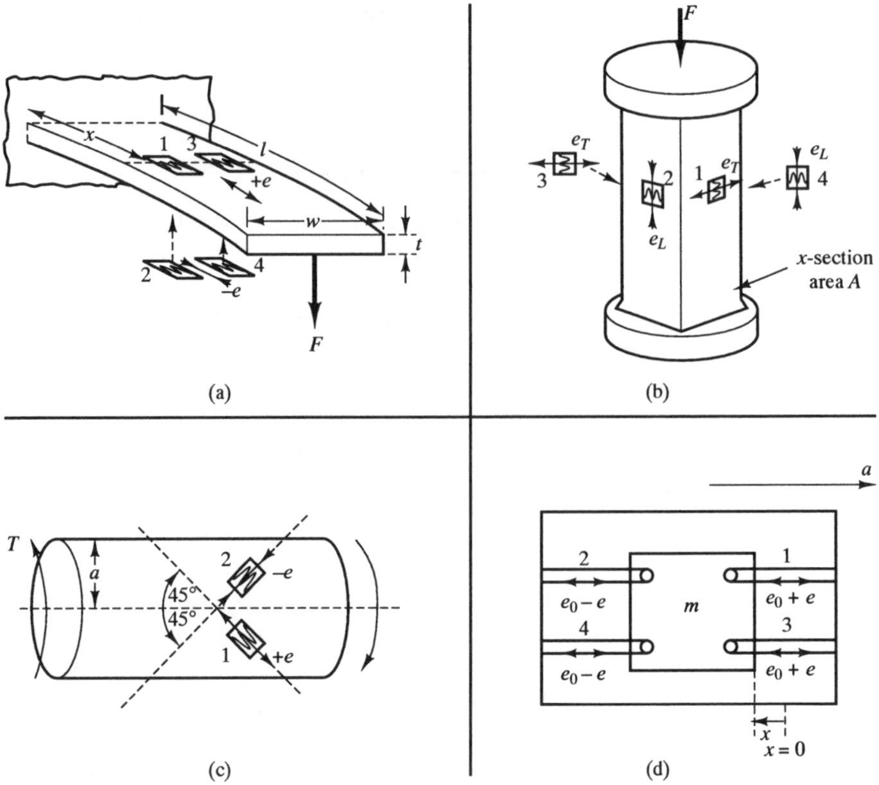

Fig. 8.13 Practical elastic sensing elements using strain gauges
(a) Cantilever load cell
(b) Pillar load cell
(c) Torque sensor
(d) Unbonded strain gauge accelerometer

the cantilever force element or load cell, the applied force F causes the cantilever to bend so that the top surface experiences a tensile strain $+e$ and the bottom surface an equal compressive strain $-e$. The magnitude of strain e is given by:

$$e = \frac{6(l - x)}{wt^2E} F \qquad [8.52]$$

where E is Young's modulus for the cantilever material and the other quantities are defined in Fig. 8.13.

Strain gauges 1 and 3 sense a tensile strain $+e$ so that their resistance increases by ΔR. Gauges 2 and 4 sense a compressive strain $-e$ so that their resistance decreases by an equal amount. From eqn [8.10],

$$\Delta R = GR_0e \qquad [8.53]$$

where G is the gauge factor and R_0 the unstrained resistance of the gauges. This gives:

$$\begin{aligned} R_1 &= R_3 = R_0 + \Delta R = R_0(1 + Ge) \\ R_2 &= R_4 = R_0 - \Delta R = R_0(1 - Ge) \end{aligned} \qquad [8.54]$$

The four gauges are connected into a deflection bridge (see Fig. 9.4(b) and eqn [9.18]).

In the pillar load cell the applied force F causes a compressive stress $-F/A$, where

A is the cross sectional area of the pillar. This produces a longitudinal, compressive strain:

$$e_L = -\frac{F}{AE} \qquad [8.55]$$

which is accompanied by a transverse, tensile strain (eqn 8.5):

$$e_T = -\nu e_L = \frac{+\nu F}{AE} \qquad [8.56]$$

where E and ν are respectively Young's modulus and Poisson's ratio for the pillar material. Strain gauges are bonded onto the pillar so that gauges 1 and 3 sense e_T and gauges 2 and 4 sense e_L; thus

$$\begin{aligned} R_1 = R_3 &= R_0 + R_0Ge_T = R_0\left(1 + \frac{G\nu F}{AE}\right) \\ R_2 = R_4 &= R_0 + R_0Ge_L = R_0\left(1 - \frac{GF}{AE}\right) \end{aligned} \qquad [8.57]$$

The output voltage of the deflection bridge is given by eqn [9.19].

Figure 8.13(c) shows a cylindrical shaft acting as a torque sensing element. The applied torque T produces a shear strain ϕ in the shaft and corresponding linear tensile and compressive strains on the shaft surface. Gauge 1 is mounted with its active axis at $+45°$ to the shaft axis, where the tensile strain has a maximum value $+e$, and gauge 2 at $-45°$ to the shaft axis where the compressive strain has a maximum value $-e$. Gauges 3 and 4 are mounted at similar angles on the other side of the shaft and experience strains $+e$, $-e$ respectively. This maximum strain is given by:

$$e = \frac{T}{\pi S a^3} \qquad [8.58]$$

where S is the shear modulus of the shaft material and a is the radius of the shaft. The resistances of the gauges are $R_1 = R_3 = R_0(1 + Ge)$ and $R_2 = R_4 = R_0(1 - Ge)$ and the corresponding bridge output voltage is given by eqn [9.18].

A simplified diagram of a practical accelerometer using four unbonded strain gauges is shown in Fig. 8.13(d). The space between the seismic mass and casing is filled with liquid to provide damping. The unbonded strain gauges are stretched fine metal wires, which provide the spring restoring force as well as acting as secondary displacement sensors. The gauges are prestressed, so that at zero acceleration each gauge experiences a tensile strain e_0 and has a resistance $R_0(1 + Ge_0)$. If the casing is given an acceleration a, then the resultant displacement of the seismic mass m relative to the casing is (eqn [8.47]):

$$x = \frac{m}{k}a = \frac{1}{\omega_n^2}a \qquad [8.59]$$

where k is the effective stiffness of the strain gauges. Gauges 1 and 3 increase in length from L to $L + x$, and gauges 2 and 4 decrease in length from L to $L - x$. The tensile strain in gauges 1 and 3 increases to $e_0 + e$, and that in gauges 2 and 4 decreases to $e_0 - e$, where:

$$e = \frac{x}{L} = \frac{a}{\omega_n^2 L} \qquad [8.60]$$

The four gauges are connected into a deflection bridge circuit with out of balance voltage given by eqn [9.20]. In order to ensure that all four gauges are kept in tension over the whole range of movement of the mass, the maximum acceleration induced strain is only one-half of the initial strain, i.e.

$$e_{MAX} = \frac{a_{MAX}}{\omega_n^2 L} = \frac{e_0}{2} \quad \text{or} \quad a_{MAX} = \frac{e_0 L \omega_n^2}{2} \qquad [8.61]$$

Thus the acceleration input span is proportional to the square of the natural frequency. A family of accelerometers of this type, using 350 Ω gauges, cover the ranges $\pm 5g$ to $\pm 500g$ with natural frequencies between 300 and 3000 Hz and a damping ratio of 0.7 ± 0.1.[8]

Figure 8.14(a) shows elastic pressure elements in current use. The bourdon tubes (7)–(9) are characterised by low stiffness, low natural frequency but large displacement sensitivity. A typical C-shaped tube of 50 mm bending diameter has a typical displacement travel of up to 4 mm, and is used with a potentiometer displacement sensor. C-shaped tubes have an approximately rectangular cross section; are usually made from brass or phosphor–bronze and cover pressure ranges from 35 k Pa up to 100 MPa. The membranes, diaphragms and capsules (1)–(5) are far stiffer devices with higher natural frequency but lower displacement sensitivity. These elements are used with capacitive (Section 8.2) or strain gauge elements which are capable of sensing small displacements of 0.1 mm of less.

The performance equations for a flat circular diaphragm, clamped around its circumference, are summarised in Fig. 8.14(b). We see that the relation between applied pressure P and centre deflection y_c is in general non-linear, but this non-

Fig. 8.14 Elastic pressure sensing elements
(a) Different types (after Neubert[9])
(b) Flat circular diaphragm

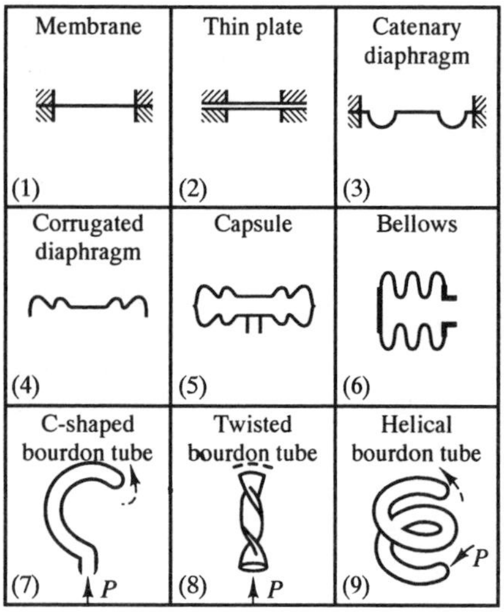

(a)

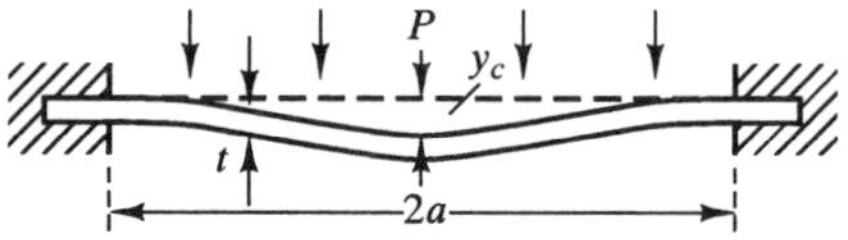

$$P = \frac{16\,Et^3}{3a^4(1-\mu^2)}\, y_c \left[1 + 0.5\left(\frac{y_C}{t}\right)^2\right] \quad \text{Accurate steady-state relation}$$

if $\left(\frac{y_C}{t}\right) \ll 1$ then

Steady-state sensitivity $K = \frac{y_C}{P} \approx \frac{3a^4(1-\mu^2)}{16\,Et^3} = \frac{1}{16\,t\gamma}$

and

Natural frequency $f_n = \frac{2.56}{\pi}\sqrt{\frac{Et^2}{3a^4(1-\mu^2)\rho}} = \frac{2.56}{\pi}\sqrt{\frac{\gamma}{\rho}}$ Hz

E = Young's modulus
μ = Poisson's ratio
ρ = density
$\gamma = \frac{Et^2}{3a^4(1-\mu^2)}$

(b)

linearity can be minimised by keeping the ratio y_c/t small. Thus if the non-linearity is to be limited to 1% maximum we require:

$$0.5\left(\frac{y_c}{t}\right)^2 \leq \frac{1}{100} \quad \text{or} \quad y_c \leq 0.14t \qquad [8.62]$$

The equations demonstrate that high steady-state sensitivity K and high natural frequency f_n cannot both be achieved. A higher sensitivity can be achieved by reducing the effective stiffness $\gamma = Et^2/[3a^4(1-\nu^2)]$, but since $f_n \alpha \sqrt{\gamma}$ this has the effect of reducing the natural frequency.

8.7 Piezoelectric sensing elements

If a force is applied to any crystal, then the crystal atoms are displaced slightly from their normal positions in the lattice. This displacement x is proportional to the applied force F: i.e., in the steady state,

$$x = \frac{1}{k}\mathrm{F} \qquad [8.63]$$

The stiffness k of the crystal is large, typically $2 \times 10^9\,\mathrm{N\,m^{-1}}$. The dynamic relation between x and F can be represented by the second-order transfer function:

$$\frac{\Delta\bar{x}}{\Delta\bar{F}}(s) = \frac{1/k}{\dfrac{1}{\omega_n^2}s^2 + \dfrac{2\xi}{\omega_n}s + 1} \qquad [8.64]$$

where $\omega_n = 2\pi f_n$ is large, typically $f_n = 10$ to 100 kHz, and ξ is small, typically $\xi \approx 0.01$.

In a piezoelectric crystal, this deformation of the crystal lattice results in the crystal acquiring a net charge q, proportional to x, i.e.

$$q = Kx \qquad [8.65]$$

Thus from [8.63] and [8.65] we have

Direct piezoelectric effect

$$q = \frac{K}{k}F = dF \qquad [8.66]$$

where $d = K/k$ coulombs N^{-1} is the charge sensitivity to force. Thus a piezoelectric crystal gives a direct electrical output, proportional to applied force, so that a secondary displacement sensor is not required. The piezoelectric effect is reversible; eqn [8.66] represents the direct effect where an applied force produces an electric charge. There is also an inverse effect where a voltage V applied to the crystal causes a mechanical displacement x, i.e.

Inverse piezoelectric effect

$$x = dV \qquad [8.67]$$

The inverse effect is important in ultrasonic transmitters (Chapter 16). The dimensions of d in eqn [8.66], i.e. coulombs N^{-1}, are identical with the dimensions $\mathrm{m\,V^{-1}}$ of d in eqn [8.67].

In order to measure the charge q, metal electrodes are deposited on opposite faces of the crystal to give a capacitor. The capacitance of a parallel plate capacitor formed from a rectangular block of crystal of thickness t (Fig. 8.5 and eqn [8.12]) is given by:

$$C_N = \frac{\epsilon_0 \epsilon A}{t} \qquad [8.68]$$

The crystal can therefore be represented as a charge generator q in parallel with a capacitance C_N, or alternatively by a Norton equivalent circuit (Section 5.1.3) consisting of a current source i_N in parallel with C_N. The magnitude of i_N is:

$$i_N = \frac{\mathrm{d}q}{\mathrm{d}t} = K\frac{\mathrm{d}x}{\mathrm{d}t} \qquad [8.69]$$

or, in transfer function form:

$$\frac{\Delta \bar{i}_N}{\Delta \bar{x}}(s) = Ks \qquad [8.70]$$

where d/dt is replaced by the Laplace operator s. We note that for a steady force F, F and x are constant with time, so that dx/dt and i_N are zero.

If the piezoelectric sensor is connected directly to a recorder (assumed to be a pure resistive load R_L) by a cable (assumed to be a pure capacitance C_C) then the complete equivalent circuit is shown in Fig. 5.11. The transfer function relating recorder voltage V_L to current i_N is shown to be:

$$\frac{\Delta \bar{V}_L(s)}{\Delta \bar{i}_N(s)} = \frac{R_L}{1 + R_L(C_N + C_C)s} \qquad [5.16]$$

The overall system transfer function relating recorder voltage V_L to input force F is:

$$\frac{\Delta \bar{V}_L}{\Delta \bar{F}}(s) = \frac{\Delta \bar{V}_L}{\Delta \bar{i}_N}\frac{\Delta \bar{i}_N}{\Delta \bar{x}}\frac{\Delta \bar{x}}{\Delta \bar{F}} \qquad [8.71]$$

From [8.64], [8.70] and [5.16] we have:

$$\frac{\Delta \bar{V}_L}{\Delta \bar{F}}(s) = \frac{R_L}{1 + R_L(C_N + C_C)s} Ks \frac{1/k}{\frac{1}{\omega_n^2}s^2 + \frac{2\xi}{\omega_n}s + 1}$$

$$= \frac{K}{k}\frac{1}{(C_N + C_C)}\frac{R_L(C_N + C_C)s}{1 + R_L(C_N + C_C)s}\frac{1}{\frac{1}{\omega_n^2}s^2 + \frac{2\xi}{\omega_n}s + 1}$$

Transfer function for basic piezoelectric force measurement system

$$\frac{\Delta \bar{V}_L}{\Delta \bar{F}}(s) = \frac{d}{(C_N + C_C)}\frac{\tau s}{(1 + \tau s)}\frac{1}{\left(\frac{1}{\omega_n^2}s^2 + \frac{2\xi}{\omega_n}s + 1\right)} \qquad [8.72]$$

where $\tau = R_L(C_N + C_C)$.

The above transfer function emphasises the two disadvantages of this basic system:

(a) The steady-state sensitivity is equal to $d/(C_N + C_C)$. Thus the system sensitivity depends on the cable capacitance C_C, i.e. on the length and type of cable.
(b) The dynamic part of the system transfer function (ignoring recorder dynamics) is:

$$G(s) = \frac{\tau s}{(\tau s + 1)} \frac{1}{\left(\frac{1}{\omega_n^2}s^2 + \frac{2\xi}{\omega_n}s + 1\right)} \qquad [8.73]$$

The second term is characteristic of all elastic elements and cannot be avoided; however, it causes no problems if the highest signal frequency ω_{MAX} is well below ω_n. The first term $\tau s/(\tau s + 1)$ means that the system cannot be used for measuring d.c. and slowly varying forces.

To illustrate this we can plot the frequency response characteristics $|G(j\omega)|$ and arg $G(j\omega)$ for a typical system (Fig. 8.15). Here

$$\text{amplitude ratio } |G(j\omega)| = \frac{\tau\omega}{\sqrt{1 + \tau^2\omega^2)}} \frac{1}{\sqrt{\left[\left(1 - \frac{\omega^2}{\omega_n^2}\right)^2 + 4\xi^2 \frac{\omega^2}{\omega_n^2}\right]}} \qquad [8.74]$$

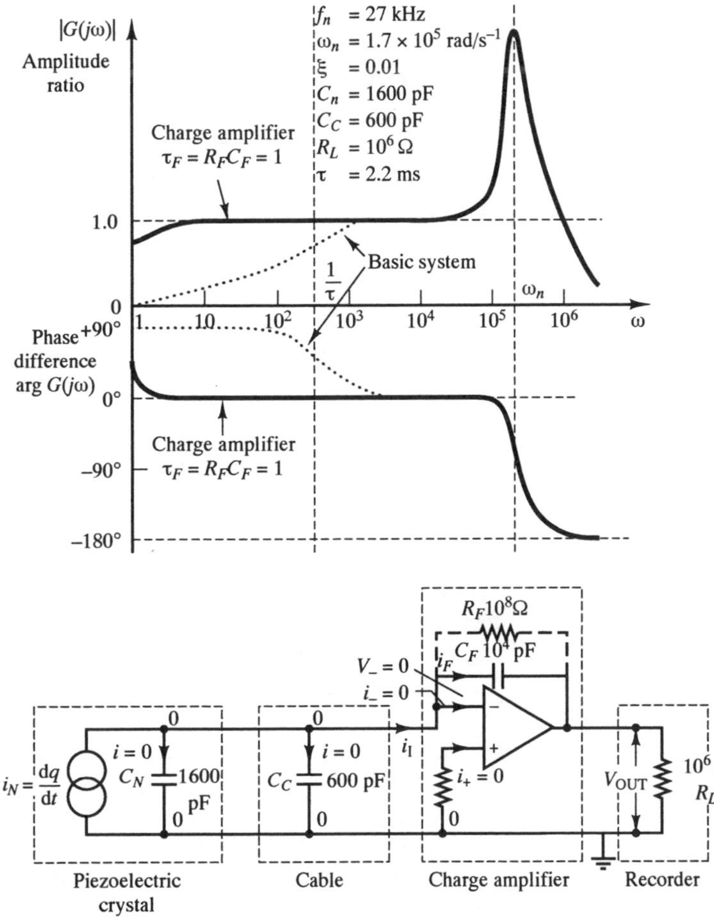

Fig. 8.15 Approximate frequency response characteristics and equivalent circuit for a piezoelectric measurement system

Phase difference

$$\arg G(j\omega) = 90° - \tan^{-1}(\omega\tau) - \tan^{-1}\left(\frac{2\xi(\omega/\omega_n)}{1 - (\omega^2/\omega_n^2)}\right).$$

The $\tau\omega/\sqrt{(1 + \tau^2\omega^2)}$ term causes a low frequency roll-off so that $|G(j\omega)| = 0$ at $\omega = 0$ and the system cannot be used for frequencies much below $1/\tau$.

Using the values of Fig. 5.11, i.e. $C_N = 1600$ pF, $C_C = 600$ pF, $R_L = 1$ MΩ, $\tau = 2.2 \times 10^{-3}$ s and $1/\tau = 455$ rad s^{-1} or 72 Hz, we find there is a useful operating range between $3/\tau$ and $0.2\omega_n$, i.e. 216 Hz and 5.4 kHz, where $0.95 \leq |G(j\omega)| \leq 1.05$ and ϕ is close to zero. This is, however, insufficient for many applications.

These disadvantages are largely overcome by introducing a charge amplifier into the system, as shown in Fig. 8.15. This is an integrator which ideally gives an output proportional to $\int i_N dt$, i.e. an output proportional to charge q, since $i_N = dq/dt$. Thus the system gives a non-zero output for a steady-force input. From Fig. 8.15 we have:

$$i_1 = i_F + i_- \qquad [8.75]$$

and the charge on feedback capacitor C_F is

$$q_F = C_F(V_- - V_{OUT}) \qquad [8.76]$$

For an ideal operational amplifier (Section 9.2) we have $i_+ = i_- = 0$ and $V_- = V_+$. In this case we have $V_- = V_+ = 0$ and $i_F = dq_F/dt$, so that:

$$i_1 = i_F = \frac{dq_F}{dt} = -C_F\frac{dV_{OUT}}{dt} \qquad [8.77]$$

Since there is zero potential drop across C_N and C_C, there is no current flow through either capacitor, so that:

$$i_1 = i_N = \frac{dq}{dt} \qquad [8.78]$$

Thus from [8.77] and [8.78] we have:

Transfer characteristics of ideal charge amplifier

$$\frac{dV_{OUT}}{dt} = -\frac{1}{C_F}\frac{dq}{dt} \quad \text{i.e.} \quad V_{OUT} = -\frac{q}{C_F} \qquad [8.79]$$

assuming that $V_{OUT} = 0$ when $q = 0$.

From eqns [8.79], [8.65] and [8.64], the overall transfer function for the force measurement system with a charge amplifier is:

Transfer function for piezoelectric system with ideal charge amplifier

$$\frac{\Delta\bar{V}_{OUT}}{\Delta\bar{F}}(s) = \frac{d}{C_F}\frac{1}{\left(\frac{1}{\omega_n^2}s^2 + \frac{2\xi}{\omega_n}s + 1\right)} \qquad [8.80]$$

The steady-state sensitivity is now d/C_F, i.e. it depends only on the capacitance C_F of the charge amplifier and is independent of transducer and cable capacitance. In the ideal case the $\tau s/(1 + \tau s)$ term is not present, so that $|G(j\omega)| = 1$ at $\omega = 0$.

Table 8.3 Properties of piezoelectric materials (adapted from Neubert, 1975[10])

	Material	Charge sensitivity d pC N^{-1}	Dielectric constant ϵ	Young's modulus $E \times 10^9$ N m^{-2}
Natural	Quartz	2.3	4.5	80
	Tourmaline	1.9,2.4*	6.6	160
Piezoelectric ceramic	Lead zirconate–titanate	265	1500	79
	Lead metaniobate	80	250	47

* Depends on direction of applied force

In practice a resistor R_F must be connected across feedback capacitor C_F to provide a path for d.c. current. This reintroduces a $\tau_F s/(1 + \tau_F s)$ term into the system transfer function, where $\tau_F = R_F C_F$. By making R_F and C_F large, the frequency response of the charge amplifier will extend down to well below 1 Hz. For example if $R_F = 10^8\,\Omega$, $C_F = 10^4$ pF, $\tau_F = 1.0$ s; then $|G(j\omega)| = 0.95$ at $\omega = 3$ rad s^{-1}, i.e. at $f \approx 0.5$ Hz.

Table 8.3 summarises the relevant properties of piezoelectric materials in common use. Crystals of materials such as quartz and tourmaline are naturally piezoelectric when properly cut, but have low sensitivities (d coefficients). Piezoelectric ceramics such as lead zirconate–titanate, with far higher d coefficients, are now preferred as sensing elements. These are man-made ceramics possessing ferroelectric properties; i.e. if an electric field is applied, and then removed, the material retains some residual polarisation and hence become piezoelectric.

Piezoelectric elements are commonly used for the measurement of acceleration and vibration. Here the piezoelectric crystal is used in conjunction with a seismic mass m. If the accelerometer casing is given an acceleration a, then there is a resulting inertia force $F = ma$ acting on the seismic mass and the piezoelectric crystal. This results in the crystal acquiring a charge $q = dF$, where d coulombs N^{-1} is the crystal charge sensitivity to force. The resulting steady-state charge sensitivity of the piezoelectric accelerometer is:

$$\frac{\Delta q}{\Delta a} = dm \quad \mathrm{C \cdot m^{-1} \cdot s^2} \qquad [8.81]$$

Usually charge sensitivity is expressed in coulomb g^{-1}, where 1 g = 9.81 metre s^{-2}, i.e.

$$\frac{\Delta q}{\Delta a} = 9.81\, dm \quad \mathrm{C\, g^{-1}} \qquad [8.82]$$

If the accelerometer is used with a charge amplifier with feedback capacitance C_F, the corresponding steady-state system voltage sensitivity is:

$$\frac{\Delta V}{\Delta a} = \frac{9.81 dm}{C_F} \quad \mathrm{V\, g^{-1}} \qquad [8.83]$$

The dynamic and frequency response characteristics of piezoelectric accelerometers, with and without charge amplifiers, are as described earlier in this section.

Figure 8.16(a) shows a compression mode accelerometer, in which a crystal disc

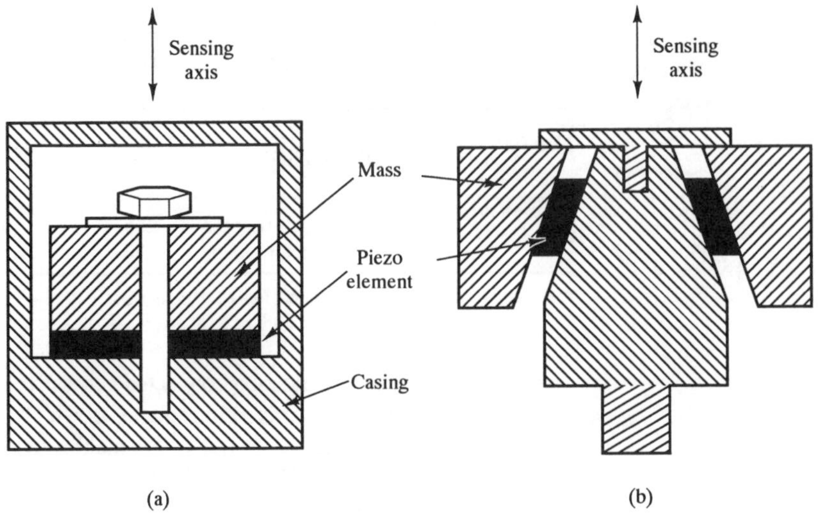

Fig. 8.16 Piezoelectric accelerometers
(a) Compression mode
(b) Shear mode (after Purdy[11])

is sandwiched between the seismic mass and the accelerometer casing. The whole is bolted together to produce a large static pre-load compression force on the crystal. Depending on direction, the acceleration to be measured causes the compression force on the crystal to be greater or less than the static pre-load value. Figure 8.16(b) shows a shear-type accelerometer; here a piezoelectric element in the shape of a truncated conical tube is sandwiched between a tapered central pillar and the outer seismic mass.[11] The acceleration to be measured causes corresponding shear forces in the piezoelectric element. A general purpose shear-mode accelerometer of this type has a charge sensitivity of 25 pC g^{-1}, a capacitance C_N of 1600 pF, a natural frequency f_n of 27 kHz and a weight of 18 g.

Consider a body, e.g. a part of a machine such as the casing of a pump or a compressor, which is executing sinusoidal vibrations with displacement amplitude $\hat{x}$ at frequency f Hz, i.e.

$$\text{displacement } x = \hat{x}\sin 2\pi ft \quad \text{m} \qquad [8.84]$$

$$\text{velocity } \dot{x} = 2\pi f\hat{x}\cos 2\pi ft \quad \text{m s}^{-1}$$

$$= \hat{v}\cos 2\pi ft, \text{ where } \hat{v} = 2\pi f\hat{x} \qquad [8.85]$$

$$\text{acceleration } \ddot{x} = -4\pi^2 f^2\hat{x}\sin 2\pi ft \quad \text{m s}^{-2}$$

$$= -\hat{a}\sin 2\pi ft, \text{ where } \hat{a} = 4\pi^2 f^2\hat{x} \qquad [8.86]$$

In the monitoring of machine vibration it is important to measure at least one of the amplitudes $\hat{x}$, $\hat{v}$, $\hat{a}$ of displacement, velocity and acceleration respectively. This is especially so where the machine has certain resonant frequencies; here the above amplitudes may be large. It is possible to use a displacement transducer, differentiate the output once to obtain velocity information and twice to obtain acceleration information. However, displacement transducers often have insufficient sensitivity, inadequate frequency response or too large a mass for these applications; moreover the operation of differentiation amplifies unwanted high frequency noise. A better solution is to use a piezoelectric accelerometer, which has sufficient sensitivity, adequate frequency response and low mass; the accelerometer output can be integrated once for velocity information and twice for displacement information. The operation of integration tends to attenuate high frequency noise.

8.8 Piezoresistive sensing elements

In Section 8.1 the piezoresistive effect was defined as the change in resistivity ρ of a material with applied mechanical strain e, and is represented by the term $(1/e)(\Delta\rho/\rho)$ in the equation for gauge factor of a strain gauge. We saw that silicon doped with small amounts of *N*- or *P*-type material exhibits a large piezoresistive effect and is used to manufacture strain gauges with high gauge factors. The traditional way of making diaphragm pressure sensors is to cement metal foil strain gauges onto the flat surface of a metal diaphragm.

In piezoresistive pressure sensors the elastic element is a flat silicon diaphragm. The distortion of the diaphragm is sensed by four piezoresistive strain elements made by introducing doping material into areas of the silicon, where the strain is greatest.

One method of introducing the doping material is **diffusion** at high temperatures; the resulting four strain gauges are connected into a deflection bridge circuit in the normal way. A typical sensor of this type has an input range of 0 to 100 kPa, a sensitivity of around 3 mV/kPa (for a 10 V bridge supply voltage), a natural frequency of 100 kHz and combined non-linearity and hysteresis of ±0.5%.[12]

Figure 8.17(a) shows a piezoresistive pressure sensor where P-type doping material is introduced into an N-type silicon diaphragm using **ion implantation** technology.[13] Four piezoresistive strain elements are thus produced (two in tension, two in compression), which are connected into a deflection bridge circuit.

Fig. 8.17 Piezoresistive pressure sensors (after Noble[13])

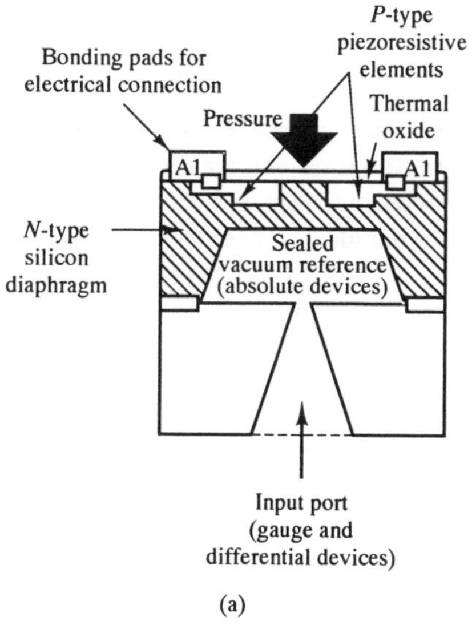

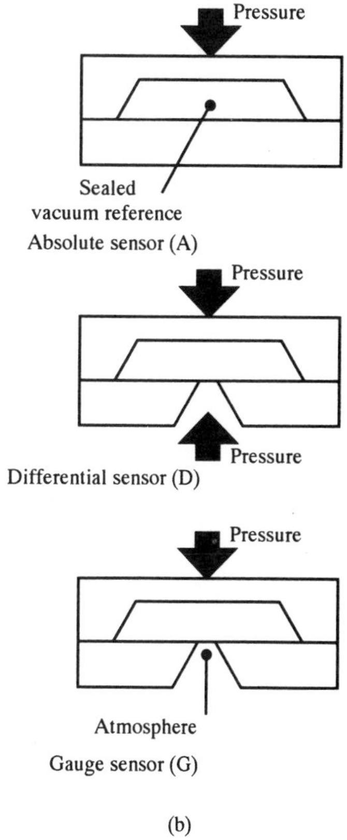

Figure 8.17(b) shows how the element may be used to measure absolute, differential and gauge pressures. A typical sensor of this type has an input range of 0 to 100 kPa, a sensitivity of around 1 mV/kPa (for a 12 V bridge supply voltage), a 10% to 90% rise time of 100 μs and typical combined non-linearity and hysteresis of $\pm 0.1\%$.

8.9 Electrochemical sensing elements

8.9.1 Ion selective electrodes

Ion selective electrodes (ISEs) are sensors which directly measure the activity or concentration of ions in solution. They could, for example, be used to measure the concentration of lead, sodium or nitrate ions in drinking water. When an ISE is immersed in a solution, a reaction takes place between the charged species in the solution and those on the sensor surface. An equilibrium is then established between these species: there is a corresponding equilibrium potential difference between the sensor and solution, which depends mainly but not entirely, on the activity of a single ion. This output signal depends also, to some extent, on the activity of other ions present in the solution, the electrodes are therefore *selective* rather than *specific*. For example, the output signal of a potassium electrode depends not only on the activity of potassium (K^+) ions in solution, but on some fraction of the sodium (Na^+) ions also present.

The Thévenin or open circuit e.m.f. of ISEs for monovalent ions is given by the modified Nernst equation:

Thévenin e.m.f. for monovalent ion ISE

$$E_{Th} = E_0 + \frac{R\theta}{F} \log_e(a_X + K_{X/Y} a_Y) \qquad [8.87]$$

where E_0 = constant depending on electrode composition
R = universal gas constant ($8.314\ JK^{-1}$)
θ = absoute temperature (K)
F = Faraday number (96 493 coulombs)
a_X = activity of ion X in solution, e.g. K^+
a_Y = activity of ion Y in solution, e.g. Na^+
$K_{X/Y}$ = selectivity coefficient of X electrode to Y ($0 < K < 1$)

The smaller the value of $K_{X/Y}$ the more selective the electrode is to X; typically $K_{K^+/Na^+} \approx 2.6 \times 10^{-3}$. The above equation is applicable to any monovalent ion, i.e. where a single electron is involved in the reaction. If n electrons are involved in the reaction then eqn [8.87] should be modified by replacing $R\theta/F$ with $R\theta/nF$.

The activity a_A of an ion A in solution depends on, but is not proportional to, the concentration C_A of that ion. The activity of an ion of particular concentration in a given solution, is a function not only of its own concentration, but also the concentration of all other ions and species. The relation between a_A and C_A is of the form:

$$a_A = f_A C_A \qquad [8.88]$$

where f_A is the activity coefficient for the ion A. In general $f_A < 1$ and is a non-linear function of C_A. At low concentrations ($< 10^{-4}$ mole litre^{-1}), $f_A \approx 1$ and $a_A \approx C_A$.

Figure 8.18(a) shows the basic system for measuring the activity or concentration of an ion in solution. Because an ion-selective electrode is only a single electrode and therefore only one half of a complete electrochemical cell, it must be used with a suitable standard reference electrode, immersed in the same test solution. Reference electrodes are electrochemical half-cells whose potential is held at a constant value by the chemical equilibrium maintained inside them. Figure 8.18(a) also shows the

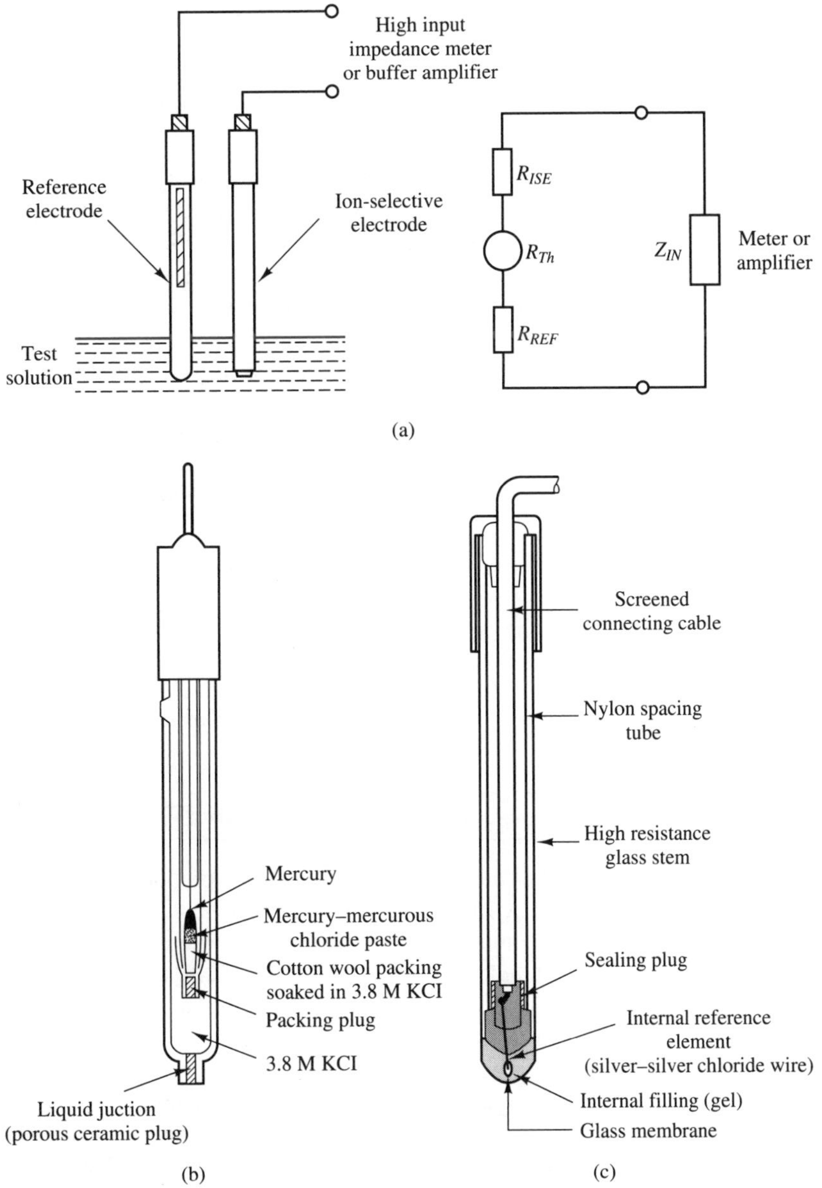

Fig. 8.18 Ion-selective electrodes
(a) Basic system for ion concentration measure and equivalent circuit
(b) Calomel reference electrode (after Bailey[14], © P.L. Bailey 1980. Reprinted by permission of John Wiley & Sons, Ltd)
(c) Practical pH sensor (after Thompson[15])

Thévenin equivalent circuit for the system; the Thévenin resistance R_{Th} is made up of the resistance of the ISE in series with the resistance of the reference electrode and can be as high as $10^9\,\Omega$. This high resistance means that the potential difference E_{Th} between ion selective and reference electrodes must be measured using either a high input impedance millivoltmeter or a buffer amplifier with $Z_{IN} \approx 10^{12}\,\Omega$ (Section 5.1). One commonly used reference electrode is the calomel electrode shown in Fig. 18.8(b).[14] This consists of a mercury–mercurous chloride (Hg/Hg_2Cl_2) electrode system which is in contact with a saturated solution of potassium chloride (KCl). The glass container has a porous ceramic plug; this provides a liquid junction, which allows ions to diffuse between the KCl and test solutions.

An important industrial application of ISEs is the measurement of the activity of hydrogen ions in solution. Acidity and alkalinity are quantified using pH where:

$$\mathrm{pH} = -\log_{10} a_{\mathrm{H}^+} = -\frac{1}{2.303}\log_e a_{\mathrm{H}^+} \qquad [8.89]$$

where a_{H^+} is the activity of the H^+ ion in solution. Hydrogen ion activity can vary from 10^0 for a strong acid, through 10^{-7} for a neutral solution (e.g. pure water), to 10^{-14} for a strong alkali. The corresponding pH values are 0, 7 and 14.

A typical pH sensor is shown in Fig. 8.18(c),[15] here the ISE is a membrane of a specially formulated glass. A thin layer (10^{-7} m thick) of hydrated silica forms on the wetted surface of the glass. An equilibrium is set up between the activities of H^+ ions in this layer and those in the test solution; an increase in acidity causes more H^+ ions to be deposited in the layer which raise its potential relative to the bulk solution. Because the H^+ ion is the only positive ion small enough to fit into the silica lattice, the glass electrode is highly selective to H^+ and the selectivity coefficient K to other positive ions can usually be assumed to be zero. The corresponding Thévenin e.m.f. for the sensor is:

$$E_{Th} = E_0 + \frac{R\theta}{F}\log_e a_{\mathrm{H}^+} \qquad [8.90]$$

Using eqn [8.89], at $\theta = 298\,\mathrm{K}$ (25 °C) we have:

$$E_{Th} = E_0 - \frac{8.314 \times 298 \times 2.303}{96\,493}\,\mathrm{pH}$$

Thévenin e.m.f. for glass ISE

$$E_{Th} = E_0 - 0.0592\,\mathrm{pH}\quad \mathrm{V} \qquad [8.91]$$

Thus the sensor output voltage is proportional to pH with a sensitivity of –59.2 mV/pH at 25°C. The constant E_0 depends on the construction of the glass electrode and the choice of the reference electrode; the sensor shown has an internal reference electrode consisting of a silver wire coated with silver chloride immersed in a potassium chloride/silver gel. The Thévenin resistance R_{Th} of the sensor is made up of the glass electrode resistance (100 to 300 MΩ) in series with the reference electrode resistance (10 kΩ).[15]

Table 8.4 gives the type and concentration range of 19 ISEs in common use.[16] Many of these are **solid-state electrodes**, where the membrane is either a single crystal or a compacted disc of a highly insoluble inorganic salt. Other ISEs use **neutral**

Table 8.4 Ion-selective electrodes (adapted from Ref. [16])

Model No.	Electrode	Type	Concentration range
ISE 301	Chloride	Solid-state	$10^0 - 5 \times 10^{-5}$ M
ISE 302	Bromide	Solid-state	$10^0 - 5 \times 10^{-6}$ M
ISE 303	Iodide	Solid-state	$10^0 - 10^{-6}$ M
ISE 304	Cyanide	Solid-state	$10^{-2} - 10^{-6}$ M
ISE 305	Sulphide	Solid-state	$10^0 - 10^{-7}$ M
ISE 306	Copper	Solid-state	$10^0 - 10^{-6}$ M
ISE 307	Lead	Solid-state	$10^{-1} - 10^{-4}$ M
ISE 308	Silver	Solid-state	$10^0 - 10^{-6}$ M
ISE 309	Cadmium	Solid-state	$10^{-1} - 10^{-5}$ M
ISE 310	Calcium	PVC Membrane	$10^0 - 10^{-5}$ M
ISE 311	Nitrate	PVC Membrane	$10^0 - 10^{-5}$ M
ISE 312	Barium	PVC Membrane	$10^0 - 10^{-4}$ M
ISE 313	Water Hardness	PVC Membrane	$10^0 - 10^{-4}$ M (as $Ca^{2+} + Mg^{2+}$)
ISE 314	Potassium	PVC Membrane	$10^0 - 10^{-5}$ M
ISE 315	Sodium	PVC Membrane	$10^{-1} - 10^{-5}$ M
ISE 321	Ammonia	Gas-permeable Membrane	$10^{-1} - 10^{-5}$ M
ISE 322	Sulphur Dioxide	Gas-permeable Membrane	$5 \times 10^{-2} - 5 \times 10^{-5}$ M
ISE 323	Carbon Dioxide	Gas-permeable Membrane	$10^{-2} - 10^{-4}$ M
ISE 324	Nitrogen Oxides	Gas-permeable Membrane	$10^{-2} - 5 \times 10^{-6}$ M

carriers as the active material, these are non-ionic organic molecules which are incorporated into an inert matrix of PVC or silicone rubber. ISEs for sensing soluble gases such as ammonia and sulphur dioxide are also available. The gas diffuses through a **gas-permeable membrane** and then dissolves in an internal filling solution; the resulting change in the activity of a given ion in the solution (e.g. H^+) is measured using an ion selective and a reference electrode in the usual way.

8.9.2 Solid-state gas sensors

Some solid-state materials give an electrochemical response to certain gases; an example is zirconium oxide which is sensitive to the presence of oxygen in a gas mixture.[17] The sensor consists of a small hollow cone of the oxide, held at a constant temperature of 640 °C, and coated on both inside and outside with porous platinum. Because of vacancies in the crystal lattice, zirconium oxide acts as an electrolytic conductor at high temperatures. This means, that if the partial pressures of oxygen on the inside and outside of the cone are different, then oxygen ions migrate through the sensor to produce a net voltage across the platinum electrodes. The Thévenin e.m.f. on the sensor is given by the Nernst equation:

Thévenin e.m.f. for zirconium oxide oxygen sensor

$$E_{Th} = E_0 + \frac{R\theta}{nF} \log_e \left(\frac{P_{O_2}^{REF}}{P_{O_2}^{SAM}} \right) \qquad [8.92]$$

where E_0 = constant

n = number of electrons involved in reaction (4)

$P_{O_2}^{REF}$ = partial pressure of oxygen in reference gas (cone inside)

$P_{O_2}^{SAM}$ = partial pressure of oxygen in sample gas (cone outside)

Provided both the sensor temperature θ and the partial pressure $P_{O_2}^{REF}$ in the air inside the sensor are held constant, then E_{Th} depends only on the partial pressure of oxygen in the sample gas outside the sensor. Typically, the sensitivity of the sensor is 60 mV/%O_2 at 640 °C.

Conclusion

This chapter has discussed the principles, characteristics and applications of a wide range of sensing elements.

References

8.1 BS 1904: 1964, *Industrial Platinum Resistance Thermometer Elements*, British Standards Institution.

8.2 Mullard Ltd. 1974 *Technical Information on Two-point NTC Thermistors.*

8.3 BENTLEY J P 1984 'Temperature sensor characteristics and measurement system design', *J. Phys. E: Scientific Instruments*, Vol. 17, pp. 430–9.

8.4 NEUBERT H K P 1975 *Instrument Transducers: An Introduction to their Performance and Design* (2nd edn), Oxford University Press, London, pp. 237–8.

8.5 BS 4937: 1974 *International Thermocouple Reference Tables*, British Standards Institution.

8.6 DOEBELIN E O 1975 *Measurement Systems: Application and Design* (2nd Edn), McGraw-Hill, New York, pp. 520–1.

8.7 LEE-INTEGER 1985 'Advanced relative humidity sensor', *Electronic Product Review*, June.

8.8 Consolidated Electrodynamics, *Bulletin 4202B/1167 on Type 4-202 Strain Gauge Accelerometer.*

8.9 NEUBERT H K P *Instrument Transducers*, p. 56 (see Ref. 4).

8.10 NEUBERT H K P *Instrument Transducers*, pp. 258, 265 (see Ref. 4).

8.11 PURDY D 1981 'Piezoelectric devices. A step nearer problem-free vibration measurement', *Transducer Technology*, Vol. 4, No. 1.

8.12 Endevco 1980 *Product Development News*, Vol. 16, issue 3.

8.13 NOBLE M 1985 'IC sensors boost potential of measurement systems', *Transducer Technology*, Vol. 8, No. 4.

8.14 BAILEY P L 1976 *Analysis with Ion Selective Electrodes*, Heyden.

8.15 THOMPSON W 'pH facts — the glass electrode', *Kent Technical Review*, pp. 16–22.

8.16 E.D.T. Research 1984 Technical Data on Ion Selective Electrodes.

8.17 Sirius Instruments Ltd 1983 in *Technical Bulletin on Zirconia Oxygen Analysers.*

Problems

8.1 A platinum resistance sensor is to be used to measure temperatures between 0 and 200 °C.

Given that the resistance $R_T\,\Omega$ at T °C is given by $R_T = R_0(1 + \alpha T + \beta T^2)$ and $R_0 = 100.0$, $R_{100} = 138.50$, $R_{200} = 175.83\,\Omega$ calculate:

(a) the values of α and β;
(b) the non-linearity at 100 °C as a percentage of full-scale deflection.

8.2 Four strain gauges are bonded onto a cantilever as shown in Fig. 8.13(a). Given that the gauges are placed halfway along the cantilever and the cantilever is subject to a downward force of 0.5 *N*, use the data given below to calculate the resistance of each strain gauge.

Cantilever data		*Strain gauge data*	
Length	$l = 25$ cm	Gauge factor	$G = 2.1$
Width	$w = 6$ cm	Unstrained resistance	$R_0 = 120\,\Omega$
Thickness	$t = 3$ mm		
Young's modulus	$E = 70 \times 10^9$ Pa		

8.3 A variable dielectric capacitive displacement sensor consists of two square metal plates, side 5 cm, separated by a gap of 1 mm. A sheet of dielectric material 1 mm thick and the same area as the plates can be slid between them as shown in Fig. 8.5. Given that the dielectric constant of air is 1 and that of the dielectric material 4, calculate the capacitance of the sensor when the input displacement $x = 0.0$, 2.5, 5.0 cm.

8.4 A variable reluctance sensor consists of a core, variable air gap and an armature. The core is a steel rod of diameter 1 cm, relative permeability 100, bent to form a semi-circle of diameter 4 cm. A coil of 500 turns is wound onto the core. The armature is a steel plate of thickness 0.5 cm and relative permeability 100. Assuming the relative permeability of air = 1.0 and the permeability of free space = $4\pi \times 10^{-7}\,\mathrm{H\,m^{-1}}$, calculate the inductance of the sensor for air gaps of 1 mm and 3 mm.

8.5 By taking a central flux path, estimate the inductance of the sensor shown in Fig. Prob. 5

(a) for zero air gap
(b) for a 2 mm air gap.

Assume the relative permeability of core and armature is 10^4 and that of air is unity.

Fig. Prob. 5

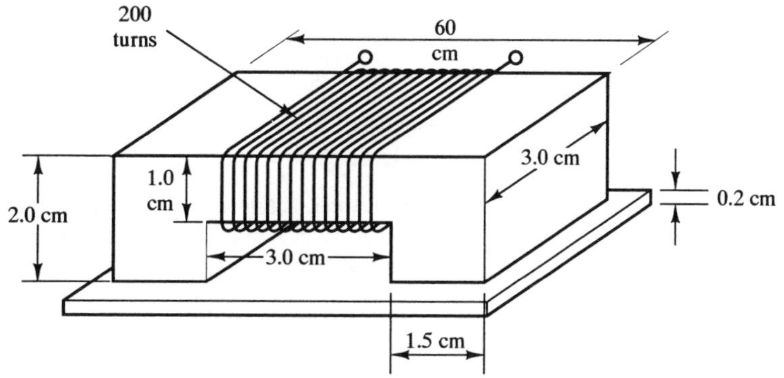

8.6 A variable reluctance tachogenerator consists of a ferromagnetic gear wheel with 22 teeth rotating close to a magnet and coil assembly. The total flux *N* linked by the coil is given by:

$$N(\theta) = 4.0 + 1.5 \cos 22\theta \text{ milliwebers}$$

where θ is the angular position of the wheel relative to the axis of the magnet. Calculate the amplitude and frequency of the output signal when the angular velocity of the wheel if 1000 and 10 000 r.p.m.

8.7 An iron v. constantan thermocouple is to be used to measure temperatures between 0 and 300 °C. The e.m.f. values are as given in Table 8.2.

(a) Find the non-linearity at 100 °C and 200 °C as a percentage of full scale.
(b) Between 100 and 300 °C the thermocouple e.m.f. is given by $E_{T,0} = a_1T + a_2T^2$; calculate a_1 and a_2.
(c) The e.m.f. is 12 500 μV relative to a reference junction of 20 °C and the corresponding reference junction circuit voltage is 1000 μV. Use the result of (b) to estimate the measured junction temperature.

8.8 An accelerometer consisting of an elastic element and a potentiometric displacement sensor has to meet the following specification:

Input range = 0 to $5g$	Damping ratio = 0.8
Output range = 0 to 10 V	Maximum non linearity = 2% of f.s.d.
Natural frequency = 10 Hz	Seismic mass = 0.005 kg.

(a) Calculate the required spring stiffness and damping constant.
(b) What should the input displacement range of the potentiometer be? (g = 9.81 m s^{-2}).
(c) If the accelerometer is to be used with a recorder of 10 kΩ resistance, what is the maximum allowable potentiometer resistance?

8.9 An accelerometer is to measure the angular acceleration of a rotating mixing vessel. The angular position of the vessel varies sinusoidally with time with amplitude 2.5 rad and period 2 s. The rotating seismic mass is equivalent to a mass of 0.1 kg on a weightless arm of length 5 cm. The stiffness of the spring is 2.5×10^{-2} Nm rad^{-1} and the damping ratio is $1/\sqrt{2}$. The angular position of the seismic mass is measured with a secondary potentiometric sensor. What should the input range of this sensor be?

8.10 A flat circular diaphragm of density 6×10^3 kg m^{-3} is to be used as a pressure sensor. The element should fulfil the following specification:

Input range = 0 to 10^4 Pa
maximum non-linearity = 1% f.s.d
Amplitude ratio to be flat with ±3% up to 100 Hz

Using the equations given in Fig. 8.14, and assuming a damping ratio of 0.01, calculate:

(a) The thickness t of the diaphragm;
(b) The output displacement range of the sensor.

8.11 A piezoelectric crystal, acting as a force sensor, is connected by a short cable of negligible capacitance and resistance to a voltage detector of infinite bandwidth and purely resistive impedance of 10 MΩ.

(a) Use the crystal data below to calculate the system transfer function and to sketch the approximate frequency response characteristics of the system.
(b) The time variation in the thrust of an engine is a square wave of period 10 ms. Explain carefully, but without performing detailed calculations, why the above system is unsuitable for this application.
(c) A charge amplifier with feedback capacitance C_F = 1000 pF and feedback resistance R_F = 100 MΩ is incorporated into the system. By sketching the frequency response characteristics of the modified system, explain why it is suitable for the application of part (b).

Crystal Data	Charge sensitivity to force	= 2 pC N^{-1}
	Capacitance	= 100 pF
	Natural frequency	= 37 kHz
	Damping ratio	= 0.01

8.12 (a) The casing of a compressor is executing sinusoidal vibrations with a displacement amplitude of 10^{-4} m and frequency 500 Hz. Calculate the amplitude of the acceleration of the casing in units of g(g = 9.81 m s^{-2}).

(b) A piezoelectric crystal accelerometer has a steady-state sensitivity of 2.0 pC/g, a natural frequency of 20 kHz, a damping ratio of 0.1 and a capacitance of 100 pF. It is connected to an oscilloscope (a resistive load of 10 MΩ) by a cable of capacitance 100 pF.

(i) without performing detailed calculations, sketch magnitude and phase frequency response characteristics for the accelerometer.
Using these characteristics:

(ii) Estimate the amplitude of the voltage measured by the oscilloscope when the accelerometer is mounted on the casing described in (a).

(iii) Explain why this system is unsuitable for measuring vibrations at 20 Hz. How should the system be modified to make these measurements possible?

9

Signal conditioning elements

As stated in Chapter 1, signal conditioning elements convert the output of sensing elements into a form suitable for further processing. This form is usually a d.c. voltage, d.c. current or a variable frequency a.c. voltage.

9.1 Deflection bridges

Deflection bridges are used to convert the output of resistive, capacitive and inductive sensors into a voltage signal.

9.1.1 Thévenin equivalent circuit for a deflection bridge

In Chapter 5 we saw that any linear network can be represented by a Thévenin equivalent circuit consisting of a voltage source E_{Th} together with a series impedance Z_{Th}. Figure 9.1 shows a general deflection bridge network. E_{Th} is the open-circuit output voltage of the bridge, i.e. when current i in $BD = 0$. Using Kirchoff's laws:

$$\text{Loop } PABCQ \quad V_S = i_1 Z_2 + i_1 Z_3 \quad \text{i.e.} \quad i_1 = \frac{V_S}{Z_2 + Z_3} \qquad [9.1]$$

Fig. 9.1 Calculation of Thévenin equivalent circuit for a deflection bridge

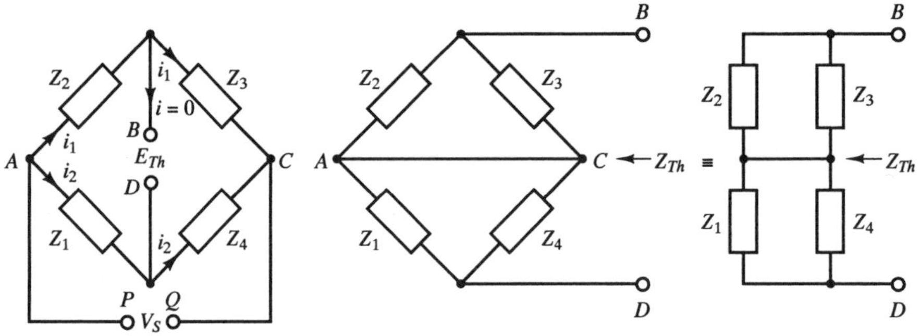

$$\text{Loop } PADCQ \quad V_S = i_2 Z_1 + i_2 Z_4 \quad \text{i.e.} \quad i_2 = \frac{V_S}{Z_1 + Z_4} \qquad [9.2]$$

Assuming Q is at earth potential, then:

$$\begin{aligned} \text{Potential at } P \text{ and } A &= V_S \\ \text{Potential at } B &= V_S - i_1 Z_2 \\ \text{Potential at } D &= V_S - i_2 Z_1 \end{aligned}$$

E_{Th} is equal to the potential difference between B and D, i.e.

$$\begin{aligned} E_{Th} &= (V_S - i_1 Z_2) - (V_S - i_2 Z_1) \\ &= i_2 Z_1 - i_1 Z_2 \end{aligned}$$

Using [9.1] and [9.2] we have

Thévenin voltage for general deflection bridge

$$E_{Th} = V_S \left\{ \frac{Z_1}{Z_1 + Z_4} - \frac{Z_2}{Z_2 + Z_3} \right\} \qquad [9.3]$$

Z_{Th} is the impedance, looking back into the circuit, between the output terminals BD, when the supply voltage V_S is replaced by its internal impedance. Assuming the internal impedance of the supply is zero, then this is equivalent to a short circuit across AC (see Fig. 9.1). We see that Z_{Th} is equal to the parallel combination of Z_2 and Z_3 in series with the parallel combination of Z_1 and Z_4, i.e.

Thévenin impedance for general deflection bridge

$$Z_{Th} = \frac{Z_2 Z_3}{Z_2 + Z_3} + \frac{Z_1 Z_4}{Z_1 + Z_4} \qquad [9.4]$$

If a load, e.g. a voltmeter or amplifier, of impedance Z_L is connected across the output terminals BD, then the current through the load is $i = E_{Th}/(Z_{Th} + Z_L)$. The corresponding voltage across the load is $V_L = E_{Th} Z_L/(Z_{Th} + Z_L)$. Thus in the limit that $|Z_L| \gg |Z_{Th}|$, $V_L \to E_{Th}$.

9.1.2 Design of resistive deflection bridges

In a resistive or Wheatstone bridge all four impedances Z_1 to Z_4 are pure resistances R_1 to R_4. From [9.3] we have

Output voltage for resistive deflection bridge

$$E_{Th} = V_S \left\{ \frac{R_1}{R_1 + R_4} - \frac{R_2}{R_2 + R_3} \right\} \qquad [9.5]$$

We first consider the case when only one of the resistances is a sensing element. Here R_1 depends on the input measured variable I, i.e. $R_1 = R_I$, and R_2, R_3, R_4 are fixed resistors. This gives

$$E_{Th} = V_S \left\{ \frac{1}{1 + R_4/R_I} - \frac{1}{1 + R_3/R_2} \right\} \qquad [9.6]$$

from which we see that to design a single element bridge we need to specify the three parameters V_S, R_4 and R_3/R_2. The individual values of R_2 and R_3 are not critical, it is their ratio which is crucial to the design. The three parameters can be specified by considering the range and linearity of the output voltage and electrical power limitations for the sensor. Thus if I_{MIN}, I_{MAX} are minimum and maximum values of the measured variable, and $R_{I_{MIN}}$, $R_{I_{MAX}}$ are the corresponding sensor resistances, then in order for the bridge output voltage to have a range from V_{MIN} to V_{MAX} the following conditions must be obeyed:

$$V_{MIN} = V_S \left\{ \frac{1}{1 + R_4/R_{I_{MIN}}} - \frac{1}{1 + R_3/R_2} \right\} \qquad [9.7]$$

$$V_{MAX} = V_S \left\{ \frac{1}{1 + R_4/R_{I_{MAX}}} - \frac{1}{1 + R_3/R_2} \right\} \qquad [9.8]$$

Often we require $V_{MIN} = 0$, i.e. the bridge to be balanced when $I = I_{MIN}$; in this case [9.7] reduces to:

Relationship between resistances in a balanced Wheatstone bridge

$$\frac{R_4}{R_{I_{MIN}}} = \frac{R_3}{R_2} \qquad [9.9]$$

A third condition is required to complete the design. One important consideration is the need to limit the electrical power $i_2^2 R_I$ in the sensor to a level which enables it to be dissipated as heat flow to the surrounding fluid; otherwise the temperature of the sensor rises above that of the surrounding fluid, thereby affecting the sensor resistance (Chapter 14). Thus if $\hat{w}$ watts is the maximum power dissipation, using [9.2] we require:

$$V_S^2 \frac{R_I}{(R_I + R_4)^2} \leq \hat{w} \qquad [9.10]$$

for $I_{MIN} \leq I \leq I_{MAX}$.

Another requirement which is often important is the need to keep the non-linearity of the overall relationship between E_{Th} and I within specified limits. Assuming $V_{MIN} = 0$, then the ideal relationship between V and I is the ideal straight line:

$$V_{IDEAL} = \frac{V_{MAX}}{I_{MAX} - I_{MIN}} I - \frac{V_{MAX}}{I_{MAX} - I_{MIN}} I_{MIN} \qquad [9.11]$$

From Chapter 2, the non-linear function $N(I) = E_{Th} - V_{IDEAL}$. Suppose we require the non-linearity as a percentage of full scale to be less than $\hat{N}$ per cent over the measurement range; we then have the condition:

$$\frac{E_{Th} - V_{IDEAL}}{V_{MAX}} \times 100 \leq \hat{N} \qquad [9.12]$$

for $I_{MIN} \leq I \leq I_{MAX}$, where E_{Th} and V_{IDEAL} are given by [9.6] and [9.11] respectively.

The value of ratio R_3/R_2 varies according to the type of resistive sensor used;

it is useful to have a graph which gives some insight into this. From [9.9] we have $R_4 = (R_3/R_2)R_{I_{MIN}}$. Substituting for R_4 in [9.6] gives:

$$\frac{E_{Th}}{V_S} = \frac{1}{1 + (R_3/R_2)(R_{I_{MIN}}/R_I)} - \frac{1}{1 + R_3/R_2}$$

or

$$v = \frac{1}{1 + r/x} - \frac{1}{1 + r} = \frac{x}{x + r} - \frac{1}{1 + r} \quad [9.13]$$

where

$$v = \frac{E_{Th}}{V_S}, \quad r = \frac{R_3}{R_2}, \quad x = \frac{R_I}{R_{I_{MIN}}}$$

Figure 9.2 shows graphs of v versus x, for x in the range 0.1 to 2.0, and for r = 0.1, 1.0, 10.0 and 100. We note that v is always zero at $x = 1$ corresponding to the bridge being balanced at $I = I_{MIN}$: also that $v(x)$ is in general non-linear, the degree of non-linearity depending on r.

For a strain gauge (Section 8.1) the change in resistance $\Delta R = R_0 G e$ is very small; this means that x is very close to 1. We require the sensitivity of the bridge to be as high as possible, i.e. $(\partial v/\partial x)_{x \approx 1}$ to be maximum. From Fig. 9.2 we see

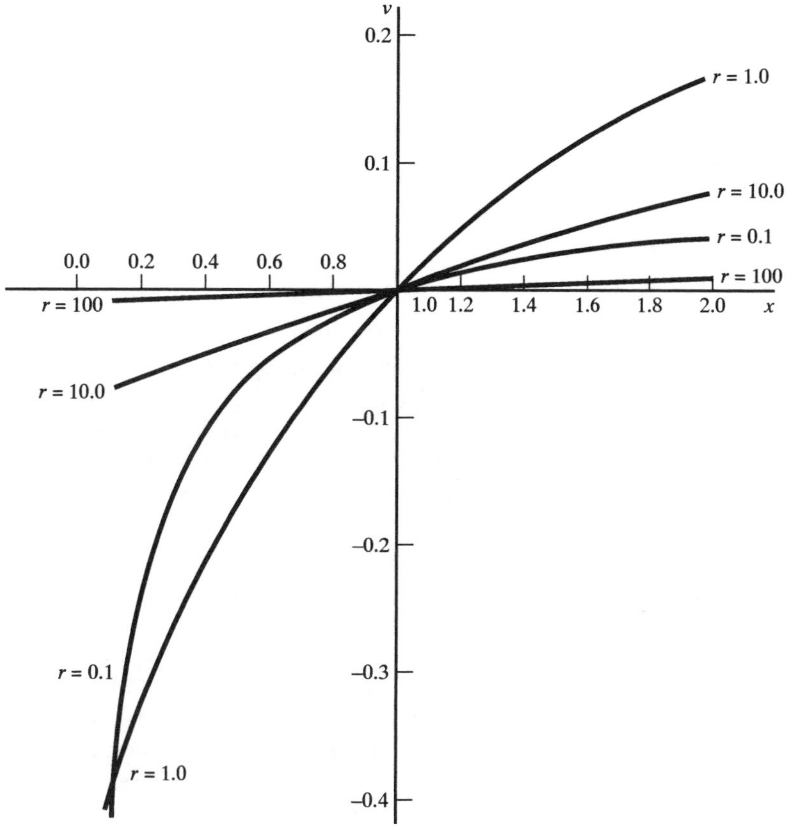

Fig. 9.2 The deflection bridge function $v(x) = x/(x+r) - 1/(1+r)$

that this occurs when $r = R_3/R_2 = 1$, i.e. $R_3 = R_2$. Using eqn [9.9], this means that $R_4 = R_{I_{MIN}} = R_0$. Thus for a bridge with a single strain gauge, we require R_2, R_3, R_4 to all equal the unstrained gauge resistance R_0. The value of V_S is determined by the power condition [9.10]. Putting $r = 1$ in [9.13] gives:

$$v(x) = \frac{x - 1}{2(x + 1)} \approx \tfrac{1}{4}(x - 1)$$

so that:

$$\frac{E_{Th}}{V_S} = \frac{1}{4}\left[\frac{R_I}{R_{I_{MIN}}} - 1\right] = \frac{1}{4}\left[\frac{R_I - R_{I_{MIN}}}{R_{I_{MIN}}}\right] = \frac{1}{4}\frac{\Delta R}{R_0} = \tfrac{1}{4}Ge$$

Thus:

Output voltage for single element strain gauge bridge

$$E_{Th} = \frac{V_S}{4}Ge \qquad [9.14]$$

i.e. the relationship between E_{Th} and e is linear.

The resistance $R_T\,\Omega$ of a metal resistance sensor, e.g. platinum, at T°C is given approximately by (Section 8.1) $R_T = R_0(1 + \alpha T)$. Typically x varies between 1 and 2 ($R_0 = 100\,\Omega$, $R_{250} \approx 200\,\Omega$). Since this device has only a small non-linearity (less than 1%) a linear bridge is required. From Fig. 9.2 we see this means a large value of r, e.g. $r \approx 100$, linearity being obtained at the expense of low sensitivity. For large r (i.e. $r \gg 1$) eqn [9.13] approximates to:

$$v \approx \frac{1}{r}(x - 1) \quad \text{i.e.} \quad \frac{E_{Th}}{V_S} = \frac{R_2}{R_3}\left(\frac{R_T}{R_{T_{MIN}}} - 1\right)$$

If $T_{MIN} = 0$ °C, then since $R_T/R_0 = 1 + \alpha T$:

Output voltage for metal resistance sensor bridge (large r)

$$E_{Th} = V_S \frac{R_2}{R_3}\alpha T \qquad [9.15]$$

i.e. a linear relationship between E_{Th} and T.

For given V_S (satisfying [9.10]), R_3/R_2 can be found from [9.15] and R_4 from [9.9]. Figure 9 Prob. 16 shows an accurate arrangement for temperature measurement involving a platinum resistance thermometer with four-lead connection to a deflection bridge. With a copper or nickel resistance sensor, this circuit provides the voltage source $E_{T_2,0}$ necessary for automatic reference junction compensation of a thermocouple e.m.f. (see Section 8.5 and Problem 9.2).

The resistance R_θ of a thermistor varies non-linearly with absolute temperature θ K according to the relationship $R_\theta = K \exp(\beta/\theta)$ (Section 8.1). A typical thermistor has a resistance of 12 kΩ at 298 K (25 °C) falling to 2 kΩ at 348 K (75 °C); i.e. x varies from 1.0 to 0.17 over this measurement range. By choosing a suitable value of r, usually between 0.25 and 0.30, it is possible to use the bridge non-linearity to partially compensate for the thermistor non-linearity; this means the overall relationship between E_{Th} and θ is reasonably linear over this range (see Figs 9.2 and 9.3). Supposing we require an output range of 0 to 1.0 V, corresponding to

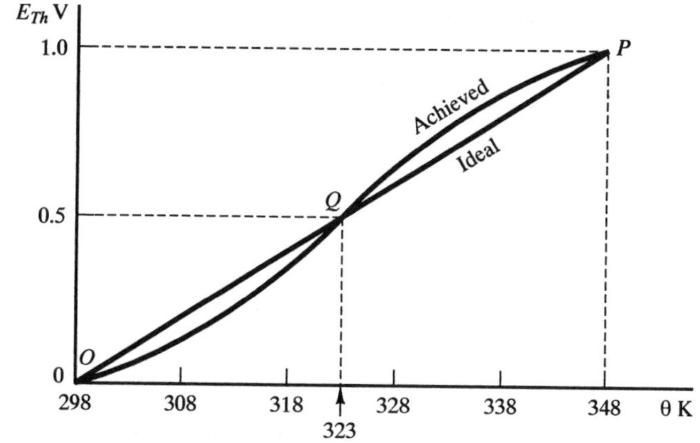

Fig. 9.3 Design conditions and characteristics for thermistor bridge

a temperature range of 298 to 348 K; then almost minimum non-linearity is obtained if we design the bridge so that $E_{Th} = 0.5$ V at $\theta = 323$ K (Fig. 9.3). The bridge is then designed by solving the following three equations for V_S, R_4 and R_3/R_2:

$$0.0 = V_S \left\{ \frac{1}{1 + R_4/R_{298}} - \frac{1}{1 + R_3/R_2} \right\} \quad \text{(point } O\text{)}$$

$$0.5 = V_S \left\{ \frac{1}{1 + R_4/R_{323}} - \frac{1}{1 + R_3/R_2} \right\} \quad \text{(point } Q\text{)} \qquad [9.16]$$

$$1.0 = V_S \left\{ \frac{1}{1 + R_4/R_{348}} - \frac{1}{1 + R_3/R_2} \right\} \quad \text{(point } P\text{)}$$

A bridge with two metal resistance thermometer elements can be designed to give an output voltage approximately proportional to temperature differences $T_1 - T_2$. The bridge incorporates one element at T_1 °C and another at T_2 °C so that $R_1 = R_0(1 + \alpha T_1)$ and $R_2 = R_0(1 + \alpha T_2)$, as shown in Fig. 9.4(a). In order to balance the bridge when $T_1 - T_2 = 0$, we require $R_4/R_1 = R_3/R_2$ when $T_1 = T_2$; this implies $R_4/R_0 = R_3/R_0$; i.e. $R_4 = R_3$. Using [9.6], we have:

$$E_{Th} = V_S \left\{ \frac{1}{1 + \frac{R_3}{R_0} \frac{1}{(1 + \alpha T_1)}} - \frac{1}{1 + \frac{R_3}{R_0} \frac{1}{(1 + \alpha T_2)}} \right\}$$

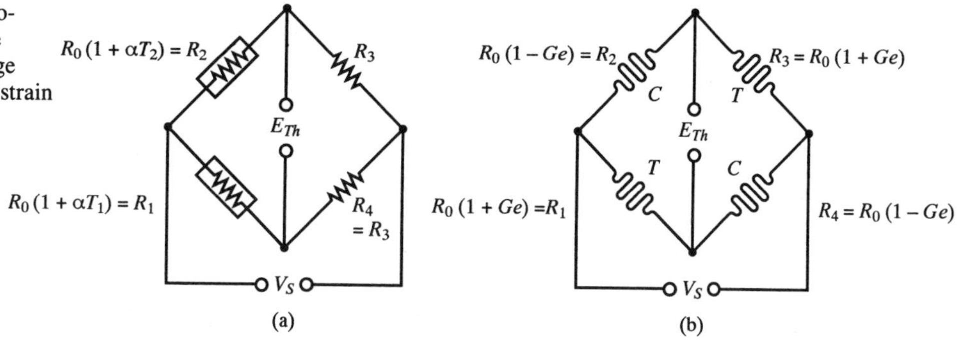

Fig. 9.4 (a) Two-element resistance thermometer bridge (b) Four-element strain gauge bridge

and if we choose R_3 so that $R_3/R_0 \gg 1$, this approximates to:

$$E_{Th} \approx V_S \left\{ \frac{(1 + \alpha T_1)}{R_3/R_0} - \frac{(1 + \alpha T_2)}{R_3/R_0} \right\}$$

i.e.

Output voltage for two element resistance thermometer bridge

$$E_{Th} = V_S \frac{R_0}{R_3} \alpha(T_1 - T_2) \qquad [9.17]$$

which has a similar form to [9.15]. This bridge has several applications, including the measurement of humidity with wet and dry bulb thermometers and as a detector in a thermal radiation measurement system (Section 15.2).

Bridges with four active strain gauges, mounted on elastic elements, are commonly used for the measurement of force, torque, acceleration and pressure. Providing that the gauges are correctly connected into the bridge, so that one opposite pair (e.g. R_1, R_3) are in tension and the other opposite pair (e.g. R_2, R_4) are in compression; then the sensitivity is greater than for a single element bridge. This bridge also compensates for changes in gauge resistance due to temperature. For metal strain gauges the effect of temperature is to multiply each gauge resistance by the factor $(1 + \alpha \mathrm{T})$; this cancels out in the output voltage eqn [9.5]. The only design decision to be made is the value of supply voltage; V_S is determined either by the maximum power condition [9.10] or by the maximum specified gauge current (Section 8.1).

For the cantilever load cell and torque element of Section 8.6 (Figs 8.13(a) and (c)), we have $R_1 = R_3 = R_0(1 + Ge)$ and $R_2 = R_4 = R_0(1 - Ge)$, where R_0 is the unstrained gauge resistance. Here

$$e = \frac{6(l - x)F}{wt^2E}$$

for the cantilever and

$$e = \frac{T}{\pi S a^3}$$

for the torque element. Using [9.5] the output voltage of the bridge (Fig. 9.4) is thus given by:

$$E_{Th} = V_S \left\{ \frac{R_0(1+Ge)}{R_0(1+Ge)+R_0(1-Ge)} - \frac{R_0(1-Ge)}{R_0(1-Ge)+R_0(1+Ge)} \right\}$$

$$= V_S \left\{ \frac{(1 + Ge)}{2} - \frac{(1 - Ge)}{2} \right\}$$

i.e.

Output voltage for cantilever and torque elements

$$E_{Th} = V_S Ge \qquad [9.18]$$

Here the output voltage is four times that of the single gauge bridge (eqn [9.14]). For the pillar load cell (Fig. 8.13(b)):

$$R_1 = R_3 = R_0\left(1 + \frac{G\nu F}{AE}\right) \quad \text{and} \quad R_2 = R_4 = R_0\left(1 - \frac{GF}{AE}\right)$$

Using eqn [9.5] and assuming $(GF/AE) \ll 1$, the approximate bridge output voltage is:

Output voltage for pillar load cell

$$E_{Th} \approx \frac{V_S}{2}(1 + \nu)\frac{GF}{AE} \qquad [9.19]$$

Finally for the unbonded strain gauge accelerometer (Fig. 8.13(d)), we have

$$R_1 = R_3 = R_0(1+G(e_0+e)] \quad \text{and} \quad R_2 = R_4 = R_0[1+G(e_0-e)]$$

where e_0 = strain due to prestressing which is present at zero acceleration and e is the acceleration-induced strain $= a/\omega_n^2 L$. The output voltage of the bridge is given by:

Output voltage for strain gauge accelerometer

$$E_{Th} = \frac{V_S G e}{(1 + G e_0)} \qquad [9.20]$$

The output voltage of these bridges is usually small and requires high amplification; typically if $V_S = 15$ V, $G = 2$ and $e = 10^{-5}$, then $E_{Th} \approx V_S G e = 300\ \mu$V. Because of the problems in amplifying low level d.c. signals (see following section), an a.c. supply voltage $V_S = \hat{V}_S \sin \omega_s t$ is often used to give an a.c. output voltage. A.C. carrier systems are discussed fully in Section 9.3.

9.1.3 Design of reactive deflection bridges

A reactive bridge has an a.c. supply voltage; two arms are usually reactive impedances and two arms resistive impedances. Figure 9.5(a) shows the bridge to be used with the capacitance level transducer of Section 8.2; here we have

$$C_h = \frac{2\pi\epsilon_0}{\log_e(b/a)}[l + (\epsilon - 1)h]$$

$Z_1 = 1/(j\omega C_0)$, $Z_2 = R_2$, $Z_3 = R_3$ and $Z_4 = 1/(j\omega C_h)$.

Fig. 9.5 (a) Bridge for capacitive level sensor (b) Bridge for inductive push–pull displacement sensor

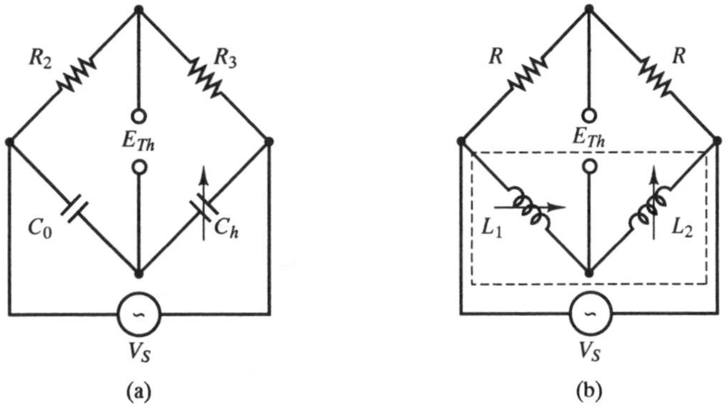

This gives

$$E_{th} = V_S \left\{ \frac{1}{1 + C_0 | C_h} - \frac{1}{1 + R_3 | R_2} \right\}$$

Thus in order to get $\hat{E}_{Th} = 0$ at minimum level h_{MIN}, we require $C_0 = C_{h_{MIN}}(R_3/R_2)$, giving:

Output voltage for capacitance level bridge

$$E_{Th} = V_S \left\{ \frac{1}{1 + \dfrac{C_{h_{MIN}}}{C_h} \dfrac{R_3}{R_2}} - \frac{1}{1 + \dfrac{R_3}{R_2}} \right\} \qquad [9.21]$$

Again if R_3/R_2 is made large compared with 1, this approximates to the linear form:

$$E_{Th} \approx V_S \frac{R_2}{R_3} \left(\frac{C_h}{C_{h_{MIN}}} - 1 \right)$$

The most common two-element reactive bridges incorporate either capacitive or inductive push-pull displacement sensors (Sections 8.2 and 8.3). The capacitive sensor (Fig. 8.5) has $C_1 = \epsilon\epsilon_0 A/(d + x)$ and $C_2 = \epsilon\epsilon_0 A/(d - x)$; if this is connected into an a.c. bridge so that $Z_1 = 1/(j\omega C_1)$, $Z_2 = Z_3 = R$, $Z_4 = 1/(j\omega C_2)$ we have:

$$E_{Th} = V_S \left\{ \frac{C_2}{C_1 + C_2} - \frac{1}{2} \right\} \qquad [9.22]$$

This gives:

Output voltage for capacitance push-pull bridge

$$E_{Th} = \frac{V_S}{2d} x \qquad [9.23]$$

i.e. the relationship between E_{Th} and x is linear and independent of frequency ω. We note that the alternative way of connecting the sensor into the bridge so that $Z_1 = 1/(j\omega C_1)$ and $Z_2 = 1/(j\omega C_2)$ gives an output voltage which is non-linearly related to x and dependent on the supply frequency ω. A similar result is obtained with the variable reluctance push-pull displacement sensor (Fig. 8.6). This has:

$$L_1 = \frac{L_0}{1 + \alpha(a - x)}, \quad L_2 = \frac{L_0}{1 + \alpha(a + x)}$$

and from Fig. 9.5(b) we have $Z_1 = j\omega L_1$, $Z_2 = Z_3 = R$, $Z_4 = j\omega L_2$, giving:

$$E_{Th} = V_S \left\{ \frac{L_1}{L_1 + L_2} - \frac{1}{2} \right\} \qquad [9.24]$$

from which

Output voltage for inductive push-pull bridge

$$E_{Th} = \frac{V_S \alpha x}{2(1 + \alpha a)} \qquad [9.25]$$

Again the relationship between E_{Th} and x is linear and frequency independent. In practice, however, these sensors will have associated resistance so that eqns [9.23], [9.25] are only approximate.

9.2 Amplifiers

Amplifiers are necessary in order to amplify low level signals, e.g. thermocouple or strain gauge bridge output voltages, to a level which enables them to be further processed.

9.2.1 The ideal operational amplifier and its applications

The operational amplifier can be regarded as the basic building block for modern amplifiers. It is a high gain, integrated circuit amplifier designed to amplify signals from d.c. up to many kHz. It is not normally used by itself but with external feedback networks to produce precise transfer characteristics which depend almost entirely on the feedback network. Usually there are two input terminals and one output terminal, the voltage at the output terminal being proportional to the difference between the voltages at the input terminals. Figure 9.6 shows the circuit symbol and a simplified equivalent circuit for an operational amplifier. Table 9.1 summarises the main characteristics of an ideal operational amplifier together with those of a typical practical amplifier.[1]

If we assume ideal behaviour, then the calculations of transfer characteristics of operational amplifier feedback networks are considerably simplified. These

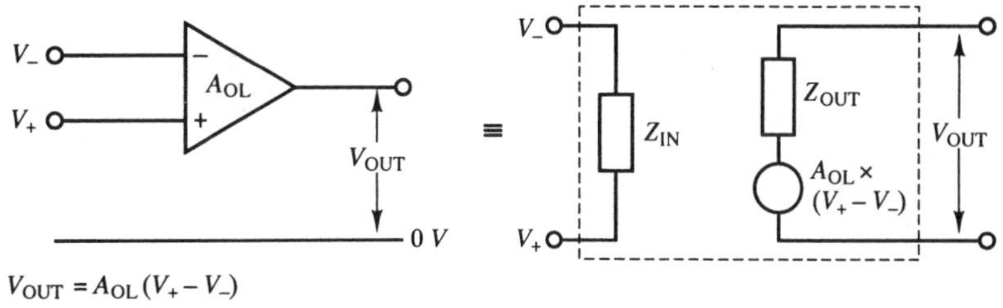

Fig. 9.6 Circuit symbol and simplified equivalent circuit for operational amplifier

Table 9.1 Ideal and typical operational amplifier characteristics

Parameter	Ideal op-amp	Typical op-amp
D.C. open loop gain A_{OL}	∞	100 dB (10^5)
Input impedance Z_{IN}	∞	2 MΩ
Output impedance Z_{OUT}	0	75 Ω
Input offset voltage V_{OS}	0	1 mV
Temp. coeff. of input offset voltage γ	0	5 μV °C^{-1}
Input bias current i_B	0	80 nA
3 dB bandwidth 0 to f_B	0 to ∞	0 to 10 Hz
Common mode rejection ratio CMMR	∞	90 dB

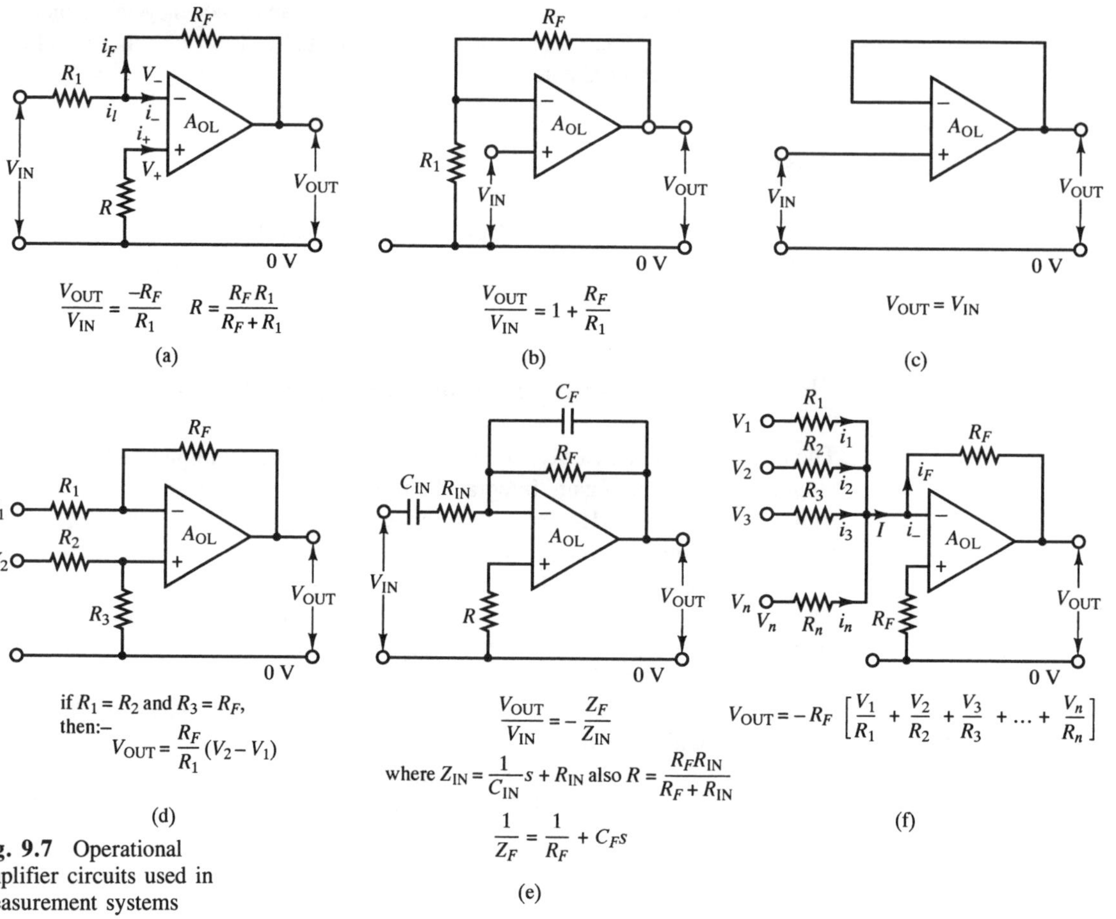

Fig. 9.7 Operational amplifier circuits used in measurement systems
(a) Inverting amplifier
(b) Non-inverting amplifier
(c) Voltage follower
(d) Differential amplifier
(e) A.c. amplifier (lead–lag)
(f) Voltage summer

Summing point conditions for ideal operational amplifier with negative feedback

calculations can then be modified if it is necessary to take into account the non-ideal behaviour of practical amplifiers.

Figure 9.7 shows six operational amplifier circuits, which are commonly used in measurement systems, together with the ideal transfer characteristics. These transfer characteristics are derived using the following two ideal conditions:

$$i_- = i_+ = 0$$
$$V_- = V_+ \qquad [9.26]$$

The second condition results from assuming that $A_{OL} \to \infty$: since $V_{OUT} = A_{OL}(V_+ - V_-)$ and V_{OUT} is finite, then the differential voltage $(V_+ - V_-) \to 0$ i.e. $V_+ = V_-$. We can illustrate the use of these conditions by deriving the transfer characteristics of the inverting amplifier (Fig. 9.7(a)). We have: $i_1 = i_- + i_F$ i.e. $i_1 = i_F$ (since $i_- = 0$), where:

$$i_1 = \frac{V_{IN} - V_-}{R_1} \quad \text{and} \quad i_F = \frac{V_- - V_{OUT}}{R_F}$$

Since $i_+ = 0$, $V_+ = 0$, this means that here $V_- = V_+ = 0$. Thus

$$\frac{V_{IN}}{R_{IN}} = -\frac{V_{OUT}}{R_F}$$

i.e.

Transfer characteristics for inverting amplifier

$$\frac{V_{OUT}}{V_{IN}} = \frac{-R_F}{R_{IN}} \qquad [9.27]$$

The transfer characteristics for the other five arrangements can be derived in a similar way.

The voltage follower (Fig. 9.7(c)) has unity voltage gain and high input impedance equal to that of the operational amplifier itself. It can therefore be used as a buffer amplifier to connect an element with high output impedance to one with low input impedance. The differential amplifier (Fig. 9.7(d)) is used in conjunction with deflection bridge circuits: V_2 is the voltage at bridge output terminal B and V_1 the voltage at output terminal D (Fig. 9.1). The differential amplifer thus amplifies the bridge out-of-balance voltage E_{Th} which is the difference between the voltages at B and D. For example a four-element strain gauge bridge with $V_S = 15$ V, $G = 2$, $e = 10^{-5}$ has $V_2 = V_B = 7.5\text{ V} + 150\,\mu\text{V}$, $V_1 = V_D = 7.5\text{ V} - 150\,\mu\text{V}$; the amplifier amplifies the differential voltage $V_2 - V_1 = 300\,\mu\text{V} = E_{Th}$. Circuits (a), (b), (c), (d) and (f) have a bandwidth from 0 to f_B Hz, where f_B depends on the frequency response characteristics of the operational amplifier itself and the closed loop gain. These amplifiers can be used to amplify low level d.c. signals but will also amplify unwanted drift and low frequency interference voltages. The a.c. amplifier (Fig. 9.7(e)) has ideally a bandwidth between f_1 and f_2 (Fig. 9.8) and can be used in an a.c. carrier system which rejects drift and interference voltages (Section 9.3).

The voltage summer (Fig. 9.7(f)) forms the basis of a digital-to-analogue converter which is in turn used in an analogue-to-digital converter (Section 10.1).

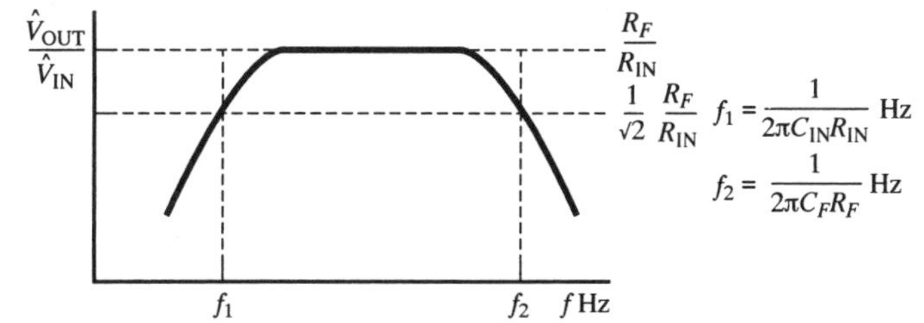

Fig. 9.8 Ideal frequency response characteristics of a.c. amplifier

9.2.2 Limitations of practical operational amplifiers

Table 9.1 summarises the characteristics of a typical practical operational amplifier. The parameters V_{OS}, γ, i_B, CMRR influence the d.c. performance of the amplifier and are discussed first.

The existence of input offset voltage V_{OS} means that V_{OUT} is unequal to zero when both V_- and $V_+ = 0$ volts, i.e.

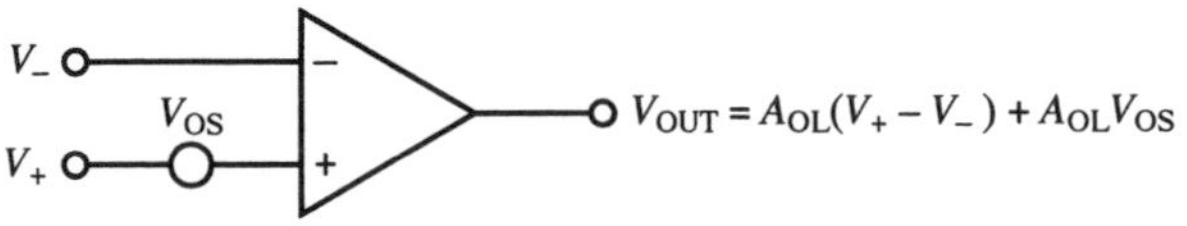

[9.28]

Most operational amplifiers have facilities for adjusting V_{OS} to zero, i.e. for obtaining $V_{OUT} = 0$ when $V_+ = V_- = 0$. However, V_{OS} is dependent on the temperature T_E °C of the amplifier environment, where γ is the appropriate temperature coefficient. Thus supposing V_{OS} is set to zero at $T_E = 15$ °C; then if T_E subsequently increases to 25 °C, the resulting input offset voltage is $\gamma(25 - 15)$ i.e. $\approx 50\,\mu$V, which causes a change of approximately $50 \times 10^5\,\mu$V, i.e. 5 V in the output of the open loop operational amplifier. The effect on a closed loop amplifier is reduced but may be still important. For an inverting amplifier we have:

$$V_{OUT} = -\frac{R_F}{R_{IN}} V_{IN} + \left(1 + \frac{R_F}{R_{IN}}\right) V_{OS} \qquad [9.29]$$

With a thermocouple temperature sensor, $V_{IN} \approx 40T\,\mu$V, i.e. $40\,\mu$V change for every 1 °C change in measured temperature T. The above variation in V_{OS} with environmental temperature therefore causes the output of the inverting amplifier to drift significantly.

Ideally the output voltage should depend only on the differential voltage $(V_+ - V_-)$ and should be independent of the common mode voltage $V_{CM} = (V_+ + V_-)/2$. For a practical operational amplifier, however, we have:

$$V_{OUT} = A_{OL}(V_+ - V_-) + A_{CM}V_{CM} \qquad [9.30]$$

where A_{CM} is the common mode gain. A more commonly used term is Common Mode Rejection Ratio (CMRR), where:

$$\text{CMMR} = \frac{A_{OL}}{A_{CM}}$$

or in decibels

$$(\text{CMMR})_{dB} = 20 \log_{10}\left(\frac{A_{OL}}{A_{CM}}\right) \qquad [9.31]$$

This gives:

$$V_{OUT} = A_{OL}\left[(V_+ - V_-) + \frac{V_{CM}}{\text{CMRR}}\right] \qquad [9.32]$$

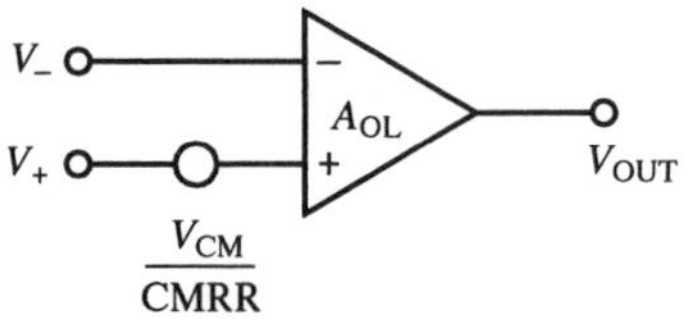

i.e. the equivalent circuit for an open loop amplifier.

Typically $(\text{CMRR})_{dB} = 90$ dB, i.e. CMRR $= 3 \times 10^4$. For a closed loop differential amplifier (Fig. 9.7(d)), we have:

$$V_{OUT} \approx \frac{R_F}{R_1}(V_2 - V_1) + \left(1 + \frac{R_F}{R_1}\right)\frac{V_{CM}}{\text{CMRR}} \qquad [9.33]$$

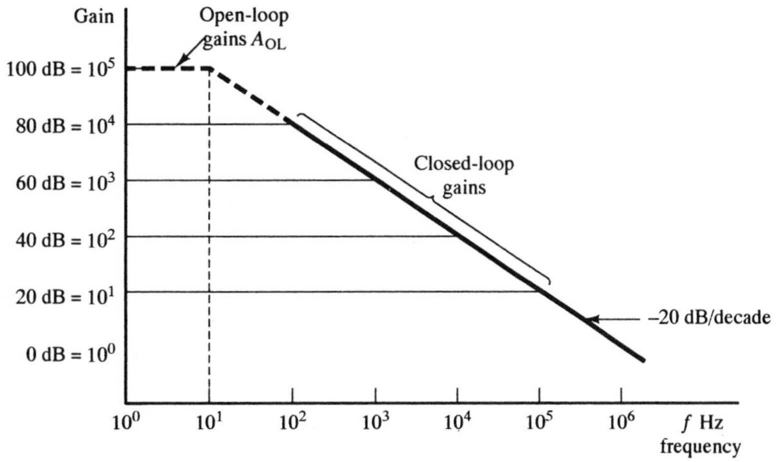

Fig. 9.9 Typical gain/frequency characteristics for operational amplifier

If the differential amplifier is used with the strain gauge bridge described earlier, we have $V_2 - V_1 = 300\,\mu V$, $V_{CM} = 7.5$ V so that $V_{CM}/\text{CMRR} = 7.5/(3 \times 10^4) = 250\,\mu V$; i.e. differential and effective common mode voltages are comparable.

The a.c. performance of a practical operational amplifier is determined by its dynamic characteristics. These are adequately represented by a first-order lag, i.e.

$$\frac{\Delta V_{OUT}}{\Delta(V_+ - V_-)}(s) = \frac{A_{OL}}{1 + \tau s} \qquad [9.34]$$

where A_{OL} is the d.c. open loop gain, τ the time constant and $\Delta(V_+ - V_-)$, ΔV_{OUT} corresponding small changes in input and output. The variation in open-loop amplifier gain with frequency f is thus given by:

Gain/frequency relation for open-loop amplifier

$$A_{OL}(f) = \frac{\hat{V}_{OUT}}{\widehat{(V_+ - V_-)}}(f) = \frac{A_{OL}}{\sqrt{1 + (f/f_B)^2}} \qquad [9.35]$$

where $f_B = 1/2\pi\tau$ is the upper bandwidth frequency.

The open-loop gain/bandwidth product $A_{OL} \times f_B$ is typically $\approx 10^5 \times 10 = 10^6$ Hz. Figure 9.9 shows the variation in gain with frequency for both open- and closed-loop amplifiers. We note that the lower the gain of the closed-loop amplifier the greater the bandwidth. This is because the gain/bandwidth product of a closed-loop amplifier is also equal to the open-loop value of $A_{OL}f_B$. The frequency response of the a.c. amplifier (Figs 9.7(e) and 9.8) depends, to some extent, on the frequency characteristics of the operational amplifier as well as those of the external components.

9.2.3 Instrumentation amplifiers[1,2]

An instrumentation amplifier is a high performance differential amplifier system consisting of several closed-loop operational amplifiers. An ideal instrumentation amplifier gives an output voltage which depends *only* on the difference of two voltages V_1 and V_2, i.e.

$$V_{OUT} = K(V_2 - V_1) \qquad [9.36]$$

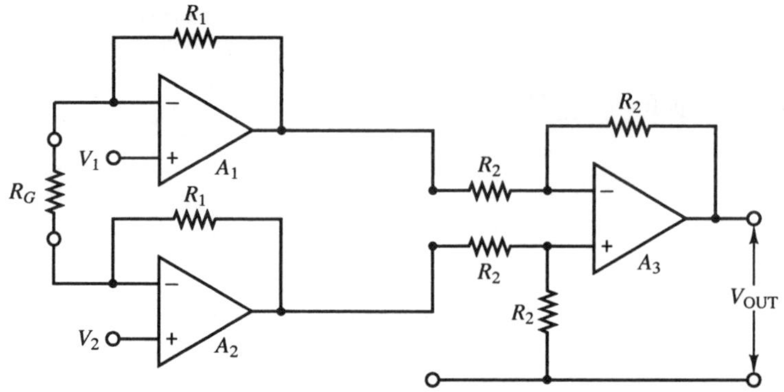

Fig. 9.10 Typical instrumentation amplifier

where the gain K is precisely known and can be varied over a wide range. A practical instrumentation amplifier should have a gain which can be set by a single external resistor and should combine:

high input impedance;
high common mode rejection ratio;
low input offset voltage;
low temperature coefficient of offset voltage.

The differential amplifier already discussed (Fig. 7.4(d)) which uses a single operational amplifier is inadequate: in order to obtain high gain, R_1 must be low. This means low input impedance and low common mode rejection.

Figure 9.10 shows a typical instrumentation amplifier system consisting of three operational amplifiers, A_1, A_2 and A_3. The two input non-inverting amplifiers A_1 and A_2 provide an overall differential gain of $(1 + 2R_1/R_G)$ and a common mode gain of unity. The output amplifier A_3 is a unity gain differential amplifier. An amplifier of this type has:

$Z_{IN} \approx 300$ to $5000\,\mathrm{M\Omega}$;
$(\mathrm{CMRR})_{dB} \approx 74$ to $100\,\mathrm{dB}$;
offset voltage $V_{OL} \approx 0.2\,\mathrm{mV}$;
temperature coefficient of offset voltage $\gamma \approx 0.25$ to $10\,\mu\mathrm{V\,°C^{-1}}$.

9.3 A.C. carrier systems

Two of the most difficult problems in conditioning low level d.c. signals from sensors are external interference in the signal circuit (Section 6.4) and amplifier drift (Section 9.2.2). These problems can be avoided if the signal is converted into a.c. form, amplified and then re-converted into d.c.

The initial or primary elements in an a.c. carrier system are usually of one of the following types:

(a) R, L, C sensors in a suitable deflection bridge (Section 9.1);
(b) an LVDT sensor (Section 8.3).

The output voltage of a four-element strain gauge bridge with a cantilever load cell is given by $E_{Th} = GV_S e$ (eqn [9.18]), i.e. the output signal is proportional to the *product* of the supply voltage V_S and the strain signal e. Similar results are obtained with other bridge circuits (eqns [9.15], [9.17], [9.19], [9.20], [9.23], [9.25]); in each case the bridge output voltage is proportional to the product of the supply voltage and the measured variable. The output voltage of the LVDT sensor (Section 8.3) is also approximately proportional to the product of supply voltage and displacement x of the core from the null point. In an a.c. carrier system the supply voltage is a.c. i.e. $V_S = \hat{V}_S \sin 2\pi f_S t$, with a frequency f_S of typically a few kHz. The measured variable is also a function of time; in the case of a step change in force on the cantilever, the strain variation $e(t)$ is a damped sine wave. In the general case $e(t)$ is a complicated function of time which can be expressed as a sum of many sine waves (Sections 4.3 and 6.2):

$$e(t) = \sum_{i=1}^{i=m} \hat{e}_i \sin 2\pi f_i t$$

Figure 9.11 shows a typical measurement signal frequency spectrum containing frequencies from $f_1 = 0$ Hz up to f_m (often only a few Hz).

The bridge output voltage is thus given by:

$$E_{Th} = \sum_{i=1}^{i=m} \hat{V}_i \sin 2\pi f_i t \sin 2\pi f_S t \qquad [9.37]$$

where $\hat{V}_i = G\hat{V}_S\hat{e}_i$. If we, for the moment, consider only the ith component:

$$V_i = \hat{V}_i \sin 2\pi f_i t \sin 2\pi f_S t$$

then we have a sine wave of frequency f_S and amplitude $\hat{V}_i \sin 2\pi f_i t$. This is amplitude modulation (AM); here the amplitude of the carrier signal frequency f_S is being altered according to the instantaneous value of the modulating signal frequency f_i (Fig. 9.11). Using the identity:

$$\sin A \sin B = \tfrac{1}{2}\cos(A - B) - \tfrac{1}{2}\cos(A + B)$$

the amplitude modulated signal

$$V_i = \hat{V}_i \sin 2\pi f_S t \sin 2\pi f_i t$$

can be expressed in the form:

$$V_i = \frac{\hat{V}_i}{2}\cos 2\pi(f_S - f_i)t - \frac{\hat{V}_i}{2}\cos 2\pi(f_S + f_i)t \qquad [9.38]$$

Thus the AM signal is the sum of two cosinusoidal signals with frequencies $f_S - f_i$, $f_s + f_i$; the corresponding frequency spectrum (Fig. 9.11) contains two lines. This form of amplitude modulation is called **balanced amplitude modulation**, because when $\hat{e}_i = 0$, $\hat{V}_i = 0$, i.e. the bridge is balanced. This means also that the frequency spectrum does not contain the carrier frequency f_S, there are only two lines at the sideband frequencies $f_S \pm f_i$. Hence the alternative name is **double sideband carrier suppressed amplitude modulation**. We see that each frequency f_i in the modulating signal spectrum contributes two lines to the AM spectrum, e.g. if $f_S = 1$ kHz, then $f_i = 1$ Hz gives lines at 999 and 1001 Hz. The AM spectrum corresponding to the total modulating signal (0 to f_m Hz) consists therefore of frequencies between $f_S - f_m$ and $f_S + f_m$, e.g. between 995 and 1005 Hz if $f_S = 1$ kHz, $f_m = 5$ Hz.

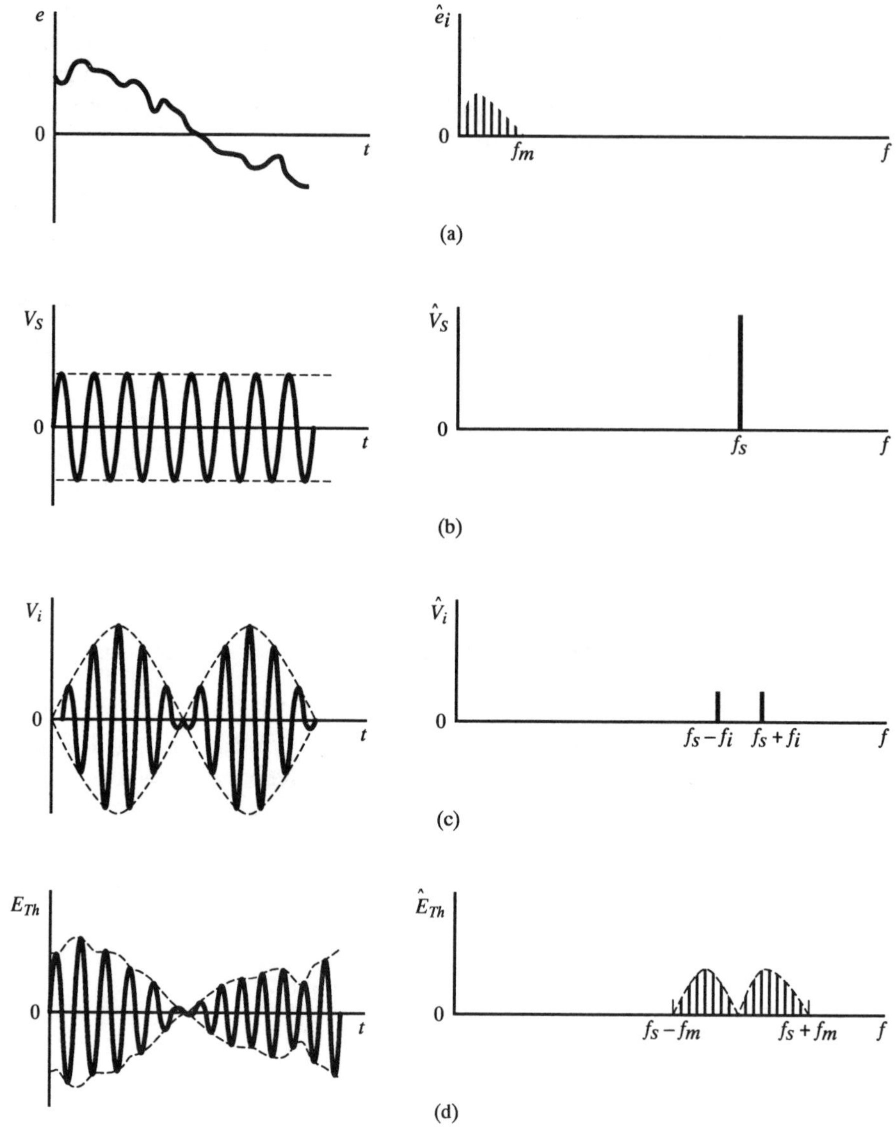

Fig. 9.11 Corresponding waveforms and frequency spectra for amplitude modulation (a) Measurement signal (b) Supply voltage (c) AM with sinusoidal modulating signal (d) Complete AM waveform

We see therefore that the operation of multiplication and thence modulation has the effect of shifting the spectrum of the measurement signal from low frequencies to higher frequencies. The amplitudes $\hat{V}_i$ are small but an a.c. amplifier (Figs 9.7(e) and 9.8) can now be used. The bandwidth frequencies f_1 and f_2 are chosen so that they include the measurement signal between $f_S - f_m$ and $f_S + f_m$ (Fig. 9.11), but exclude low frequency drift voltages and 50 Hz mains interference.

In the above example, an amplifier with $f_1 = 900$ Hz and $f_2 = 1100$ Hz will successfully amplify the measurement signal and reject drift and interference. It is important to realise that this system only rejects drift and interference signals which are introduced *after* the modulator; for example drift in bridge output voltage due to drift in bridge supply voltage will be regarded as part of the measurement signal.

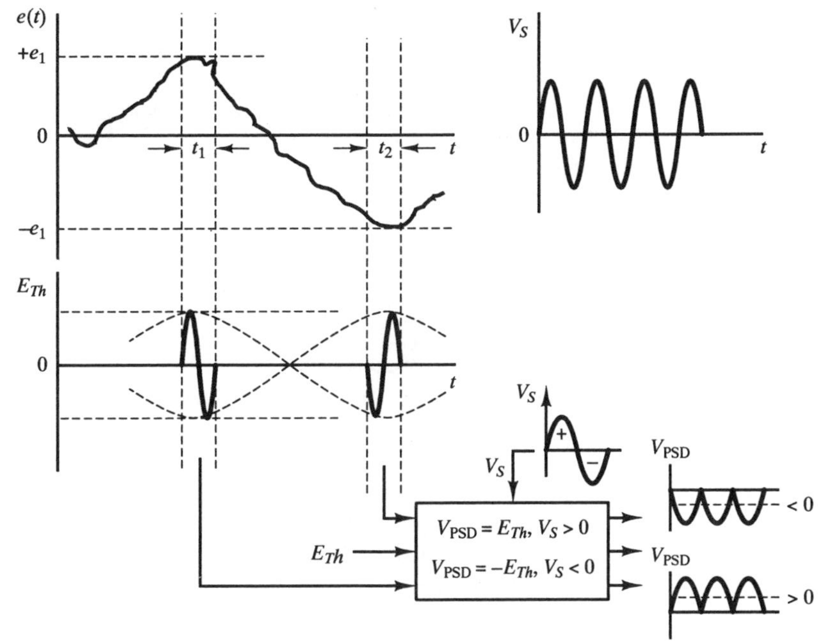

Fig. 9.12 Phase-sensitive demodulation

The AM signal must now be demodulated in order to obtain an output voltage which is proportional to the strain signal $e(t)$. Figure 9.12 shows corresponding waveforms for $e(t)$ and supply voltage $V_S(t)$; we see that at time interval t_1, $e(t_1) = +e_1$ and at time interval t_2, $e(t_2) = -e_1$. Thus at time t_1, the cantilever is below the neutral position and gauges 1 and 3 are in tension. At time t_2 the cantilever is above the neutral position and gauges 1 and 3 are in compression. The corresponding bridge output voltages are:

$$E_{Th} = +Ge_1\hat{V}_S \sin 2\pi f_S t \quad \text{at } t_1$$

and

$$E_{Th} = -Ge_1\hat{V}_S \sin 2\pi f_S t = Ge_1\hat{V}_S \sin(2\pi f_S t + 180) \quad \text{at } t_2$$

i.e. the signals have equal amplitude but are 180° out of phase (Fig. 9.12).

Thus in order to distinguish between the situations at t_1 and t_2, we need a demodulator which detects this phase difference and gives a positive output voltage when e is positive and a negative voltage when e is negative. Such a demodulator is called a **phase-sensitive demodulator** (PSD). The simplest form of PSD is an amplifier with a gain of ± 1 according to the sign of the supply voltage; that is

$$\left.\begin{aligned} V_{PSD} &= E_{Th} \quad \text{when} \quad V_S > 0 \\ V_{PSD} &= -E_{Th} \quad \text{when} \quad V_S < 0 \end{aligned}\right\} \qquad [9.39]$$

The action of this element is shown in Fig. 9.12. We see that a low-pass filter (Section 6.5) is required to give the mean value of the output signal. The bandwidth of this filter should be less than f_S, in order to filter out carrier frequencies; but greater than f_m, in order not to filter out measurement signal frequencies. The complete a.c. carrier system is shown in Fig. 9.13.

Fig. 9.13 Complete a.c. carrier system

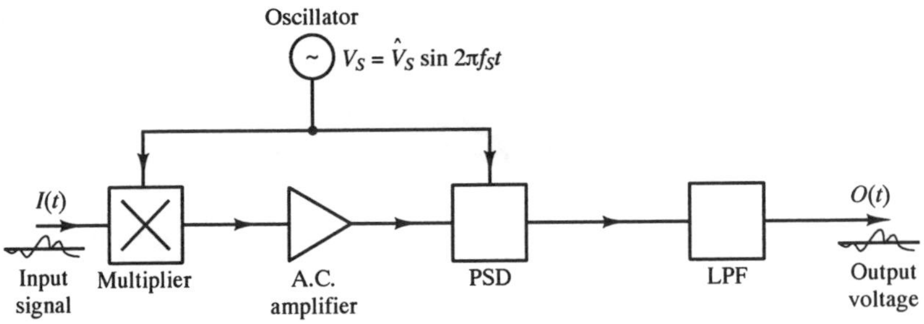

We have seen above that an a.c. bridge can be used to generate an a.c. voltage from a resistive, capacitive or inductive sensor. This raises the question of how to convert a low-level d.c. voltage signal, e.g. a thermocouple e.m.f., into a.c. form. The usual solution here is to sample the signal at high speed, around 1000 times per second, using a transistor switch or 'chopper' driven by an oscillator. The resulting 'chopped' signal is a series of pulses whose amplitudes correspond to the instantaneous value of the d.c. voltage; this is pulse amplitude modulation (PAM). The PAM signal can be processed by an a.c. amplifier and phase-sensitive demodulator so that the chopper simply replaces the multiplier in Fig. 9.13. The resulting **chopper-stabilised amplifier** has excellent drift characteristics.

9.4 Current transmitters

In Section 5.1 we saw that sensing elements could be represented either by a Thévenin voltage source or a Norton current source. In Section 6.3 we saw that a current transmission system rejects series interference more efficiently than a voltage transmission system. In this section we discuss transmitters used in the process industries to give an output d.c. current signal in the standard range 4 to 20 mA proportional to the input measured variable. The output has a **live zero** of 4 mA: this enables a distinction to be made between a zero input situation (output = 4 mA) and a fault situation (output = 0 mA).

These transmitters were initially **closed-loop systems** with **high-gain negative feedback** (Section 3.3). Here the overall transmitter sensitivity depends mainly on the sensitivity of the elements in the feedback path and is largely unaffected by changes in the characteristics of forward path elements, for example amplifier drift. The zero and span of these transmitters can be adjusted but, in the case of force balance systems, the adjustments have to be made mechanically, for example by adjusting the position of a pivot. Closed-loop transmitters have now been largely replaced by **open-loop transmitters**. Here the overall transmitter sensitivity depends on the sensitivity of every element in the system; these element sensitivities can change due to modifying inputs and/or non-linear effects. However by using precision mechanical components and high performance electronic integrated circuits, the above effects are minimised and open-loop transmitters are now widely used. They have the advantage that zero and span adjustments can be made electronically which is more convenient than mechanical adjustment. The most recent development is that of **intelligent** or **'smart' transmitters**, here a microcomputer is incorporated into an open-loop transmitter.

This results in an improvement in overall accuracy and enhanced flexibility and versatility.

9.4.1 Closed-loop transmitters

Figure 9.14(a) shows a simplified block diagram of a closed-loop differential pressure transmitter operating on the force balance or more accurately torque balance principle. The diaphragm force $(P_1 - P_2)A_D$ produces a clockwise torque $A_D(P_1 - P_2)a$ on the lever arm which is supported by a clockwise torque F_0b due to the zero

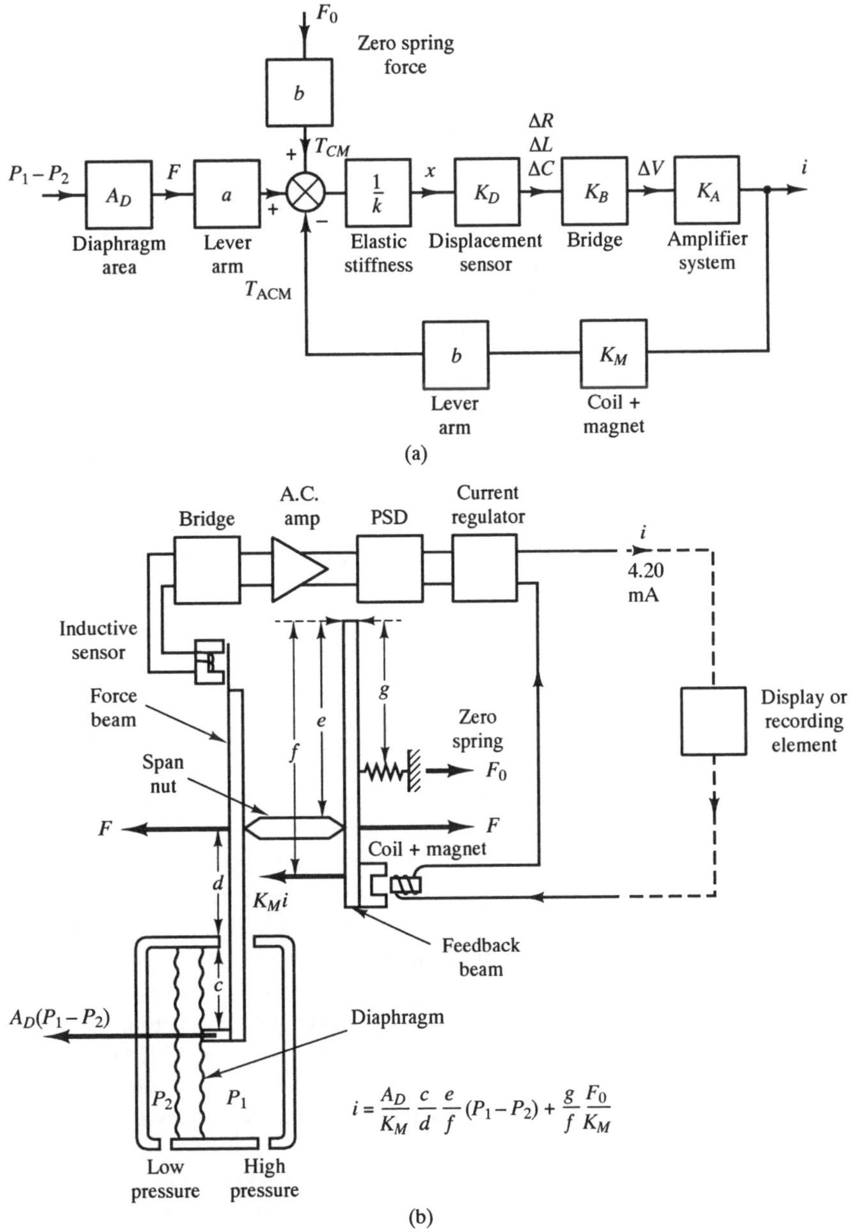

Fig. 9.14 Closed loop differential pressure transmitters
(a) Simplified block diagram
(b) Simplified schematic diagram

spring. the output current i is fed back to a coil, inside a permanent magnet; this produces a force $K_M i$ on the lever arm and a corresponding anticlockwise torque $K_M ib$. Any imbalance of these torques causes the lever to rotate; the resulting displacement x is measured by a sensor-bridge-amplifier system. The resulting change in output current is fed back to the coil and magnet to adjust the anticlockwise torque until an approximate torque balance is again obtained.

Inspection of Fig. 9.14(a) give

$$i = \frac{K_D K_B K_A / k}{1 + K_D K_B K_A b K_M / k} [aA_D(P_1 - P_2) + bF_0] \qquad [9.40]$$

If the gain K_A of the amplifier system is high so that $\frac{K_D K_B K_A b K_M}{k}$ is very much greater than 1, then eqn [9.40] reduces to:

$$i = \frac{a}{b}\frac{A_D}{K_M}(P_1 - P_2) + \frac{F_0}{K_M} \qquad [9.41]$$

From eqn [9.41] we see that transmitter sensitivity $= (a/b)(A_D/K_M)$ and zero bias $= (F_0/K_M)$. The closed-loop transmitter has the advantage, therefore, that the overall sensitivity is independent of diaphragm stiffness, displacement sensor sensitivity and amplifier gain, so that once set it should remain constant and not drift. The main disadvantage of this type of transmitter is that sensitivity (input span) and zero adjustments must be made mechanically; the input span is adjusted by altering the ratio a/b and the zero by altering F_0. For example, suppose $A_D = 10^{-2}\,m^2$ (circular diaphragm approx. 11 cm in diameter) and $K_M = 10^3\,N\,A^{-1}$. Then in order to obtain an output range of 4 to 20 mA, corresponding to 0 to 10^4 Pa (0 to 1 metre water), we require $a/b = 0.16$, $f_0 = 4$ N.

Figure 9.14(b) shows a simplified schematic diagram of a two lever torque balance differential pressure transmitter using an inductive displacement sensor.

9.4.2 Open-loop transmitters

Open-loop differential pressure transmitters

Figure 9.15(a) shows a general simplified block diagram for an open-loop differential pressure transmitter. The input differential pressure $(P_1 - P_2)$ acting over the area A_D of the sensing diaphragm produces a deflecting force $A_D(P_1 - P_2)$ which is opposed by the elastic spring force kx. The resulting diaphragm deformation x is small ($\approx 10^{-5}$ m) and approximately proportional to $(P_1 - P_2)$. This displacement can be measured using variable capacitance, variable reluctance or resistive strain gauge sensors. Another possibility is to use a piezoresistive sensor consisting of a silicon diaphragm with diffused P- or N-type material to give four strain gauges. With strain gauge elements, a d.c. 4-element strain gauge bridge is used; the amplifier system consists of an instrumentation amplifier and output current regulator. With capacitive or inductive sensors, an a.c. bridge is used with a carrier amplifier system consisting of an a.c. amplifier, phase sensitive demodulator, d.c. amplifier and output current regulator. Span and zero adjustments are made electronically via the amplifier system.

Figure 9.15(b) shows a **differential capacitance displacement sensor**. The input differential pressure causes the sensing diaphragm to be displaced a small distance

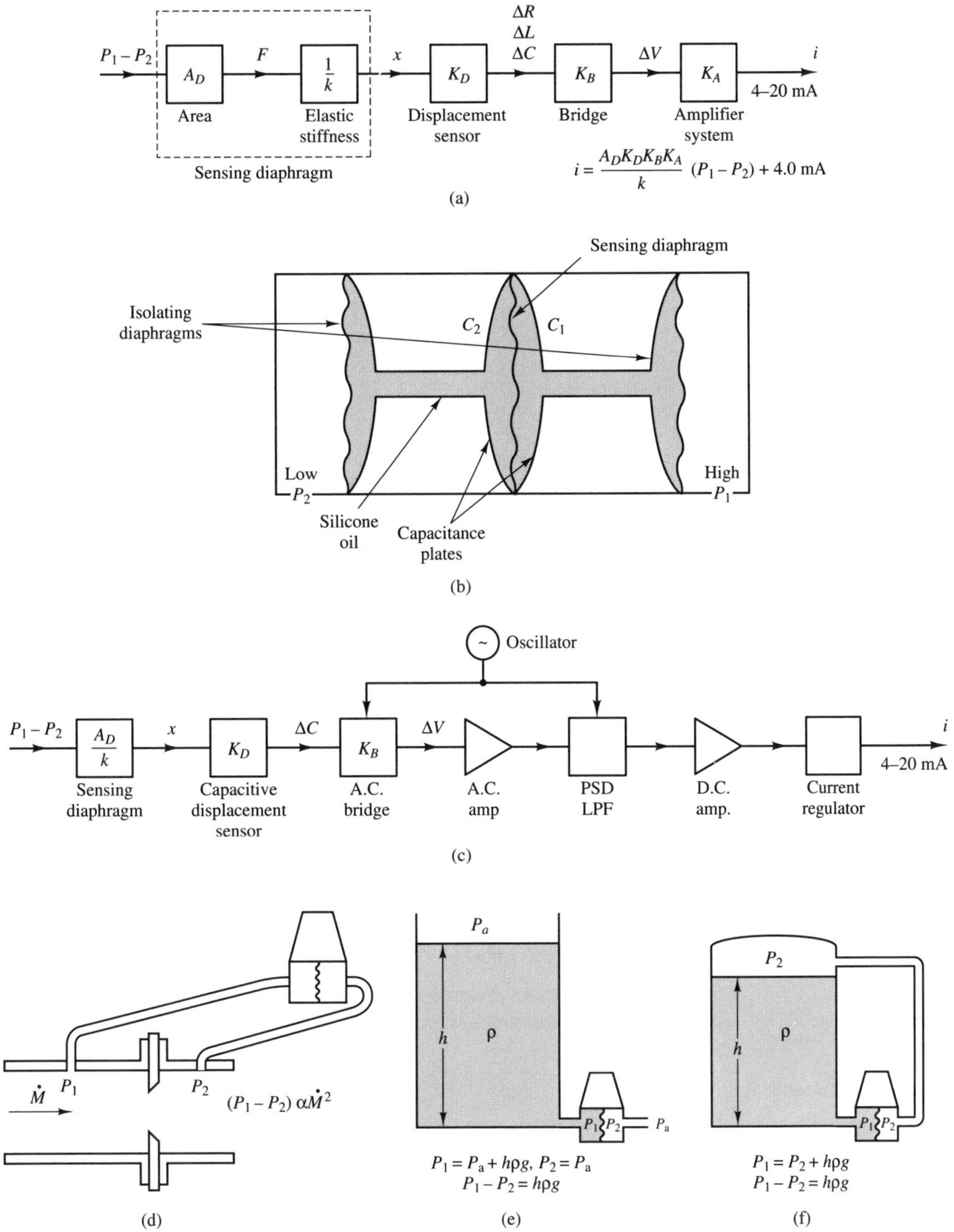

Fig. 9.15 Open-loop differential pressure transmitters: (a) General block diagram; (b) Differential capacitance — principle; (c) Differential capacitance — block diagram; (d) Application — flow measurement; (e) Application — level measurement open vessel; (f) Application — level measurement closed vessel

x to the left. This in turn causes C_1 to fall from C_0 to $C_0 - \Delta C$ and C_2 to increase from C_0 to $C_0 + \Delta C$ where ΔC is the change in capacitance proportional to $(P_1 - P_2)$. The capacitances are incorporated in an a.c. bridge circuit, similar to Figure 9.5(a) which has:

$$Z_1 = \frac{1}{j\omega C_1},\ Z_2 = Z_3 = R,\ Z_4 = \frac{1}{j\omega C_2}$$

The bridge output voltage is therefore given by:

$$\Delta V = V_S \left\{ \frac{C_2}{C_1 + C_2} - \frac{1}{2} \right\} \qquad [9.42]$$

$$= V_S \left\{ \frac{C_0 + \Delta C}{C_0 - \Delta C + C_0 + \Delta C} - \frac{1}{2} \right\}$$

which simplifies to:

$$\Delta V = \frac{V_S}{2C_0} \Delta C \qquad [9.43]$$

Thus ΔV is proportional to ΔC and the bridge sensitivity K_B is equal to $V_S/2C_0$. Figure 9.15(c) shows a block diagram of the complete transmitter, typically the overall accuracy is $\pm 0.25\%$.

This type of transmitter is widely used in the process industries for measuring pressure, flow rate and liquid level. The transmitter can be used for measuring gauge pressure (pressure relative to atmosphere) by opening the low pressure chamber to atmosphere, i.e. setting P_2 equal to atmospheric pressure P_a. By evacuating and sealing the low pressure chamber, i.e. setting P_2 equal to zero, the transmitter can be used for measuring absolute pressure (pressure relative to vacuum).

Figure 9.15(d) shows the use of the transmitter with a differential pressure flowmeter to measure mass flow rate $\dot{M}$ of a fluid in a pipe, here differential pressure $(P_1 - P_2)$ is proportional to $\dot{M}^2$ (Section 12.3.1). Figures 9.15(e) and (f) show the application of the transmitter to the measurement of liquid level in open and closed vessels, in both cases differential pressure is proportional to the height h of liquid.

Open-loop temperature transmitters

Figure 9.16 shows block diagrams for two types of open-loop transmitters. Both use an a.c. carrier amplifier system (Section 9.3) and give a 4–20 mA d.c. current output.

In the millivolt input type, Figure 9.16(a), the input is a thermocouple e.m.f. E_{T_1,T_2}. The automatic reference junction circuit provides an e.m.f. equal to $E_{T_2,0}$; this is added to E_{T_1,T_2} to give a compensated e.m.f. $E_{T_1,0}$ which is relative to a fixed reference junction temperature of 0 °C (Section 8.5). This e.m.f. is then 'chopped' to give the a.c. voltage necessary for the carrier system. This type of transmitter can be located close to the process so that the signal is transmitted to a remote control room as a 4–20 mA current using normal leads. This is simpler and less prone to external interference than the alternative of transmitting the basic millivolt e.m.f. E_{T_1,T_2} using extension leads as in Figure 8.10(d).

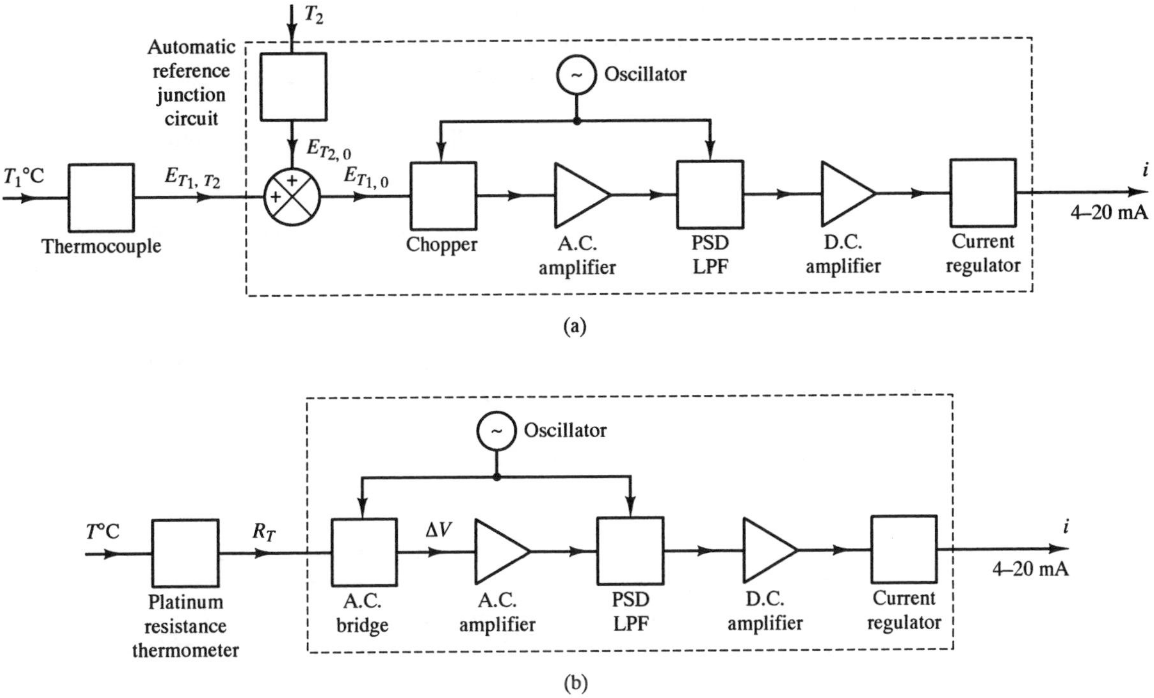

Fig. 9.16 Open-loop temperature transmitters
(a) Millivolt input
(b) Resistance input

In the resistance input type, Figure 9.16(b), the input is the resistance R_T of a resistance thermometer. This is incorporated into a deflection bridge circuit which gives a voltage output ΔV proportional to temperature T (Section 9.1.1 and eqn [9.15]). Problem 9.16 shows a four-lead resistance thermometer with two sensor leads and two dummy leads, incorporated into a deflection bridge circuit. Here the bridge output voltage is independent of the lead resistance R_C and therefore the temperature of the plant environment. The a.c. bridge output voltage necessary for the carrier system is obtained either by using an a.c. bridge supply or using a d.c. bridge supply and 'chopping' the resulting d.c. bridge output voltage.

9.4.3 Intelligent or smart transmitters

By incorporating a microcomputer into the basic open-loop transmitter not only is the accuracy of the transmitter increased but flexibility and versatility are enhanced. An intelligent or smart transmitter can perform the following functions.

Computer calculation of measured value (Section 3.3). Here the microcomputer solves the **inverse equation** to calculate an accurate value of the measured variable from the sensor output signal. This means that non-linear, hysteresis and environmental effects in the sensor can be compensated for. The overall accuracy of a smart transmitter is therefore typically $\pm 0.1\%$ of span compared to $\pm 0.25\%$ for a conventional transmitter. This accurate measured value in engineering units is then converted into a standard 4–20 mA current output.

Remote diagnosis of faults. Signals at different points in the transmitter are continuously monitored and the actual values compared with expected values. This way both the existence and nature of a fault can be detected.

Remote re-ranging. Span and zero adjustments can be made remotely from the control room rather than at the transmitter itself.

Remote identification. The transmitter identification number and maintenance records are stored in the computer memory, this information can again be accessed remotely.

Digital communications. A digital communications signal can be superimposed on the 4–20 mA analogue current signal.

Local indication of measured value. An analogue indication or digital display of measured value is available at the transmitter.

Figure 9.17 is a block diagram of a smart differential pressure transmitter.[3] The differential capacitance pressure sensor of Figure 9.15(b) gives a change in capacitance ΔC which depends on input differential pressure $(P_1 - P_2)$. This is converted into a d.c. voltage V_1 by the signal conditioning system of Figure 9.15(c).

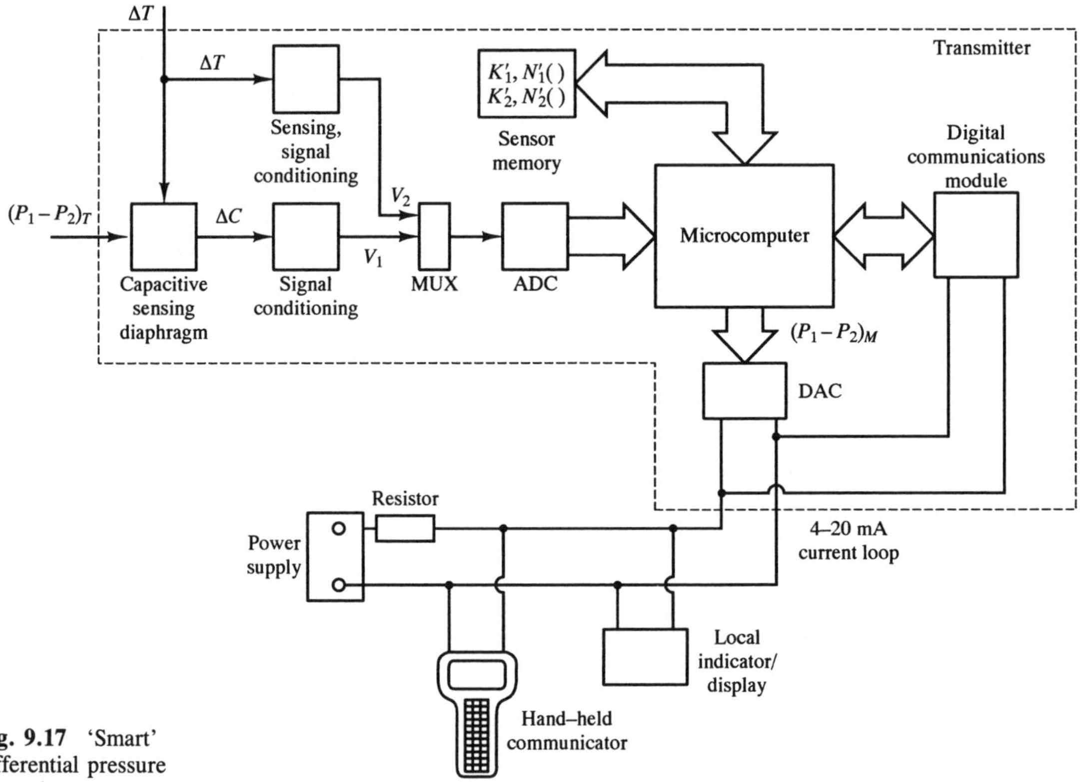

Fig. 9.17 'Smart' differential pressure transmitter

The relationship between $(P_1 - P_2)$ and V_1 is non-linear and is also affected by changes in the temperature of the sensor. ΔT is the deviation in temperature from standard conditions. Calibration of a given sensor/signal conditioning element gives corresponding values of $(P_1 - P_2)$, V_1 and ΔT and an inverse equation of the form (equation [3.17]):

$$P_1 - P_2 = K_1'V_1 + N_1'(V_1) + a_1' + K_M'\Delta TV_1 + K_I'\Delta T \qquad [9.44]$$

The coefficients K_1', a_1' etc. are specific to an individual element. Similarly ΔT is measured with a temperature sensor and appropriate signal conditioning element to give a d.c. output voltage V_2. Here the inverse equation is:

$$\Delta T = K_2'V_2 + N_2'(V_2) + a_2' \qquad [9.45]$$

where again K_2', $N_2(\ )$, a_2' are specific to an individual element. The coefficients K_1', $N_1'(\ )$, K_2', $N_2'(\ )$ etc. are stored in a sensor memory which is identified with the individual differential pressure and temperature sensors.

The voltages V_1 and V_2 are then input to a **time division multiplexer** (Section 18.1) which transfers them one at a time to an analogue-to-digital converter (ADC) see Section 10.1. The ADC gives a parallel digital output signal proportional to V_1 or V_2. This is fed to a microcomputer along with the inverse equation coefficients from the sensor memory. The microcomputer solves equations [9.45] and [9.44] to give an accurate measured value of the differential pressure $(P_1 - P_2)$ in engineering units. If the transmitter is being used to measure flow rate or liquid level, the measured value of flow rate or level is calculated in engineering units from $(P_1 - P_2)$. This measured value is then output from the microcomputer to a digital-to-analogue converter (DAC) see Section 10.1. The DAC forms a 4–20 mA current loop with a d.c. power supply and a series resistor; an analogue indicator and/or recorder is connected into the loop to display the measured value of $(P_1 - P_2)$.

The measured value of $(P_1 - P_2)$ is also transferred from the computer to a digital communications module. This gives a serial digital signal corresponding to $(P_1 - P_2)$ using **frequency shift keying** (FSK) see Section 18.6. This is a sinusoidal signal at two distinct frequencies, one corresponding to a binary '1' the other to binary '0'. The FSK signal (typical amplitude of 0.5 mA) is then superimposed on the 4–20 mA analogue signal; this means that no d.c. component is added to the 4–20 mA signal (Section 18.7). Communication between the transmitter and the outside world is via a hand-held communicator which is connected into the current loop. There is two way communication between the communicator and the microcomputer via the digital communications module. Thus a technician can transmit information requesting a range change and receive information back confirming that the change has been implemented.

9.5 Oscillators and resonators

In Section 9.3 we studied a.c. carrier systems, where the amplitude of a sinusoidal voltage depends on the magnitude of the measured variable — **amplitude modulation** (AM); these systems reject external interference added after the modulator. An alternative is **frequency modulation** (FM) where the frequency of a sinusoidal voltage

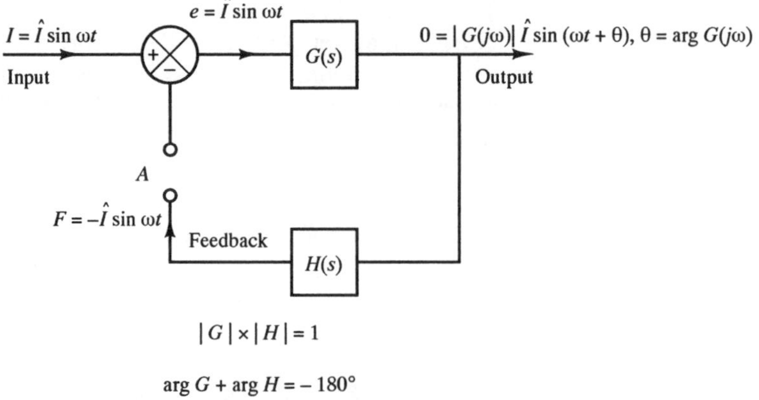

Fig. 9.18 Principle of oscillator/resonator

depends on the magnitude of the measured variable. FM systems also reject interference and have the following advantages over AM systems:

(a) The external interference affects signal amplitude more than it affects signal frequency; FM is therefore inherently more resistant to interference than AM.
(b) By counting pulses over a fixed time interval, a frequency signal can easily be converted to digital form (Section 10.1.4).

Examples of sensors which directly generate frequency signals are the variable reluctance tachogenerator (Section 8.4), the turbine flowmeter (Section 12.3.2) and the vortex flowmeter (Section 12.3.3). Because of the advantages of frequency signals however, we need general methods for producing them. These are *oscillators*, which are purely electrical, and *resonators* which are electromechanical. Oscillators and resonators are feedback systems which rely on the dynamic property of a closed-loop system, consisting of elements with dynamic characteristics, to sustain continuous oscillations, if certain conditions are satisfied.

Figure 9.18 shows a closed-loop system consisting of a forward path element with transfer function $G(s)$ and a feedback path element with transfer function $H(s)$. Suppose that an input signal $I = \hat{I}\sin\omega t$ is applied to the system with the feedback link broken at A. The corresponding feedback signal F is also sinusoidal and is given by (Section 4.2):

$$F = |G(j\omega)|\,|H(j\omega)|\hat{I}\sin(\omega t + \phi) \qquad [9.46]$$

where:

$$\phi = \arg G(j\omega) + \arg H(j\omega) \qquad [9.47]$$

If we make the product of magnitudes $|G| \times |H|$ equal to 1 and the sum of arguments (phase angles) equal to $-180°$, then $F = -\hat{I}\sin\omega t$, i.e. F has the same amplitude as the input signal but is 180° out of phase. If the feedback link is now closed, then $e = 2\hat{I}\sin\omega t$, but if $I(t)$ is reduced to zero, then $e = \hat{I}\sin\omega t$ and the system remains in continuous oscillation at frequency ω. The conditions for continuous oscillation at ω are therefore:

Conditions for continuous oscillation at ω

$$|G(j\omega)| \times |H(j\omega)| = 1$$
$$\arg G(j\omega) + \arg H(j\omega) = -180° \qquad [9.48]$$

Under these conditions there is a sinusoidal output signal at frequency ω for no input signal.

9.5.1 Oscillators

Here the feedback element $H(s)$ is an L—C—R circuit and $G(s)$ is an amplifier (maintaining amplifier). In Fig. 9.19(a) the L—C—R circuit consists of an inductive sensor (pure inductance with series loss resistance) in series with a fixed pure capacitance; in Fig. 9.19(b) the L—C—R circuit consists of a capacitive sensor (pure capacitance with parallel loss resistance) in series with a fixed pure inductance. From Fig. 9.19 we see that each circuit is described by a second-order differential equation and a corresponding second-order transfer function of the form:

$$H(s) = \frac{1}{\dfrac{1}{\omega_n^2}s^2 + \dfrac{2\xi}{\omega_n}s + 1} \qquad [9.49]$$

Fig. 9.19 Oscillators (a) Inductive sensor (b) Capacitive sensor

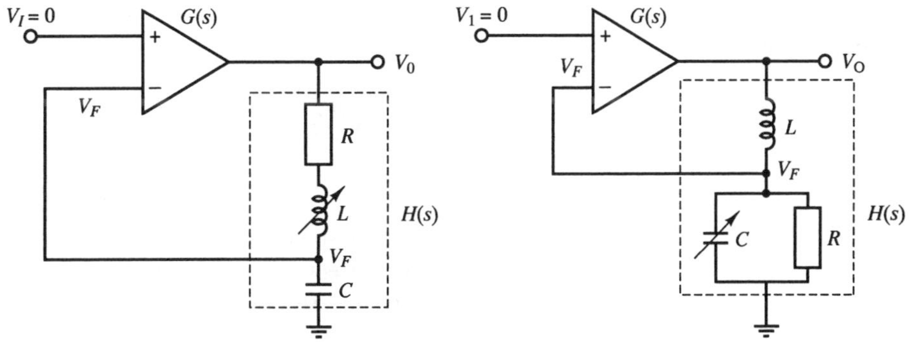

(a)

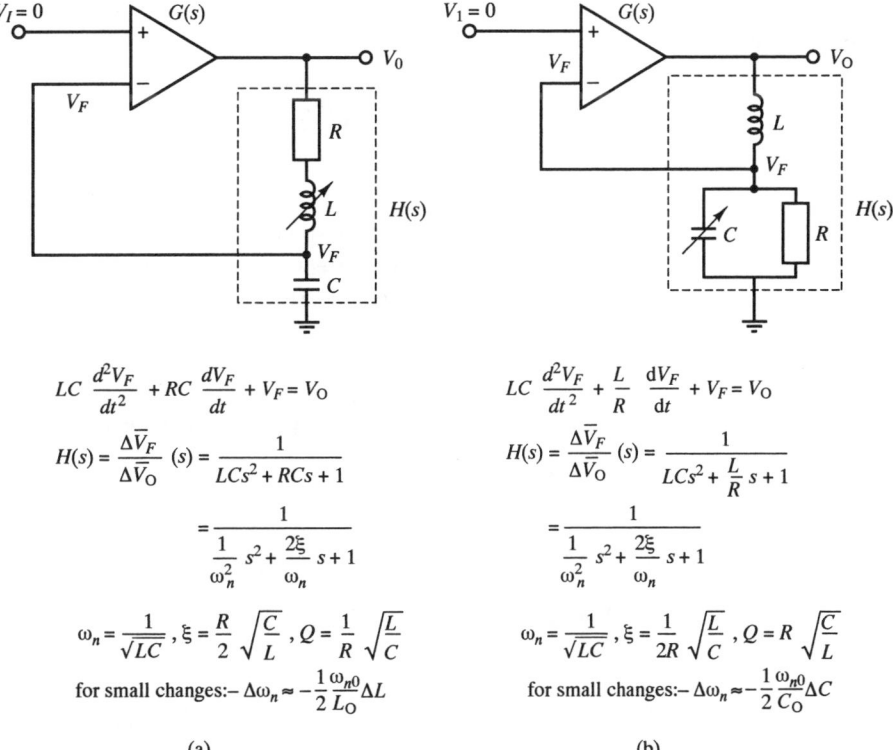

$$LC\frac{d^2V_F}{dt^2} + RC\frac{dV_F}{dt} + V_F = V_O$$

$$H(s) = \frac{\Delta\bar{V}_F}{\Delta\bar{V}_O}(s) = \frac{1}{LCs^2 + RCs + 1} = \frac{1}{\dfrac{1}{\omega_n^2}s^2 + \dfrac{2\xi}{\omega_n}s + 1}$$

$$\omega_n = \frac{1}{\sqrt{LC}},\ \xi = \frac{R}{2}\sqrt{\frac{C}{L}},\ Q = \frac{1}{R}\sqrt{\frac{L}{C}}$$

for small changes:– $\Delta\omega_n \approx -\dfrac{1}{2}\dfrac{\omega_{n0}}{L_O}\Delta L$

(b)

$$LC\frac{d^2V_F}{dt^2} + \frac{L}{R}\frac{dV_F}{dt} + V_F = V_O$$

$$H(s) = \frac{\Delta\bar{V}_F}{\Delta\bar{V}_O}(s) = \frac{1}{LCs^2 + \dfrac{L}{R}s + 1} = \frac{1}{\dfrac{1}{\omega_n^2}s^2 + \dfrac{2\xi}{\omega_n}s + 1}$$

$$\omega_n = \frac{1}{\sqrt{LC}},\ \xi = \frac{1}{2R}\sqrt{\frac{L}{C}},\ Q = R\sqrt{\frac{C}{L}}$$

for small changes:– $\Delta\omega_n \approx -\dfrac{1}{2}\dfrac{\omega_{n0}}{C_O}\Delta C$

The figure also gives expressions for the natural frequency ω_n, the damping ratio ξ and the Q factor for the circuits. Since $\omega_n = 1/\sqrt{(LC)}$ for both circuits, a change in L for the inductive sensor or a change in C for the capacitive sensor both result in a change in circuit natural frequency. We therefore design the system to oscillate at ω_n. From eqn [9.49] we have:

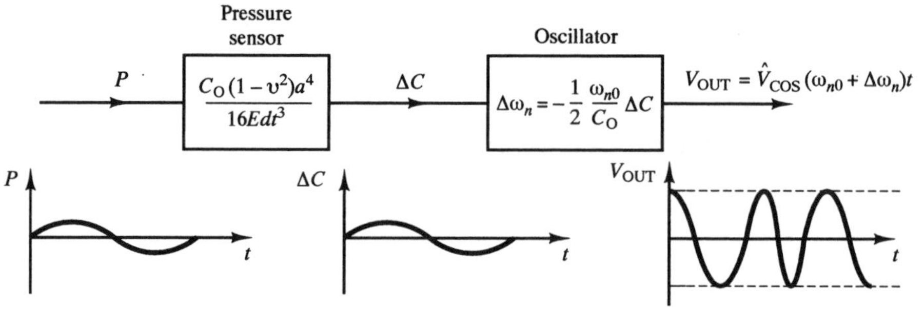

Fig. 9.20 Frequency pressure measurement system

$$H(j\omega) = \frac{1}{\left(1 - \frac{\omega^2}{\omega_n^2}\right) + j2\xi\frac{\omega}{\omega_n}} \quad \text{and} \quad H(j\omega_n) = \frac{1}{j2\xi} \qquad [9.50]$$

From eqns [9.48], in order to sustain oscillations at $\omega = \omega_n$, the gain and phase of the maintaining amplifier must satisfy the conditions:

Gain and phase conditions for maintaining amplifier in oscillator

$$|G(j\omega_n)| = \frac{1}{|H(j\omega_n)|} = 2\xi = \frac{1}{Q} \qquad [9.51]$$
$$\arg G(j\omega_n) = -180° - \arg H(j\omega_n) = -90°$$

over the range of ω_n corresponding to the range of variation of either L or C.

Figure 9.20 shows a frequency system for measuring sinusoidal variations in pressure. The capacitive pressure sensor of Fig. 8.5 is used; for small changes, the increase in capacitance ΔC is proportional to the applied pressure P (eqn [8.17]). The sensor is incorporated in the oscillator circuit of Fig. 9.19(b); the change $\Delta\omega_n$ in circuit natural frequency is approximately proportional to ΔC. The frequency of the oscillator output signal varies sinusoidally with time giving rise to the **frequency modulated signal** shown.

9.5.2 Resonators

Figure 9.21 shows the block diagram for a resonator.[4] The forward path element $G(s)$ is again an amplifier; the feedback element $H(s)$ consists of:

a drive element to convert voltage to force
an elastic force sensor to convert force to displacement
a displacement sensor to convert displacement into voltage.

From Fig. 9.21 we have:

$$H(s) = \frac{K_s K_D}{k} \frac{1}{\frac{1}{\omega_n^2}s^2 + \frac{2\xi}{\omega_n}s + 1} \qquad [9.52]$$

Fig. 9.21 Block diagram of resonator

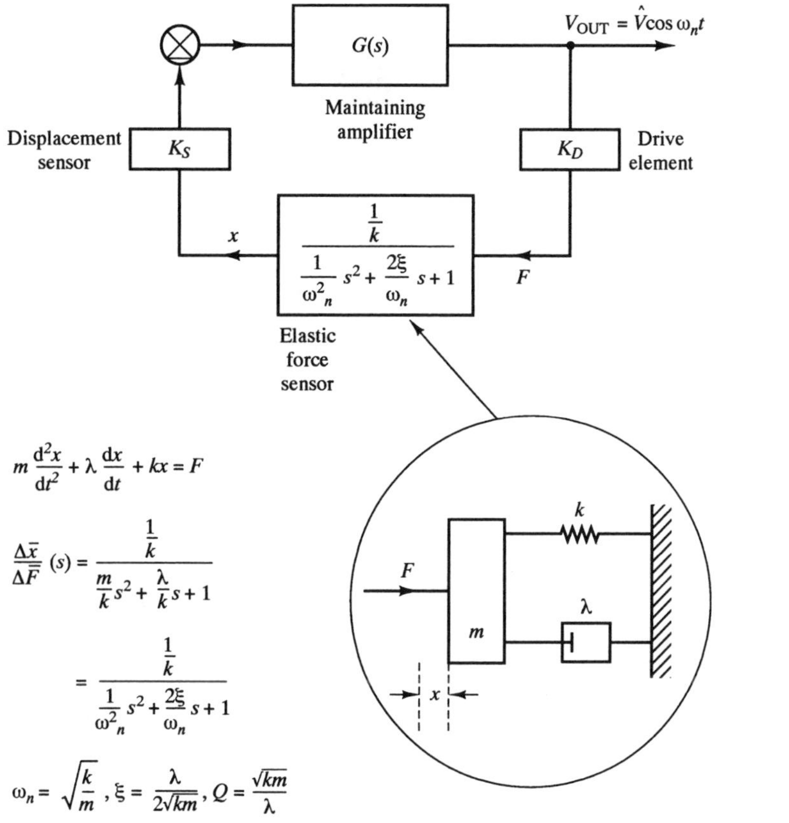

the figure also gives expressions for natural frequency ω_n, damping ratio ξ and Q factor. Since $\omega_n = \sqrt{(k/m)}$; changes in either effective stiffness k or effective mass m cause corresponding changes in ω_n. Thus providing the system is designed to oscillate at ω_n; it could be used, for example, to sense force variations (affecting k) or density variations (affecting m). From eqns [9.48] and [9.52], in order to sustain oscillations at $\omega = \omega_n$, the gain and phase of the maintaining amplifier must satisfy the conditions:

Gain and phase conditions for maintaining oscillations in a resonator

$$|G(j\omega_n)| = \frac{1}{|H(j\omega_n)|} = \frac{2\xi k}{K_s K_D} = \frac{\lambda}{K_s K_D}\sqrt{\frac{k}{m}} \qquad [9.53]$$

$$\arg G(j\omega_n) = -180° - \arg H(j\omega_n) = -90°$$

over the range of ω_n, corresponding to the range of variation in either m or k.

Figure 9.22(a) shows a wire length l, mass per unit length σ, held under tension T between two rigid supports and positioned in a transverse magnetic field B. If

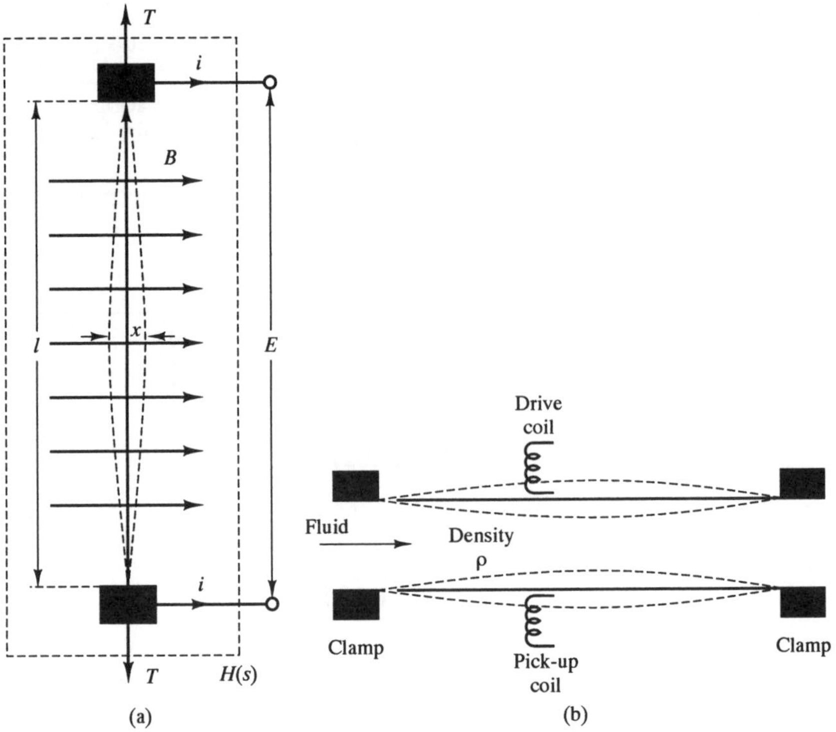

Fig. 9.22 Examples of resonators
(a) Vibrating wire element
(b) Vibrating tube element

the maintaining amplifier passes a current i through the wire, the corresponding deflecting force F on the wire is:

$$F = Bli \tag{9.54}$$

The transfer function relating the transverse deflection x of the wire centre to F is:

$$\frac{\Delta\bar{x}}{\Delta\bar{F}} = \frac{\frac{1}{k}}{\frac{1}{\omega_n^2}s^2 + \frac{2\xi}{\omega_n}s + 1} \tag{9.55}$$

where the fundamental frequency of transverse vibrations is:

$$\omega_n = 2\pi f_n = \frac{\pi}{l}\sqrt{\frac{T}{\sigma}} \tag{9.56}$$

The induced e.m.f. across the ends of the wire is:

$$E = Bl\dot{x} \tag{9.57}$$

where $\dot{x}$ is the velocity of the wire. The overall feedback transfer function is:

$$H(s) = \frac{\Delta\bar{E}}{\Delta\bar{i}}(s) = \frac{\Delta\bar{E}}{\Delta\bar{\dot{x}}}\,\frac{\Delta\bar{\dot{x}}}{\Delta\bar{F}}\,\frac{\Delta\bar{F}}{\Delta\bar{i}} \tag{9.58}$$

$$= \frac{(Bl)^2 \frac{1}{k}s}{\frac{1}{\omega_n^2}s^2 + \frac{2\xi}{\omega_n}s + 1}$$

By choosing $G(s)$ so that conditions [9.48] are obeyed at $\omega = \omega_n$, the system oscillates at the wire natural frequency:

$$f_n = \frac{1}{2l}\sqrt{\frac{T}{\sigma}}.$$

This vibrating wire principle is used in differential pressure transmitters[5] and load cells. A change in either input differential pressure or force causes T and f_n to change, producing a corresponding change in the frequency of the transducer output signal.

Figure 9.22(b) shows a vibrating thin-walled tube element, inside which a fluid is flowing.[6] The tube is clamped at each end and a transverse force is applied to the tube using a drive coil connected to the maintaining amplifier. This causes transverse deflection of the tube centre which is sensed by a pick-up coil and fed back to the amplifier. By correct choice of amplifier gain and phase characteristics, the tube is continuously maintained in transverse oscillation at a fundamental frequency $f_n = (1/2\pi)(\sqrt{k/m})$. The total vibrating mass m and therefore f_n depend on the fluid density ρ, as well as the mass of the tube itself. The frequency f_n of the resonator output signal depends on fluid density ρ according to the following non-linear relation:

$$\rho = \frac{A}{f_n^2} + \frac{B}{f_n} + C \qquad [9.59]$$

where A, B, C, are constants. This tube arrangement is particularly useful for liquids and liquid-solid mixtures.

Conclusion

This chapter has studied methods of converting the output of sensing elements into a form suitable for further processing, this is usually a d.c. voltage, d.c. current or variable frequency a.c. voltage. The chapter begins by examining the principles of **bridge circuits** which convert a change in resistance, capacitance or inductance into a d.c. or a.c. voltage. It then goes on to study the principles and characteristics of ideal and practical **operational amplifiers**. The third section looks at **a.c. carrier systems** where **amplitude modulation** is used to reject external interference. In the fourth section three types of **current transmitter** are studied, **closed-loop, open-loop** and **intelligent**, these give a standard 4 to 20 mA output signal. The final section examines **oscillators** and **resonators** which give a variable frequency signal, i.e. **frequency modulation**.

References

9.1 WONG Y J and OTT G E 1977 *Function Circuits — Design and Applications*, Burr Brown Electronics series, McGraw-Hill, New York.

9.2 SMITH J L 1971 *Modern Operational Amplifier Design*, Wiley, New York.
9.3 Rosemount Ltd. 1989 *Technical Information on Model 3051C Smart Differential Pressure Transmitter*.
9.4 LANGDON R M 1985 'Resonator sensors — a review', *J. Phys. E: Scientific Instruments*, vol. 18, pp. 103–15.
9.5 The Foxboro Company 1981 *Instruction manual on 823 DP—F resonant wire differential pressure transmitter*.
9.6 Sarosota Automation Instrumentation Division 1985 *Technical information on FD 800 series density meters*.

Problems

9.1 A platinum resistance thermometer has a resistance of 100 Ω at 0 °C and a temperature coefficient of resistance of $4 \times 10^{-3}\,°C^{-1}$. Given that a 15 V supply is available, design a deflection bridge giving an output range of 0 to 100 mV for an input range of 0 to 100 °C:

(a) using the procedure summarised by eqns [9.7] and [9.8];
(b) using the linear approximation of eqn [9.15).

Given values for all circuit components and assume a high impedance load.

(c) How should the circuit be altered if the input range is changed to 50 to 150 °C?

9.2 A thermocouple has an e.m.f. of 4.1 mV at 100 °C and 16.4 mV at 400 °C relative to a cold junction of 0 °C. A deflection bridge incorporating a nickel metal resistance thermometer is to be used as the voltage source $E_{T_2,0}$ necessary for automatic reference compensation of the thermocouple e.m.f. The nickel thermometer has a resistance of 10 Ω at 0 °C and a temperature coefficient of resistance of $6.8 \times 10^{-3}\,°C^{-1}$. Design the deflection bridge assuming a 1 V supply is available (*Hint*: eqns [8.44] and [9.15]).

9.3 The resistance R_θ kΩ of a thermistor at θK is given by:

$$R_\theta = 1.68 \exp\left[3050\left(\frac{1}{\theta} - \frac{1}{298}\right)\right]$$

The thermistor is incorporated into the deflection bridge circuit shown in Fig. Prob. 3.

(a) Assuming that V_{OUT} is measured with a detector of infinite impedance
 (i) Calculate the range of V_{OUT} corresponding to an input temperature range of 0 to 50 °C.
 (ii) The non-linearity at 12 °C as a percentage of full-scale deflection.
(b) Calculate the effect on the range of V_{OUT} of reducing the detector impedance to 1 kΩ.

Fig. Prob. 3

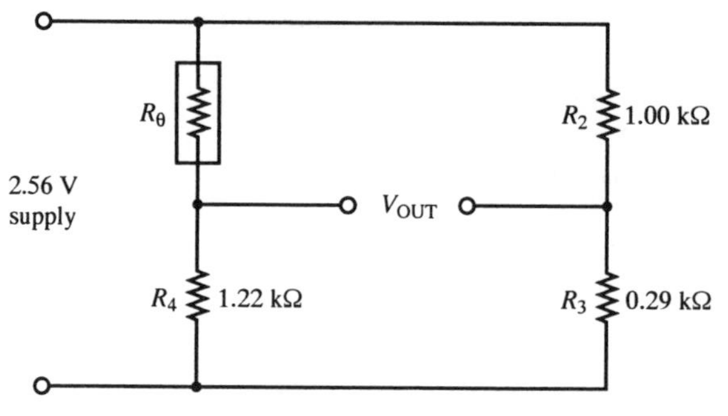

9.4 The resistance R_θ Ω of a thermistor varies with temperature θ K according to the following equation:

$$R_\theta = 0.0585 \exp\left(\frac{3260}{\theta}\right)$$

Design a deflection bridge, incorporating the thermistor, to the following specification:

(a) input range 0 to 50 °C;
(b) output range 0 to 1.0 V;
(c) relationship between output and input to be approximately linear.

9.5 Four strain gauges, with specification given below, are available to measure the torque on a cylindrical shaft 4 cm in diameter connecting a motor and load.

(a) Draw clearly labelled diagrams showing:
  (i) the arrangement of the gauges on the shaft
  (ii) the arrangement of the gauges in the bridge circuit, for optimum accuracy and sensitivity.

(b) Calculate the maximum achievable bridge out-of-balance voltage for an applied torque T of 10^3 N m given the following:

Tensile and compressive strains = $\pm T/\pi S a^3$ where $S = 1.1 \times 10^{11}$ N m^{-2} is the shear modulus of the shaft material and a is the radius of the shaft in metres.

Strain gauge data:	resistance	= 120 Ω
	gauge factor	= 2.1
	maximum current	= 50 mA

9.6 A load cell consists of a domed vertical steel cylinder 20 cm high and 15 cm in diameter. Four flat surfaces, at right angles to each other, are cut on the vertical surface so as to form 10 cm squares. Resistance strain gauges are attached to these flat surfaces so that two gauges (on opposite faces) suffer longitudinal compression and two gauges (on the other pair of opposite faces) suffer transverse tension. The strain gauges have the following specification:

Resistance = 100 Ω
Gauge factor = 2.1
Maximum gauge current = 30 mA

The gauges are connected in a temperature compensated bridge and the out-of-balance signal is input to a differential amplifier. Calculate the minimum amplifier gain if the amplifier output voltage is to be 1 V for a compressive force of 10^5 N

Young's modulus for steel = 2.1×10^{11} N m^{-2}
Poisson's ratio for steel = 0.29

9.7 The unbonded strain gauge accelerometer of Fig. 8.13 and Section 9.1.2 has a natural frequency of 175 Hz and an input range of 0 to 5 g. Given that the strain due to prestressing is twice the maximum acceleration induced strain, use the strain gauge data given below to calculate the range of the bridge output voltage.

Strain gauge data:	Resistance	= 120 Ω
	Gauge factor	= 2.1
	Maximum current	= 50 mA
	Length at zero acceleration	= 2.3 cm.

9.8 The capacitance level transducer of Section 8.2 and Fig. 8.5 is to be used to measure the depth h of liquid in a tank between 0 and 7 m. The total length l of the transducer is 8 m and the ratio b/a of the diameters of the concentric cylinders is 2.0. The dielectric constant ϵ of the liquid is 2.4 and the permittivity of free space ϵ_0 is 8.85 pF m^{-1}. the transducer is incorporated into the deflection bridge of Fig. 9.5(a) with $R_2 = 100\ \Omega$, $R_3 = 10\ \text{k}\Omega$, $\hat{V}_s = 15$ V.

(a) Calculate the value of C_0 so that the amplitude $\hat{E}_{Th}$ is zero when the tank is empty.
(b) Using this value of C_0 calculate $\hat{E}_{Th}$ at maximum level
(c) Explain why the relationship between $\hat{E}_{Th}$ and h is non-linear and calculate the non-linearity at $h = 3.5$ m as a percentage of full-scale deflection.

9.9 Figure Prob. 9 shows a variable reluctance force sensor which is incorporated into the bridge circuit of Fig. 9.5(b). When the applied force is zero the armature is positioned along the centre line *AB*.

Fig. Prob. 9

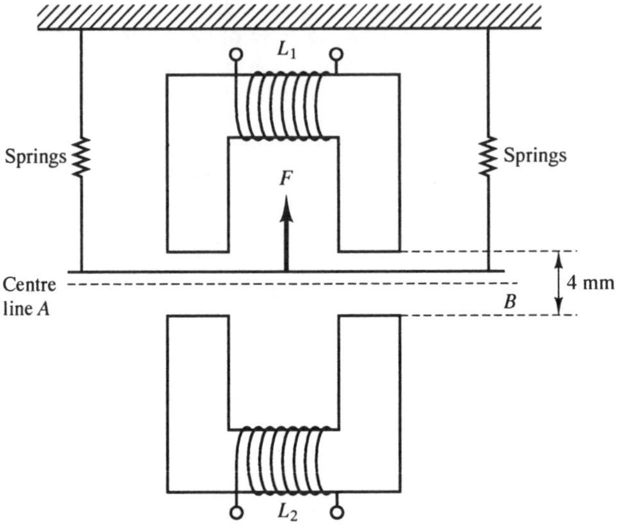

(a) Explain why the sensor would be suitable for measuring force signals containing frequencies between 0 and 10 Hz.
(b) Sketch the frequency spectrum of the bridge output voltage.
(c) Use the data given to calculate the form of the bridge output voltage when $F = +1.0$ N and $F = -1.0$ N.
(d) Using the results of (c) explain how to demodulate the bridge output voltage.

Data. Overall spring stiffness $= 10^3$ N m^{-1}
Effective mass of spring and armature $= 25 \times 10^{-3}$ kg
Damping ratio $= 0.7$

$$\text{Inductance of each coil} = \frac{20}{1 + 2d}\ \text{mH} \quad (d = \text{air gap in mm})$$

Amplitude of bridge supply = 1 V
Frequency of bridge supply = 1000 Hz.

9.10 Use eqn [9.26] to derive the ideal transfer characteristics of the operational amplifier circuits shown in Figs 9.7(b) to 9.7(f).

9.11 Design an a.c. amplifier system which incorporates an ideal operational amplifier, to meet the following specification.

Midband input impedance = 10 kΩ

Midband gain = 100
Bandwidth = 100 to 1000 Hz

9.12 The electronic torque balance D/P transmitter of Fig. 9.14(a) is to have an output range of 4 to 20 mA for input ranges between 0 to 0.5 m of water and 0 to 5.0 m of water. The diaphragm has a diameter of 10 cm and the maximum values of *a/b* is 2.0.

(a) Complete the design by calculating the electromagnetic force constant and the zero spring force;

(b) Find the values of a/b corresponding to input ranges of 0 to 1 m and 0 to 5 m.

$g = 9.81\ \text{m s}^{-2}$, density of water $= 10^3\ \text{kg m}^{-3}$

9.13 The variable reluctance displacement transducer of problem 8.4 is incorporated into an electrical oscillator circuit with a fixed capacitance of 500 pF. Find the variation in frequency of the oscillator output signal corresponding to a variation in air gap of between 1 and 3 mm.

9.14 The natural frequency f_n Hz of a thin-walled tube executing circumferential vibrations in a fluid of density ρ kg m^{-3} is given by:

$$f_n^2 = \frac{1.25 \times 10^{10}}{(\rho + 350)}$$

The stiffness of the tube is $10^9\ \text{N m}^{-1}$ and the damping ratio $\xi = 0.1$. The tube is incorporated in a closed loop system which also includes a drive coil of sensitivity $10^4\ \text{N V}^{-1}$ and a displacement sensing coil of sensitivity $10^3\ \text{V m}^{-1}$. The system is required to give a sinusoidal voltage output signal whose frequency changes with fluid density in the range $\rho = 250$ to $1500\ \text{kg m}^{-3}$.
Find the gain and phase characteristics of the maintaining amplifier. What is the effect of changes in fluid viscosity on the system?

9.15 A solid state capacitive humidity sensor has a capacitance given by:

$$C = 1.7\,\text{RH} + 365\ \text{pF}$$

where RH is the percentage relative humidity. The sensor has an associated parallel resistance of 100 kΩ. The sensor is incorporated into a feedback oscillator system with a pure inductance and a maintaining amplifier. The oscillator is to give a sinusoidal output voltage at the natural frequency of the L—C circuit for a relative humidity between 5 and 100%.

(a) Draw a diagram of a suitable oscillator system.

(b) If the frequency of the output signal is to be 100 kHz at RH = 100% calculate the required inductance.

(c) Find the gain and phase characteristics of the maintaining amplifier.

9.16 Figure Prob. 16 shows a four-lead bridge circuit; R_c is the resistance of the leads connecting

Fig. Prob. 16

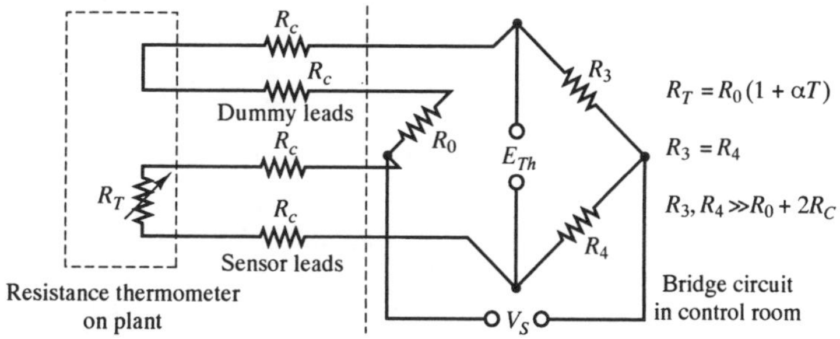

the sensor to the bridge circuit. Show that $E_{Th} \approx V_s(R_0/R_3)\alpha T$, i.e. the bridge output voltage is unaffected by changes in R_c.

9.17 (a) A stretched thin wire, clamped at each end, has length l, effective mass M, elastic stiffness k and damping constant λ. The wire can be displaced along a direction perpendicular to its axis, within a mutually perpendicular magnetic field of flux density B. Derive the transfer function relating the voltage E induced across the ends of the wire to the current i applied to the wire.

(b) The above wire is to be incorporated into a resonator system for force measurement. The system maintains the wire in continuous transverse oscillation at the fundamental natural frequency f_n, which depends on the tensile force F applied to the wire. The system is required to give a sinusoidal output signal, with variable frequency, for a force input F, between 1×10^5 and 10×10^5 N. Use the data given below to calculate:

(i) the frequency range of the resonator output signal

(ii) the gain and phase characteristics of the maintaining amplifier.

wire mass M	$= 0.05$ kg
wire length l	$= 0.1$ m
wire stiffness k	$= \pi^2(F/l)$ N m^{-1}
Damping constant λ	$= 10$ Ns m^{-1}
Flux density B	$= 10^3$ Wb m^{-2}

10

Signal Processing Elements and Software

The output signal from the conditioning elements is usually in the form of a d.c. voltage, d.c. current or variable frequency a.c. voltage. In many cases calculations must be performed on the conditioning element output signal in order to establish the value of the variable being measured. Examples are the calculation of temperature from a thermocouple EMF signal, the calculation of total mass of product gas from flow rate and density signals. These calculations are referred to as **signal processing** and are usually performed digitally using a microcomputer. The first section of this chapter discusses the principles of **analogue-to-digital conversion** and the operation of typical analogue-to-digital converters. The following section explains the structure and operation of typical **microcomputer systems**. The various forms of microcomputer **software**, both low- and high-level languages, are discussed in the next section. The final section looks at the two main types of signal processing calculation in measurement systems; steady-state calculation of measured value and dynamic digital **compensation** and **filtering**.

10 Analogue-to-digital (A/D) conversion

This section commences by discussing the three operations involved in A/D conversion; these are **sampling, quantisation** and **encoding**. The first operation is performed by a **sample-and-hold device**; the second and third are combined in an **analogue-to-digital converter**.

10.1.1 Sampling

In Chapter 6 we saw that a continuous signal $y(t)$ could be represented by a set of samples y_i, $i = 1, \ldots, N$, taken at discrete intervals of time ΔT (sampling interval). The operation is shown in Fig. 10.1; the switch is closed f_S times per second, where **sampling frequency** $f_S = 1/(\Delta T)$. In order for the sampled signal $y_S(t)$ to be an adequate representation of $y(t)$, f_S should satisfy the conditions of the Nyquist sampling theorem, which can be stated as follows:

A continuous signal can be represented by, and reconstituted from, a set

Fig. 10.1 Time waveform and frequency spectrum of sampled signal

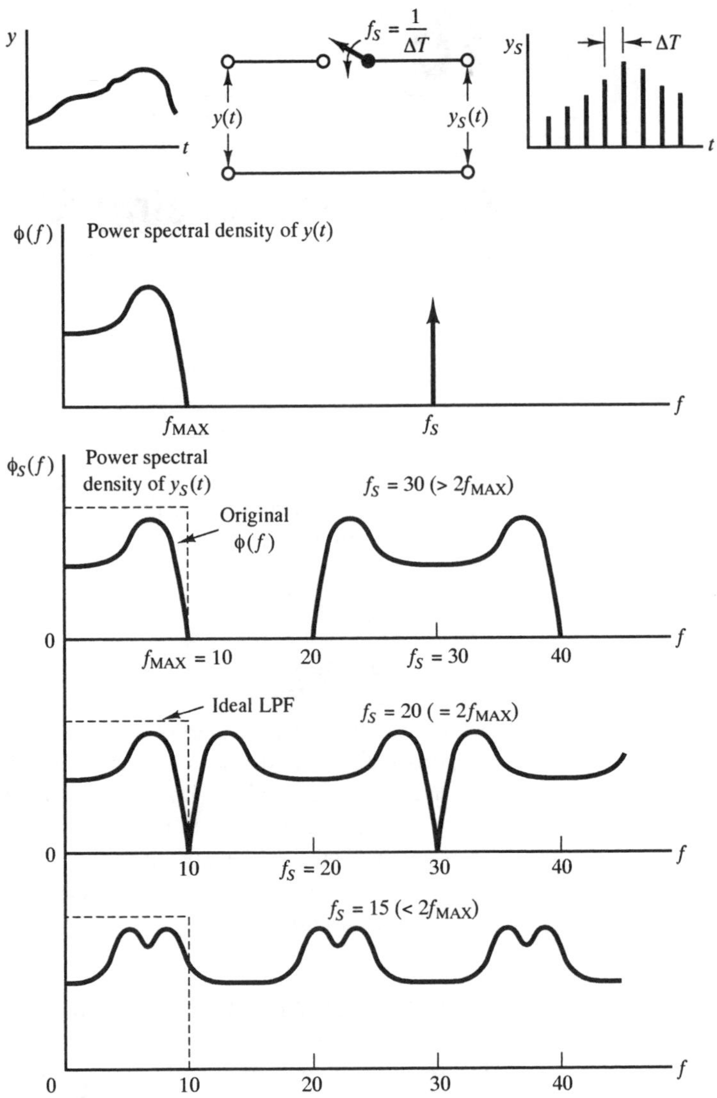

of sample values providing that the number of samples per second is at least twice the highest frequency present in the signal.

Mathematically we require:

Nyquist sampling theorem

$$f_S \geq 2f_{MAX} \quad [10.1]$$

where f_{MAX} is the frequency beyond which the continuous signal power spectral density $\phi(f)$ becomes negligible (Chapter 6 and Fig. 10.1).

The above result can be explained by examining the power spectral density $\phi_S(f)$ of the sampled signal. Because the sampled signal is a series of sharp pulses, $\phi_S(f)$ contains additional frequency components which are centred on multiples of the

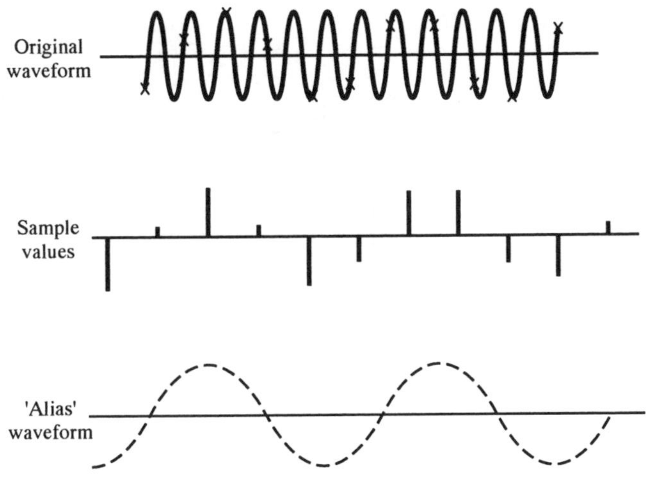

Fig. 10.2 Aliasing

sampling frequency. Figure 10.1 shows $\phi_S(f)$ in the three situations $f_S > 2f_{MAX}$, $f_S = 2f_{MAX}$ and $f_S < 2f_{MAX}$. If $f_S > 2f_{MAX}$, then the additional frequency components can easily be filtered out with an ideal low-pass filter of bandwidth 0 to f_{MAX} and the original signal reconstituted. If $f = 2f_{MAX}$, it is just possible to filter out the sampling components and reconstitute the signal. If $f < 2f_{MAX}$, the sampling components occupy the same frequency range as the original signal and it is impossible to filter them out and reconstitute the signal.

The effect of sampling at too low a frequency is shown in Fig. 10.2. Here a sine wave of period 1 s, i.e. frequency 1 Hz, is being sampled approximately once every second, i.e. the sampling frequency is below the Nyquist minimum of 2 samples/second. The diagram shows that it is possible to reconstruct an entirely different sine wave of far lower frequency from the sample values. This is referred to as the 'alias' of the original signal; it is impossible to decide whether the sample values are derived from the original signal or its alias. the phenomenon of two different signals being constructed from a given set of sample values is referred to as **aliasing**.

The operation of analogue-to-digital conversion can take up to a few milliseconds; it is necessary therefore to hold the output of the sampler constant at the sampled value while the conversion takes place. This is done using a **sample-and-hold device** as shown in Fig. 10.3. In the **sample** state the output signal follows the input signal: in the **hold** state the output signal is held constant at the value of the input signal at the instant of time the **hold** command is sent. The sample-and-hold waveform shown is ideal; in practice errors can occur due to the finite time for the transition between sample-and-hold states (aperture time) and reduction in the hold signal (droop).

10.1.2 Quantisation

Although the above sample values are taken at discrete intervals of time, the values y_i can take any value in the signal range y_{MIN} to y_{MAX} (Fig. 10.4). In quantisation

Fig. 10.3 Sample and hold

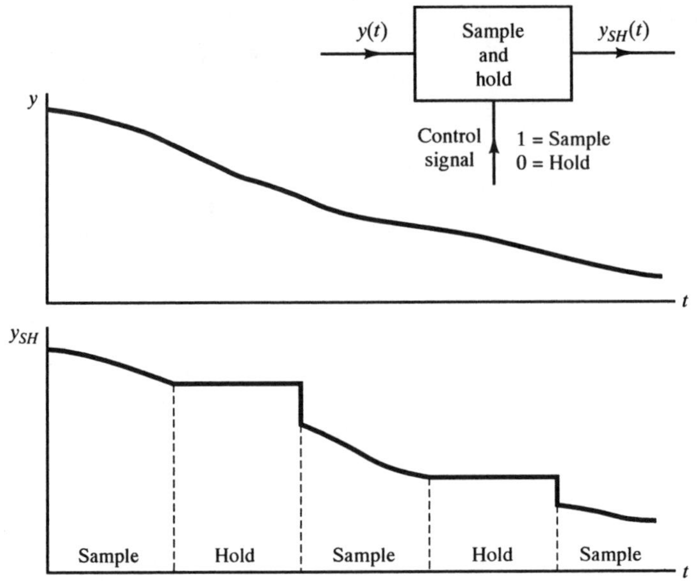

Fig. 10.4 Quantisation

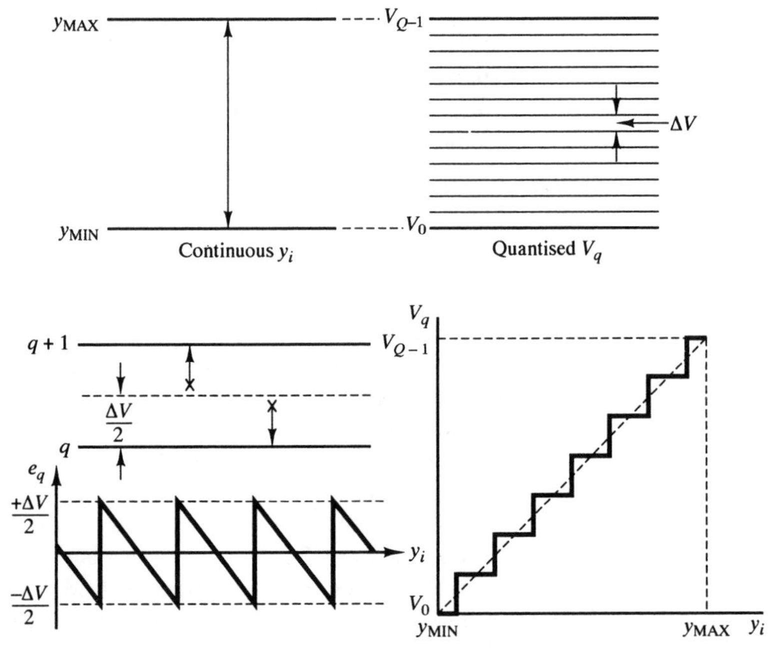

the sample voltages are rounded either up or down to one of Q quantisation values or levels V_q, where $q = 0, 1, 2, \ldots, Q - 1$. These quantum levels correspond to the Q decimal numbers 0, 1, 2, ... $Q - 1$. If $V_0 = y_{\text{MIN}}$ and $V_{Q-1} = y_{\text{MAX}}$, then there are $(Q - 1)$ spacings occupying a span of $y_{\text{MAX}} - y_{\text{MIN}}$. The spacing width or quantisation interval ΔV is therefore:

$$\Delta V = \frac{y_{\text{MAX}} - y_{\text{MIN}}}{Q - 1} \qquad [10.2]$$

The operation of quantisation produces an error $e_q = V_q - y_i$ termed the **quantisation error**. Normally if y_i is above the halfway point between two levels q, $q + 1$ it is rounded up to V_{q+1}; if y_i is below half-way it is rounded down to V_q (Fig. 10.4). The maximum quantisation error e_q^{MAX} is therefore $\pm\Delta V/2$ or expressed as a percentage of span $y_{MAX} - y_{MIN}$:

Maximum percentage quantisation error

$$e_q^{MAX} = \pm \frac{\Delta V}{2(y_{MAX} - y_{MIN})} \times 100 \text{ per cent} = \pm \frac{100}{2(Q-1)} \text{ per cent} \qquad [10.3]$$

We see that the relationship between V_q and y_i is characterised by a series of discrete steps or jumps, this is an example of resolution errors discussed in Chapter 2. Obviously the greater the number of levels Q, the lower the quantisation errors.

10.1.3 Encoding

The encoder converts the quantisation values V_q into a parallel digital signal corresponding to a binary coded version of the decimal numbers 0, 1, 2, . . ., $Q - 1$. The commonly used decimal or denary number system uses a **base** or **radix** of 10, so that any positive integer is expressed as a series of powers of 10 (decades):

$$d_n \times 10^n + d_{n-1} \times 10^{n-1} + \cdots + d_i \times 10^i + \cdots + d_1 \times 10^1 + d_0 \times 10^0 \qquad [10.4]$$

where the d_i are the respective weights or digits which take the values 0 to 9. In digital computers the **binary number system** is used. This has a base of 2 that any positive integer can be expressed as a series of powers of 2:

$$\underset{\substack{\uparrow \\ \text{MSB} \\ \text{(Most significant bit)}}}{b_n \times 2^n} + b_{n-1} \times 2^{n-1} + \cdots + b_i \times 2^i + \cdots + b_1 \times {}^1 + \underset{\substack{\uparrow \\ \text{LSB} \\ \text{(Least significant bit)}}}{b_0 \times 2^0} \qquad [10.5]$$

where the weightings b_i are referred to as bits or digits. A bit can take only the values 0 or 1 so that calculations on binary numbers are easily performed by logic circuits which distinguish between two states — on or off, true or false. In order to convert the decimal number 183 to binary, we express it in the form:

$$1 \times (128) + 0 \times (64) + 1 \times (32) + 1 \times (16) + 0 \times (8) + 1 \times (4) + 1 \times (2) + 1 \times (1)$$

i.e.

$$1 \times 2^7 + 0 \times 2^6 + 1 \times 2^5 + 1 \times 2^4 + 0 \times 2^3 + 1 \times 2^2 + 1 \times 2^1 + 1 \times 2^0$$

giving the 8-bit binary number 10110111. To produce an electrical signal corresponding to this number we require 8 wires in parallel, the voltage on each wire being typically 5 V for a '1' and 0 V for a '0'; this signal is an 8-bit parallel digital signal. The number of binary digits n required to encode Q decimal numbers is given by:

$Q = 2^n$ i.e.

Table 10.1

Analogue input (V)	Decimal number	Digital output
0	0	00000000
1.2	$\frac{1.2}{5.0} \times 255 = 61.2 \approx 61$	00111101
3.7	$\frac{3.7}{5.0} \times 255 = 188.7 \approx 189$	10111101
5	255	11111111

Number of digits in binary code

$$n = \log_2 Q = \frac{\log_{10} Q}{\log_{10} 2} \qquad [10.6]$$

Thus if $Q = 200$, $n = \log_{10} 200/\log_{10} 2 = 2.301/0.301 = 7.64$. Since, however, n must be an integer, we require 8 bits which corresponds to $Q = 2^8 = 256$. From eqn [10.3], the corresponding maximum quantisation error is $\pm\ 100/2(255)\% = +\ 0.196\%$. If the input range of the converter is 0 to 5 V, then the corresponding analogue input, decimal numbers and digital output signals are as shown in Table 10.1.

Other commonly used codes are as follows:

Binary coded decimal (b.c.d.)

Here each decade of the decimal number is *separately* coded into binary. Since $2^3 = 8$ and $2^4 = 16$, four binary digits DCBA are required to encode the 10 numbers 0 to 9 in each decade. In 8:4:2:1 b.c.d. $A \equiv 2^0 = 1$, $B \equiv 2^1 = 2$, $C \equiv 2^2 = 4$, $D \equiv 2^3 = 8$; thus the decimal number 369 becomes

DCBA	DCBA	DCBA
0011	0110	1001
3	6	9

The number of decades p of b.c.d. required to encode Q decimal numbers is given by: $Q = 10^p$, i.e. $p = \log_{10} Q$ and the corresponding total number of binary digits is:

Number of digits in b.c.d.

$$4p = 4 \log_{10} Q \qquad [10.7]$$

The input signal to alphanumeric displays (Chapter 11) is normally in b.c.d. form; since the signal is already separated into decades the conversion into seven segment or hexadecimal dot code is easier than with pure binary.

Octal code

Here a base of 8 is used so that the weights take the values 0 to 7. In order to convert 183 decimal, i.e. $(183)_{10}$, into octal we express it in the form:

$$2(64) + 6(8) + 7(1)$$

i.e.

$$2 \times 8^2 + 6 \times 8^1 + 7 \times 8^0$$

i.e.

267 octal or $(267)_8$

A binary number is easily coded into octal by arranging the digits in groups of 3, where for each group $A \equiv 2^0 = 1$, $B \equiv 2^1 = 2$, $C \equiv 2^2 = 4$. Thus 010110111 (binary code for $(183)_{10}$) becomes:

CBA	CBA	CBA	
010	110	111	
2	6	7	i.e. 267 octal

Hexadecimal code (hex)

Here a base of 16 is used and the digits are the 10 numbers 0 to 9 and the 6 letters ABCDEF. Since $2^4 = 16$, each hexadecimal digit corresponds to 4 binary digits:

Binary	0000	0001	0010	0011	0100	0101	0110	0111
Hex	0	1	2	3	4	5	6	7
Binary	1000	1001	1010	1011	1100	1101	1110	1111
Hex	8	9	A	B	C	D	E	F

Some corresponding decimal, binary and hexadecimal numbers are

Decimal	Binary	Hexadecimal
94	01011110	5E
167	10100111	A7
238	11101110	EE

Floating point representation

Here a number is expressed in the form aN^b where a is the **mantissa**, b the **exponent** and N the **base** of the number system. Thus for a denary or decimal system with $N = 10$, we have:

$$-43\,700 = -.437 \times 10^{+5}$$
$$+0.000\,259 = +.259 \times 10^{-3}$$

Both the mantissa and the exponent, each with a corresponding sign bit, can then be separately represented inside a computer using either binary, b.c.d. or hexadecimal code. Thus if 3-decade b.c.d. is used for the mantissa and 4-digit binary for the exponent, with 0 corresponding to − and 1 to +, the above numbers can be repesented by:

1	0101	0	0100	0011	0111
0	0011	1	0010	0101	1001
sign bit	exponent	sign bit	mantissa		

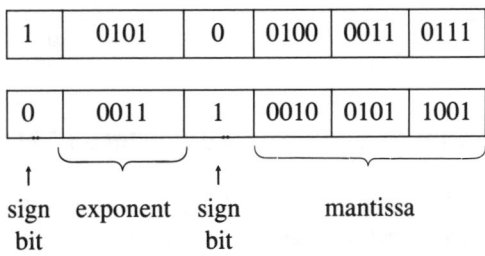

This representation is used when the range of numbers in a given computation is very wide. For example the above format covers the range from 0.001×10^{-15} to $0.999 \times 10^{+15}$. Arithmetic operations are performed on exponent and mantissa separately.

10.1.4 Frequency to digital conversion

In several cases the output signal from primary sensing or signal conditioning elements is an a.c. voltage with a frequency which depends on the measured variable. Examples are:

the variable reluctance tachogenerator (Section 8.4);
electrical and electromechanical oscillators for displacement, density and differential pressure measurement (Section 9.4);
turbine, vortex and Doppler flow meters (Sections 12.2 and 16.4).

Figure 10.5 shows how a frequency signal can be converted into parallel digital form. The Schmitt trigger converts the sinusoidal signal to a 0–5 V square wave signal which is the clock input to an 8-bit binary counter. The number of clock pulses during a fixed time interval ΔT is counted, the total count N is then proportional to the frequency f of the signal, i.e.

$$N = f\Delta T \qquad [10.8]$$

The operation of the counter is controlled using **stop count** and **reset** (to zero) logic signals; this system provides the basis of the mirocomputer-based speed measurement system discussed in Section 10.4.1.

Fig. 10.5 Frequency to digital conversion

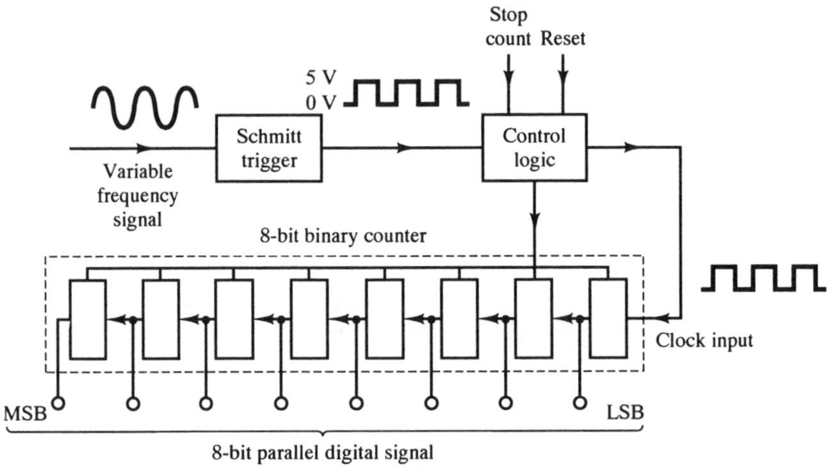

10.1.5 Digital-to-analogue converters (DACs)

A DAC gives an analogue output voltage which is proportional to an input parallel digital signal, e.g. an 8-bit binary singal $b_7b_6 \cdots b_1b_0$. In Fig. 10.6 an operational

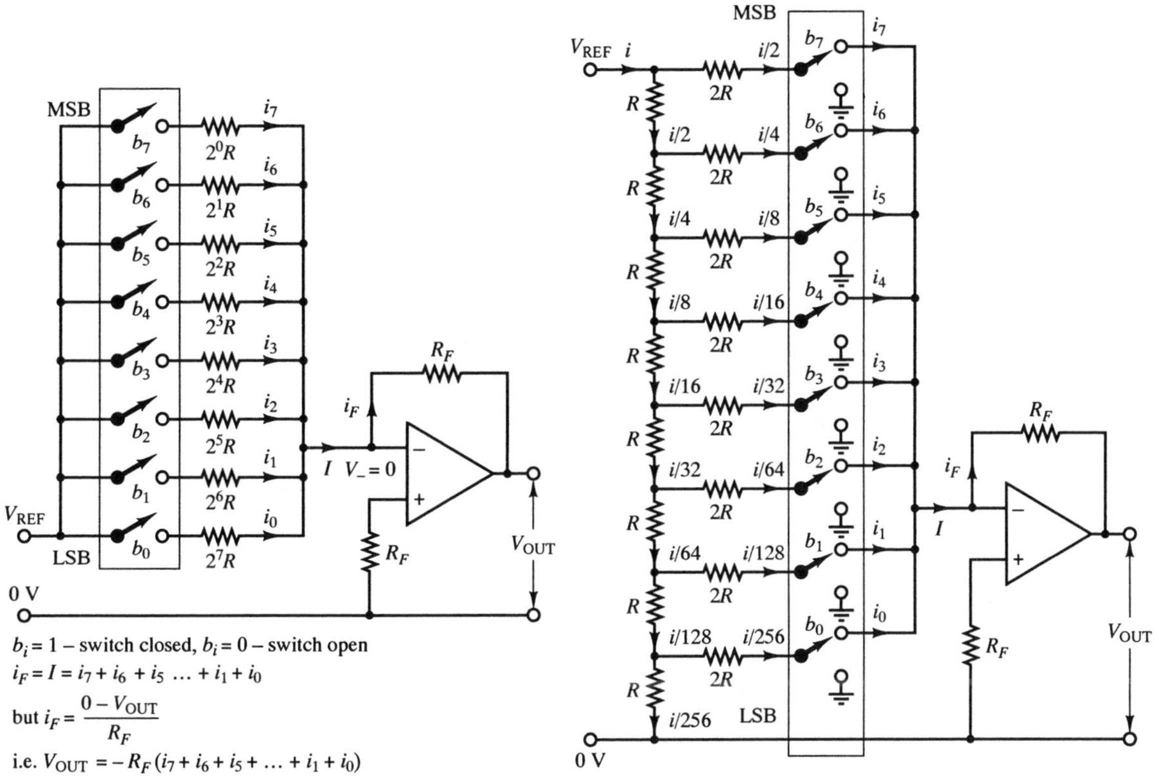

$b_i = 1$ – switch closed, $b_i = 0$ – switch open

$i_F = I = i_7 + i_6 + i_5 \ldots + i_1 + i_0$

but $i_F = \dfrac{0 - V_{OUT}}{R_F}$

i.e. $V_{OUT} = -R_F(i_7 + i_6 + i_5 + \ldots + i_1 + i_0)$

$$i_7 = \frac{V_{REF}}{2^0 R} b_7,\ i_6 = \frac{V_{REF}}{2^1 R} b_6, \ldots, i_1 = \frac{V_{REF}}{2^6 R} b_1,\ i_0 = \frac{V_{REF}}{2^7 R} b_0$$

$$V_{OUT} = \frac{-R_F}{R} V_{REF} \left[\frac{b_7}{2^0} + \frac{b_6}{2^1} + \frac{b_5}{2^2} + \ldots \frac{b_1}{2^6} + \frac{b_0}{2^7}\right]$$

(a)

$b_i = 1$ – switch up $b_i = 0$ – switch down

$i_7 = i/2\, b_7,\ i_6 = i/4\, b_6,\ i_5 = i/8\, b_5, \ldots\ i_1 = i/128\, b_1,\ i_0 = i/256\, b_0$

$$V_{OUT} = \frac{-R_F i}{2} \left[\frac{b_7}{2^0} + \frac{b_6}{2^1} + \frac{b_5}{2^2} + \ldots \frac{b_1}{2^6} + \frac{b_0}{2^7}\right]$$

(b)

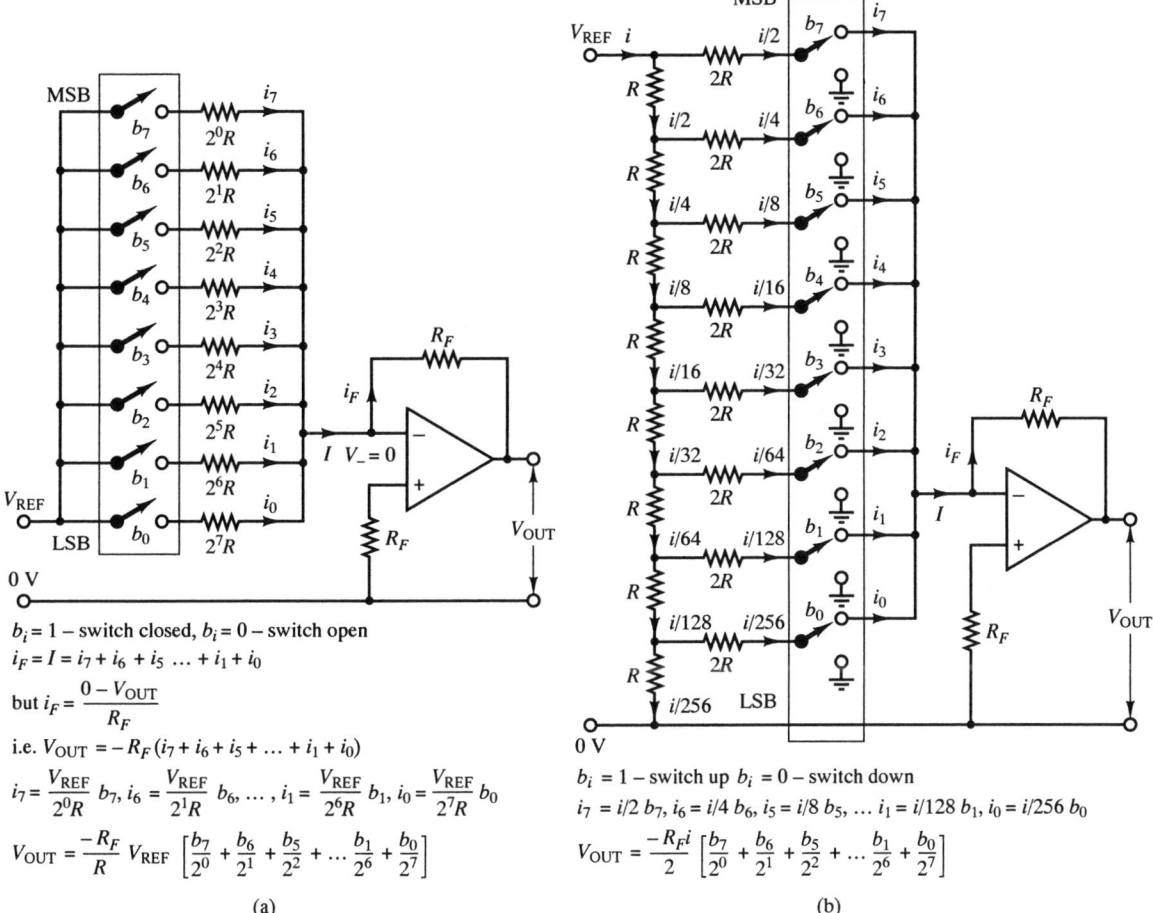

Fig. 10.6 Summing amplifier DACs
(a) Binary weighted resistor network
(b) *R*–2*R* ladder network

amplifier is used to sum a number of currents which are either zero or non-zero depending on whether the coresponding bit is 0 or 1. The current corresponding to the most significant bit is twice that corresponding to the next significant bit and so on. This is achieved in Fig. 10.6(a) by using a network of binary weighted resistors 2^0R, 2^1R, 2^2R, ..., 2^7R. The problem with this arrangement is that a very large range of resistance values is required. A better alternative is the ladder network shown in Fig. 10.6(b); the advantage of this circuit is that the required current distribution can be obtained with only two values of resistance R, $2R$.

10.1.6 Analogue-to-digital converters (ADCs)

Several types of ADC use a digital-to-analogue converter (DAC) in a closed loop arrangement. One common type is the **counter-ramp converter** shown in Fig. 10.7(a).[1] On receipt of the leading edge of the **initiate conversion** pulse the counter is reset to zero on receipt of the trailing edge the counter starts to count

Fig. 10.7 Analogue to digital converters
(a) Counter-ramp converter
(b) Successive approximation converter

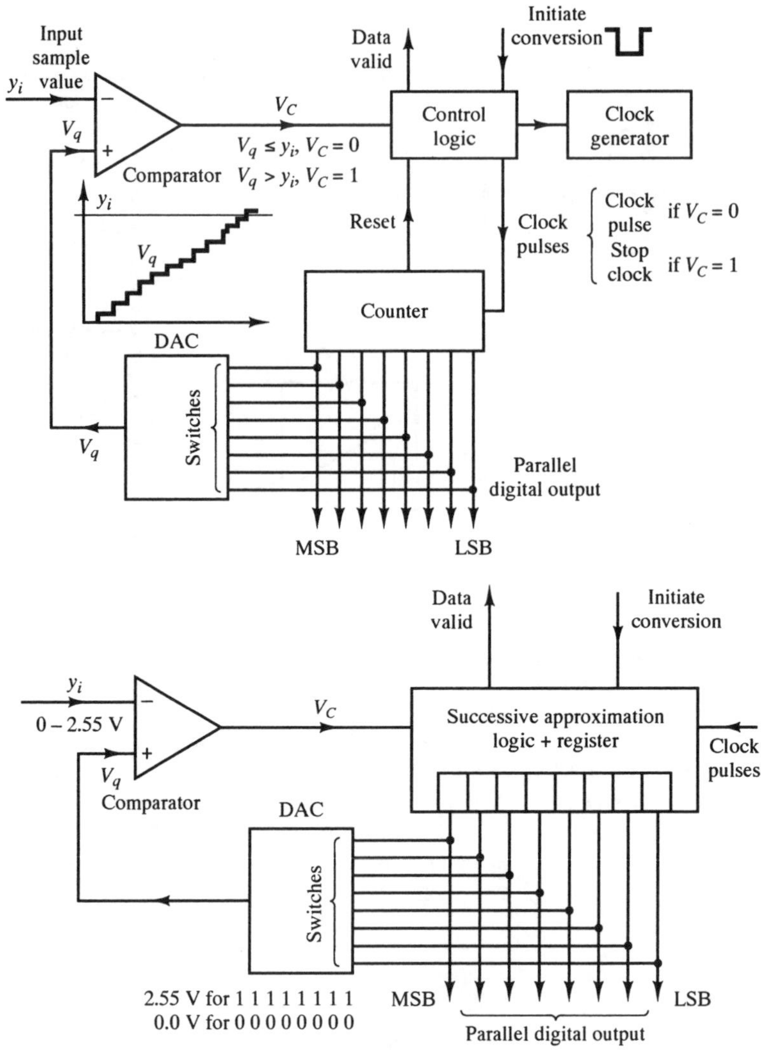

clock pulses. As the binary count and parallel digital signal increase, the appropriate switches in the DAC are closed, causing the converter output voltage V_q to increase in a staircase manner. While V_q is less than or equal to the input sample voltage y_i, the comparator output is 0 and the counting continues. When V_q becomes greater than y_i the comparator output switches to 1, the clock pulses are inhibited so that the count is stopped, and the **data valid** signal changes state to indicate that the conversion is complete. The parallel digital output signal corresponds to the final binary count and is held constant until the next initiate conversion pulse. The quantisation error $V_q - y_i$ should be within $\pm\frac{1}{2}$LSB. Typically the conversion time is around 1 ms.

Figure 10.7(b) shows a schematic diagram of a **successive approximation** analogue-to-digital converter.[1] This method involves making successive guesses at the binary code corresponding to the input voltage y_i. The trial code is converted

Table 10.2 Typical sequence of guesses in successive approximation ADC

	Input voltage y_i = 0.515 V				
	Clock pulse	DAC input	DAC V_q output volts	Comparator output V_C	Result
Initiate conversion →	1 Clear register	00000000	0	0	
	2 First guess	01111111 $(127)_{10}$	1.27	1 HIGH	$b_7 = 0$
	3 Next guess	00111111 $(63)_{10}$	0.63	1 HIGH	$b_6 = 0$
	4	00011111 $(31)_{10}$	0.31	0 LOW	$b_5 = 1$
	5	00101111 $(47)_{10}$	0.47	0 LOW	$b_4 = 1$
	6	00110111 $(55)_{10}$	0.55	1 HIGH	$b_3 = 0$
	7	00110011 $(51)_{10}$	0.51	0 LOW	$b_2 = 1$
	8	00110101 $(53)_{10}$	0.53	1 HIGH	$b_1 = 0$
Data valid →	9 Final guess	00110100 $(52)_{10}$	0.52	1 HIGH	$b_0 = 0$
	Output digital signal = 00110100				

into an analogue voltage using a DAC and a comparator is used to decide whether the guess is too high or too low. On the basis of this result another guess is made, and the process repeated until V_q is within half a quantisation interval of y_i.

Table 10.2 shows a series of guesses for an 8-bit binary converter with an input range of 0 to 2.55 V. The first guess is always 01111111 corresponding to $(127)_{10}$ i.e. approximately half full scale: this guess is high so that b_7 is set to 0, if the guess had been low b_7 would be set to 1. This next guess is 0011111 corresponding to $(63)_{10}$, i.e. approximately one-quarter full scale, this guess is also high so that b_6 is confirmed as 0. The process continues until all the remaining bits have been confirmed, the DATA VALID signal then changes state.

Figure 10.8 shows a general schematic diagram of a **flash or parallel analogue-to-digital converter**.[1] In any n-digit binary ADC there are Q quantisation voltage levels V_0 to V_{Q-1}, where $Q = 2^n$. In a flash ADC there are $Q-1$ comparators in parallel and $Q-1$ corresponding voltage levels V_1 to V_{Q-1}. There is no need to provide the V_0 voltage level. In each comparator q, the input sample value y_i is compared with the corresponding voltage level V_q. If y_i is less than or equal to V_q, the output is zero corresponding to 0. If y_i is greater than V_q, the output is non-zero corresponding to a 1, i.e.

$$\begin{aligned} V_q^c &= 0, \quad y_i \le V_q \\ V_q^c &= 1, \quad y_i > V_q \end{aligned} \qquad [10.9]$$

Thus if y_i lies between V_q *and* V_{q+1}, i.e. $V_q < y_i \le V_{q+1}$, the output of the lowest

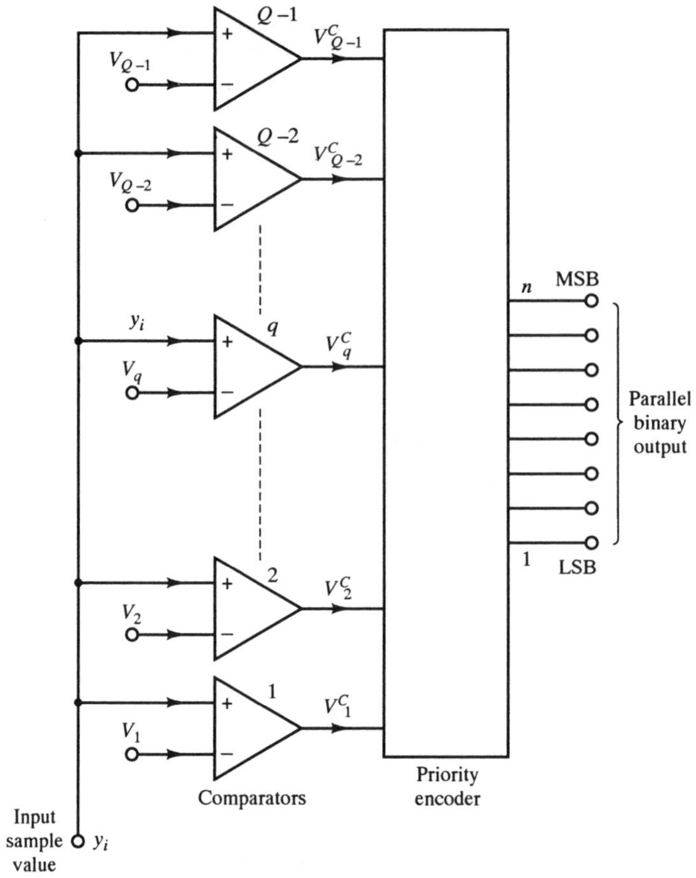

Fig. 10.8 Flash analogue-to-digital converter

q comparators 1 to q will all be 1 and the output of the remaining comparators $q + 1$ to $Q - 1$ will all be 0. Thus the comparators provide a $Q - 1$ digit parallel input code to a **priority encoder** which generators an n digit binary parallel output code corresponding to the value of q. The main advantage of the flash converter is the short conversion time: the main disadvantage is that the large number of comparators required to give acceptable resolution mean that it is relatively expensive.

10.2 Typical microcomputer system

10.2.1 Structure of typical microcomputer system

Figure 10.9 shows the structure of a typical microcomputer system,[2,3] an 8-bit system is shown but 16- and 32-bit systems are also available. The system consists of several elements: **microprocessor**, **read only memory** (ROM), **random access memory** (RAM), and **imput/out interface** which are interconnected by a **data bus**, **address bus** and **control lines**.

Input data (from a counter or ADC) enters the computer at the input parallel interface as an 8-bit parallel digital signal. A string of 8 bits is referred to as an 8-bit word or byte. The data is then carried to the different elements using a set of 8 parallel wires called an 8-bit data bus. The data bus will carry, for example,

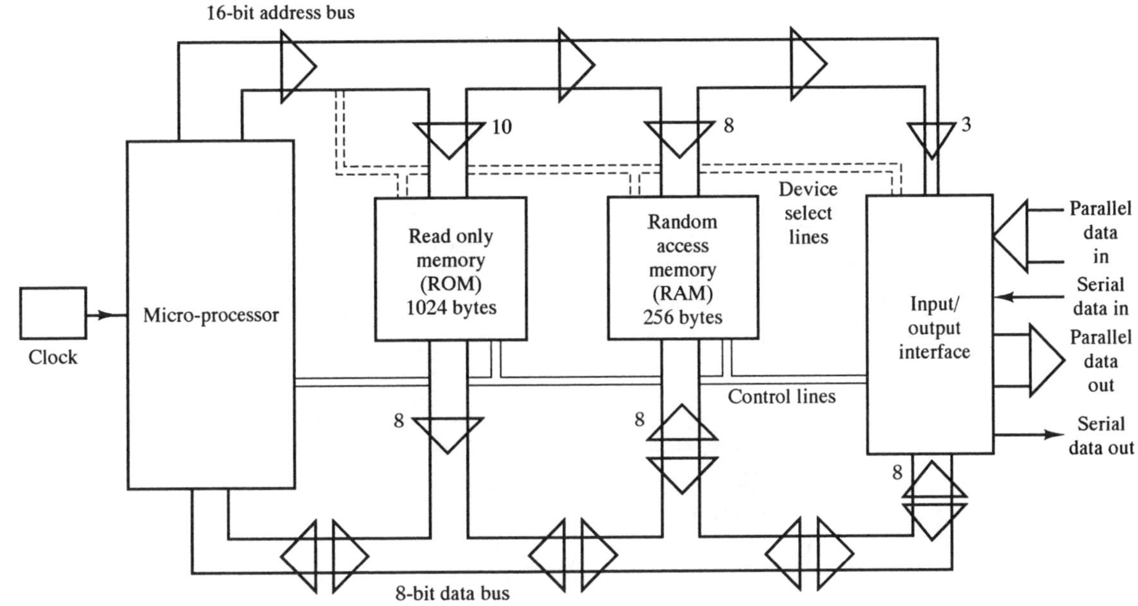

Fig. 10.9 Simplified schematic diagram of typical 8-bit microcomputer system

data from the input interface or RAM to the microprocessor, data from the microprocessor to the RAM, data from the microprocessor or RAM to the output interface. Data leaves the computer at the parallel or serial output interface to pass to a data presentation element such as an alphanumeric display, CRT display or digital recorder.

The computer system is sequential, i.e. it operates by executing a sequence of steps. Each step is performed as the result of an **instruction**; the sequence of instruction is called a **program**. Instructions are normally stored in the ROM and both instructions and data in the RAM. The data bus is used to carry instructions from these memory devices to the processor. Both types of memory devices have many storage elements, each with an associated **address**. This is a binary or hexadecimal number which specifies the location of the storage element within the memory. **Reading** is an operation whereby a copy of a data word or instruction is transferred from a given storage location to another device without changing the contents of the store. **Writing** is an operation whereby a data word or instruction is placed in a given storage location.

Information can be written into or read out of any part of the RAM; it is, however, **volatile memory**; i.e. the stored information is lost when the power supply is switched off. The ROM is a permanent memory, but information can only be read out of the ROM. In order to carry out these **read/write** operations the storage location must be **addressed**; here the microprocessor generates a parallel digital signal which corresponds to the address of the location and presents it to the memory device.

The system of Fig. 10.9 has a 16-bit address bus which is capable of addressing up to 2^{16} different locations. In this system the ROM has a storage capacity of 1024 8-bit bytes and requires a 10-bit address signal. The RAM has a storage capacity of 256 8-bit bytes and requires an 8-bit address signal. The address bus will carry, for example, the address of the next instruction to be executed to the ROM, the address of a data word to the RAM or the address or number of an input or output

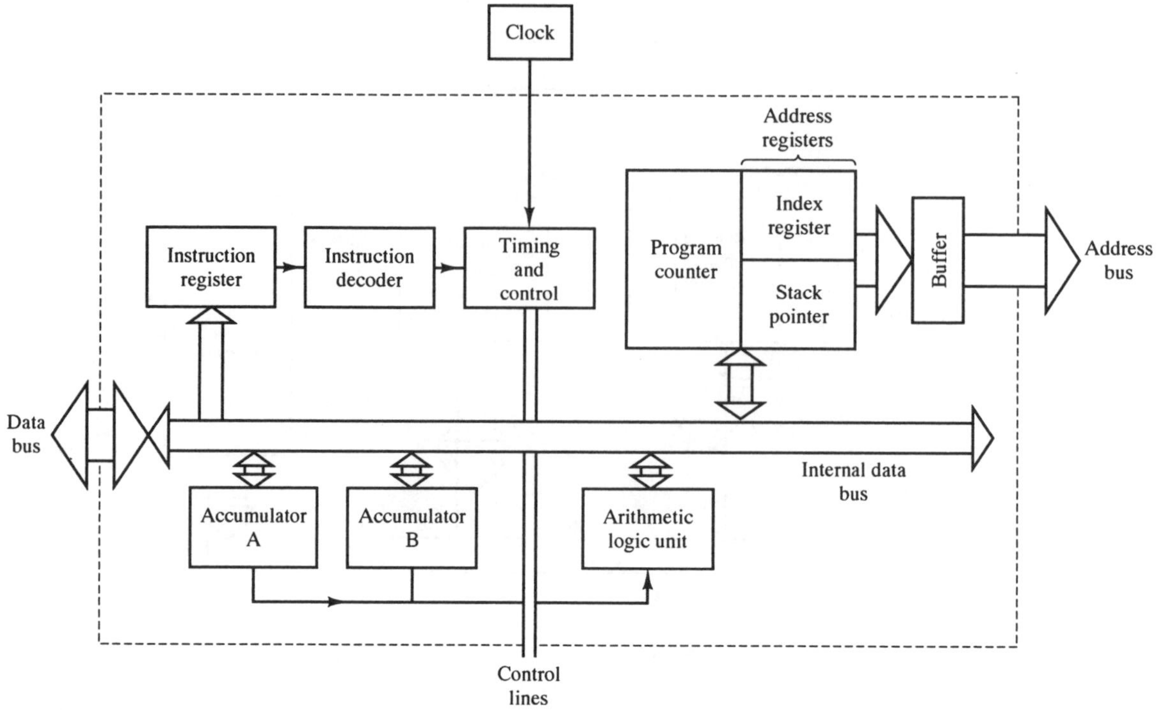

Fig. 10.10 Typical 8-bit microprocessor

device to the input/output (I/O) interface.

The **control lines** carry clock signals and control signals to each element in the computer to ensure the coordination and synchronisation necessary for the information transfers mentioned above. Input and output data can be in both parallel and serial form. Serial digital signals are discussed in Chapter 18.

Figure 10.10 shows a typical 8-bit microprocessor. The **program counter** is a register which contains the address of the instruction to be executed next. This instruction address passes to the **address registers** (index register and stack pointer) which hold the address prior to passing it to the address bus. The corresponding instruction is received from the data bus and is passed to the **instruction register**. Here it is held while it is decoded in the **instruction decoder** and executed. The address registers also pass data addresses to the address bus.

The **accumulators** are data registers. These receive data words from the data bus (prior to the execution of an instruction), and transmit data words back to the data bus (after the execution of an instruction). The **arithmetic–logic unit** (ALU) is a logic circuit which can perform arithmetic (addition, subtraction) and logical (AND, OR) operations on one or two data words held in accumulators. The result of the operation is held in an accumulator prior to passing to the data bus. Advanced microprocessors have circuits which also multiply and divide.

10.2.2 Typical 16-bit microprocessor

Figure 10.11 is a schematic diagram of a typical 16-bit microprocessor. This has a 16-bit external data bus and a 24-bit external address bus; this address bus provides

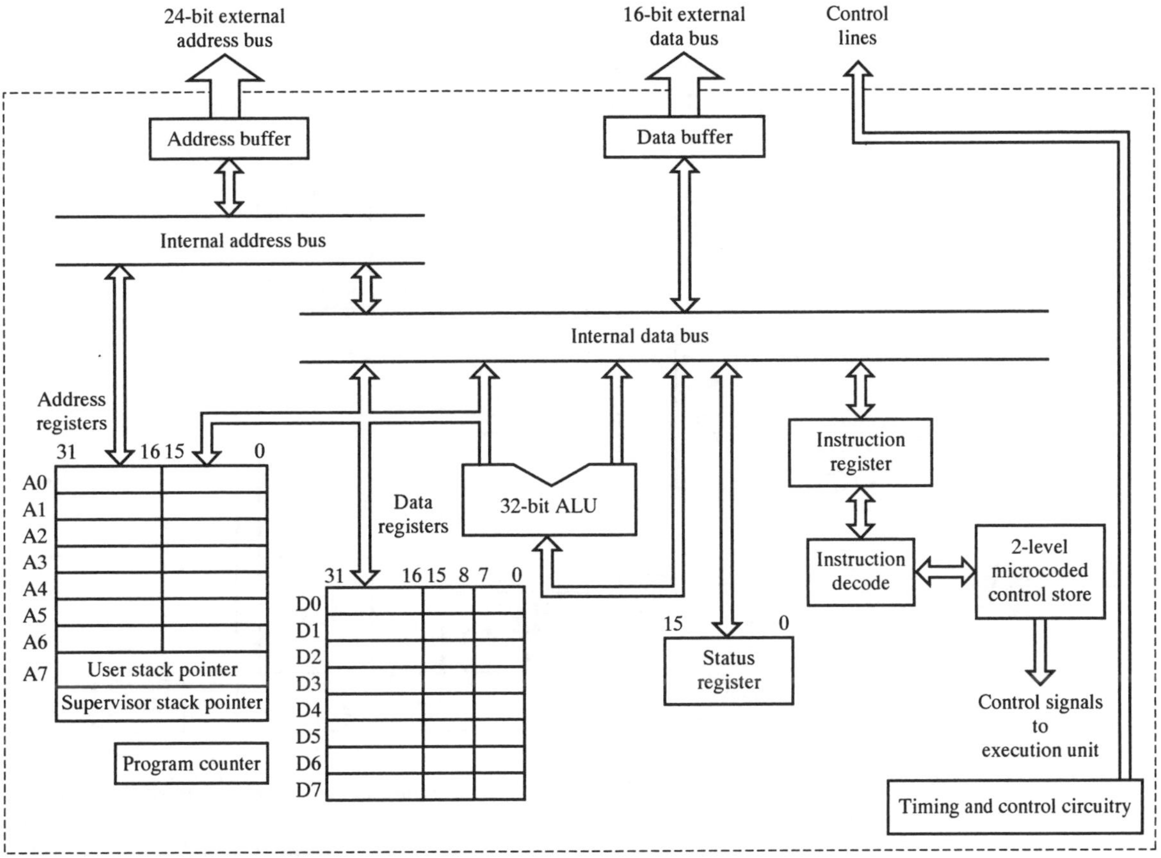

Fig. 10.11 Schematic diagram of typical 16-bit microprocessor

access to 16 Megabytes of external memory. The internal address and data buses are predominantly 32 bits wide. There are no dedicated accumulators but instead eight general-purpose data registers each of which is 32 bits wide. Each register can handle four basic types of data:

Eight 4-bit b.c.d. nibbles
Four 8 bit bytes
Two 16-bit words
One 32-bit double word

There are also no dedicated index registers but instead eight general-purpose address registers. Register 7 of this set is actually two sets of 32-bit registers implemented as stack pointers, although only one can ever be in operation at any one time.

10.2.3 Typical microcontroller

Microcontrollers are now widely used for measurement system signal processing. Here large scale integration means that several functions, for example the processor, ROM, RAM, ADC, DAC, serial port, parallel port, are all brought together on

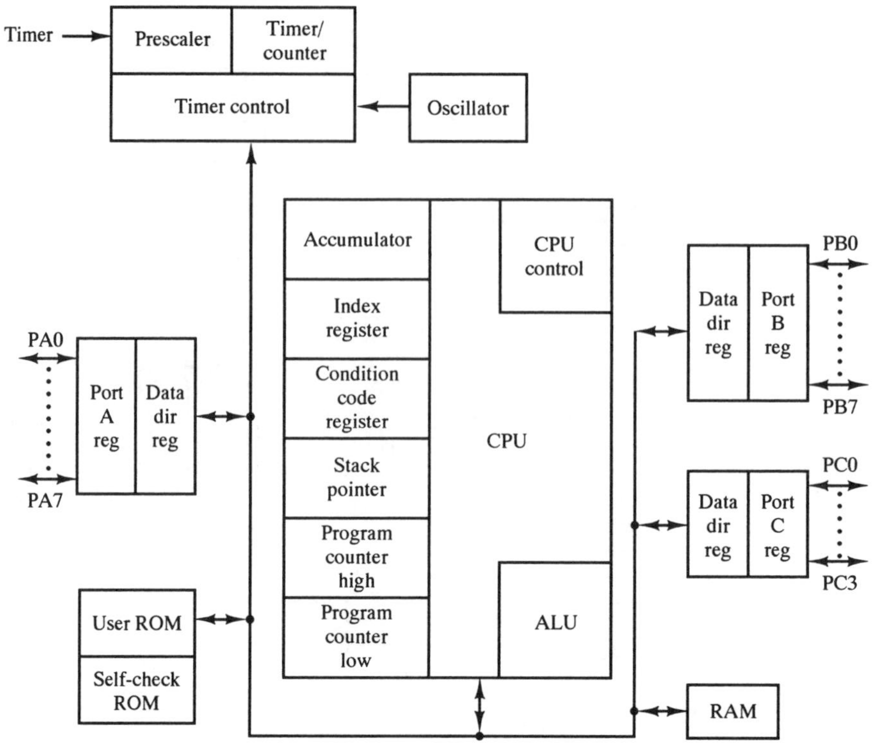

Fig. 10.12 Schematic diagram of MC6805 microcontroller

a single silicon chip. Figure 10.12 shows the internal architecture of the Motorola MC6805 microcontroller. This has a central processing unit (CPU), internal ROM, internal RAM, 2 bi-directional 8-bit input/output ports, 1 bi-directional 4-bit I/O port and an internal 8-bit timer with oscillator. Typically there are 3776 bytes of internal ROM and 112 bytes of internal RAM.

10.3 Microcomputer software

10.3.1 Assembly language and machine code

Each basic instruction in a microcomputer program is of the form:

operator operand

i.e. an operator operating on an operand. The operand can be either a data word or an address. In an 8-bit computer the basic data word is an 8-bit byte corresponding to 0 to 255 decimal or 00 to FF hexadecimal. However, if two numbers are added or multiplied together the result may lie outside this range, e.g. if 157 is added to 201 i.e. the result is 358 decimal or 166 hexadecimal. To accommodate this a 2-byte operand is used corresponding 0 to 65535 decimal or 0000 to FFFF hexadecimal. A computer with a 16-bit address bus can then address up to $2^{16} = 65536$ different locations. This means that a single-byte operand is used to address locations 0 to 255 (00 to FF) and a 2-byte operand to address locations 256 to 65535 (0100 to FFFF).

Table 10.3 Possible address map for microcomputer

Device	Address bits																Address range
	15	14	13	12	11	10	9	8	7	6	5	4	3	2	1	0	
RAM	.	.	.	.	0	0	.	.	x	x	x	x	x	x	x	x	(0 to 255, 0000 to 00FF)
ROM	.	.	.	.	1	1	x	x	x	x	x	x	x	x	x	x	(3072 to 4095, 0C00 to 0FFF)
I/O Interface	.	.	.	.	0	1	.	.	.	.	.	.	.	x	x	x	(1024 to 1029, 0400 to 0405)

x ≡ addres bit
0 or 1 ≡ device select bit
. ≡ irrelevant

The microprocessor has to send address signals to the ROM, RAM, **I/O interface** and any other memory or data storage device in the system. It is necessary therefore to have an **address map** which specifies the address of each device and the address of each storage location or register within the device. Table 10.3 shows a possible address map for the system of Fig. 10.9 which has a 256-byte RAM, 1024-byte ROM and an I/O interface.

The table shows that bits 10 and 11 are used for selecting which device is to be addressed; bits 12 to 15 are not used. The I/O interface is allocated an address range of 0400 to 0405 to enable different interfaces to be addressed. The system can be expanded by adding extra units of 1024 byte (1 kbyte) ROM and 256 byte RAM.

The **operator** specifies the operation to be carried out on the operand, e.g. add, subtract, clear, increment, load and store. In **assembly language** each operator is represented by a 3- or 4-letter mnemonic statement which suggests as closely as possible the operation to be carried out. A typical microprocessor is capable of executing between 50 and 100 different instructions or operations; examples are given in Table 10.4. Each mnemonic statement has a hexadecimal code associated with it, this is referred to as **op-code** or **machine code**. The program must be translated from mnemonic to machine code form before it can be executed.

Table 10.4 shows assembler and machine code versions of a simple program for adding two numbers. The program begins by clearing accumulators A and B, then loads the contents of storage location 2C into accumulator A and the contents of location 7B into accumulator B. The contents of accumulator B are added to those of accumulator A, the result is placed in accumulator A and then transferred to storage location BF. In the machine code version the computer is told the address of every operator and operand. Thus for the instruction LDB 7B, the operator code 1B is stored in location 9B and the operand code 7B is stored in the next location 9C.

10.3.2 High-level languages

Assembly language programs generally take less time to execute and require less memory than those written in other languages; they also support any type of input, output and peripheral device. However, assembly programming is far more difficult; we can see from the example in Table 10.4 that several lines of code are necessary to perform a simple addition operation. Several types of error are possible in assembly programming which are much less likely in a high-level language. If we assume that it takes about the same time, per line of code to write, debug, test, execute

Table 10.4 Typical assembly and machine code instructions

Operator		
Mnemonic	Op-code	Description
ADAB	01	Adds contents of accumulator B to contents of accumulator A and places result in accumulator A
SBAB	03	Subtracts contents of accumulator B from contents of accumulator A and places result in accumulator A
LDB	1 B	Loads the contents of memory into accumulator B
STRA	29	Stores the contents of accumulator A in memory
INCA	33	Increments contents of accumulator A by 1
DIX	3 F	Decrements contents of index register by 1
SHRA	4 D	Shifts contents of accumulator A one place to right
CLRB	5 A	Clears accumulator B
LDX	5 E	Loads contents of memory into index register
JMP	63	Jump to instruction stored in operand address

Assembly instructions		Machine code instructions	
Operator	Operand	Hex address	Hex op-code
CLRA		97	59
CLRB		98	5 A
LDA	2 C	99	1 A
		9 A	2 C
LDB	7 B	9 B	1 B
		9 C	7 B
ADAB		9 D	01
STRA	BF	9 E	29
		9 F	BF

and comment irrespective of the language, then assembly language implementation will be far more expensive. Another disadvantage of assembly language is that it is specific to one type of processor and is therefore not **portable** to other microprocesses or microcomputers. Assembly code written for one type of processor cannot be moved to another processor with a different instruction set.

High-level languages have the following characteristics:

(1) They include **procedures**: a procedure is a sequence of operations which defines exactly how a task is to be performed.
(2) The control of program flow is determined by the way the program is **structured**, i.e. by the sequential ordering of operations or by explicit linkages between them.
(3) They include **assignment statements**; an assignment statement assigns a new value to a variable, usually by computing the numerical value of an arithmetic expression specified in the statement. Examples are:

$$A = B + C$$
$$Y = 5.0 * \sin T$$
$$Y = \exp(-7.2 * X)$$

Table 10.5 Typical BASIC statements and operators

Statements	*Operators*
BAUD	ADD (+)
CALL	DIVIDE (/)
CLEAR	EXPONENTIATION (**)
CLEAR(S&I)	MULTIPLY (*)
CLOCK(1&0)	SUBTRACT (—)
DATA	LOGICAL AND (.AND.)
READ	LOGICAL OR (.OR.)
RESTORE	LOGICAL X-OR (.XOR.)
DIM	LOGICAL NOT (.OR.)
DO-WHILE	ABS()
DO-UNTIL	INT()
END	SGN()
FOR-TO-STEP	SQR()
NEXT	RND
GOSUB	LOG()
RETURN	EXP()
GOTO	SIN()
ON-GOTO	COS()
ON-GOSUB	TAN()
IF-THEN-ELSE	ATN()
INPUT	=, >, >=, <, <=, <>
LET	ASC()
ONERR	CHR()
ONEX1	CBY()
ONTIME	DBY()
PRINT	XBY()
PRINT#	GET

(4) Several **data types** are possible, examples are integer, real, Boolean. High-level languages are processor-independent and use a **compiler** to translate the complete high-level language program into the assembly code for a given processor. They therefore require more memory (the compiler is itself a large program) and take more time to execute. However, high-level programs are easier to program and correct.

BASIC is a very popular, easy-to-learn language. It began as an interpreted rather than a compiled language. This means that it is translated and executed statement-by-statement, using a program called an **interpreter**, instead of being fully translated into machine code before execution. There are several 'dialects' of BASIC which are specific to a given processor, this makes it less portable than other high-level languages. Table 10.5 shows some typical BASIC statements and operators, these include input/output statements, program sequence control statements and arithmetic operators.[4] Control of flow is obtained by attaching a numerical label (line number) to each statement and then executing the resulting program in an ascending sequence of these statement labels. An assignment statement would therefore read:

```
10 LET I = J + 50
```

Subroutines are called with GOSUB N statements, where N is the line number of the first statement of the subroutine. Return to the main program is then via a RETURN statement at the end of the subroutine. Program sequence is controlled by IF, THEN, ELSE, GOTO N statements. This reliance on line numbers can lead to a lack of program structure making programs difficult to follow and can lead to errors if the wrong line number is given.

'C' is widely used as a language for microcomputers in measurement systems; it is a relatively low-level language operating fairly close to assembly instructions. It has a highly modular simple structure which uses **functions** as sub-programs; a function is a self-contained unit of code. Before any variable is used as a function, it has to be **declared**, i.e. given a name and type. The general term for a name is an **identifier**. C permits only a few data types, these are:

float	floating point	3.4×10^{38} to 1.2×10^{-38}
int	integer	−32,768 to +32,767
char	up to 256 characters (e.g. upper/lower case letters)	

The **assignment operator** = assigns to the variable on the left-hand side a value given by the expression on the right-hand side. Thus to convert Celsius temperature into Fahrenheit we have:

```
temp _ f = temp _ c * 9.0/5.0 + 32
```

The basic **arithmetic operators** + (add), − (subtract), * (multiply), / (divide) are available; higher level arithmetic is provided by **standard library functions**. **Relational operators** pose a question to which the answer is either yes or no. These are:

a > b, Is *a* greater than *b*?
a < b, Is *a* less than *b*?
a = = b, Is *a* equal to *b*?
a > = b, Is *a* greater than or equal to *b*?

There are also the **logical operators** && for AND and || for OR. The **increment operator** i ++ increases the value of variable i by 1, the **decrement operator** i − − decreases it by 1. Functions are defined by enclosing the code that the function is to contain in braces { }. Standard library functions are used to provide high-level facilities e.g. graphics, input/output, arithmetic, examples are:

clrscr ( ) — clear screen

cprint f("message") — print "message"

The "message" contained in the circular brackets is termed the argument. Program sequence is controlled by the **if-else**, **while** and **do-while** constructs. For example the if-else construct has the form:

```
if (test condition)
  {
    Block of code to be executed if test condition is TRUE
  }
else {
    Block of code to be executed if test condition is FALSE
    }
```

In C, loops can be implemented using the **for** construct, which has the form:

```
for (initialisation; test condition; action at end)
  {
    code to be executed while test condition is true
  }
```

Figure 10.13 shows a typical C program, this displays a table of Fahrenheit temperatures for the Celsius temperatures 0, 1, 2, 3, . . . , 20. C is a highly portable, easy to learn, cost-effective language which can handle any input/output devices using specially-written driver functions.

Fig. 10.13 C program to display Celsius and Fahrenheit temperatures

```
main ( )
      {
      float temp_c = 0.0;
      float temp_f;
      int i;
      clrscr ( );
      for (i = 0; i < = 20; i + +)
      {
      temp_c = (float) i;
      temp_f = temp_c * 9.0/5.0 + 32.0;
      cprint f ("%5.1 f degrees C is % 5.1 f degrees F
          |r|n", temp_c, temp_f);
      }
wait ( );
      }
```

10.4 Signal processing calculations

10.4.1 Steady-state compensation

Computer calculation of measured value in the steady state has already been discussed in Section 3.3. Here the output U of the uncompensated system (normally comprising sensing, signal conditioning and ADC element) is input to the microcomputer. The computer then solves the steady-state inverse model equation:

$$I = K'U + N'(U) + a' + K'_M I_M U + K'_I I_I \qquad [10.10]$$

which obtains an estimate I' of model input I and thus compensates for non-linearity and environmental effects in the uncompensated system. The computer requires estimates I'_M, I'_I of the modifying and interfering inputs I_M, I_I and values of the model parameters K', $N'(\ )$, a', etc. We now consider examples of the computer solution of the inverse equation in both linear and non-linear cases.

Linear case

In this case the system input I is exactly proportional to the output U, i.e.

$$I = K'U \qquad [10.11]$$

As an example of the solution of this equation we consider a speed measurement system (Figure 10.14) using an 8-bit microcomputer programmed in assembly language.

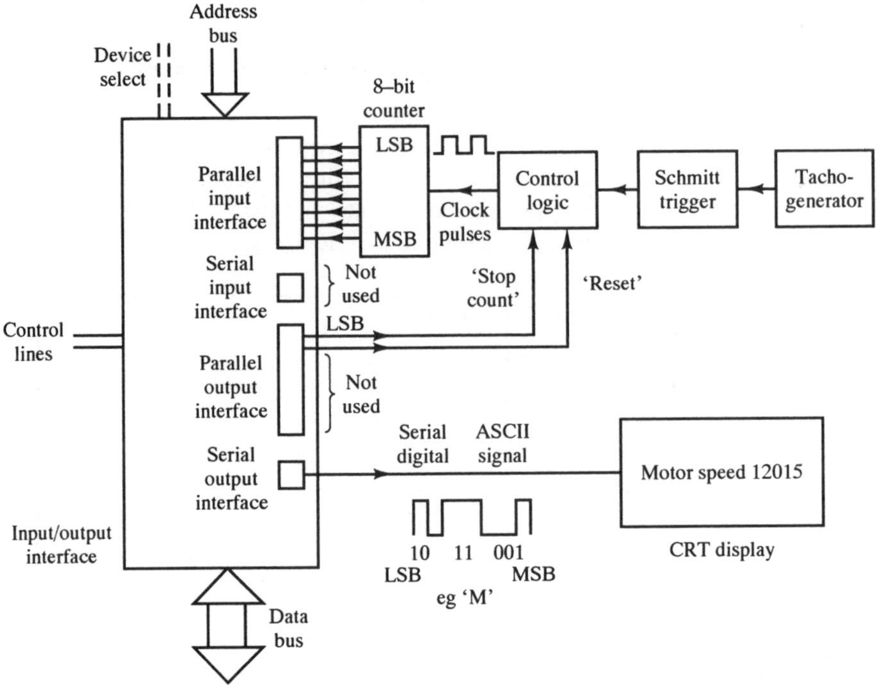

Fig. 10.14 Speed measurement system and input/output interface

In this system the angular velocity of a motor is sensed by a variable reluctance tachogenerator and the measured speed in r.p.m. displayed on a cathode ray tube display or VDU (visual display unit). The tachogenerator (Section 8.4) gives an a.c. output voltage whose frequency is proportional to motor angular velocity. This signal is input to the frequency-to-digital converter of Fig. 10.5, i.e. a Schmitt trigger followed by an 8-bit binary counter which gives a 8-bit parallel digital output signal. This is connected to the parallel input interface of the microcomputer as shown in Fig. 10.14. The output digital signal to the CRT display (Section 11.5) must be in serial form (Section 18.3) and in ASCII code (see later notes); this is provided by the serial output interface. The parallel output interface provides the STOP COUNT (LSB) and RESET logic control signals for the counter.

The frequency f Hz of the tachogenerator output voltage is given by:

$$f = \frac{m\omega_r}{2\pi} \qquad [8.37]$$

where ω_r rad s^{-1} is the angular velocity of the wheel and m the number of teeth. If $\dot{n}$ is the angular speed in r.p.m., then:

$$\frac{\dot{n}}{60} = \frac{\omega_r}{2\pi}$$

i.e.

$$f = \frac{m\dot{n}}{60} \qquad [10.12]$$

Thus if $m = 8$ and the range of $\dot{n}$ is 0 to 18 750 r.p.m., then the corresponding frequency range is 0 to 2500 Hz. Since the range of the counter is 0 to 255, a counting interval $\Delta T = 0.1$ sec was chosen (maximum count = 250). From eqns [10.8] and [10.12] the average speed $\dot{n}$ r.p.m. over time interval ΔT is:

$$\dot{n} = \frac{60N}{m\Delta T} = 75N \qquad [10.13]$$

where N is the count registered over 0.1 s. The equation has the form of [10.11]. Thus in order to measure $\dot{n}$ the pulses are counted over 0.1 s and the count multiplied by 75. The counter is then reset to zero and the process repeated. The main stages in the program are therefore:

Stage A — Control of counter for 0.1 sec count
Stage B — Multiplication of count by 75
Stage C — Conversion of hexadecimal count to decimal
Stage D — Conversion of decimal to ASCII for display

Each of these stages will now be discussed in more detail.

Stage A. The computer first sets RESET = 1, STOP COUNT = 1 to reset the counter to zero. It then sets RESET = 0, STOP COUNT = 0 to start the count. The count proceeds for an interval of 0.1 s (10^5 μs). The count is then stopped by setting RESET = 0, STOP COUNT = 1. The time interval of 10^5 μs is obtained by using two nested delay loops, the outer loop gives 100 passes of the inner loop, which has a time delay of 10^3 μs.

Stage B. Here the final count is multiplied by 75 to give the speed in r.p.m. Since the speed range is 0 to 18 750 r.p.m. and numbers can only be expressed in 8-bit integer form, two 8-bit bytes are required to represent the speed data. Two bytes represent the numbers 0 to 65 535 ($2^{16} - 1$) decimal, 0000 to FFFF hexadecimal, and require two storage locations in RAM or ROM. Since no multiplication instruction is available in assembler language, the multiplication is performed by successively adding the count to itself 75 − 1 = 74 times. The more significant byte is then stored in a given location and the less significant byte in an adjacent location.

Stage C. The above two-byte number is equivalent to a 4-digit hexadecimal number PQRS, this must be converted into a 5-digit decimal number JKNVW before it can be displayed. This involves initially division by 100 decimal (64 hex) to obtain the number of hundreds and then division of both quotient and remainder by 10 decimal (A hex) to find the number of thousands, hundreds, tens and units. The resulting five digits J, K, N, V and W are then stored in five separate locations.

Stage D. The CRT or VDU display (Section 11.5) accepts input data in American Standard Code for Information Interchange (**ASCII**). This code is in either 7-digit binary (an eighth digit is added as a parity check) or hexadecimal form and is shown in Table 10.6.

There are 128 possible codes of which 96 are used for the **display** of characters, these include the numerals 0 to 9, the letters A to Z (upper and lower case), punctuation marks and other symbols. The remaining 32 codes are used for **control** of the display; these include LF (line feed) which moves the display to the next line

Table 10.6 American Standard Code for Information Interchange (ASCII)

	ASCII character set (7-bit code)							
	More significant							
	0	1	2	3	4	5	6	7
Less significant	000	001	010	011	100	101	110	111
0 0000	NUL	DLE	SP	0	@	P	’	p
1 0001	SOH	DC1	1	1	A	Q	a	g
2 0010	STX	DC2	″	2	B	R	b	r
3 0011	ETX	DC3	#	3	C	S	c	s
4 0100	EOT	DC4	$	4	D	T	d	t
5 0101	ENQ	NAK	%	5	E	U	e	u
6 0110	ACK	SYN	&	6	F	V	f	v
7 0111	BEL	ETB	′	7	G	W	g	w
8 1000	BS	CAN	(	8	H	X	h	x
9 1001	HT	EM	)	9	I	Y	i	y
A 1010	LF	SUB	*	:	J	Z	j	z
B 1011	VT	ESC	+	;	K	[	k	{
C 1100	FF	FS	,	<	L	\	l	:
D 1101	CR	GS	—	=	M	]	m	}
E 1110	SO	RS	•	>	M	↑	n	~
F 1111	SI	VS	/	?	O	↓	o	DEL

and CR (carriage return) which moves the display to the first position on the same line.

From Table 10.6 we see that the decimal numbers 0 to 9, which are represented by the 4-digit binary codes 0000 to 1001, can be converted by adding 48 decimal, i.e. adding the three more significant digits 011, to give the 7-digit ASCII code. Each of the five decimal numbers JKNVW is thus converted into ASCII and this information together with caption information passed to the serial output interface. Here it is converted into binary **serial** form, ie a series of 1's and 0's, which are transmitted, one bit at a time along a single wire to the display unit. Figure 10.14 shows the transmission of the letter ‘M’ in the binary ASCII coded form 1001101.

The above calculation would have been significantly simpler using floating paint numbers and a high-level language with multiplication and ASCII conversion operations.

Non-linear case

If the effects of environmental inputs can be neglected equation [10.10] simplifies to:

$$I = K'U + N'(U) + a' \qquad [10.14]$$

In an eight-bit microcomputer programmed in assembly language numbers will be in eight-bit integer form rather than floating point and the multiplication operation will not be available. This means that the non-linear function $N'(U)$ cannot be realised using a polynomial in U and a **look-up table** is normally used. We will consider the example of temperature measurement in the range 0 to 250 °C using a copper–constantan thermocouple.

Figure 10.15 shows the graph of the inverse relation between temperature T °C and e.m.f. E mV (relative to a reference junction at temperature 0 °C). Here the vertical axis corresponds to the dependent variable T and the horizontal axis to the independent variable E. The table shows some corresponding values of E, T, X and Y, where X, Y are integer variables scaled to take values between 0 and 250 and defined by the equations:

$$X = \frac{250}{12} E, \; \mathrm{Y} = \mathrm{T} \qquad [10.15]$$

thus a change in X of 1 corresponds to a change in E of 0.048 mV and a change in Y of 1 corresponds to a change in T of 1 °C. There are 251 pairs of values of X, Y and 251 adjacent storage locations are used to store the value of Y corresponding to each value of X. The computer reads the value of input X and adds to it the address of the first storage location. This gives the address of the location where the corresponding value of Y is stored, this value is then fetched.

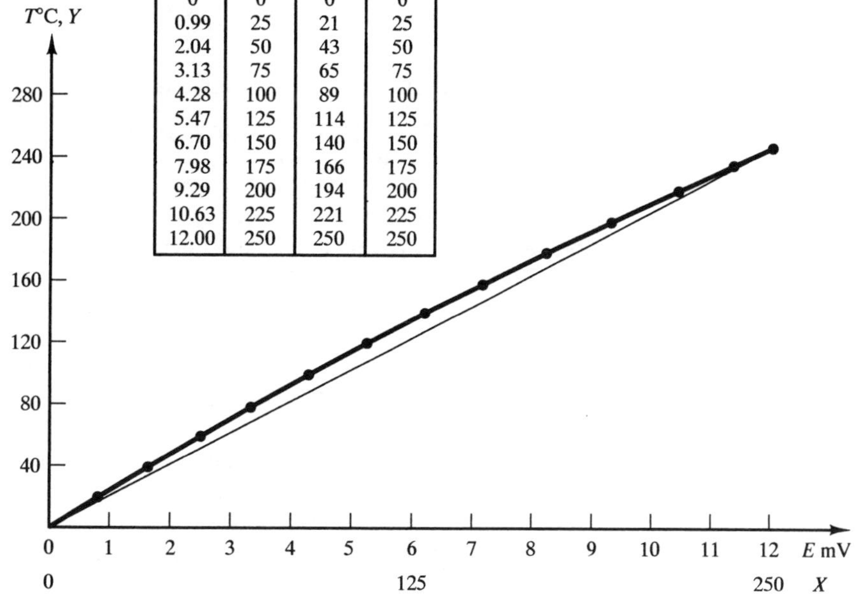

Fig. 10.15 Temperature/E.M.F. relationship and look-up table for copper–constantan thermocouple

E mV	T°C	X	Y
0	0	0	0
0.99	25	21	25
2.04	50	43	50
3.13	75	65	75
4.28	100	89	100
5.47	125	114	125
6.70	150	140	150
7.98	175	166	175
9.29	200	194	200
10.63	225	221	225
12.00	250	250	250

The importance of using scaled variables can be illustrated by the example of a 16-bit microcomputer where high-level arithmetic operations are available but the numbers are in 16-bit integer form rather than floating point. Here a square root operation is available which is used to calculate the volume flow rate $Q\,\text{m}^3/\text{hr}$ of liquid in a pipe from a measurement of the differential pressure ΔP created across an orifice plate flowmeter (Section 12.3.1). The range of integer variables is 0 to 65 535 but since $\sqrt{65\,535} = 256$, the square root operation could limit the number of calculated values of Q to 256. The problem is again solved by introducing scaled variables. Suppose the ranges of Q and ΔP are respectively 0 to $10\,\text{m}^3/\text{hr}$ and 0 to 10^4 Pa and they are related by the equation:

$$Q = 0.1\sqrt{\Delta P} \qquad [10.16]$$

We now introduce integer variables X and Y scaled to take values between 0 and 65 535 and defined by the equations:

$$X = \frac{65\,535}{10^4}\Delta P, \; Y = \frac{65\,535}{10}Q \qquad [10.17]$$

The computer then solves the corresponding scaled equation:

$$Y = 256\sqrt{X} \qquad [10.18]$$

In cases where both high-level arithmetic operations and floating point representation of numbers are available then the non-linear function can be realised using a polynomial and the inverse equation solved without using scaled variables. If we consider again the example of a copper–constantan thermocouple, this time measuring temperature T between 0 and 400 °C, the corresponding range of e.m.f. E will be 0 to 21 mV. If a 12-bit ADC is used, then the input to the computer can be regarded as a 12-bit integer decimal number X proportional to E, with range between 0 and 4095. Since 4095 corresponds to 21 mV, E is first calculated using:

$$E = \frac{21}{4095}X = 0.5128 \times 10^{-2}X \qquad [10.19]$$

T is then calculated using the inverse equation:

$$T = 0.2555 \times 10^{2}E - 0.5973 \times 10^{0}E^2 + 0.2064 \times 10^{-1}E^3 - 0.3205 \times 10^{-3}E^4 \qquad [10.20]$$

The coefficients in both equations can be expressed in floating point form and the operations of exponentiation, multiplication and addition used to solve equation [10.20].

10.4.2 Dynamic digital compensation and filtering

Dynamic compensation has already been discussed in Section 4.4. Given an uncompensated element with transfer function $G_u(s)$, a compensating element $G_c(s)$ is introduced into the system such that the overall transfer function $G(s) = G_u(s)G_c(s)$ satisfies a given frequency response specification. The compensating transfer function $G_c(s)$ can be implemented using analogue signal conditioning elements; for example a lead-lag circuit (Figure 9.7(e)) can be used to extend the frequency response of a thermocouple. In this section we see how a microcomputer can be used as a **digital signal processor** to give a compensating z-transfer function $G_c(z)$. Also in section 6.5.5 we saw how filtering can be used to reduce the effects of noise and/or interference on a signal. This improvement in signal-to-noise ratio can only be achieved if the signal spectrum occupies a frequency range which is substantially different from that of the noise/interference spectrum. In Section 9.2.1 we saw how an **analogue filter** $G(j\omega)$ for continuous signals can be implemented using an operational amplifier circuit. For example the a.c. amplifier of Figure 9.7(e) is a band-pass filter. In this section we see how a **digital filter** $G(z)$ for sampled signals can be implemented using a digital signal processor.

Principles of digital signal processing[1]

Figures 10.16(a)(b) shows a continuous signal $x(t)$ and the corresponding sampled signal $x_s(t)$; the sample interval is ΔT so that the ith sample occurs at time $i\Delta t$. Figure 10.16(c) shows an infinite train d(t) of unit inpulses, the ith impulse occurs at time $i\Delta T$ and is denoted by $\delta(t - i\Delta T)$ where $\delta(\ )$ is the **Dirac δ function**. We therefore have:

$$\mathrm{d}(t) = \sum_{i=0}^{i=\infty} \delta(t - i\Delta T) \qquad [10.21]$$

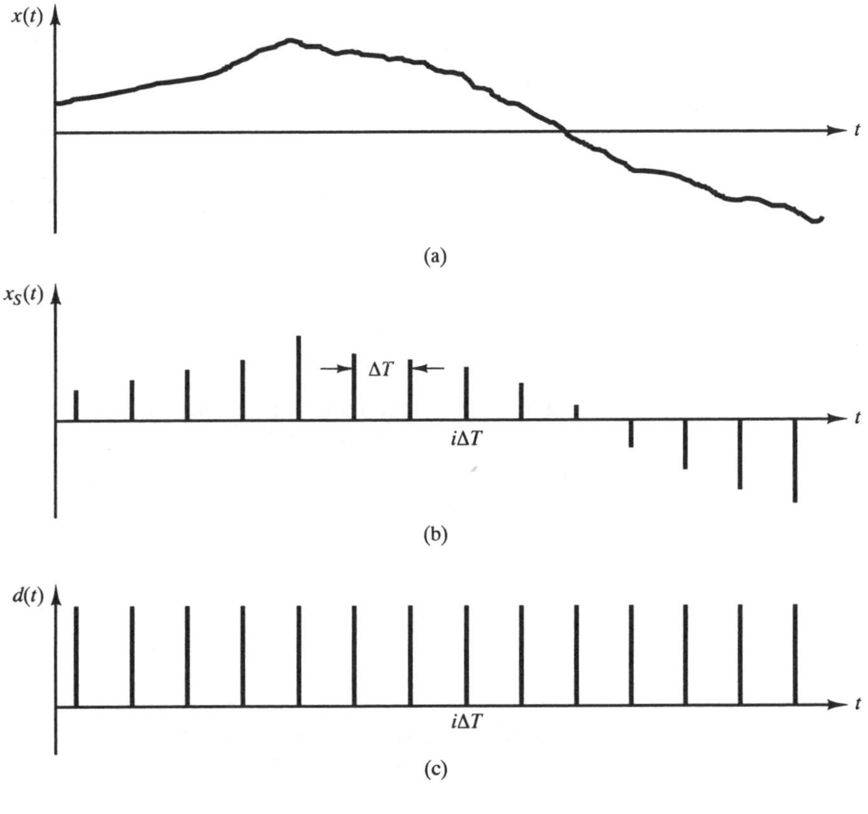

Fig. 10.16 Signal conversion
(a) Continuous signal
(b) Sampled signal
(c) Infinite impulse train

We can now regard $x_s(t)$ as the product of $x(t)$ and $d(t)$ ie.:

$$x_s(t) = \sum_{i=0}^{i=\infty} x(t)\delta(t - i\Delta T) \qquad [10.22]$$

The products $x(t)\delta(t - i\Delta T)$ are only non-zero at times $i\Delta T$ so that we have:

$$x_s(t) = \sum_{i=0}^{i=\infty} x(i\Delta T)\delta(t - i\Delta T) \qquad [10.23]$$

We now calculate the **Laplace Transform** $\bar{x}_s(s)$ of $x_s(t)$ (Section 4.1.1)

$$\bar{x}_s(s) = \mathcal{L}\left\{\sum_{i=0}^{i=\infty} x(i\Delta T)\delta(t - i\Delta T)\right\}$$

$$= \sum_{i=0}^{i=\infty} x(i\Delta T)\mathcal{L}\{\delta(t - i\Delta T)\}$$

But

$$\mathcal{L}\{\delta(t - i\Delta T)\} = \exp(-i\Delta Ts)$$

(Laplace Transform of a delayed unit impulse) giving:

Laplace Transform of sampled signal

$$\bar{x}_s(s) = \sum_{i=0}^{i=\infty} x(i\Delta T)\exp(-i\Delta Ts) \qquad [10.24]$$

We now introduce the **z Transform** $\bar{x}_s(z)$ of $x_s(t)$ by defining:

$$z = \exp(\Delta Ts) \qquad [10.25]$$

so that $z^{-1} = \exp(-\Delta Ts)$ corresponds to a time delay of 1 sampling interval and $z^{-i} = \exp(-i\Delta Ts)$ corresponds to a time delay of i sampling intervals. We therefore have:

z-transform of sampled signal

$$\bar{x}_s(z) = \sum_{i=0}^{i=\infty} x(i)z^{-i} \qquad [10.26]$$

The **frequency spectrum** of any signal $x(t)$ can be found by calculating the **Fourier Transform** $\bar{x}(j\omega)$. To convert the Laplace Transform $\bar{x}(s)$ into $\bar{x}(j\omega)$ we simply replace s by $j\omega$ (Section 4.2.2). Correspondingly to convert the z-transform $\bar{x}(z)$ into $\bar{x}(j\omega)$ we replace z by $\exp(j\omega\Delta T)$. The fourier Transform of the sampled signal $x_s(t)$ is therefore:

Fourier transform of sampled signal

$$\bar{x}_s(j\omega) = \sum_{i=0}^{i=\infty} x(i\Delta T)\exp(-j\omega i\Delta T) \qquad [10.27]$$

In section 4.1.1 we defined the **transfer function** $G(s)$ of an element as the ratio of the Laplace transform of the output signal to that of the input signal. Here we

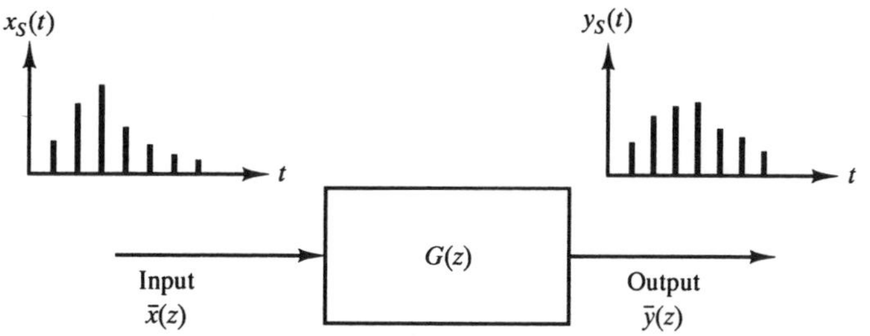

Fig. 10.17 Processing of sampled signals

can similarly define the z-transform $G(z)$ of a digital signal processor as the ratio of the z-transform of the output signal $\bar{y}(z)$ to that of the input signal $\bar{x}(z)$, i.e.

z-transfer function of an element

$$G(z) = \frac{\bar{y}(z)}{\bar{x}(z)} \qquad [10.28]$$

so that

$$\bar{y}(z) = G(z)\bar{x}(z).$$

Figure 10.17 shows a digital signal processor with transfer function $G(z)$ converting a sampled signal $x_s(t)$, with z-transform $\bar{x}(z)$, into a sampled signal $y_s(t)$ with z-transform $\bar{y}(z)$.

Types of digital filter

There are two main types of digital filter.
Non-recursive filter: here $G(z)$ is of the form:

Non-recursive filter

$$G(z) = g_0 + g_1 z^{-1} + g_2 z^{-2} + \ldots + g_n z^{-n} \qquad [10.29]$$

so that

$$\begin{aligned} \bar{y}(z) &= G(z)\bar{x}(z) \\ &= [g_0 + g_1 z^{-1} + g_2 z^{-2} + \ldots + g_n z^{-n}]\bar{x}(z) \end{aligned}$$

The **inverse transform** of $\bar{y}(z)$ is the value y_i of $y_s(t)$ at the ith sampling interval $i\Delta T$. This can be found by recalling that the operator z^{-1} corresponds to a time delay of one sampling interval. This means the inverse transform of $z^{-1}\bar{x}(z)$ is $x(i-1)$, ie. the value of $x_s(t)$ one sampling interval earlier; similarly the inverse transform of $z^{-2}\bar{x}(z)$ is $x(i-2)$ etc. Thus we have:

Inverse transform of non-recursive filter

$$y_i = g_0 x(i) + g_1 x(i-1) + g_2 x(i-2) + \ldots + g_n x(i-n) \qquad [10.30]$$

The inverse transform is often referred to as a **difference equation**. From equation [10.30] we see that in a non-recursive filter, a given output sample value depends on input sample values only, i.e. it does depend on output sample values so there is no feedback from the output.

Fig. 10.18
Implementation of non-recursive filter

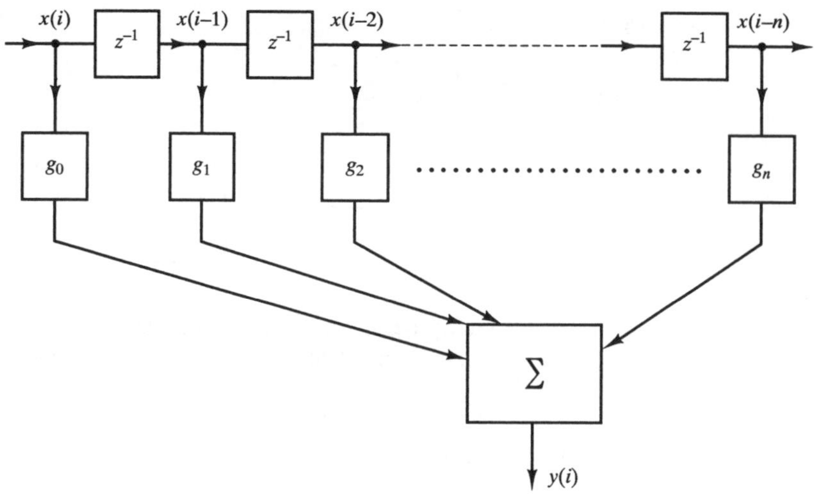

Figure 10.18 shows the implementation of a non-recursive filter using the operations of time shifting, coefficient multiplication and addition.

Recursive filter: here $G(z)$ is of the form:

Recursive filter

$$G(z) = \frac{a_0 + a_1 z^{-1} + a_2 z^{-2} + \ldots + a_n z^{-n}}{1 + b_1 z^{-1} + b_2 z^{-2} + \ldots + b_m z^{-m}} = \frac{\bar{y}(z)}{\bar{x}(z)} \qquad [10.31]$$

i.e.

$$(a_0 + a_1 z^{-1} + a_2 z^{-2} + \ldots + a_n z^{-n})\bar{x}(z)$$
$$= (1 + b_1 z^{-1} + b_2 z^{-2} + \ldots b_m z^{-m})\bar{y}(z)$$

Using the same procedure as above, the corresponding inverse transform or difference equation is:

$$y(i) + b_1 y(i-1) + b_2 y(i-2) + \ldots + b_m y(i-m)$$
$$= a_0 + a_1 x(i-1) + a_2 x(i-2) \ldots + a_n x(i-n)$$

giving:

Inverse transform of recursive filter

$$y_i = [a_0 x(i) + a_1 x(i-1) + \ldots + a_n x(i-n)]$$
$$- [b_1 y(i-1) + b_2 y(i-2) + \ldots b_m y(i-m)] \qquad [10.32]$$

From equation [10.32] we see that in a recursive filter, a given output sample value depends on both input and output sample values meaning there is feedback from the output. Figure 10.19 shows the implementation of a recursive filter.

Examples of digital filters and compensators

Integration. There are several situations where a signal must be integrated in order to obtain the required measured value. Two common examples are the calculation

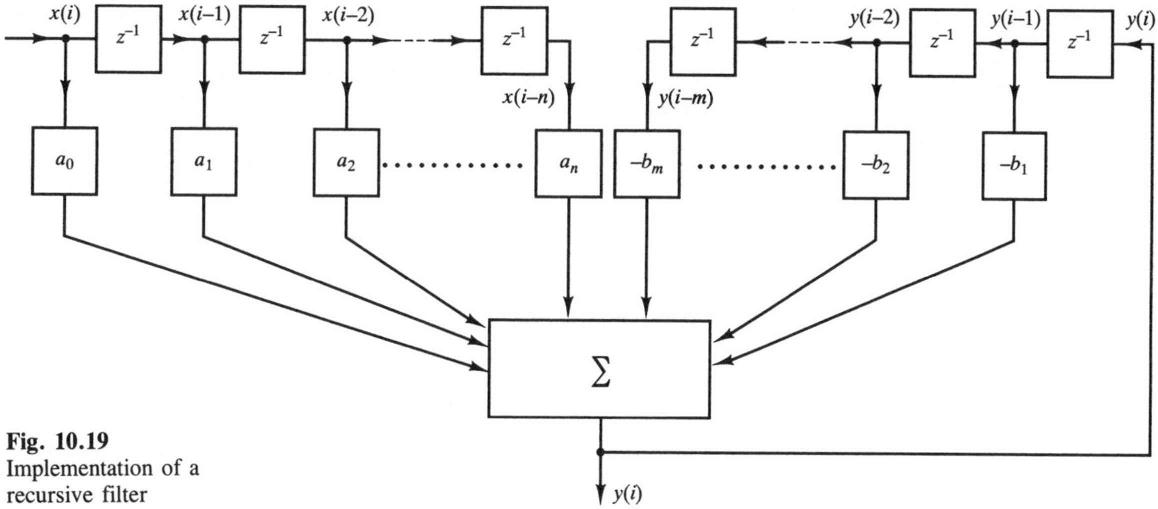

Fig. 10.19 Implementation of a recursive filter

Fig. 10.20 Approximate evaluation of definite integral

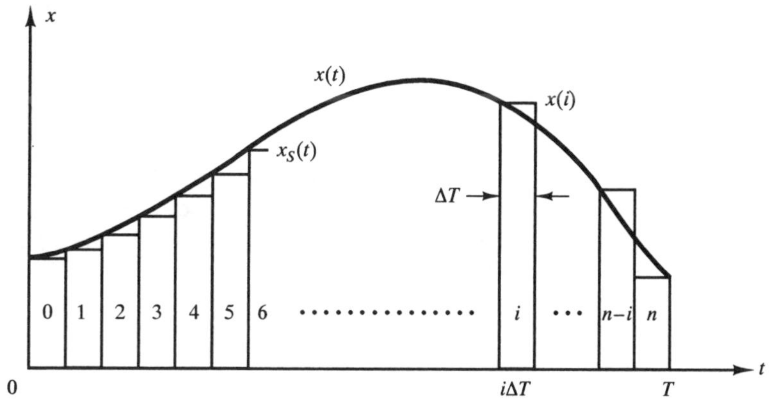

of the total mass of fluid delivered in a given time interval and the determination of the composition of a gas from the areas under the peaks in a gas chromatograph. Figure 10.20 shows how the definite integral $\int_0^T x(t)\mathrm{d}t$, the area under the continuous function $x(t)$, may be approximated by the area under the sampled signal $x_s(t)$. The area of the rectangle associated with the ith sample value is $x(i)\Delta T$. Thus the integral:

$$y(T) = \int_0^T x(t)\mathrm{d}t$$

may be approximated by the difference equation:

Sampled approximation to integral

$$y(T) \approx y(n) = \Delta T \sum_{i=0}^{i=n} x(i) \qquad [10.33]$$

where $(n + 1)\Delta T = T$.

Differentiation. There are also situations where a signal must be differentiated to obtain the required measured value. Examples are the calculation of the rate of change of temperature in a reactor or the rate of change of flow rate in a pipe. In some situations the derivative:

$$\dot{x}(t) = \frac{dx}{dt}$$

may be approximated by the difference equation:

Sampled approximation to derivative

$$\dot{x}(i) = \frac{1}{\Delta T}[x(i+1) - x(i)] \quad [10.34]$$

However, for this method to be valid, there should be sufficient binary digits to ensure a large number of quantisation levels and small quantisation errors.

Band-pass filter. We consider the example of a band-pass filter which is used to improve the signal-to-noise ratio for a sinusoidal vortex flowmeter signal affected by wideband random turbulence (Section 12.3.3). A suitable filter has the z-transform:

$$G(z) = \frac{1 - z^{-2}}{1 - 2r\cos(\omega_0 \Delta T)z^{-1} + r^2 z^{-2}} \quad [10.35]$$

where ω_0 rad/s is the centre frequency of the filter, ΔT s the sampling interval and r a dimensionless parameter between 0 and 1 which determines the bandwidth of the filter. If we require f_0 to be 40 Hz, use a sampling frequency f_s of 300 samples/s and choose r to be 0.95, we have:

$$\omega_0 = 2\pi \times 40 \text{ rad/s}$$

$$\Delta T = \frac{1}{f_s} = 3.33 \times 10^{-3} \text{ s}$$

$$\omega_0 \Delta T = 0.837$$

$$\cos \omega_0 \Delta T = 0.670$$

giving

$$G(z) = \frac{\bar{y}(z)}{\bar{x}(z)} = \frac{1 - z^{-2}}{1 - 2.272z^{-1} + 0.903z^{-2}} \quad [10.36]$$

The frequency response $G(j\omega)$ of the filter can be found by replacing z^{-1} by $e^{-j\omega\Delta T}$ and z^{-2} by $e^{-2j\omega\Delta T}$ to give:

$$G(j\omega) = \frac{1 - e^{-j2\omega\Delta T}}{1 - 1.272e^{-j\omega\Delta T} + 0.903e^{-j2\omega\Delta T}} \quad [10.37]$$

If ω_s rad/s is the angular sampling frequency then $\Delta T = 2\pi/\omega_s$ giving:

$$G(j\omega) = \frac{1 - \exp(-j4\pi\omega/\omega_s)}{1 - 1.272\exp(-j2\pi\,\omega/\omega_s) + 0.903\exp(-j4\pi\omega/\omega_s)} \quad [10.38]$$

Fig. 10.21 Band-pass digital filter
(a) Frequency response
(b) Implementation

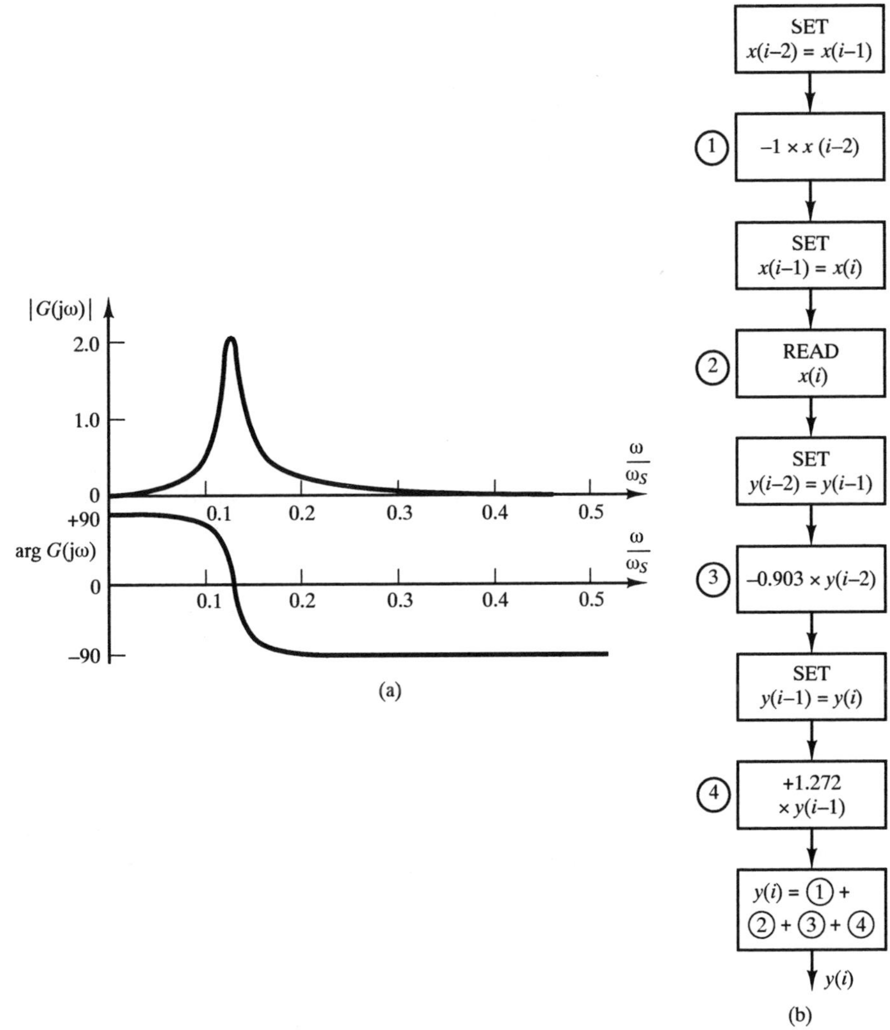

Figure 10.21(a) shows the magnitude $|G(j\omega)|$ and argument arg $G(j\omega)$ of $G(j\omega)$ plotted against normalised frequency ω/ω_s.

Rearranging equation [10.38] gives:

$$(1 - 1.272z^{-1} + 0.903z^{-2})\bar{y}(z) = (1 - z^{-2})\bar{x}(z)$$

and taking the inverse transform gives the difference equation:

$$y(i) - 1.272y(i-1) + 0.903y(i-2) = x(i) - x(i-2)$$

i.e.

$$y(i) = x(i) - x(i-2) + 1.272y(i-1) - 0.903y(i-2) \qquad [10.39]$$

Figure 10.21(b) shows the flowsheet for a program which solves equation [10.39] to implement the digital filter.

Conclusion

This chapter has discussed the conversion of analogue signals into digital form; also microcomputer structure, operation and software and how microcomputers perform steady-state and dynamic digital calculations to establish the measured value of a variable.

References

10.1 OWENS A R 1982, Digital signal conditioning and conversion, *J. Phys E: Scientific Instruments*, Vol. 15, No. 8, pp. 789–806.

10.2 Motorola Semiconductors, 1975, *Microprocesor Applications Manual*, McGraw-Hill, New York.

10.3 Intel Corporation, 1976, *Technical information on Intel 8080 microprocessors.*

10.4 Intel Corporation, 1989, *MCS BASIC—52 Users Manual.*

Problems

10.1 An analogue-to-digital converter has an input range of 0 to 5 V and incorporates a 12-bit encoder.

a) Assuming a binary encoder find:
   (i) the maximum percentage quantisation error;
   (ii) the digital output signals corresponding to input voltages of 0.55 V and 2.63 V.

(b) Convert the binary codes of part a(ii) into hexadecimal form.

(c) Repeat part (a) assuming a 3-decade, 8:4:2:1 binary coded decimal (b.c.d.) encoder.

10.2 The digital-to-analogue converter of Fig. 10.6(a) is required to give an output voltage in the range 0 to 2.55 V, corresponding to an 8-bit digital input signal 00000000 to 11111111.

(a) Assuming $V_{\text{REF}} = -15$ V, $R = 1$ kΩ, calculate the value of R_F required.

(b) Find the output voltage corresponding to an input signal of 11000101.

10.3 An analogue first-order low-pass filter has the transfer function:

$$G(s) = \frac{1}{1 + \tau s}$$

where τ is the time constant of the filter.

(a) What is the 3 dB bandwidth of the analogue filter?

(b) Use the bilinear transformation:

$$s = \frac{2}{\Delta T} \frac{(1 - z^{-1})}{(1 + z^{-1})}$$

where ΔT is the sampling interval, to show that an equivalent digital filter has the z-transfer function:

$$G(z) \frac{\bar{y}(z)}{\bar{x}(z)} = \frac{a_0 + a_1 z^{-1}}{1 - b\, z^{-1}}$$

where

$$a_0 = a_1 = \frac{1}{\left(\frac{2\tau}{\Delta T} + 1\right)}, \quad b_1 = \frac{\left(\frac{2\tau}{\Delta T} - 1\right)}{\left(\frac{2\tau}{\Delta T} + 1\right)}$$

(c) Hence show that the corresponding difference equation is:

$$y(i) = a_0 x(i) + a_1 x(i-1) + b_1 y(i-1)$$

(d) In the special case that $2\tau/\Delta T = 1$ show that:

$$G(z) = \tfrac{1}{2}(1 + z^{-1})$$

and that:

$$|G(j\omega)| = \cos\frac{\omega\Delta T}{2}, \quad \arg G(j\omega) = -\tfrac{1}{2}\omega\Delta T$$

What is the 3 dB bandwidth of this filter?

11

Data Presentation Elements

The data presentation element is the final element in the measurement system; its function being to communicate the measured value of the variable to a human observer. It is important that the measured value is presented as clearly and easily as possible, otherwise the value registered by the observer may be different. Consider an accurate flow measurement system where the true value of flow rate is $11.3\,m^3\,hr^{-1}$ and the measured value $11.5\,m^3\,hr^{-1}$; i.e. a **measurement system error** of $11.5 - 11.3 = 0.2\,m^3\,hr^{-1}$. If the observed value is $12.0\,m^3\,hr^{-1}$, then the **observation error** is $12.0 - 11.5 = 0.5\,m^3\,hr^{-1}$. This is greater than the measurement error and means that the high system accuracy is wasted.

Observation error depends on many factors:

the distance of the element from observer;
ambient lighting;
the eyesight, patience and skill of the observer.

However, a clear presentation is of major importance. This chapter begins by reviewing **data presentation** elements in current use, and goes on to examine their principles and characteristics in detail.

11.1 Review and choice of data presentation elements

Figure 11.1 lists elements in wide current use and classifies them according to whether they are indicators or recorders, analogue or digital. In choosing an element the first decision to be made is whether or not a permanent record is required. A record would be required for example in the following situations:

(a) A high speed event; e.g. a human heart beat, which is too fast to be followed by an observer.
(b) Large amounts of data; e.g. measurements of temperature, pressure, flow rates in a chemical reactor, which require subsequent quantitative analysis and calculation, possibly by computer. Similarly in the transfer of chemical products from producer to customer by pipeline a record of flow rates is essential for costing purposes.

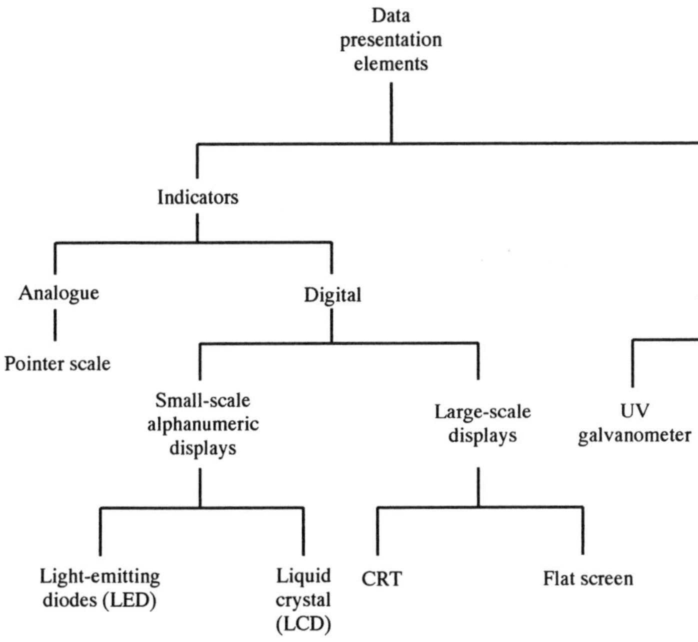

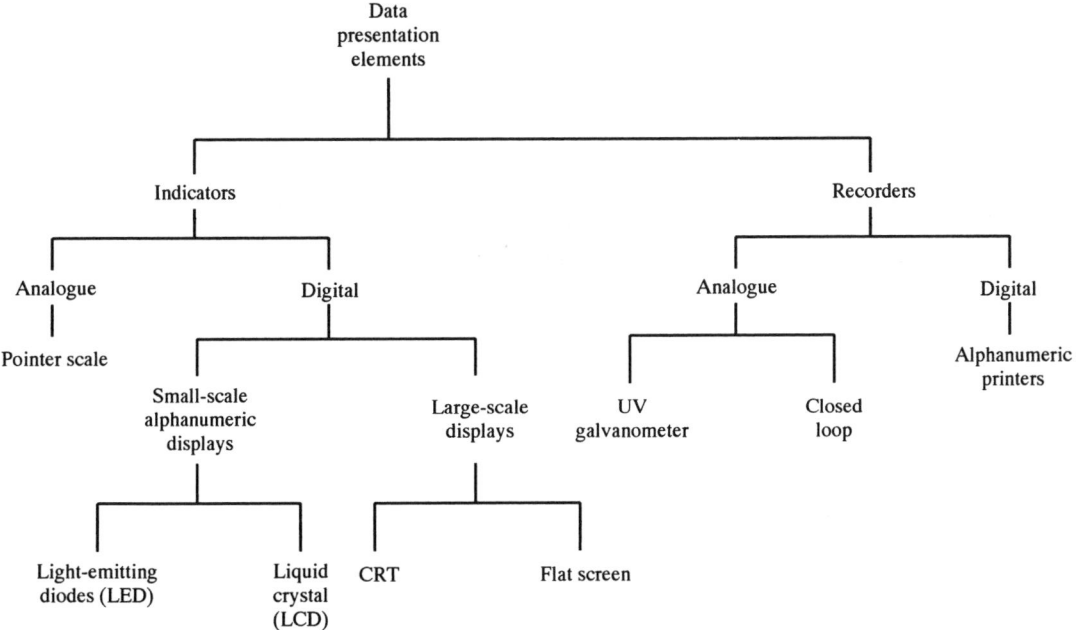

Fig. 11.1 Classification of data presentation elements

(c) In fault situations; e.g. an aircraft crash, where it is essential to establish the exact sequence of events so that the real cause of the fault can be found.

In situation (a), an analogue recording, i.e. a continuous graph of blood pressure versus time, on a paper chart is ideal. However, in situation (b), a digital recording with the measured value of each variable typed on a strip of paper is required. An analogue chart recorder can be directly connected to an analogue voltage source such as a deflection bridge or amplifier. A digital recorder requires a serial or parallel digital signal as input. The ultraviolet (UV) galvanometer chart recorder is a delicate, low-impedance, sensitive, high-bandwidth device, especially suited to the laboratory measurement of high-speed events. The potentiometric or closed-loop servo recorder is a far more rugged device with high impedance, high accuracy but low bandwidth. It is commonly used in monitoring industrial processes where variables are changing relatively slowly.

In applications where indication is sufficient, a choice must first be made between analogue pointer-scale indicators and digital displays. With a pointer-scale indicator the observer must **interpolate** if the pointer lies between two scale marks. Thus if the pointer lies between 9 and 10, the observer must decide whether the measured value is say 9.4, 9.5 or 9.6. Thus an observation error of up to ± 0.5 units is possible. This problem is avoided with a 3-decade alphanumeric display which presents the measured variable directly as 9.5. The pointer-scale indicator can, however, be directly connected to an analogue voltage source, whereas the digital display requires a parallel, digital signal in b.c.d. form. The small-scale alphanumeric devices are well suited to displaying the measured value of a single variable. If we wish to display a large amount of information, i.e. descriptions and measured values of a large number of variables, then large-scale displays, either CRT or flat screen, should be used.

11.2 Pointer-scale indicators

Figure 11.2(a) shows simplified diagrams and an equivalent circuit for a moving coil indicator connected to a Thévenin signal source E_{Th}, R_{Th}. The coil is situated in a radial magnetic field of flux density B, so that a current i through the coil produces a deflecting torque:

$$T_D = BnAi \qquad [11.1]$$

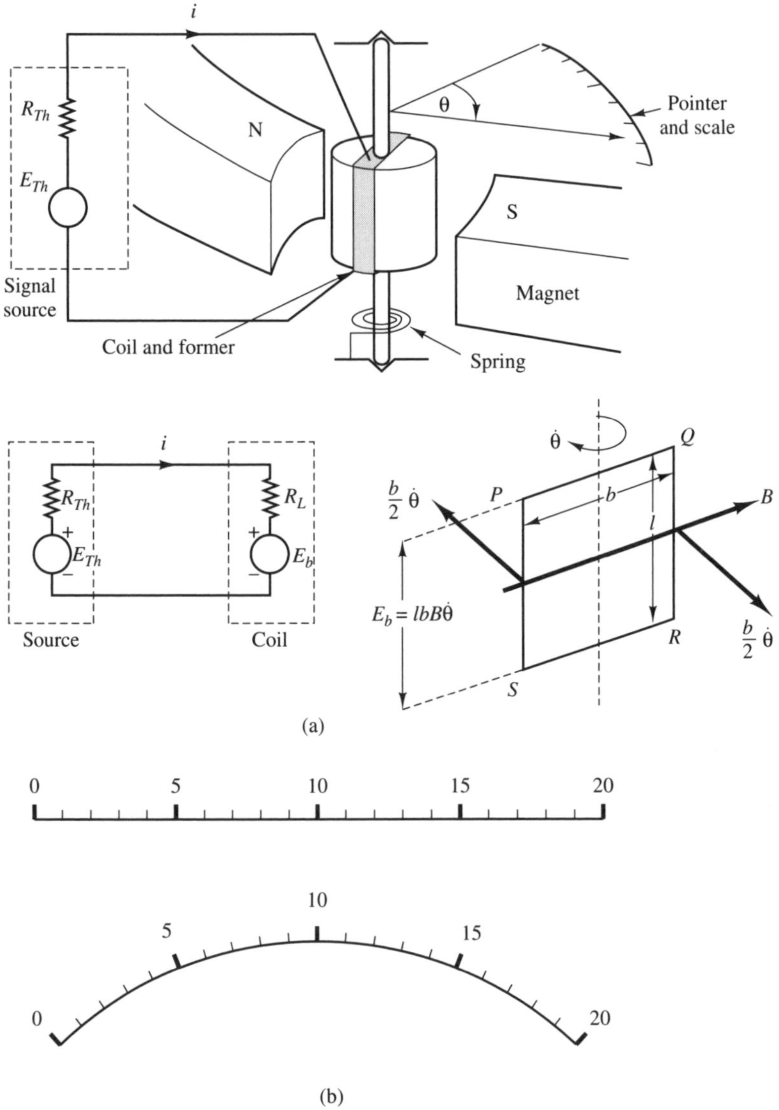

Fig. 11.2 Pointer–scale indicators: principle and recommended scale format
(a) Mechanical arrangement and circuit
(b) Scales (after British Standards Institution[2]

where A is the cross sectional area of the coil and n the number of turns. This deflecting torque is opposed by the spring restoring torque:

$$T_R = c\theta \qquad [11.2]$$

where c is the spring stiffness and θ the angular deflection. Assuming negligible frictional torque, the resultant unbalanced torque on the coil is $T_D - T_R$. This is equal to the product of moment of inertia I and angular acceleration $d^2\theta/dt^2$, i.e.

$$BnAi - c\theta = I\frac{d^2\theta}{dt^2} \qquad [11.3]$$

The current i is given by:

$$i = \frac{E_{Th} - E_b}{R_{Th} + R_L} \qquad [11.4]$$

where R_L is the resistance of the coil and E_b is the back e.m.f. induced in the coil due to its motion in the magnetic field. E_b can be calculated using Faraday's law of electromagnetic induction as shown in Fig. 11.1(a). If we consider one turn $PQRS$ of the coil, then the voltage induced in PS, with length l moving with translational velocity $(b/2)(d\theta/dt)$ perpendicular to the magnetic field B, is $Bl(b/2)(d\theta/dt)$. An equal and opposite voltage is induced in QR, so that the resultant voltage in $PQRS$ is $Blb(d\theta/dt)$, i.e. $BA(d\theta/dt)$ since $A = lb$. The total back e.m.f. for a coil of n turns is:

$$E_b = nAB\frac{d\theta}{dt} \qquad [11.5]$$

From [11.3]–[11.5] we have:

$$BnA\left[\frac{E_{Th} - nAB(d\theta/dt)}{R_{Th} + R_L}\right] - c\theta = I\frac{d^2\theta}{dt^2} \qquad [11.6]$$

i.e.

Differential equation for pointer–scale indicator

$$\frac{I}{c}\frac{d^2\theta}{dt^2} + \frac{(nAB)^2}{c(R_{Th} + R_L)}\cdot\frac{d\theta}{dt} + \theta = \frac{nAB}{c(R_{Th} + R_L)}E_{Th} \qquad [11.7]$$

By defining $\Delta\theta$, ΔE_{Th} to be deviations from initial steady conditions $\theta(0-)$, $E_{Th}(0-)$ (Section 4.1.2) we can derive the transfer function for the indicator. This can be expressed in the standard second order form:

Transfer function for pointer-scale indicator

$$\frac{\Delta\bar{\theta}}{\Delta\bar{E}_{Th}}(s) = \frac{K}{\dfrac{1}{\omega_n^2}s^2 + \dfrac{2\xi}{\omega_n}s + 1} \qquad [11.8]$$

where

$$\text{Steady-state sensitivity} \quad K = \frac{nAB}{c(R_{Th} + R_L)}\ \text{rad V}^{-1}$$

Natural frequency $\omega_n = \sqrt{\frac{c}{I}}$ rad s^{-1}

Damping ratio $\xi = \frac{(nAB)^2}{2\sqrt{(cI)}(R_{Th} + R_L)}$

From eqn [11.8] we see that the steady-state sensitivity K depends on magnetic flux density B, spring stiffness c and total circuit resistance $R_{Th} + R_L$. Reducing c gives a larger K but a lower ω_n; similarly increasing n and A also increases K but this gives a larger moment of inertia I and again reduced ω_n. The value of K is affected by stray magnetic fields, ferrous instrument panels (altering B), temperature and ageing (altering c). Departures from the ideal steady-state relation $\theta = (nAB/c)i$ are quantified using error bands (Section 2.1); there are 9 classes of accuracy, ranging from ± 0.05 to ± 5 per cent of f.s.d.[1]

A typical meter has a coil resistance of 75 Ω, ±2.5% error bands, input range 0 to 1 mA, and output range 0 to $\pi/2$ radians, i.e. a current sensitivity of 1.57×10^3 rad A^{-1}. The moment of inertia I of the coil/former/pointer assembly is fairly large, so that ω_n is small, typically $f_n \approx 0.5$ to 1.5 Hz. The damping ratio ξ depends on the total circuit resistance $R_{Th} + R_L$; a ξ of around 0.7 is ideal (Section 4.3). If ξ is greater than 0.7 (with a given R_{Th} and R_L), it can be reduced by connecting in an additional resistance in *series* with R_L. This causes a corresponding reduction in sensitivity. If ξ is less than 0.7, it can be increased by connecting a resistance in *parallel* with R_L. This causes a corresponding increase in sensitivity. The scale corresponds to the range of the measured variable, e.g. 0 to 200 °C, 0 to 50 m^3 hr^{-1}. In order to reduce observation error it is important that the scale should be legible, clear and not overcrowded with scale marks. Recommendations for the design of scales are given in BS 89.[2].

Experiments show that observers are able to mentally subdivide a scale division into five equal parts with reasonable accuracy. To achieve an observation error of 1% of f.s.d., each part should be about 1% of the scale range. This means a scale with 20 minor divisions, usually grouped into four or five major divisions as shown in Fig. 11.2(b).

11.3 Analogue chart recorders

11.3.1 Ultraviolet galvanometer recorders

The UV galvanometer recorder has the same moving coil principle as the pointer-scale indicator, but a rather more delicate construction (Fig. 11.3). The moving coil is suspended by a thin torsion strip which provides the spring restoring torque. A lightweight mirror is also attached to the assembly; this reflects an incident beam of ultraviolet light to a point $P(x, y)$ on a moving strip of photosensitive paper. Since the paper moves at a constant speed, x is proportional to time t; y depends on the angular deflection θ of the coil and thus the signal e.m.f. E_{Th}. Once the paper is developed a permanent record is obtained. Equations [11.7] and [11.8] specify the transfer function between θ and E_{Th}.

The relationship between y and θ can be found from the plan views shown in Fig.

Fig. 11.3 UV galvanometer recorder and deflection geometry

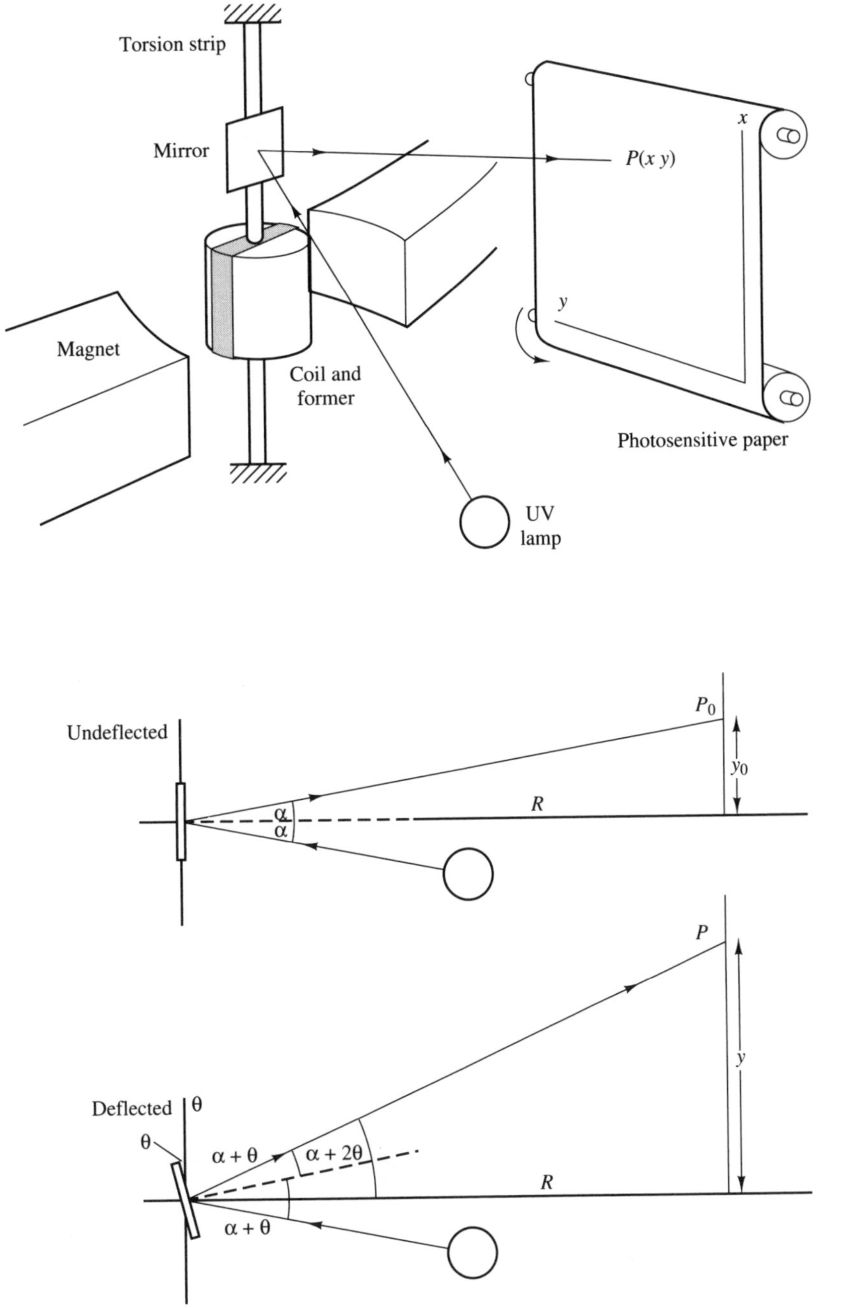

11.3. When the coil is undeflected i.e. $\theta = 0$, the angles of incidence and reflection are α. This gives:

$$y_0 = R \tan \alpha \tag{11.9}$$

where R is the distance of mirror from the paper. When $\theta \neq 0$ the angles of incidence and reflection are $\alpha + \theta$, thus:

$$y = R \tan(\alpha + 2\theta) \qquad [11.10]$$

indicating a non-linear relation between y and θ. However, for small θ, i.e. $|\theta| < 10°$, the following linear approximation is valid:

$$y = y_0 + \Delta y = R \tan \alpha + 2R\sec^2 \alpha\Delta\theta \qquad [11.11]$$

The overall steady-state voltage sensitivity of the source/recorder system is, from [11.8] and [11.10]:

Stead-state sensitivity of recorder

$$\frac{\Delta y}{\Delta E_{Th}} = \frac{\Delta y}{\Delta \theta}\frac{\Delta \theta}{\Delta E_{Th}} = \frac{2R\sec^2\alpha\; nAB}{c(R_{Th} + R_L)} \qquad [11.12]$$

The use of an optical pointer means that the pointer length R is large, giving high steady-state sensitivity, but moment of inertia I is small, giving high natural frequency ω_n.

A typical 'low frequency' galvanometer[3] has a natural frequency f_n of 100 Hz, a coil resistance R_L of 80 Ω, maximum safe current of 10 mA and a voltage sensitivity of 5 cm mV^{-1}. By connecting in an external damping resistance of 250 Ω (R_{Th} and additional series or parallel resistance) a damping ratio ξ of 0.64 is obtained. This ensures that the amplitude ratio is flat within ±5% (i.e. $0.95 \le |G(j\omega)| \le 1.05$) up to 60 Hz.

A typical 'high frequency' galvanometer[3] has a natural frequency of 5 kHz, coil resistance of 42 Ω, maximum safe current of 50 mA, and a voltage sensitivity of 1.5×10^{-3} cm mV^{-1}. This type of galvanometer is damped by immersing the torsion strip in silicone fluid. The fluid viscosity is such that a damping ratio ξ of 0.64 is obtained without the need for an external damping resistance. The galvanometer assembly is housed in a cylindrical pencil. 6, 12 or 25 galvanometer 'pencils' can be inserted in the same magnet block to give simultaneous recording of several measured variables.

11.3.2 Closed-loop recorders

Figure 11.4(a) shows a simplified schematic diagram of a typical closed-loop chart recorder using an a.c. position servomechanism. This device is more robust than the galvanometer type, has a sensitivity independent of signal circuit resistance, and is often preferred for industrial applications. It is essentially a voltage balance device,

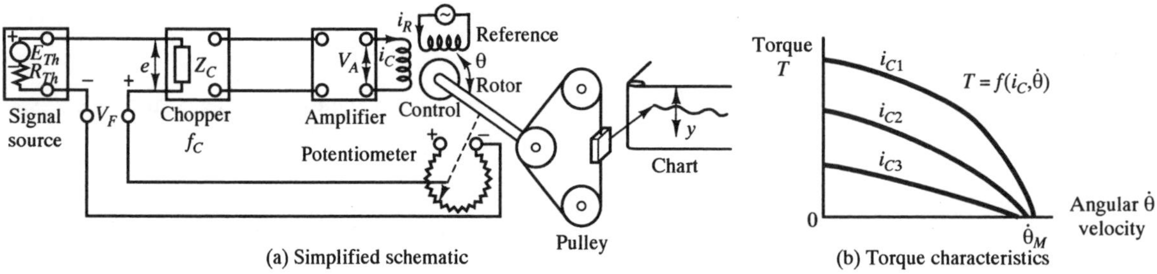

Fig. 11.4 Closed loop recorder using a.c. position servo
(a) Simplified schematic
(b) Torque characteristics (after Healey[4])

Fig. 11.5 Approximate equations and block diagram for a.c. servo recorder

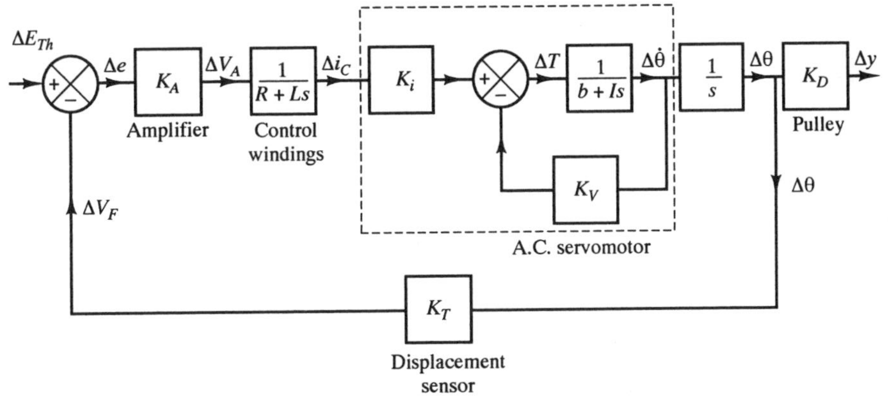

where the signal voltage E_{Th} is opposed by a feedback voltage V_F derived from a displacement sensor such as a potentiometer slidewire or variable inductance. The d.c. error voltage e is chopped at frequency f_C Hz, and the resulting a.c. signal amplified both in voltage and power. The amplifier output voltage produces a variable amplitude current $\hat{i}_C \sin 2\pi f_C t$ in the control stator winding of the two-phase a.c. servomotor. A fixed amplitude current $\hat{i}_R \cos 2\pi f_C t$, 90° out of phase, flows in the reference stator windings. Under these conditions the torque T acting on the rotor is approximately proportional to control winding current i_C, but also depends on rotor angular velocity $\dot{\theta}$ ($d\theta/dt$), as shown in Fig. 11.4(b).[4] This torque adjusts the rotor angular position θ and recorder pen position y until sensor output voltage V_F is as close as possible to E_{Th}, i.e. error voltage $e \approx 0$.

Figure 11.5 gives the system equations and the corresponding approximate transfer function block diagram for the recorder. The control winding current i_C is determined by amplifier output voltage V_a, winding resistance R and winding inductance L. Equation [11.16] is a linear approximation to the motor torque/speed/current characteristics suitable for small deviations ΔT, Δi_C, $\Delta\dot{\theta}$ from steady-state operating conditions: K_i is the slope of the torque/current curve and K_V the slope of the torque/speed curve. The rotor torque T is opposed by a mechanical damping torque $b\dot{\theta}$ to give a resultant accelerating torque $T - b\dot{\theta}$; this is equal to the product $I(d\dot{\theta}/dt)$ of total moment of inertia and angular acceleration (eqn [11.17]. K_T is the sensitivity of the displacement sensor, and K_D the sensitivity of the pulley system which converts angular rotation θ into the translational displacement y of the recorder pen. System equations are:

Error voltage $e = E_{Th} - V_F$ [11.13]

Amplifier output $V_A = K_A e$ [11.14]

Control winding $V_A = Ri_C + L\, di_C/dt$ [11.15]

Rotor torque $\Delta T = K_i \Delta i_C - K_V \dot{\theta}$ [11.16]

Rotor motion $I\, d\dot{\theta}/dt = T - b\dot{\theta}$ [11.17]

Displacement sensor $V_F = K_T \theta$ [11.18]

Pen position $y = K_D \theta$ [11.19]

By successive reduction[4] of Fig. 11.5 the transfer function relating small changes in input E_{Th} and output y can be shown to be:

Transfer function for closed loop servo recorder

$$\frac{\Delta \bar{y}}{\Delta \bar{E}_{Th}}(s) = \frac{K_D K_A K_i}{LIs^3 + [RI + L(K_V + b)]s^2 + R(K_V + b)s + K_T K_A K_i} \qquad [11.20]$$

The steady-state sensitivity of the recorder can be found by putting $s = 0$ in the transfer function; i.e.

$$K = \frac{\Delta \bar{y}}{\Delta \bar{E}_{Th}}(0) = \frac{K_D K_A K_i}{K_T K_A K_i} = \frac{K_D}{K_T} \qquad [11.21]$$

and depends only on the sensitivity of the displacement sensor and pulley system.

By careful design of these elements highly ideal steady-state performance is possible, e.g. error bands within $\pm 0.25\%$ of the ideal straight line. In some modern recorders the rotary servomotor and pulley system are replaced by a linear servomotor which consists of a coil moving along a cylindrical rod bridging two permanent magnets.

A typical recorder, with a chart 100 mm wide, has a potentiometer slide wire fed by an adjustable but stabilised constant current source. By adjusting this current, K_T, K and hence the input range of the recorder can be altered; typically from a minimum of 0 to 2 mV ($K = 50\ \text{mm mV}^{-1}$) to a maximum of 0 to 100 mV ($K = 1\ \text{mm mV}^{-1}$). Since the recorder transfer function is third-order, the standard second-order parameters ω_n, ξ are not applicable, and dynamics are normally specified using **bandwidth** (Section 4.4) and **rise time** (time interval between 10% and 90% of response to 100% step input).

Mainly because of the high moment of inertia, the dynamic response of this type of recorder is slow compared with the galvanometer type; a rise time of 1 to 2 sec and bandwidth of ≈ 1 Hz is typical. Transfer function [11.20] is only applicable to small deviations from steady conditions. Large deviations result in **velocity saturation** effects, where $\dot{\theta}$ cannot be increased beyond a maximum value $\dot{\theta}_M$ set by the motor torque/speed characteristic (Fig. 11.4).

The input impedance Z_{IN} of the recorder depends on the error voltage and chopper input impedance Z_C, i.e.

$$Z_{IN} = \frac{\Delta E_{Th}}{\Delta i_{IN}} \quad \text{and} \quad \Delta i_{IN} = \frac{e}{Z_C}$$

giving:

$$Z_{IN} = \frac{\Delta E_{Th}}{e} Z_C \qquad [11.22]$$

Z_{IN} is large, typically increasing from 30 kΩ, at maximum out-of-balance, towards infinity at perfect balance. Various techniques for marking the chart paper are available; these include inking and heat-sensitive, pressure-sensitive, electrolytic and photographic papers. Simultaneous recording of several measured variables (up to 6 or 12) can be obtained by switching the corresponding voltage signals in turn onto the position servomechanism. Each time a given signal is switched in, a dot is printed on the chart in a colour corresponding to the signal; the record thus consists of several traces of different colours each made up of a series of dots.

11.4 Small-scale alphanumeric displays

The word 'alphanumeric' is a contraction of 'alphabetic' and 'numeric' and alphanumeric devices are able to display the numerals 0 to 9, some or all of the letters A to Z, decimal points and other simple symbols. These devices accept a serial or parallel digital signal, either in binary coded decimal (8:4:2:1 b.c.d.) or ASCII form. Thus in a temperature measurement system, where the true temperature is 537 °C, the input signal to the 3-decade display may have the b.c.d. form 0101 0011 1001, so that the displayed measured temperature is 539 °C.

11.4.1 Character formats

Figure 11.6 shows two character formats in widespread use: seven-segment and 7 × 5 dot-matrix. Figure 11.6(a) shows 7 segments *a* to *g* arranged in a figure **8** configuration and the corresponding character set. This is limited to the 10 numerals and 9 upper-case letters. The 7 × 5 dot-matrix format enables a far larger set of typically 192 characters to be obtained; this includes all the numerals, upper and lower case letters (Fig. 11.6(b)), together with Greek letters and other symbols. A 9 × 7 dot-matrix format gives a better representation of lower-case letters.

For both character formats, each individual dot or segment can be switched 'on' or 'off' independently of the others; to display a given character the correct combination of elements must be switched 'on'. As an example, suppose we wish to display the decimal numbers 0 to 9, given the 4 logic signals DCBA representing one decade of 8:4:2:1 b.c.d., using the seven-segment format. A logic system, or decoder, is therefore required to convert the input DCBA code into the required 7-segment a b c d e f g code.

The conversion table shown in Fig. 11.7 shows the coresponding denary number, b.c.d. code, 7-segment code and the displayed numeral. The same principles are used in 7 × 5 dot matrix displays, in both cases the logic is implemented using a read only memory (ROM).

The segment or dot elements used in character generation are either light-emitting diodes or liquid crystal displays.

11.4.2 Light-emitting diode (LED) displays[5]

When a semiconductor diode is forward biased, as shown in Fig. 11.8(a), a current i_F flows, which depends exponentially on the forward voltage V_F:

$$i_F = i_s \exp\left(\frac{qV_F}{k\theta}\right) \qquad [11.23]$$

where i_s is the reverse saturation current, q the electron or hole charge, k Boltzmann's constant and θ K the absolute temperature. Taking natural logarithms we have:

$$\log_e i_F = \frac{q}{K\theta} V_F + \log_e i_s \qquad [11.24]$$

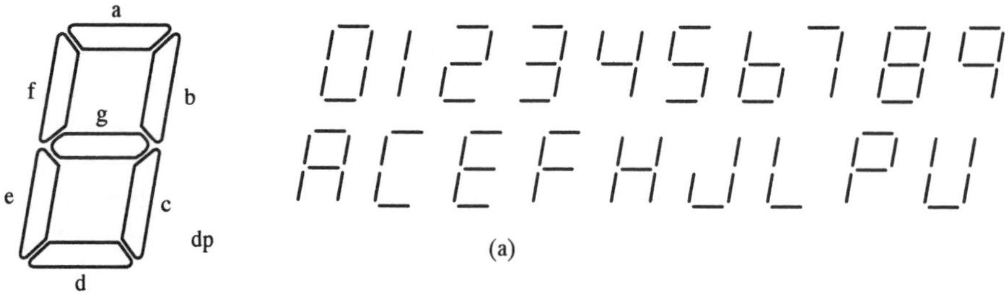

(a)

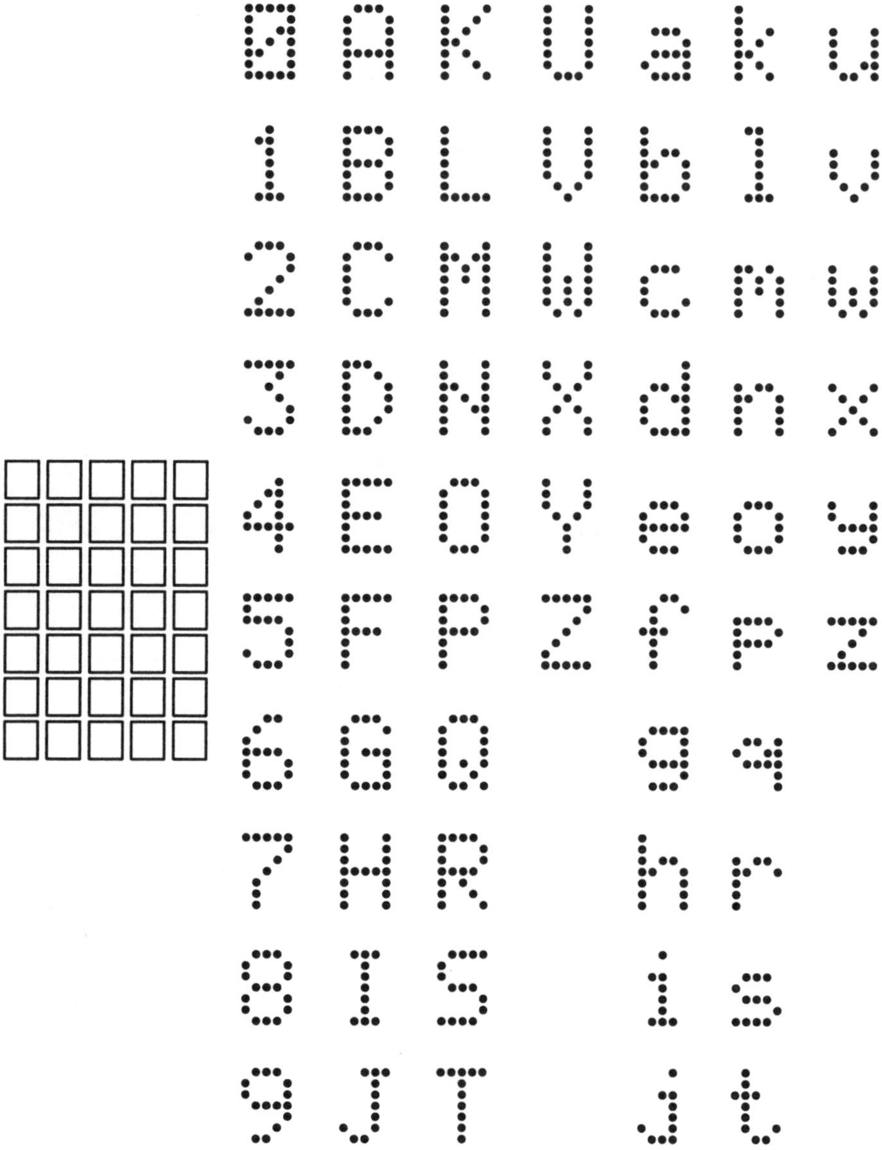

(b)

Fig. 11.6 Character formats for displays
(a) 7-segment character format
(b) 7 × 5 dot-matrix character format

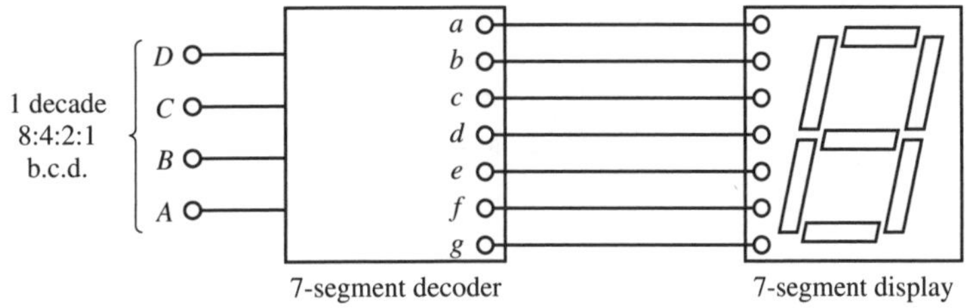

Fig. 11.7 Display of numerals using seven-segment format

Denary number	b.c.d. code D C B A	7-segment code *a b c d e f g*	Displayed numeral
0	0 0 0 0	1 1 1 1 1 1 0	
1	0 0 0 1	0 1 1 0 0 0 0	
2	0 0 1 0	1 1 0 1 1 0 1	
3	0 0 1 1	1 1 1 1 0 0 1	
4	0 1 0 0	0 1 1 0 0 1 1	
5	0 1 0 1	1 0 1 1 0 1 1	
6	0 1 1 0	0 0 1 1 1 1 1	
7	0 1 1 1	1 1 1 0 0 0 0	
8	1 0 0 0	1 1 1 1 1 1 1	
9	1 0 0 1	1 1 1 0 0 1 1	

i.e. there is an approximately linear relation between $\log_e i_F$ and V_F as shown in Fig. 11.8(b).

Light-emitting diodes have the special property that when forward biased they emit electromagnetic radiation over a certain band of wavelengths. Two commonly-used LED materials are gallium arsenide phosphide (GaAsP), which emits red light, and gallium phosphide (GaP), which emits green or yellow light.[5] In both cases the luminous intensity I_V of the diode light source increases with current i_F; for GaAsP diodes the relationship is approximately linear (Fig. 11.8(c)). The light emitted by a GaAsP (red) diode is distributed over a narrow band of wavelengths centred on 0.655 μm. Figure 11.8(d) shows the relationship between relative luminous intensity and wavelength λ. Similarly the light emitted by a GaP (green) diode is distributed over a narrow band of wavelengths centred on 0.560 μm. The human eye is far more sensitive to green light than red, so a green LED of low luminous intensity may appear as bright as a red LED of much higher luminous intensity. The response of LEDs to step changes in i_F is extremely fast; turn-on and turn-off times of 10 ns are typical.

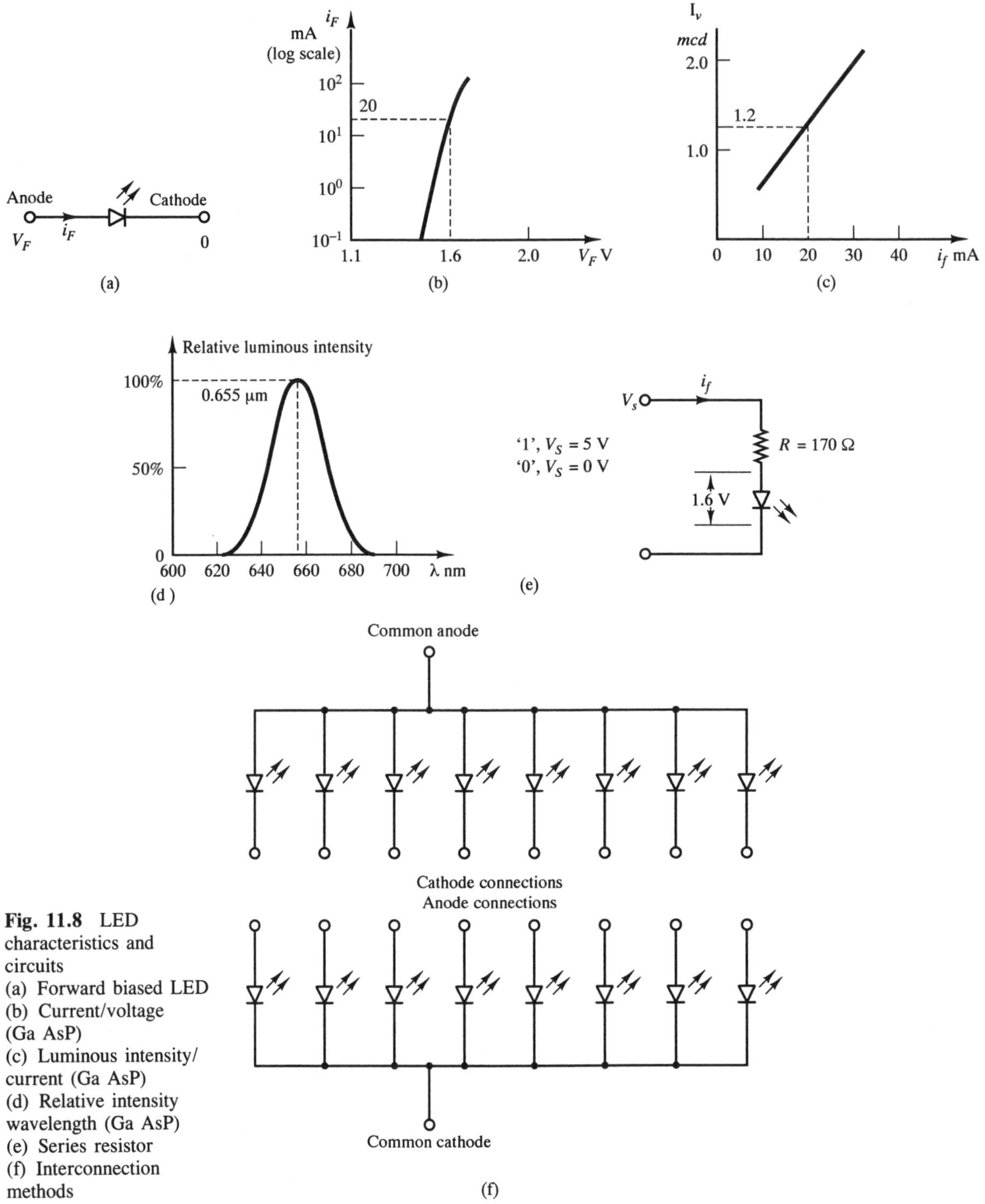

Fig. 11.8 LED characteristics and circuits
(a) Forward biased LED
(b) Current/voltage (Ga AsP)
(c) Luminous intensity/current (Ga AsP)
(d) Relative intensity wavelength (Ga AsP)
(e) Series resistor
(f) Interconnection methods

When switched 'on' a typical GaAsP diode requires a forward current i_F of around 20 mA corresponding to a luminous intensity I_v of 1.2 mcd (millicandela), and a forward voltage V_F of 1.6 V. When used as a display the diode should be switched 'on' by a logic signal in the'1' state, and switched 'off' when the signal is in the '0' state. Figure 11.8(e) shows a simple circuit for achieving this, using a series resistor R of 170 Ωm. For a '1' input, $V_s \approx 5$ V, $i_F = (5 - 1.6)/170 =$ 20 mA and the diode is 'on'. For a '0' input, $V_s \approx 0$ V, i_F is negligible and the diode is 'off'. Alternatively a 20 mA constant current source may be switched in and out by the logic signal.

A 7-segment LED display device consists of 8 individuals LEDs, one for each segment and one for the decimal point. There are two possible methods of interconnection, common anode or common cathode (Fig. 11.8(f)).

11.4.3 Liquid crystal displays (LCD)[6][7]

Liquid crystal displays do not emit light but use light incident on them from other sources. The passage of light through the device is modified to create dark areas and thus characters. Liquid crystals are materials which retain a crystal-like structure in the liquid phase, over a certain temperature range. In a normal liquid the orientation of molecules is random and optical effects, such as reflection, refraction and polarisation, average out so that no effect predominates in a given direction. However, in a liquid crystal some molecular order remains so that optical effects can predominate in certain directions, also both crystal structure and optical effects may be modified by applied electric fields.

Figures 11.9(a) and (b) show the principle of an LCD display using plane polarised light. A thin layer of liquid crystal is sandwiched between two glass plates. An x polarising filter is placed above the upper plate; this only transmits light with an electromegnetic component along the x direction. A y polarising filter is placed below the lower plate; this only transmits light with an electromagnetic component along the y direction. When the applied electric field is zero, the liquid crystals rotate the plane of polarisation of light from x to y so that it is able to pass through the lower filter. A reflector at the base of the device reflects the light back through the system to an observer, and the surface of the device appears pale grey in colour. When an electric field is applied, the crystals align themselves with the field and can no longer rotate the plane of polarisation of the light from x to y. The light is now completely absorbed by the lower filter and the surface of the device appears black to an observer. Figure 11.9(c) shows the construction of a 7-segment LCD display.[8]

11.5 Large-scale displays

The small-scale alphanumeric displays are well suited to displaying the measured values of a few variables. If a process has several measured variables then it is best to display all the process information together on a single large screen. A large screen also has the advantage that a 'picture' (mimic diagram) of the process can be displayed on the screen together with alphanumeric information on the measured values.

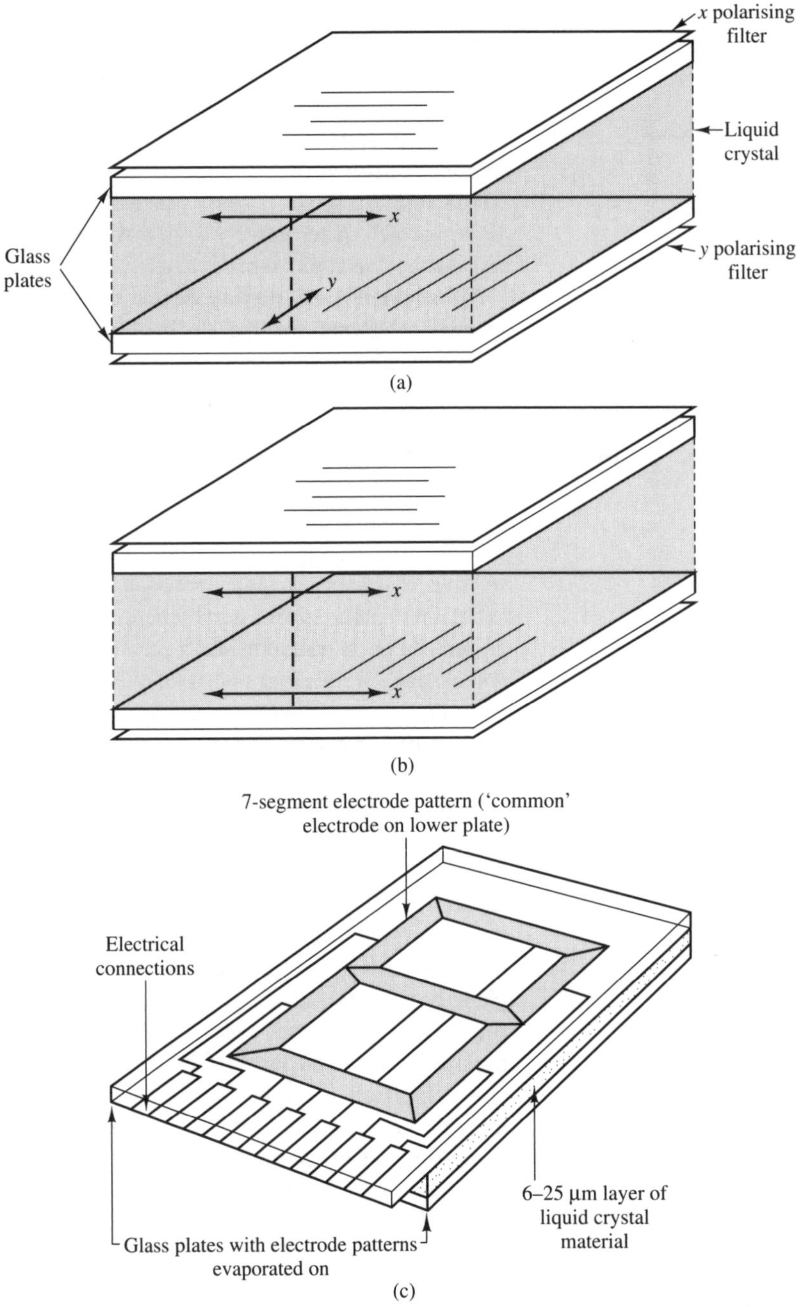

Fig. 11.9 Liquid crystal devices
(a) Principle — electric field = 0
(b) Principle — electric field ≠ 0
(c) Physical construction (after Open University T292 Instrumentation[8])

The display screen consists of a large number of small picture elements or **pixels** each of which can be rapidly switched on or off. For example a standard format for the presentation of alphanumeric data is 80 characters per line of display and 25 lines of characters, i.e. a total of 2000 characters. If each character is represented by a 7 × 5 dot-matrix pattern and if a space equivalent to one pixel is necessary to separate each character from the next and each line from the next then each character space occupies 8 × 6 pixels. The total number of pixels required on the complete screen is 96 000. If a 9 × 7 dot-matrix pattern is used, then each character space occupies 10 × 8 pixels and the total number of pixels is 160 000.

11.5.1 Monitors

A monitor is a large-scale display device based on the cathode-ray tube (CRT). (A visual display unit (VDU) is a combination of a monitor and a keyboard.)

Figure 11.10 shows a basic CRT: electrons are emitted at the cathode and accelerated towards the anode. A third electrode, called a grid or modulator, is placed between cathode and anode: by altering the potential of the modulator the number of electrons in the beam, i.e. the beam current, can be adjusted. The beam then passes through a focusing system followed by X and Y deflection systems: the focusing and deflection systems can be electromagnetic (EM), or electrostatic (ES) as shown in the diagram. The electron beam is brought to a focus on the inside surface

Fig. 11.10 Cathode ray construction and waveforms

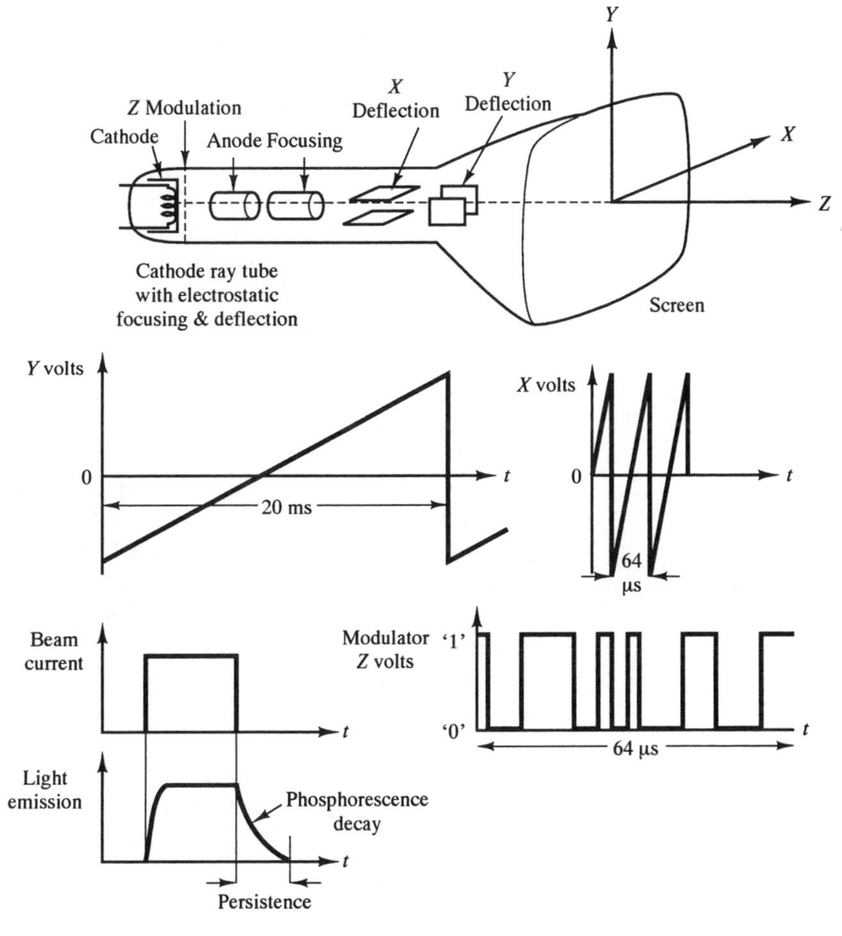

of the screen which is coated with a large number of **phosphor dots**. These dots form the pixels. Phosphors are semiconductor materials which emit visible radiation in response to the impact of electrons: a spot of light therefore appears on the screen. In response to a pulse change in beam current, i.e. a sudden increase followed by a sudden decrease, the light emission does not fall instantaneously but there is a gradual reduction called **phosphorescence decay** (Fig. 11.10). The corresponding decay time is called **persistence** of the phosphor; phosphors with a wide range of persistences are available, ranging from less than 1 μs (very short) to greater than 1 sec (very long). In **refresh displays** the phosphors must be 'refreshed' or re-energised every time the phosphorescence decays to a certain level: this is necessary to obtain a stationary pattern on the screen with minimum **flicker**. In the more expensive **storage displays**, the display is retained on the screen and refreshment is not necessary.

A **monochrome monitor** is a non-storage CRT with a standard 312-line **raster display**. In a raster-type display constant amplitude sawtooth deflection voltages are applied to both X and Y plates (Fig. 11.10). The period of the Y voltage, i.e. the time taken for the display to move from the top to bottom of the screen, is relatively long. A typical period is 20 ms corresponding to a refreshment rate of 50 frames or pictures per second. This is suitable for a phosphor with a medium persistence of say 50 ms. The period of the X voltage i.e. the time for each traverse across the screen is 312 times shorter, i.e. 64 μs. The resulting motion of the spot on the screen is shown in Fig. 11.11. 312 horizontal 'lines' are traced out during the movement of the spot from top to bottom of the screen.

Suppose we wish to generate a fixed format of 2000 alphanumeric characters arranged in 25 lines of 80 characters per line. If a 9×7 character format is used, then each character space occupies 10×8 pixels. Thus there are $25 \times 10 = 250$ horizontal lines of pixels with $80 \times 8 = 640$ pixels on each horizontal line. To create the characters on the screen, the electron beam is switched on and off to produce the required pattern. A high frequency pulse waveform is applied to the modulator electrode which causes the electron beam current to be switched on and off many times during the 64 μs required for each horizontal traverse (**z modulation**). During each traverse the electron beam moves across a single line of pixels, causing them to be switched either 'on' or 'off' depending on whether the modulator signal is a '1' or a '0'. Since there are 9 lines of pixels for each character, 9 complete horizontal traverses are required to build up each complete line of characters. Figure 11.11 shows how the characters build up in a 7×5 format. The entire display of 25 lines, using a character space of 10×8, is completed in 20 ms using 250 horizontal line scans. This is less than the theoretical number of 312, partly because of the time required for vertical flyback, and partly because the top and bottom of the CRT screen are not used to avoid image distortion.

The input measurement data to the monitor is usually in serial ASCII digital form (Chapter 10) and must be converted into the serial pulse video signal required by the modulator to energise the pixels. Each horizontal sweep takes 64 μs but only about 40 μs is available for character generation; since 640 pixels must be energised during this time, the video signal has 16×10^6 pulses or bits per second. The monitor is normally operated under computer control and an observer/operator can enter information and instructions via a keyboard. This facility enables the operator to request different display formats; e.g. the values of the most important measured

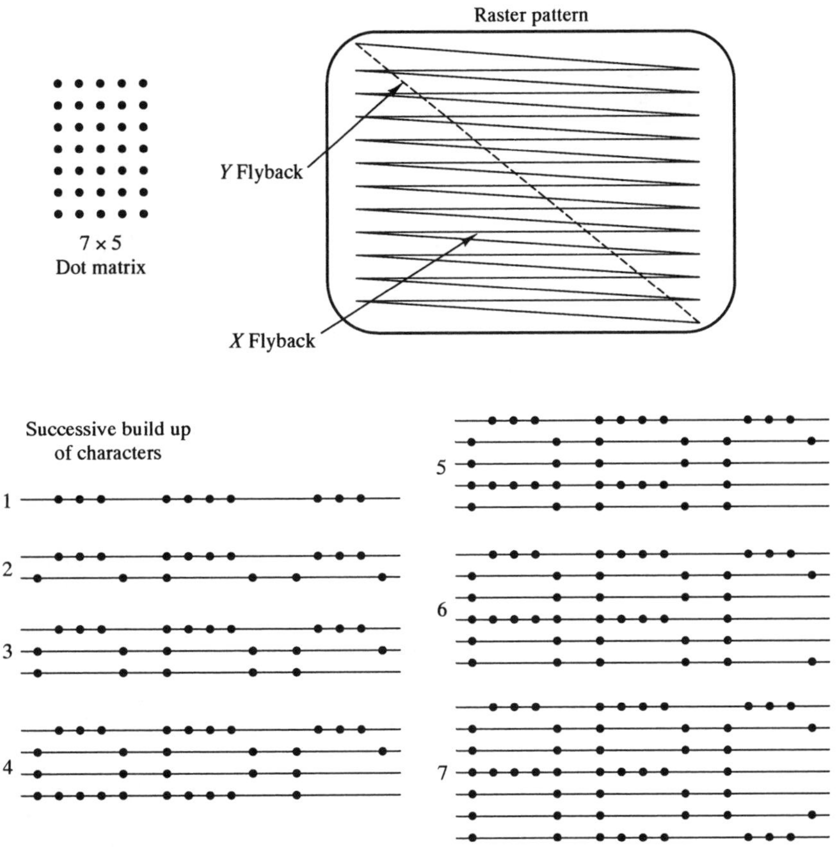

Fig. 11.11 Raster display and character generation using 7 × 5 dot-matrix format

values for the complete process, or the values of all the measured variables associated with a particular part of the process.

A **colour monitor** produces images containing a wide range of colours. The screen of a colour CRT is coated with dots of three different types of phosphor: one type of phosphor emits red light, the second green light, the third blue light. Dots of each type are arranged in equilateral triangles called triads (Fig. 11.12(a)). The monitor has three electron guns one for each type of phosphor. The corresponding electron beams are deflected horizontally and vertically to produce a raster display as in a monochrome monitor. As the beams traverse the screens, the intensity of each beam is varied according to the voltage applied to the corresponding modulator electrode. This creates varying colour intensities at the triads and colour images on the screen.

A colour monitor can be used to create **graphic displays**. In a fixed format alphanumeric display characters occupy fixed positions in a display and each character is built up using a fixed 7 × 5 or 9 × 7 dot-matrix format. In a graphic display the screen contains a full matrix of pixels, each one of which can be turned on or off to produce graphical images or pictures, alphanumeric characters at any position on the screen or a combination of both. Figure 11.12(b) shows, in black-and-white, a colour line diagram of part of a chemical plant; alphanumeric data such as

Fig. 11.12 Colour displays
(a) Phosphor dot triads
(b) Chemical plant line diagram

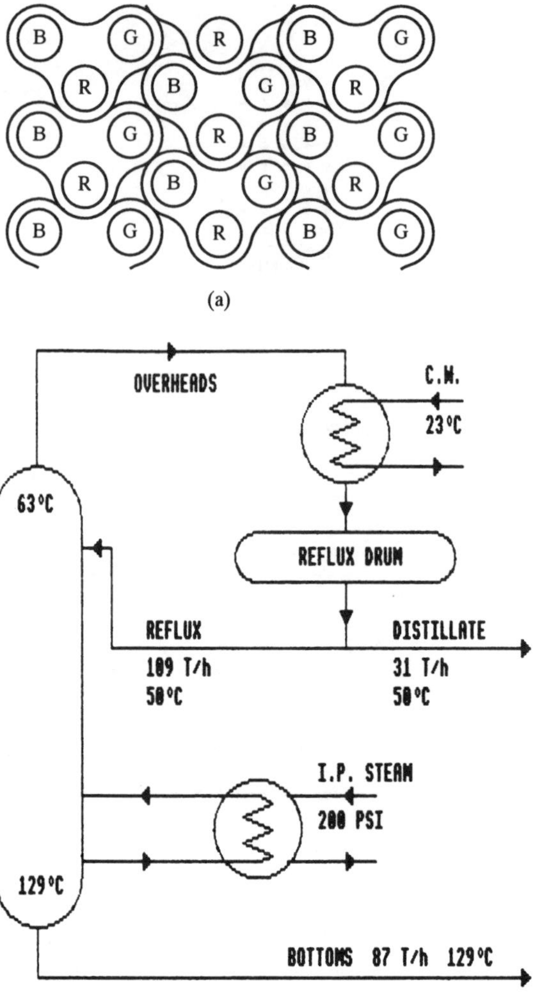

(a)

(b)

descriptions of process variables and their measured values can be displayed at any appropriate point on the diagram.

11.5.2 Flat screen displays

The CRT has the disadvantage of being both bulky and fragile. For this reason four technologies are currently being developed to provide both alphanumeric and graphic displays on a flat screen. The technologies are:

(a) Liquid crystal
(b) Electroluminescene
(c) Gas plasma discharge
(d) Vacuum fluorescence

In all cases the image is generated using a matrix of pixels. A full account of these technologies is given in.[8]

11.6 Digital printers

The analogue chart recorders discussed in Section 11.3 give a record in the form of a continuous trace or graph. This is ideal for **qualitative analysis** of high-speed events, such as a human heart beat, or complex events, such as the breakdown of a piece of machinery such as a compressor. In the first example, any irregularities or abnormal behaviour show up immediately as changes in the waveform. In the second example, the exact sequence of events — drop in lubricating oil pressure, rise in bearing temperature, drop in delivery pressure — can be found and the cause of the failure established. There are, however, applications in which a record in the form of printed numbers is required, so that a **quantitative analysis** can be performed. One example is the calculation of the total weekly consumption of fuel oil in a furnace, from flow rate data taken at hourly intervals. Another is the plotting of temperature profiles in a nuclear reactor, using many temperature measurements distributed over the reactor core. Numerical data can only be extracted from a chart record by **interpolation**, i.e. estimating the position of the trace in relation to two scale markings; an inaccurate and time-consuming process.

Alphanumeric printers create data records suitable for quantitative analysis by printing numbers, letters and special characters on paper. Figure 11.13(a) shows the layouts of a 14-column record from a strip printer, together with an expanded extract of a typical record of several temperatures in a reactor. The record has 14 characters on each line which specify the channel number, the measured value and the units of measurement. By printing '>' or '<', the record indicates that the measurement is either greater than a specified upper limit (say 200 °C) or less than a specified lower limit (say 100 °C). The input data to a printer is in parallel or serial ASCII digital form from a microcomputer (Section 10.3 and Table 10.5).

Figure 11.13(a) gives the 14 binary ASCII input code words corresponding to each of the 14 characters on the first line of the record. Each column of the record has a print hammer that strikes the paper against an inking ribbon and a typeface. The type for all possible characters in a given column is arranged on a disc. The discs are rotated and the print hammer strikes as the selected character passes the paper.

More modern methods of printing use a single print head which moves horizontally across the paper as each character is printed. In a **golf-ball printer** the print head is a sphere and the required character is selected by rotational movement of the head about two axes. In a **daisy-wheel printer** the print head is a disc consisting of a large number of vanes radiating out from the centre. There is a character at the end of each vane. The required character is selected by rotating the print head, the selected character is then struck by a print hammer onto an inking ribbon.

In a **dot-matrix printer** the print head consists of either seven (for 7 × 5 format) or nine (for 9 × 7 format) pins arranged in a vertical line (Fig. 11.13(b)). Each time a pin is 'fired', it shoots forward, strikes the inking ribbon and presses it against the paper to produce a dot. A character is created by moving the head horizontally across the paper between 7 different positions, in each position the required pattern

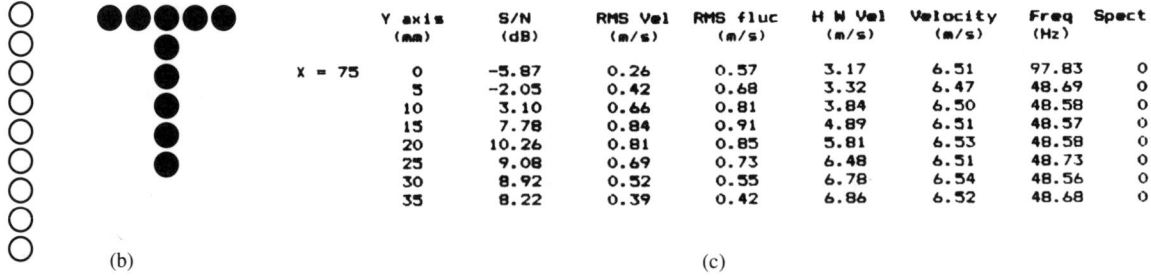

Channel number		Blank	Alarm	Blank	Polarity	Measured value data 5 digits + decimal point						Blank	Units
1	2	3	4	5	6	7	8	9	10	11	12	13	14
1	0		<			0	8	5	.	0	0		C
1	1					1	7	1	.	0	0		C
1	2					1	3	4	.	0	0		C
1	3		>			2	0	7	.	0	0		C
1	4					1	0	5	.	0	0		C
1	5					1	6	3	.	0	0		C
1	6		>			2	1	9	.	0	0		C
1	7					1	5	5	.	0	0		C
1	8		<		–	0	0	2	.	0	0		C
1	9					1	9	8	.	0	0		C
2	0		<			0	9	5	.	0	0		C

ASCII input codes for the first record:

0110001 (1)
0110000 (0)
0100000 (SP)
0111100 (<)
0100000 (SP)
0100000 (SP)
0110000 (0)
0111000 (8)
0110101 (5)
0101110 (•)
0110000 (0)
0110000 (0)
0100000 (SP)
1000011 (C)

(a)

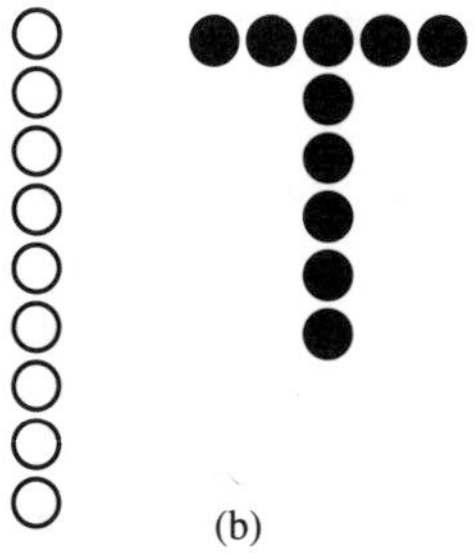

(b)

	Y axis (mm)	S/N (dB)	RMS Vel (m/s)	RMS fluc (m/s)	H W Vel (m/s)	Velocity (m/s)	Freq (Hz)	Spect
X = 75	0	-5.87	0.26	0.57	3.17	6.51	97.83	0
	5	-2.05	0.42	0.68	3.32	6.47	48.69	0
	10	3.10	0.66	0.81	3.84	6.50	48.58	0
	15	7.78	0.84	0.91	4.89	6.51	48.57	0
	20	10.26	0.81	0.85	5.81	6.53	48.58	0
	25	9.08	0.69	0.73	6.48	6.51	48.73	0
	30	8.92	0.52	0.55	6.78	6.54	48.56	0
	35	8.22	0.39	0.42	6.86	6.52	48.68	0

(c)

Fig. 11.13
Alphanumeric printers
(a) Layout, typical record and ASCII input codes for 14-column strip printer
(b) Dot-matrix (9 × 7 format) printing
(c) Extract from 80 column dot-matrix print record

of dots is produced by firing certain pins. For example to print a capital **T**; the head fires the top pin, moves, fires the top pin again, moves, fires the top seven pins, moves, fires the top pin, moves, fires the top pin once more to finish the letter.[9] Figure 11.13(c) shows an extract from a 80-column dot-matrix print record of a wind tunnel test on a vortex flowmeter, in which the dots are discernable under close examination.

Conclusion

The chapter first reviewed the types of **data presentation elements** in current use and the factors influencing choice. The principle and characteristics of **pointer-scale indicators** were first discussed followed by **analogue chart recorders**. The following section examined **small-scale alphanumeric displays**, character format and implementation using both **light emitting diodes** (LEDs) and **liquid crystal displays** (LCDs). **Large scale displays** and their implementation using **monitors** were then discussed. The final section looked at **digital printers**.

References

11.1 British Standards Institution, BS 89; 1977, *Direct Acting Indicating Electrical Measuring Instruments and Their Accessories.*

11.2 British Standards Institution BS 3693 (Part 1: 1964 and Part 2: 1969), *The Design of Scales and Indexes.*

11.3 S.E. Laboratories, *Technical Information on Galvanometers.*

11.4 HEALEY M 1967 *Principles of Automatic Control*, English Universities Press, London, p. 285 and pp. 30–44.

11.5 R.S. Components, 1986 *Technical Data on Optoelectronics/Indicators.*

11.6 Hamlin Electronics 1982 *Technical Data on seven-segment LCD displays.*

11.2 Densitron Corporation 1983 *Technical Data on Dot Matrix LCD Modules.*

11.8 The Open University 1986 Instrumentation T 292 Block 6, part 2 Displays.

11.9 Epson Corporation 1985 *Operating Manual for LX-80 Printer.*

Problems

11.1 A moving coil indicator is connected to a Thévenin signal source of resistance 125 Ω. Using the data given below,

(a) calculate the steady-state sensitivity (rad V^{-1}), natural frequency and damping ratio for the system;

(b) what additional resistance must be connected into the circuit to give a damping ratio of 0.7? What is the sensitivity of the modified system?

Data for Indicator

Number of turns on coil = 100
Coil resistance = 75 Ω
Coil area = 10^{-4} m^2
Coil moment of inertia = 2.5×10^{-5} kg m^2
Magnetic Flux density = 150 Wb m^{-2}
Spring stiffness = 10^{-3} Nm rad^{-1}

11.2 A UV galvanometer recorder has a moving coil with a resistance of 80 Ω and a maximum safe current of 10 mA. The coil system has a damping ratio of 2.64 when the resistance of the external circuit is zero. The coil is connected into a circuit with the signal source and an additional resistance so that an overall damping ratio of 0.64 is obtained.

(a) Find the additional resistance required with
  (i) a source of resistance 100 Ω;
  (ii) a source of resistance 1000 Ω.

(b) What is the maximum safe source voltage in a(i)? (Assume steady-state conditions.)

11.3 A closed loop chart recorder incorporates an a.c. position servomechanism which is described by eqns [11.13]–[11.19]. Use the data given below to calculate the following recorder characteristics:

(a) steady-state sensitivity;
(b) transfer function;
(c) amplitude ratio at 1 Hz.

Data

Amplifier gain $K_A = 10^2$

Control winding resistance $R = 10\,\Omega$
Control winding inductance $L = 100$ mH
Moment of inertia of rotor and pulley $I = 10^{-4}$ kg m^2
Mechanical damping constant $b = 5 \times 10^{-4}$ Nms rad^{-1}
Slope of torque/speed curve $K_V = 5 \times 10^{-4}$ Nms rad^{-1}
Slope of torque/current curve $K_i = 8 \times 10^{-4}$ Nm A^{-1}
Sensitivity of displacement sensor $K_T = 0.5$ V rad^{-1}
Sensitivity of pulley system $K_D = 10$ m rad^{-1}

11.4 It is desired to display the numerals 0 to 9 and the letters ACEFJH using the 7-segment character format shown in Fig. 11.5(a). Write down a table for the conversion of a 4-input DCBA (8:4:2:1 b.c.d.) to the 7-segment output a b c d e f g code to drive the display.

Part C

Specialised measurement systems

12

Flow measurement systems

Measurement of the rate of flow of material through pipes is extremely important in a wide range of industries, including chemical, oil, steel, food and public utilities. There are a large and bewildering number of different flowmeters on the market and the user is faced with the problem of choice outlined in Chapter 7. This chapter explains the principles and characteristics of the more important flowmeters in current use. The chapter is divided into five sections: essential principles of fluid mechanics, measurement of velocity at a point in a fluid, volume flow rate, mass flow rate, and flow measurement in difficult situations.

12.1 Essential principles of fluid mechanics

12.1.1 Shear stress and viscosity

There are three states of matter: solid, liquid and gas. Liquids and gases are different in many respects but behave in the same way under the action of a deforming force. Liquids and gases, i.e. fluids, **flow** under the action of a deforming force, whereas a solid retains its shape. The effect is illustrated in Fig. 12.1(a) which shows the effect of a shear force F on a rectangular block. The corresponding **shear stress** τ is the force per unit area F/A, where A is the area of the base of the block. The effect of τ is to deform the block as shown and the resulting *shear strain* is quantified by the angle ϕ. In a solid ϕ will be constant with time and of magnitude proportional to τ. In a fluid ϕ will increase with time and the fluid will flow. In a Newtonian fluid the rate of change of shear strain $\phi/\Delta t$ is proportional to τ, i.e. τ = constant $\times$ $(\phi/\Delta t)$ where Δt is the time interval in which ϕ occurs. If ϕ is small and in radians, we have $\phi = \Delta x/y$, also if v is the velocity of the top surface of the block relative to the base $v = \Delta x/\Delta t$. This gives:

$$\tau = \text{constant} \times \frac{v}{y}$$

where the constant of proportionality is the **dynamic viscosity** η of the fluid. Replacing the velocity gradient term v/y by its differential form dv/dy we have

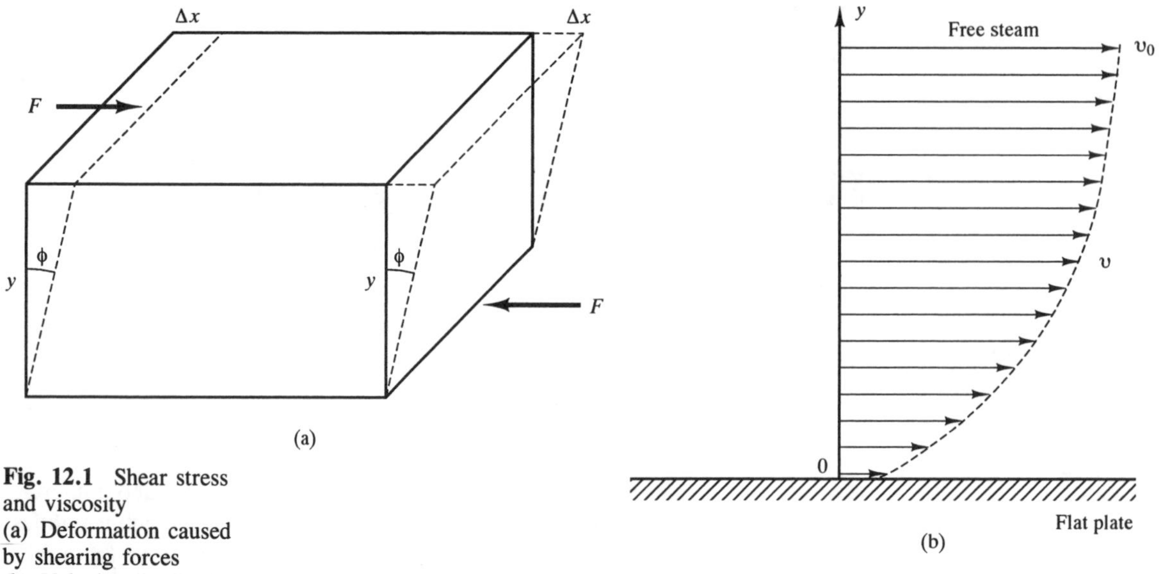

Fig. 12.1 Shear stress and viscosity
(a) Deformation caused by shearing forces
(b) Velocity distribution in boundary layers

Newton's Law of Viscosity

$$\tau = \eta \frac{dv}{dy} \qquad [12.1]$$

Figure 12.1(b) shows a fluid flowing over a solid boundary, e.g. a flat plate. The fluid in contact with the plate surface at $y = 0$ has zero velocity. As we move away from the plate, i.e. as y increases, the velocity v of the layers increases, until well away from the plate the layers have the free stream velocity v_0. The layers between the free stream and the boundary are called *boundary layers* and are characterised by a velocity gradient dv/dy. From eqn [12.1] we see that frictional shear stresses are present in these boundary layers.

12.1.2 Liquids and gases

Although liquids and gases have the common properties of fluids they have distinctive properties of their own. A liquid is difficult to compress, i.e. there is a very small decrease in volume for a given increase in pressure and may be regarded as incompressible, i.e. density ρ is independent of pressure (but will depend on temperature).

Gases are easy to compress and density depends on both pressure and temperature, for an ideal gas we have:

Equation of state for ideal gas

$$PV = mR\theta \text{ i.e. } P = \rho R\theta \qquad [12.2]$$

where P = Absolute pressure (Pa)
θ = Absolute temperature (Kelvin)

V = Volume (m^3)
ρ = Density ($kg\,m^{-3}$)
R = constant for the gas ($J\,kg^{-1}\,K^{-1}$)

For real gases the above equation must be corrected by introducing an experimental compressibility factor or gas law deviation constant.

The amount of heat required to raise the temperature of a gas by a given amount depends on whether the gas is allowed to expand, i.e. to do work, during the heating process. A gas therefore has two specific heats: specific heat at constant volume C_V and specific heat at constant pressure C_P. If the expansion or contraction of a gas is carried out **adiabatically**, i.e. no heat enters or leaves the system, the process is accompanied by a change in temperature and the corresponding relationship between pressure and volume (or density) is:

$$PV^{\gamma} = \frac{P}{\rho^{\gamma}} = \text{constant} \qquad [12.3]$$

where γ is the specific heat ratio C_P/C_V.

12.1.3 Laminar and turbulent flow: Reynolds number

Experimental observations have shown that two distinct types of flow can exist. The first is **laminar flow**, or viscous or streamline flow; this is shown for a circular pipe in Fig. 12.2(a). Here the particles move in a highly ordered manner retaining the same relative positions in successive cross sections. Thus laminar flow in a circular pipe can be regarded as a number of annular layers: the velocity of these layers increases from zero at the pipe wall to maximum at the pipe centre with significant viscous shear stresses between each layer. Figure 12.2(a) shows the resulting velocity profile; this is a graph of layer velocity v versus distance r of layer from centre, it is parabolic in shape.

The second type of flow, **turbulent flow**, is shown in Fig. 12.2(b). This is highly disordered; each particle moves randomly in three dimensions and occupies different relative positions in successive cross sections. As a result, the velocity and pressure at a given point in the pipe are both subject to small random fluctuations with time about their mean values. The viscous friction forces which cause the ordered motion in laminar flow are much smaller in turbulent flow. Figure 12.2(b) shows a typical velocity profile for turbulent flow in a circular pipe. It is obtained by taking a point r in the pipe and measuring the time average v of the velocity component, along the direction of flow at that point.

Reynolds number tells us whether the flow in a given situation is laminar or turbulent. It is the dimensionless number:

Reynolds number

$$\text{Re} = \frac{vl\rho}{\eta} \qquad [12.4]$$

where l is a characteristic length of the situation, e.g. pipe diameter. Re represents the ratio of inertial forces (proportional to $vl\rho$) to viscous forces (proportional to

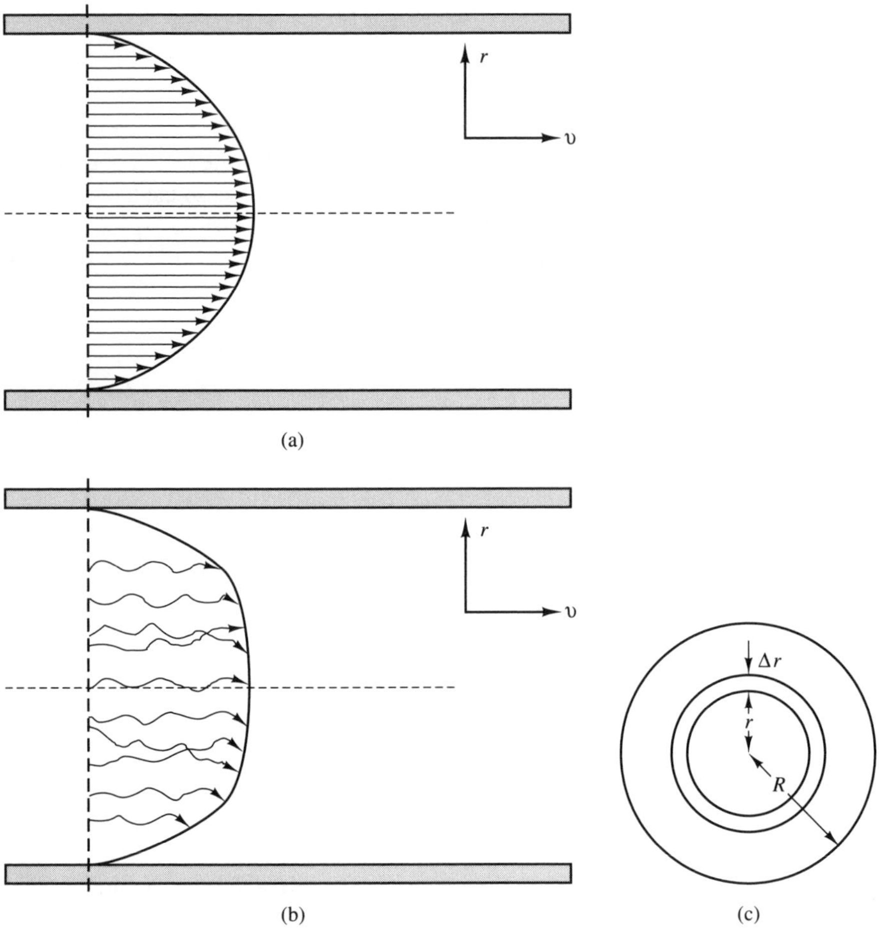

Fig. 12.2 Types of flow and velocity profiles in a circular pipe
(a) Laminar
(b) Turbulent
(c) Calculation of volume flow rate

η), thus a low value of Re implies laminar flow and a high value turbulent flow. The following is an approximate guide:

$$\begin{aligned} \mathrm{Re} &< 2 \times 10^3 && \text{— Laminar flow} \\ 2 \times 10^3 < \mathrm{Re} &< 10^4 && \text{— Transition region} \\ \mathrm{Re} &> 10^4 && \text{— Turbulent flow} \end{aligned}$$

12.1.4 Volume flow rate, mass, flow rate and mean velocity

Figures 12.2(a) and 12.2(b) show the velocity profiles $v(r)$ for both laminar and turbulent flow. If we consider an annular element radius r, thickness Δr, then this will have area $2\pi r\Delta r$ (Fig. 12.2(c)). The corresponding volume flow rate ΔQ through the element is given by:

$$\begin{aligned} \Delta Q &= \text{Area of element} \times \text{velocity} \\ &= 2\pi r \Delta r \times v(r) \end{aligned}$$

Hence the total **volume flow rate** through a circular pipe of radius R is:

Volume flow rate in a circular pipe

$$Q = 2\pi \int_0^R v(r) r \, dr \qquad [12.5]$$

In many problems the variation in velocity over the cross-sectional area can be neglected and assumed to be constant and equal to the mean velocity $\bar{v}$ which is defined by:

Mean velocity

$$\bar{v} = \frac{Q}{A} \qquad [12.6]$$

Here A is the cross-sectional area of the fluid normal to the direction of flow. Finally the corresponding mass flow rate $\dot{M}$ is given by:

Mass flow rate

$$\dot{M} = \rho Q = \rho A \bar{v} \qquad [12.7]$$

12.1.5 Continuity: conservation of mass and volume flow rate

Figure 12.3 shows a streamtube through which there is a steady flow; since conditions are steady the principle of conservation of mass means that:

mass of fluid entering in unit time = mass of fluid leaving in unit time

i.e. mass flow rate in = mass flow rate out

Using eqn [12.7] we have:

Conservation of mass flow rate

$$\rho_1 A_1 \bar{v}_1 = \rho_2 A_2 \bar{v}_2 = \dot{M} \qquad [12.8]$$

If the fluid can be considered incompressible then $\rho_1 = \rho_2$ and eqn [12.8] reduces to the volume flow rate conservation equation:

Conversation of volume flow rate

$$A_1 \bar{v}_1 = A_2 \bar{v}_2 = Q \qquad [12.9]$$

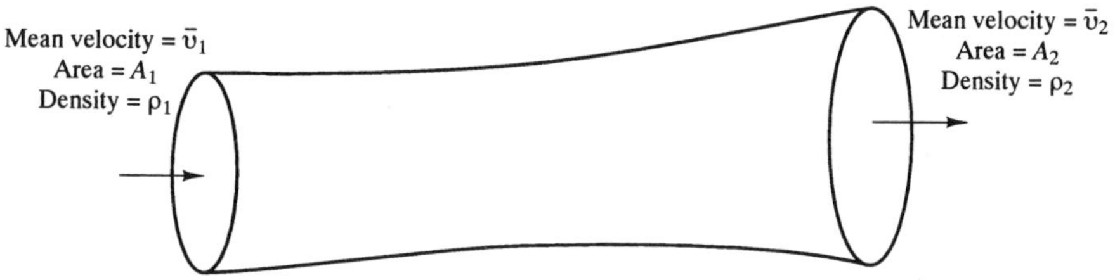

Fig. 12.3 Conservation of mass flow rate in a streamtube

12.1.6 Total energy and conservation of energy

Figure 12.4 shows an element of an incompressible fluid flowing in a streamtube.

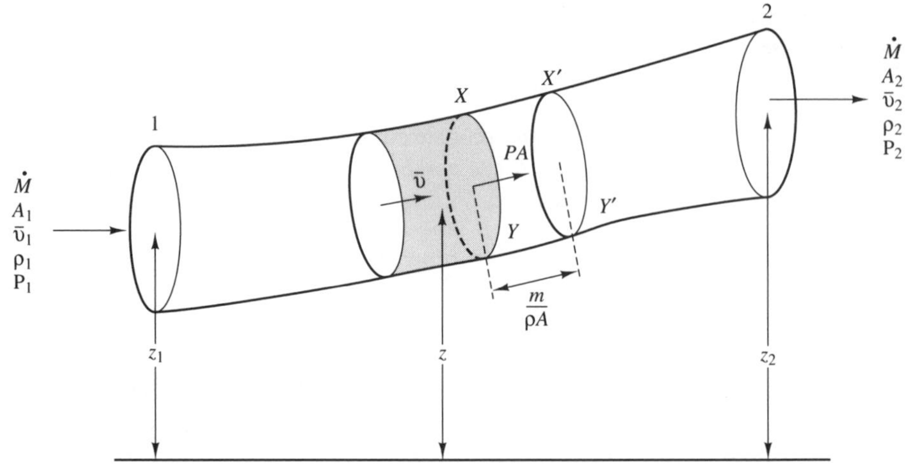

Fig. 12.4 Total energy and energy conservation in a flowing fluid

Since the element is at a height z above the datum level it possesses **potential energy** given by:

$$\text{potential energy} = mgz$$

where m is the mass of the element. The element is moving with a mean velocity $\bar{v}$ and therefore also possesses **kinetic energy** given by:

$$\text{kinetic energy} = \tfrac{1}{2}m\bar{v}^2$$

In addition the element of fluid can also do work because it is flowing under pressure. If the pressure acting over cross-section XY is P and the area of the cross-section is A, then:

$$\text{Force exerted on XY} = PA$$

If the entire element moves to occupy volume XX′Y′Y then the magnitude of this volume is m/ρ, where ρ is the density of the fluid. The corresponding distance moved XX′ is given by $m/\rho A$ and the work done by the fluid is:

$$\text{Flow Work} = \text{Force} \times \text{distance} = PA \times m/\rho A = mP/\rho$$

Flow work is often referred to as **pressure energy**, this is the energy possessed by a fluid when moving under pressure as part of a continuous stream.

The total energy of a flowing fluid is the sum of pressure, kinetic and potential energies, so that:

Fluid energy

$$\text{Total Energy/Unit Mass} = \frac{P}{\rho} + \frac{1}{2}\bar{v}^2 + gz \qquad [12.10]$$

Thus if we consider cross sectional areas 1 and 2 in Fig. 12.4, the principle of conservation of energy means that the total energy/unit mass is the same at both sections. This assumes that there is no energy inflow or outflow between Sections 1 and 2, for example no energy lost in doing work against friction. Using eqn [12.10] we have:

Conservation of energy — incompressible fluid

$$\frac{P_1}{\rho_1} + \frac{1}{2}\bar{v}_1^2 + gz_1 = \frac{P_2}{\rho_2} + \frac{1}{2}\bar{v}_2^2 + gz_2 \quad [12.11]$$

Equations [12.10] and [12.11] only apply to incompressible fluids where density ρ is independent of pressure P. For adiabatic expansion/contraction of a gas described by P/ρ^γ = constant, the flow work or pressure energy term must be modified to $\gamma/(\gamma - 1)(P/\rho)$ so that:

Conservation of energy-compressible fluid

$$\frac{\gamma}{\gamma - 1}\frac{P_1}{\rho_1} + \frac{1}{2}\bar{v}_1^2 + gz_1 = \frac{\gamma}{\gamma - 1}\frac{P_2}{\rho_2} + \frac{1}{2}\bar{v}_2^2 + gz_2 \quad [12.12]$$

12.2 Measurement of velocity at a point in a fluid

This is important in investigational work, such as studies of the velocity distribution around an aerofoil in a wind tunnel, or measurement of the velocity profile in a pipe prior to the installation of a permanent flowmeter. There are two main methods.

12.2.1 Pitot-static tube

At the impact hole part of the fluid is brought to rest; this part has therefore no kinetic energy, only pressure energy. At the static holes the fluid is moving and therefore has both kinetic and pressure energy. This creates a pressure difference $P_I - P_S$ which depends on velocity v.

Incompressible flow

Assuming energy conservation and no frictional or heat losses, the sums of pressure, kinetic and potential energies at the impact and static holes are equal. Since kinetic energy at the impact hole is zero:

$$\frac{P_I}{\rho} + 0 + gz_I = \frac{P_S}{\rho} + \frac{v^2}{2} + gz_s \quad [12.13]$$

where z_I, z_s are the elevations of the holes above a datum line and $g = 9.81\ \text{ms}^{-2}$. If $z_I = z_s$ then,

Pitot tube–incompressible flow

$$v = \sqrt{\frac{2(P_I - P_S)}{\rho}} \quad [12.14]$$

Compressible flow

The above assumes that the fluid densities at the impact and static holes are equal.

Since $P_I > P_S$ a compressible fluid has $\rho_1 > \rho$. The energy balance equation is now:

$$\frac{\gamma}{\gamma - 1}\frac{P_I}{\rho_I} + 0 = \frac{\gamma}{\gamma - 1}\frac{P_S}{\rho} + \frac{v^2}{2} \qquad [12.15]$$

where γ = ratio of specific heats at constant pressure and volume = C_P/C_V. Assuming the density changes are adiabatic we have:

$$\frac{P_I}{\rho_I^\gamma} = \frac{P_S}{\rho^\gamma} \qquad [12.16]$$

giving:

$$v = \sqrt{\left\{2\frac{\gamma}{\gamma - 1}\frac{P_S}{\rho}\left[\left(\frac{P_I}{P_S}\right)^{(\gamma-1)/\gamma} - 1\right]\right\}}$$

or, in terms of the pressure difference $\Delta P = P_I - P_S$:

Pitot tube—compressible flow

$$v = \sqrt{\left\{2\frac{\gamma}{\gamma - 1}\frac{P_S}{\rho}\left[\left(1 + \frac{\Delta P}{P_S}\right)^{(\gamma-1)/\gamma} - 1\right]\right\}} \qquad [12.17]$$

Characteristic and systems

From the incompressible eqn [12.14] we have $\Delta P = (1/2)\rho v^2$, i.e. there is a square law relation between ΔP and v (see Fig. 12.5). Applying the incompressible equation to air at standard temperature (20 °C) and pressure ($P_S = 10^5$ Pa), with $\rho = 1.2\ \text{kg m}^{-3}$, gives $\Delta P = 0.6\,v^2$. Thus at $v = 5\ \text{m s}^{-1}$ we have: $\Delta P = 15$ Pa, $\Delta P/P_S = 1.5 \times 10^{-4}$ and at $v = 100\ \text{m s}^{-1}$, $\Delta P = 6 \times 10^3$ Pa, $\Delta P/P_S = 6 \times 10^{-2}$. The small $\Delta P/P_S$ ratio means that for $v < 100\ \text{m s}^{-1}$, the difference in density between the air at the impact and static holes is negligible; the error introduced by

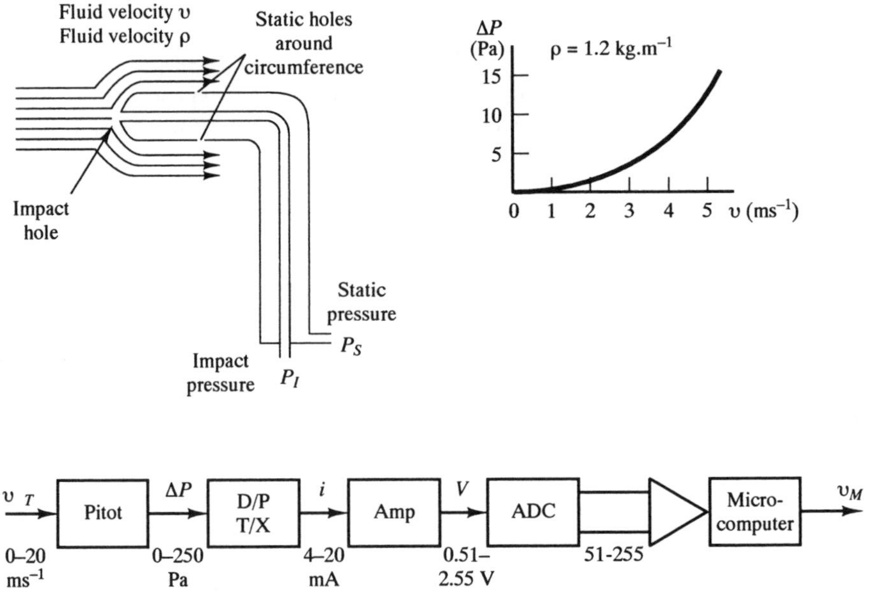

Fig. 12.5 Pitot-static tube

using the incompressible equation is within 1%. Close examination of the compressible eqn [12.17] shows that it reduces to the incompressible eqn [12.14] if $\Delta P/P_S \ll 1$.

The above very low differential pressures mean that special pressure transmitters must be used. One such transmitter uses a linear variable differential transformer to sense the deformation of a diaphragm capsule with a large area; this gives a 4 to 20 mA current output proportional to input differential pressure in the range 0 to 250 Pa.[1] Figure 12.5 shows a computer-based measurement system incorporating this transmitter for measuring air velocities in the range 0 to 20 m s^{-1}. The amplifier converts the transmitter output to a voltage signal between 0.51 and 2.55 V. The analogue-to-digital converter gives an 8-bit parallel digital output signal corresponding to decimal numbers D between 51 and 255. The computer reads D and calculates ΔP using:

$$\Delta P = 1.2255(D - 51)$$

and the measured velocity v_M using $v_M = \sqrt{1.67\,\Delta P}$.

The above system is only suitable for measuring the time average of the velocity at a point in a fluid. The system frequency response is insufficient for it to measure the rapid random velocity fluctuations present in turbulent flow.

12.2.2 Hot-wire and film anemometers

These are capable of measuring both average velocity and turbulence. A full account is given in Sections 14.2 and 14.3.

12.3 Measurement of volume flow rate

12.3.1 Differential pressure flowmeters

These are the most common industrial flowmeters for clean liquids and gases. Here a constriction is placed in the pipe and the differential pressure developed across the constriction is measured. The main problem is to accurately infer volume flowrate from the measured differential pressure (D/P).

Theoretical equation for incompressible flow through a D/P meter

The constriction causes a reduction in the cross-sectional area of the fluid. Figure 12.6(a) shows this reduction and defines relevant quantities. The following assumptions enable a theoretical calculation of pressure difference to be made.

1. Frictionless flow — i.e. no energy losses due to friction, either in the fluid itself or between the fluid and the pipe walls.
2. No heat losses or gains due to heat transfer between the fluid and its surroundings.
3. Conservation of total energy (pressure + kinetic + potential)

$$E_1 = \frac{P_1}{\rho_1} + \frac{1}{2}v_1^2 + gz_1 = E_2 = \frac{P_2}{\rho_2} + \frac{1}{2}v_2^2 + gz_2 \qquad [12.18]$$

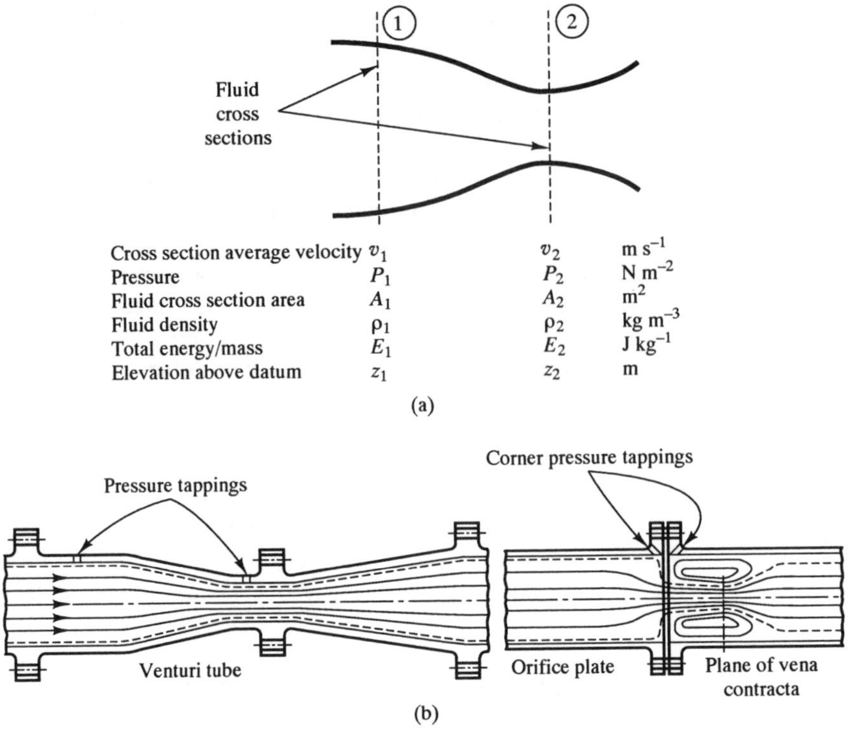

Fig. 12.6
(a) Principle of differential pressure flowmeter
(b) Effect of meter geometry on fluid cross sectional area

4. Incompressible fluid, i.e. $\rho_1 = \rho_2 = \rho$
5. Horizontal pipe, i.e. $z_1 = z_2$. This means that eqn [12.18] reduces to:

$$\frac{v_2^2 - v_1^2}{2} = \frac{P_1 - P_2}{\rho} \qquad [12.19]$$

6. Conservation of volume flow rate, i.e.

$$Q_1 = Q_2 = Q \quad \text{where } Q_1 = A_1 v_1 \quad \text{and} \quad Q_2 = A_2 v_2 \qquad [12.20]$$

Since $A_2 < A_1$, it follows from the conservation of volume flow rate [12.20], that $v_2 > v_1$; i.e. fluid velocity and kinetic energy are greater at the constriction. Since total energy is conserved [12.19] then the pressure energy at the constriction must be reduced, i.e. $P_2 < P_1$. From [12.19] and [12.20] we have:

Theoretical equation for incompressible flow through a differential pressure flowmeter

$$Q_{Th} = \frac{A_2}{\sqrt{\left[1 - \left(\frac{A_2}{A_1}\right)^2\right]}} \sqrt{\left[\frac{2(P_1 - P_2)}{\rho}\right]} \qquad [12.21]$$

Practical equation for incompressible flow

The above equation is not applicable to practice flowmeters for two main reasons:

(a) Assumption (1) of frictionless flow is not obeyed in practice. It is approached most closely by well-established turbulent flows in smooth pipes, where friction losses are small and constant but non-zero. Well-established turbulence is characterised by a Reynolds number greater than around 10^4. Reynolds number specifies the ratio between inertial forces and viscous friction forces and is given by $\mathrm{Re_D} = vD\rho/\eta$ where D is the pipe diameter and η the fluid viscosity.

(b) A_1 and A_2 are the cross-sectional areas of the fluid, which cannot be measured and which may change with flow rate. The cross sectional area of the pipe is $\pi D^2/4$ and the cross-sectional area of the meter is $\pi d^2/4$ where D and d are the respective diameters. At cross-section ① we have $A_1 = \pi D^2/4$ if the fluid fills the pipe. At cross-section ② we have $A_2 \approx 0.99\pi d^2/4$ for a Venturi, this being a gradual constriction which the fluid can follow (Fig. 12.6(b)). However, the orifice plate is a sudden constriction, which causes the fluid cross-sectional area to have a minimum value $0.6\pi d^2/4$ at the vena contracta.

For these reasons, the theoretical equation is corrected for practical use by introducing a correction factor termed the coefficient of discharge C. The modified equation is:

Practical equation for incompressible flow through a differential pressure flowmeter

$$Q_{\mathrm{ACT}} = CEA_2^M \sqrt{\frac{2(P_1 - P_2)}{\rho}} \qquad [12.22]$$

where C = discharge coefficient
E = velocity of approach factor = $1/\sqrt{(1 - \beta^4)}$
β = flowmeter-pipe diameter ratio = d/D
A_2^M = flowmeter cross sectional area = $\pi d^2/4$

Values of C depends upon:

(a) type of flowmeter, e.g. orifice plate or Venturi
(b) Reynolds number Re_D
(c) Diameter ratio β

i.e. $C = F(\mathrm{Re}_D, \beta)$ for a given flowmeter.

Values of C have been measured experimentally, for several types of flowmeters, over a wide range of fluid conditions. Corresponding measurements of Q_{ACT} and $(P_1 - P_2)$ are made for a given fluid, pipe and meter. If d, D, ρ are known, C can be found from [12.22]. Important sources are: BS 1042: Section 1.1: 1981[2] and ISO 5167: 1980.[3] These are identical and give discharge coefficient data for orifice plates, nozzles and Venturi tubes (Fig. 12.7) inserted in circular pipes which are running full. The C data is given both in table form and as regression equations which are ideal for computer use. Table 12.1 summarises the data for orifice plates.

The table is in three parts; part (a) shows the Stolz equation which expresses C in terms of Re_D and β. The equation also involves the parameters L_1 and L_2'. These parameters have different values (part (b)) for the three different recommended types of tappings (Fig. 12.7). Part (c) summarises the conditions which d, D, β and Re_D must satisfy if the Stolz equation is to be valid. The conditions on Re_D are especially complicated; the allowable range of Reynolds number depends on the value of β.

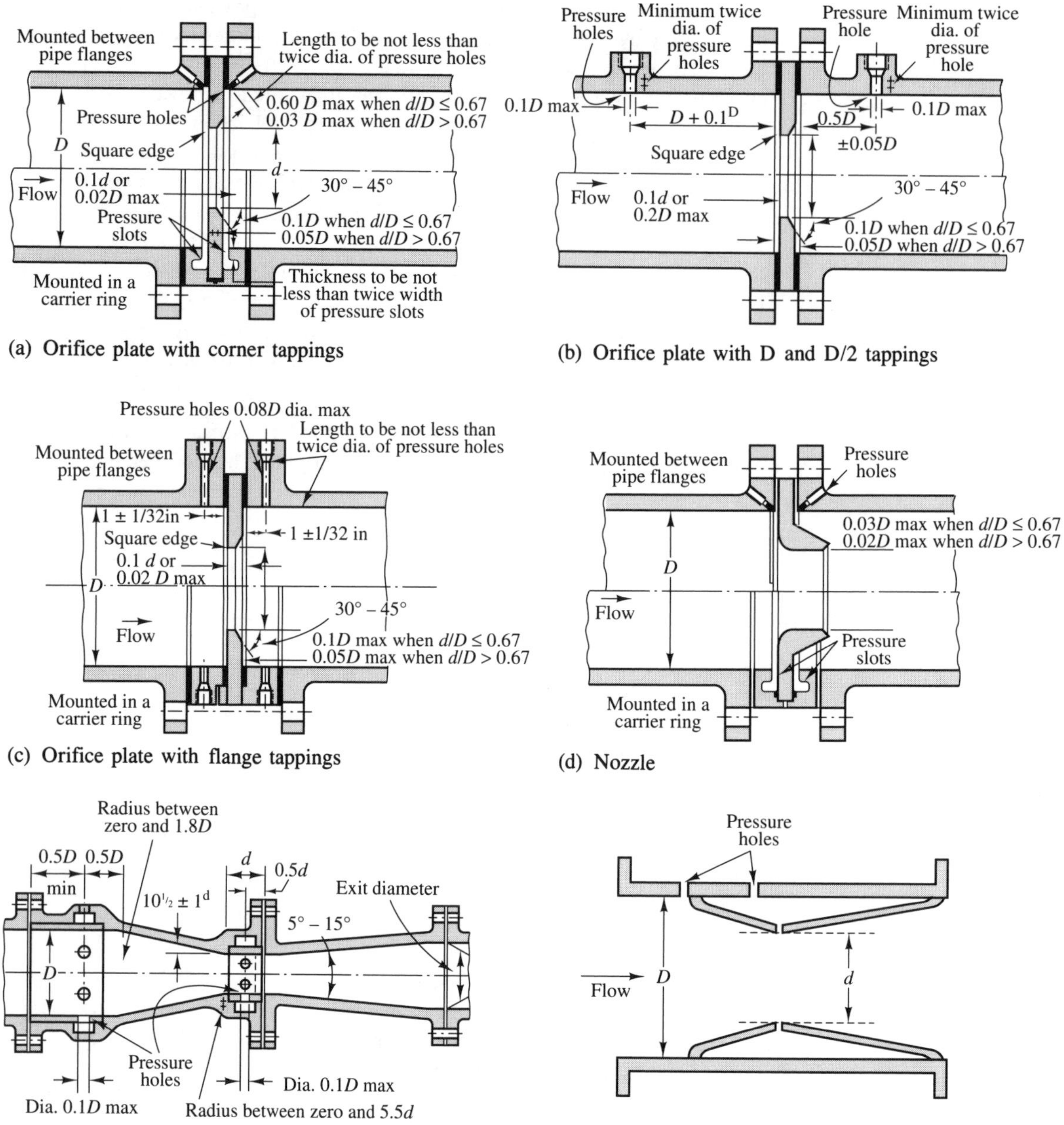

(a) Orifice plate with corner tappings

(b) Orifice plate with D and D/2 tappings

(c) Orifice plate with flange tappings

(d) Nozzle

(e) Venturi

(f) Dall tube

Fig. 12.7 Differential pressure flowmeters ((a)–(e) after British Standards Institution[2])

Compressible fluids

In order to accurately represent the behaviour of gases as well as liquids, restriction (4) above that $\rho_1 = \rho_2$ must be removed. Assuming adiabatic pressure/volume changes between cross-sections (1) and (2) we have:

Table 12.1 Discharge coefficient data for orifice plate (from BS 1042: Section 1.1: 1981‡)

(a) The Stolz equation

$$C = 0.5959 + 0.0312\beta^{2.1} - 0.184\beta^{8} + 0.0029\beta^{2.5}\left(\frac{10^6}{Re_D}\right)^{0.75} + 0.0900L_1\beta^4(1-\beta^4)^{-1} - 0.0337L_2'\beta^3$$

NOTE

If $L_1 \geq \frac{0.0390}{0.0900}$ (= 0.4333)

use 0.0390 for the coefficient of $\beta^4(1-\beta^4)^{-1}$

(b) Values of L_1 and L_2'

Corner tappings	$L_1 = L_2' = 0$
D and $D/2$ tappings	$L_1 = 1^*$. $L_2' = 0.47$
flange tappings	$L_1 = L_2' = 25.4/D$†

(c) Conditions of validity

	Corner taps	Flange taps	D and $D/2$ taps
d(mm)	≥ 12.5	≥ 12.5	≥ 12.5
D(mm)	$50 \leq D \leq 1000$	$50 \leq D \leq 760$	$50 \leq D \leq 760$
β	$0.23 \leq \beta \leq 0.80$	$0.2 \leq \beta \leq 0.75$	$0.2 \leq \beta \leq 0.75$
Re_D	$5000 \leq Re_D \leq 10^8$ for $0.23 \leq \beta \leq 0.45$; $10\,000 \leq Re_D \leq 10^8$ for $0.45 < \beta \leq 0.77$; $20\,000 \leq Re_D \leq 10^8$ for $0.77 \leq \beta \leq 0.80$	$\geq 1260\beta^2$D† $\leq 10^8$	$\geq 1260\beta^2$D† $\leq 10^8$

* Hence coefficient of $\beta^4(1-\beta^4)^{-1}$ is 0.0390

† Where D is expressed in mm

‡ Extracts from BS 1042: Section 1.1: 1981 are reproduced by permission of BSI. Complete copies can be obtained from them at Linford Wood, Milton Keynes, MK14 6LE

$$\frac{P_1}{\rho_1^\gamma} = \frac{P_2}{\rho_2^\gamma} \qquad [12.23]$$

where γ = specific heat ratio C_P/C_V: since $P_1 > P_2$, $\rho_1 > \rho_2$. The energy balance eqn [12.7] must be modified to:

$$\frac{\gamma}{\gamma-1}\frac{P_1}{\rho_1} + \frac{1}{2}v_1^2 = \frac{\gamma}{\gamma-1}\frac{P_2}{\rho_2} + \frac{1}{2}v_2^2 \qquad [12.24]$$

Since $\rho_2 < \rho_1$, i.e. the fluid expands, $Q_2 > Q_1$ and volume flow rate is not conserved. However, there is conservation of mass flowrate $\dot{M}$, i.e.

$$\dot{M}_1 = v_1 A_1 \rho_1 = \dot{M}_2 = v_2 A_2 \rho_2 \qquad [12.25]$$

This underlines the greater significance of mass flow in gas metering. From [12.23]–[12.25] we have

Theoretical equation for compressible flow

$$\dot{M}_{Th} = \epsilon \frac{A_2}{\sqrt{\left\{1 - \left(\frac{A_2}{A_1}\right)^2\right\}}} \sqrt{2\rho_1(P_1 - P_2)} \qquad [12.26]$$

where:

ϵ = expansibility factor

$$= \sqrt{\left\{\left(\frac{\gamma}{\gamma-1}\right)\left(\frac{P_2}{P_1}\right)^{2/\gamma}\left[\frac{1-(P_2/P_1)^{(\gamma-1)/\gamma}}{1-(P_2/P_1)}\right]\left[\frac{1-(A_2/A_1)^2}{1-(A_2/A_1)^2(P_2/P_1)^{2/\gamma}}\right]\right\}}$$

i.e. ϵ is a function of the 3 dimensionless groups P_2/P_1, γ, A_2/A_1 or $\Delta P/P_1$, γ, β; i.e. $\epsilon = f(\Delta P/P_1, \gamma, \beta)$ where $\Delta P = P_1 - P_2$. The above equation for ϵ is never used in practice. In BS 1042 and ISO 5167, ϵ is given as a regression equation. for orifice plates this is:

$$\epsilon = 1 - (0.41 + 0.35\beta^4)\frac{1}{\gamma}\frac{\Delta P}{P_1} \left(\text{if } \frac{P_2}{P_1} \geq 0.75\right) \qquad [12.27]$$

Note ϵ = 1.0 for a liquid.

Summarising, the following equation is applicable to any practical differential pressure flowmeter, metering any clean liquid or gas:

General, practical equation for differential pressure flowmeter

$$\dot{M}_{\text{ACT}} = CE\,\epsilon\, A_2^M \sqrt{2\rho_1(P_1 - P_2)} \qquad [12.28]$$

General characteristics

The following general characteristics of differential pressure flowmeters should be borne in mind when deciding on the most suitable meter for a given application.

1. No moving parts, robust, reliable and easy to maintain; widely established and accepted.
2. There is always a permanent pressure loss (ΔP_P) due to frictional effects (Fig. 12.8). The cost of the extra pumping energy may be significant for large installations.
3. These devices are non-linear i.e. $Q \propto \sqrt{\Delta P}$ or $\Delta P \propto Q^2$. This limits the useful range of a meter to between 25 and 100% of maximum flow. At lower flows the differential pressure measurement is below 6% of full scale and is clearly inaccurate.
4. Can only be used for clean fluids, where there is well-established turbulent flow i.e. $Re_D > 10^4$ approximately. Not generally used if solids are present, except for Venturis with dilute slurries.
5. A typical flowmeter system (Fig. 12.8) consists of the differential pressure sensing element, differential pressure transmitter (Chapter 9), interface circuit and microcomputer. For a transmitter giving a d.c. current output signal (typically 4 to 20 mA) the interface circuit consists of an amplifier acting as a current-to-voltage converter and an analogue-to-digital converter. For a

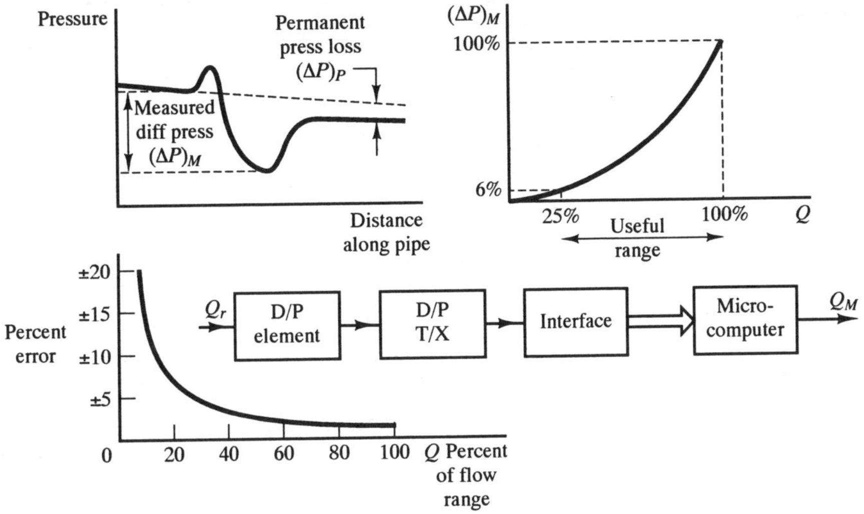

Fig. 12.8 Characteristics of differential pressure flowmeters and typical system

resonator transmitter (Section 9.4) giving a sinusoidal output of variable frequency the interface circuit consists of a Schmitt trigger and a binary counter (Fig. 10.5). The computer reads the input binary number, converts it into differential pressure ΔP and then calculates the measured flow rate Q_M using eqn [12.22]. The calculation is based on values of ρ_1, C, β, etc. stored in memory. The system measurement error $E = Q_M - Q_T$ is determined by the transmitter accuracy, quantisation errors and uncertainties in the values of the above parameters. A graph of percentage error versus flowrate is shown in Fig. 12.8.

6. Considerable care must be taken with the installation of the meter. The standards give detailed information on:
   (a) The geometry of the flowmeter itself (Fig. 12.7). C values are only applicable to meters with the prescribed geometry.
   (b) Minimum lengths of straight pipe upstream and downstream of the meter.
   (c) The arrangement of the pressure pipes connecting the flowmeter to the differential pressure device.

Types of differential pressure flowmeter

The four elements in current use are the orifice plate, Venturi, Dall tube and nozzle (see Fig. 12.7). Of these the orifice plate is by far the most widely used. It is cheap and available in a very wide range of sizes. The main disadvantages are the limited accuracy ($\pm 1.5\%$ at best) and the high permanent pressure loss $(\Delta P)_P$. There are three recommended arrangements of pressure tappings to be used with orifice plates; corner, flange and $D - D/2$ (Fig. 12.7). The values of C are given by Table 12.1 and are different for each arrangement. Table 12.2 summarises the main parameters of the four elements. The Dall tube combines a high measured differential pressure $(\Delta P)_M$ (like the orifice plate), with a low permanent pressure loss $(\Delta P)_P$ (better than Venturi).

Table 12.2

Parameter \ Meter	Venturi	Nozzle	Dall tube	Orifice Plate
Approx C	0.99	0.96	0.66	0.60
Relative values of measured diff. pressure $(\Delta P)_M$	1.0	1.06	2.25	2.72
Permanent ΔP as % of $(\Delta P)_M$ i.e. $\dfrac{(\Delta P)_P}{(\Delta P)_M} \times 100\%$	10–15%	40–60%	4–6%	50–70%

Calculation of orifice plate hole diameter d

To calculate d we need accurate values of C, E and ϵ. Since all three quantities are functions of d (via $\beta = d/D$), an iterative calculation is required. Figure 12.9 is a flowsheet for one possible calculation method; this is suitable for manual or computer implementation.

Notes on flowsheet

1. The following input data are required;
   - $\dot{M}_{MAX}$ Maximum mass flow rate (kg s^{-1})
   - ΔP_{MAX} Differential pressure at maximum flow (Pa)
   - P_1 Upstream pressure (Pa)
   - ρ_1 Fluid density at upstream conditions (kg m^{-3})
   - η Dynamic viscosity of fluid (Pa s)
   - γ Specific heat ratio
   - δ Machining tolerance (m)
   - D Pipe diameter (m)
   - i fluid index: $i = 0$ for liquid, $i = 1$ for gas
   - j Tappings index: $j = 0$ for corner, $j = 1$ for flange, $j = 2$ for $(D - D/2)$.
2. Calculation of Reynolds number uses $\dot{M}_{MAX}$ and pipe diameter D. Since $\text{Re}_D = \rho(vD/\eta)$ and since $v = \dot{M}_{MAX}/(\rho\pi D^2/4)$, $\text{Re}_D = 4\dot{M}_{MAX}/(\pi D\eta)$.
3. The values of d, D, β and Re_D must be checked against the conditions of validity of the Stolz equation (Table 12.1). An initial approximate check that Re_D is greater than 10^4 is followed by a final accurate check once β is established.
4. There is no point in calculating d more accurately than the tolerance δ to which the hole can be machined. This provides a criterion for either continuing or concluding the calculation. Thus if d_{n-1}, d_n are respectively the $(n - 1)$th and nth guesses for d, then:

   if $|d_n - d_{n-1}| > \delta$ continue calculation
   if $|d_n - d_{n-1}| \leq \delta$ conclude calculation

5. Since final values of C, ϵ and E will be close to the initial guesses, the calculation should require no more than about 6 iterations.
6. Since Venturi and Dall tubes are sold in standard sizes, a different approach is required. An approximate calculation will give the most suitable size, then an accurate calculation of $(\Delta P)_{MAX}$ is carried out for the size chosen.

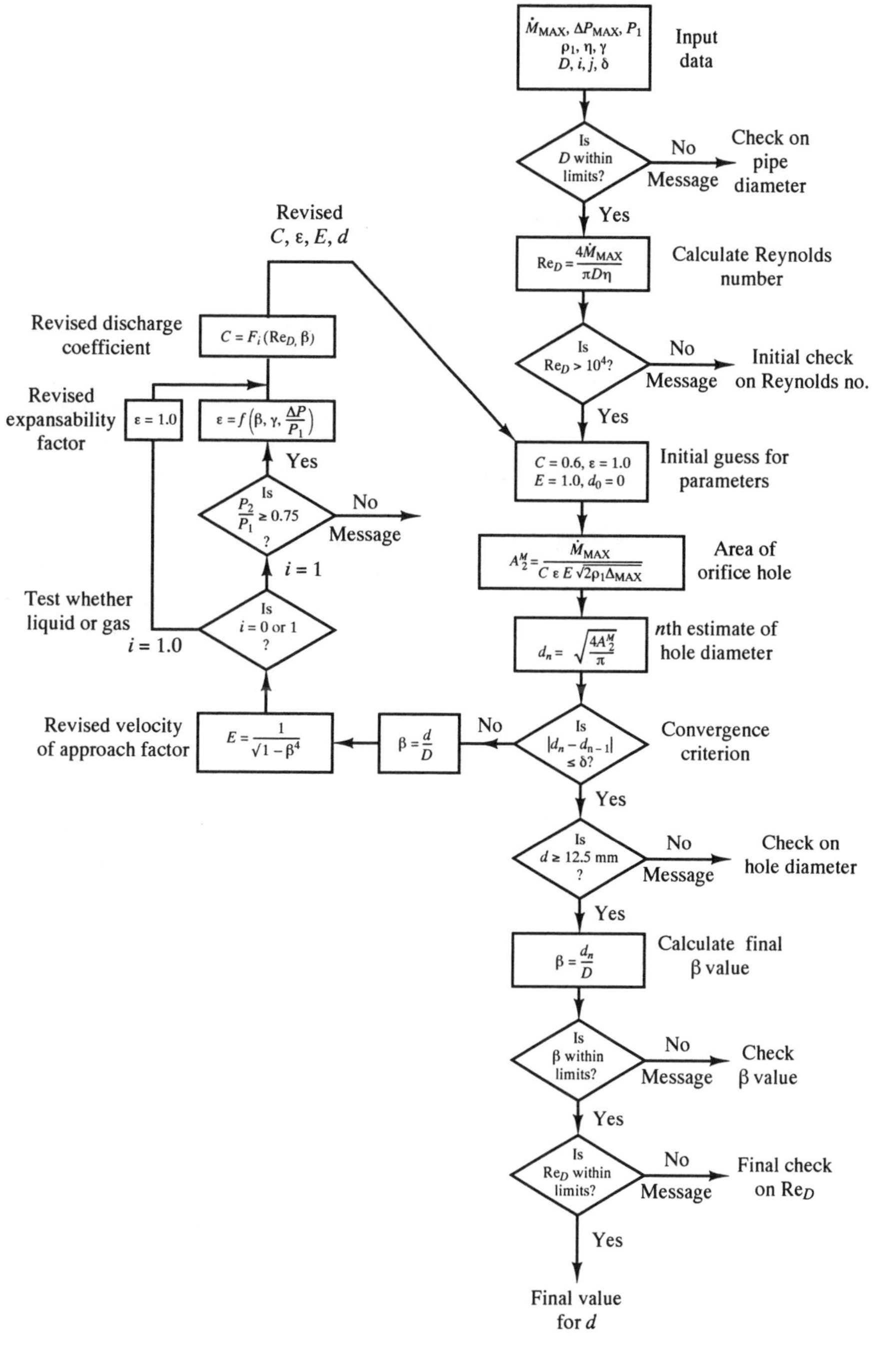

Fig. 12.9 Flowsheet for orifice plate sizing

12.3.2 Mechanical flowmeters

A mechanical flowmeter is a machine which is placed in the path of the flow, and made to move by the flow. The number of machine cycles per second f is proportional to volume flowrate Q, i.e. $f = KQ$, so that measurement of f yields Q. However, mechanical flowmeters are often used to measure the total volume of fluid $V = \int_0^T Q\mathrm{d}t$ that has been delivered during a time interval T. The total number of machine cycles during T is:

$$N = \int_0^T f\mathrm{d}t = \mathrm{K}\int_0^T Q\,\mathrm{d}t = KV$$

i.e. the total count is proportional to volume. A large number of mechanical flowmeters have been developed, but the most commonly used is the **axial flow turbine flowmeter**.

Principle of turbine flowmeter

A turbine flowmeter (see Figs 12.10 and 12.11) consists of a multi-bladed rotor suspended in the fluid stream; the axis of rotation of the rotor is parallel to the direction of flow. The fluid impinges on the blades and causes them to rotate at an angular velocity approximately proportional to flow rate. The blades, usually between four and eight in number, are made of ferromagnetic material and each blade forms a magnetic circuit with the permanent magnet and coil in the meter housing. This gives a variable reluctance tachogenerator (Section 8.4); the voltage induced in the coil is a sine wave whose frequency is proportional to the angular velocity of the blades.

Assuming that the drag torque due to bearing and viscous friction is negligible, the rotor angular velocity ω_r is proportional to Q, i.e.

$$\omega_r = kQ \qquad [12.29]$$

where k is a constant which depends on the geometry of the blade system. An approximate value for k can be evaluated using Fig. 12.10. If Q is the volume flow rate through the meter then the corresponding mean velocity $\bar{v}$ is:

$$\bar{v} = \frac{Q}{A} \qquad [12.30]$$

where A is the cross-sectional area of the fluid. Assuming the fluid fills the pipe then:

$$A = \text{Area of pipe} - \text{Area of hub} - \text{Area of blades}$$

$$= \frac{\pi}{4}D^2 - \frac{\pi}{4}d^2 - m\left(R - \frac{d}{2}\right)t \qquad [12.31]$$

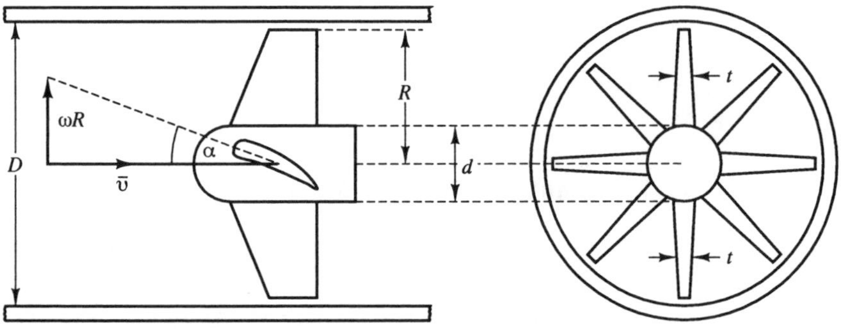

Fig. 12.10 Turbine flowmeter — principles

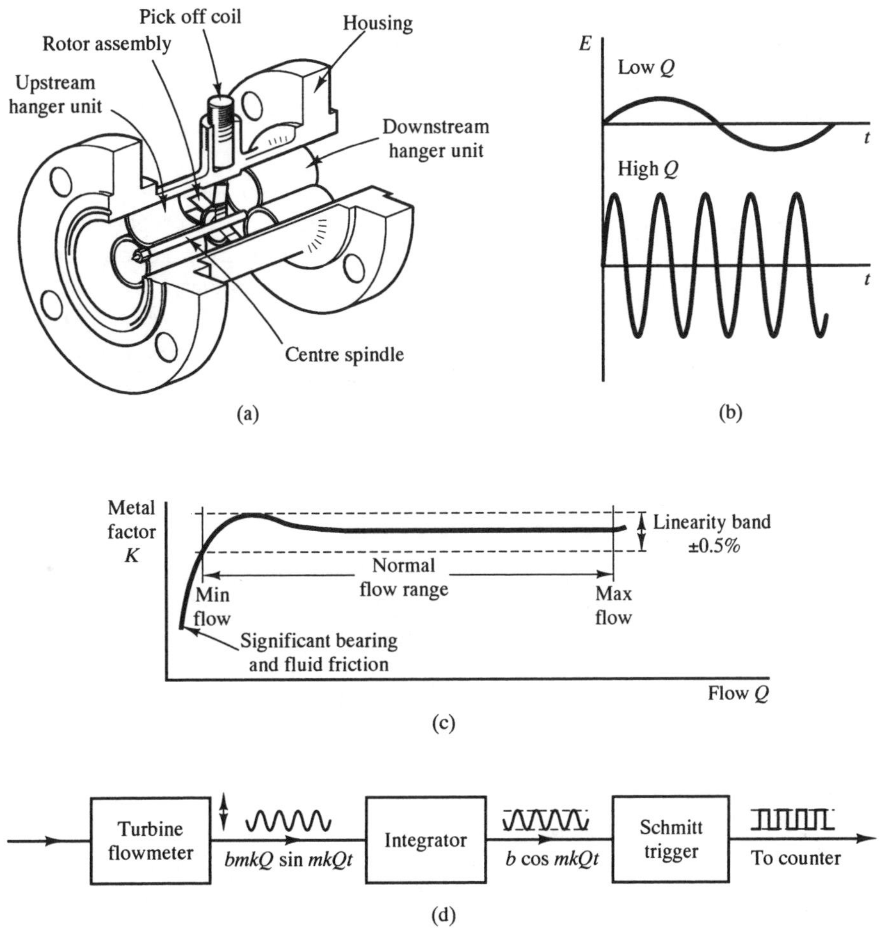

Fig. 12.11 Turbine flowmeter
(a) Construction (after Lomas, Kent Instruments, 1977, Institute of Measurement and Control Conference 'Flow-Con 77')
(b) Signals
(c) Characteristics
(d) System

where m is the number of blades and t their average thickness. From the inlet velocity triangle shown in Fig. 12.10 we have:

$$\frac{\omega_r R}{\bar{v}} = \tan \alpha$$

where ω_r is the angular velocity of the blades, $\omega_r R$ is the velocity of the blade tip perpendicular to the direction of flow and α is the inlet blade angle at tip. From eqns [12.29] and [12.30] we have:

$$k = \frac{\omega_r}{Q} = \frac{\tan \alpha}{AR} \qquad [12.32]$$

The principle of the variable reluctance tachogenerator was explained in Section 8.4; from eqn [8.37] the voltage induced in the coil is:

$$E = bmkQ \sin mkQt \qquad [8.37]$$

where m is the number of blades and b is the amplitude of the angular variation in magnetic flux. Thus:

Output signal from turbine flowmeter

$$E = bmkQ \sin mkQt \qquad [12.33]$$

This is a sinusoidal signal of amplitude $\hat{E} = bmkQ$ and frequency $f = (mk/2\pi)Q$, i.e. both amplitude and frequency are proportional to flow rate (see Fig. 12.11).

The flowmeter signal E is usually passed through an integrator and a Schmitt trigger circuit (Fig. 12.11). The output is thus a constant amplitude, square wave signal of variable frequency f which can be successfully transmitted to a remote counter, even in the presence of considerable noise and interference. Since $f = (mk/2\pi)Q = KQ$, where K is the linear sensitivity or meter factor, the total count N is proportional to total volume V. Using eqn [12.32] the meter factor is given by $K = (m \tan \alpha)/(2\pi AR)$.

Characteristics

The meter factor K for a given flowmeter is found by direct calibration: typical results are shown in Fig. 12.11. The normal flow range is usually from about 10% up to 100% of maximum rated flow. Over this range, the deviation of meter factor K from mean value is usually within $\pm 0.5\%$ and may be only $\pm 0.25\%$. Below 10% of maximum, bearing and fluid friction become significant and the relationship between f and Q becomes increasingly non-linear. Repeatability is quoted to be within $\pm 0.1\%$ of actual flow. A 5-inch meter with a range between 45 and 450 $m^3\,hr^{-1}$ and a meter factor of 1.17×10^3 pulses m^{-3} will give an output of between 15 and 150 pulses s^{-1} (with water).

Turbine flowmeters are available to fit a wide range of pipe diameters, typically between 5 and 500 mm. They tend to be more delicate and less reliable than competing flowmeters; blades and bearings can be damaged if solid particles are present in the fluid. They are relatively expensive, and there is a permanent pressure drop, typically between 0.1 and 1.0 bar, at maximum flow (with water). Modern turbine flowmeters have 'thrust compensation' where the thrust of the fluid impinging on the rotor is balanced by an opposing thrust, thus reducing bearing wear. Flow straightening vanes are positioned upstream of the meter to remove fluid swirl which would otherwise cause the rotor angular velocity to be too high or too low, depending on direction.

12.3.3 Vortex flowmeters

Principle

The operating principle of the vortex flowmeter is based on the natural phenomenon of **vortex shedding**. When a fluid flows over any body, boundary layers of slower moving fluid move over the surface of the body (Section 12.1.1). If the body is **streamlined**, as in an aerofoil (Fig. 12.12(a)), then the boundary layers can follow the contours of the body and remain attached to the body over most of its surface. The boundary layers become detached at the separation points S which are well to the rear of the body and the resulting wake is small. If, however, the body is

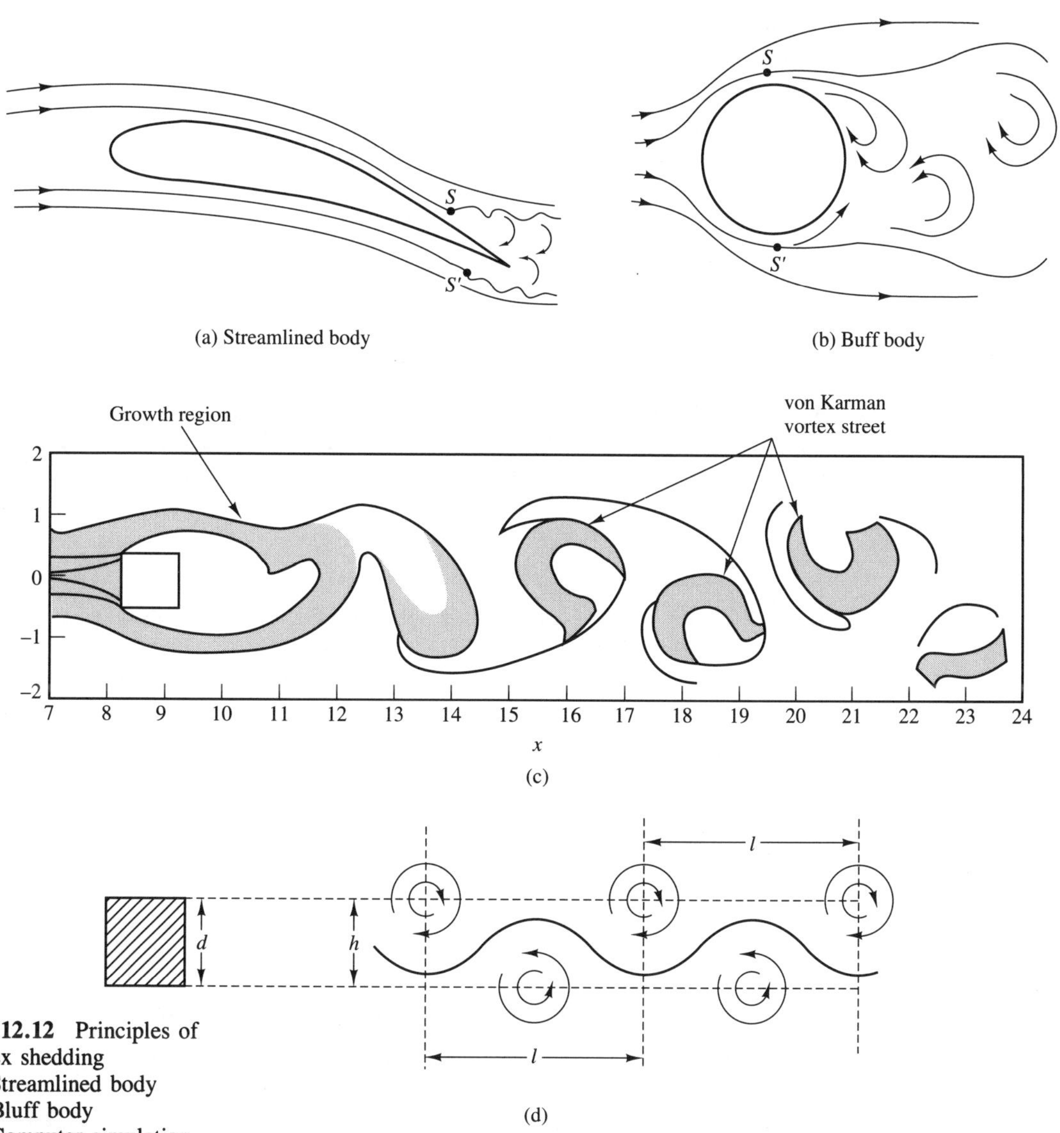

Fig. 12.12 Principles of vortex shedding
(a) Streamlined body
(b) Bluff body
(c) Computer simulation of flow behind rectangular bluff body
(d) Dimensions of idealised vortex street

unstreamlined i.e. **bluff**, e.g. a rectangular, circular or triangular cylinder (Fig. 12.12(b)), then the boundary layers cannot follow the contours and become separated much further upstream (points S'). The resulting wake is now much larger.

Figure 12.12(c) shows a computer simulation of the detailed flow around a rectangular bluff body. The boundary layers become separated at the upstream corners and roll up into **vortices** in the low pressure region immediately downstream of the body. These vortices are produced alternatively from the top and bottom surfaces. There is a growth region just downstream of the bluff where the vortices develop. Further downstream the vortices break away and move downstream forming a wake known as a **von Karman vortex street**. This consists of two rows of vortices moving downstream, parallel to each other, at a fixed velocity. The distances l between each vortex and h between the rows are constant and a vortex in one row occurs half-way between two vortices in the other row (Fig. 12.12(d)). If d is the width of the bluff body then

$$h \approx d \quad \text{and} \quad l \approx 3.6h$$

The frequency of vortex shedding f is the number of vortices produced from each surface of the bluff per second. This is given by:

Frequency of vortex shedding

$$f = S\frac{v_1}{d} \qquad [12.34]$$

where v_1 is the mean velocity at the bluff body, d the width of the bluff and S is a dimensionless quantity called the Strouhal Number. Since S is practically constant, f is proportional to v_1 thus providing the basis of a flowmeter.

Derivation of meter factor (sensitivity)

If

Q = volume flow rate $(m^3\,s^{-1})$

A, A_1 = cross sectional areas of the flow, upstream and at the bluff body respectively (m^2)

v, v_1 = velocities of the flow, upstream and at the bluff body respectively (ms^{-1})

D = pipe diameter (m)

Then assuming the flow is incompressible, conservation of volume flow rate gives

$$Q = Av = A_1 v_1 \qquad [12.35]$$

The frequency f of vortex shedding is given by

$$f = S\frac{v_1}{d} \qquad [12.36]$$

where S is the Strouhal number and d is the width of the bluff body (Fig. 12.13 shows a rectangular bluff body). Assuming the fluid fills the pipe:

$$A = \frac{\pi}{4}D^2 \qquad [12.37]$$

The fluid cross section at the bluff body is approximately given by:

$$A_1 \approx \frac{\pi}{4}D^2 - Dd = \frac{\pi}{4}D^2\left[1 - \frac{4}{\pi}\frac{d}{D}\right] \qquad [12.38]$$

From [12.35]–[12.38], the theoretical equation for the meter factor f/Q is:

$$\frac{f}{Q} = \frac{4S}{\pi D^3}\,\frac{1}{\dfrac{d}{D}\left[1 - \dfrac{4}{\pi}\dfrac{d}{D}\right]} \qquad [12.39]$$

This equation is corrected, for practical use, by introducing a bluffness coefficient k[4] thus:

Practical equation for meter factor

$$\frac{f}{Q} = \frac{4S}{\pi D^3} \cdot \frac{1}{\dfrac{d}{D}\left[1 - \dfrac{4}{\pi} k \dfrac{d}{D}\right]} \qquad [12.40]$$

k has different values for different bluff body shapes: e.g. $k = 1.1$ for a circle, and 1.5 for a rectangle and equilateral triangle.

Characteristics

1. Investigations show[4,5] that the Strouhal number S is a constant for a wide range of Reynolds number (Fig. 12.13). This means that, for a given flowmeter in a given pipe, i.e. fixed D, d and k the meter factor f/Q is practically independent of flow rate, density and viscosity. Here Re is the body Reynolds number $vd\rho/\eta$.
2. Considerable experimental work has been performed[4] to find the optimum bluff body shape and blockage ratio d/D. Results show that, for a given shape, the most regular and highest amplitude shedding is obtained at a d/D ratio which causes the meter factor f/Q to have a minimum value (Fig. 12.13). By differentiating the function:

$$F\left(\frac{d}{D}\right) = \frac{1}{\dfrac{d}{D}\left(1 - \dfrac{4}{\pi} k \dfrac{d}{D}\right)}$$

it can be shown that the minimum occurs at $d/D = \pi/(8k)$ i.e. $d/D = 0.26$ for

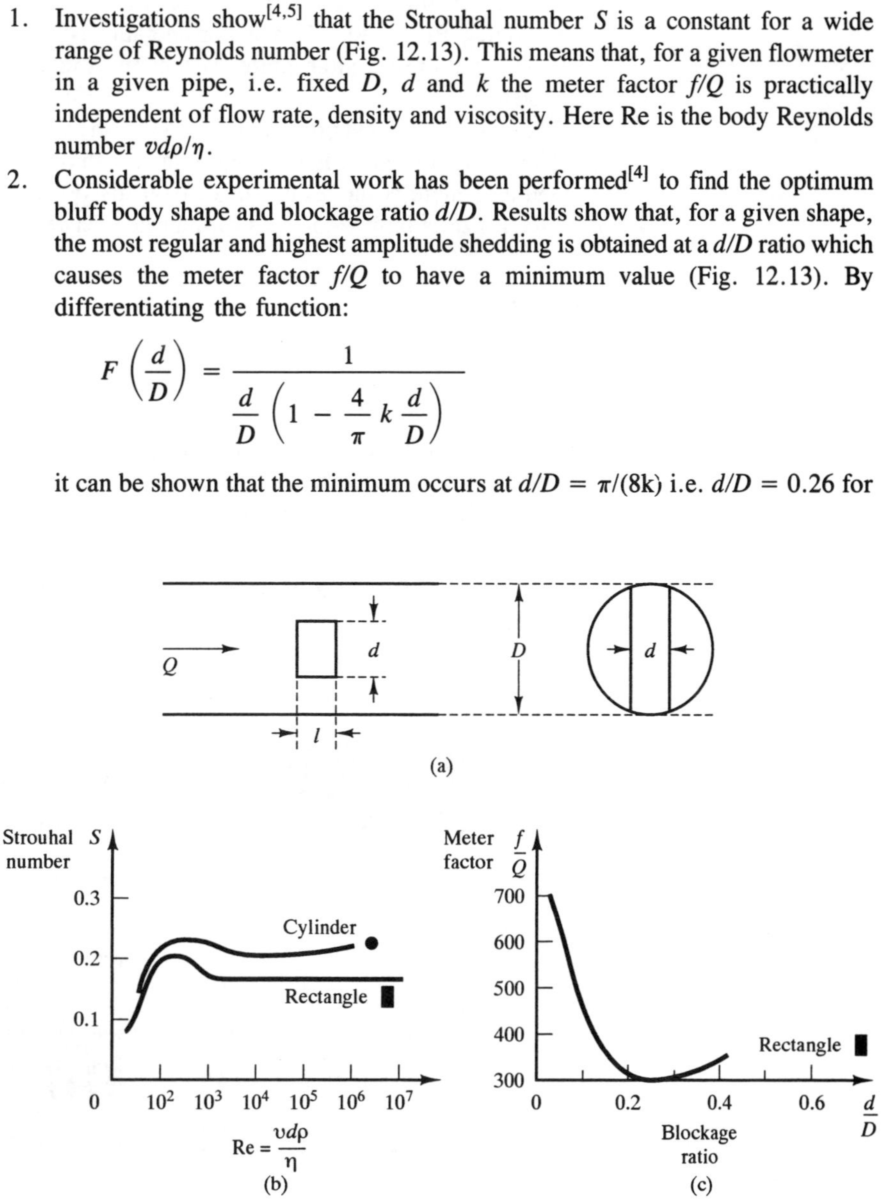

Fig. 12.13 Vortex flowmeter (after Cousins[5])
(a) Geometry
(b) Strouhal number
(c) Meter factor

a rectangle. This work also indicates that a rectangle of dimensions d and l given by:

$$\frac{d}{D} = 0.26 \text{ and } \frac{l}{d} = 0.66 \qquad [2.41]$$

produces stronger shedding than a circle, triangle or any other rectangle. However, flowmeters with more complex bluff body shapes have now been developed in order to enhance vortex formation (Fig. 12.15).

3. The vortex flowmeter gives a frequency output proportional to flow like the turbine flowmeter, but has no moving parts and is therefore more reliable.
4. Typical accuracy is $\pm 0.75\%$ of actual flow for liquids and $\pm 1.5\%$ for gases, providing $Re > 10^4$. This is considerably better than an orifice plate with differential pressure transmitter system but is inferior to the turbine flowmeter system.
5. There is a permanent pressure loss as with orifice plate and turbine flowmeters.
6. The sizing procedure is far simpler than with orifice plates (e.g. eqn [12.41], and also there is far better range, typically down to 8% of maximum.
7. Meter size is limited to pipes of diameter between 5 and 20 cm (2 and 8 in).

Vortex detection systems

Vortex shedding is characterised by approximately sinusoidal changes in fluid velocity and pressure in the vicinity of the bluff body. Figures 12.14(b), (c), (d) show typical vortex signals at different flow rates. Figure 12.14(a) shows the signal obtained without the bluff body; it has no regular pattern and corresponds to the random velocity fluctuations associated with turbulent flow.

Figure 12.15 shows four commonly used bluff body shapes, these use three different methods of vortex detection.[6–9]

(a) *Piezoelectric* (Section 8.7). Figure 12.15(a) shows a T-shaped bluff body; part of the tail is not solid but fluid filled. Flexible diaphragms in contact with the

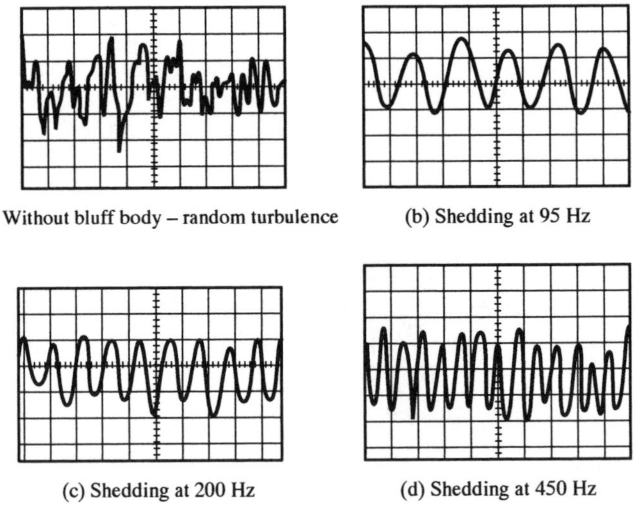

(a) Without bluff body – random turbulence (b) Shedding at 95 Hz

(c) Shedding at 200 Hz (d) Shedding at 450 Hz

Fig. 12.14 Typical vortex flowmeter signals

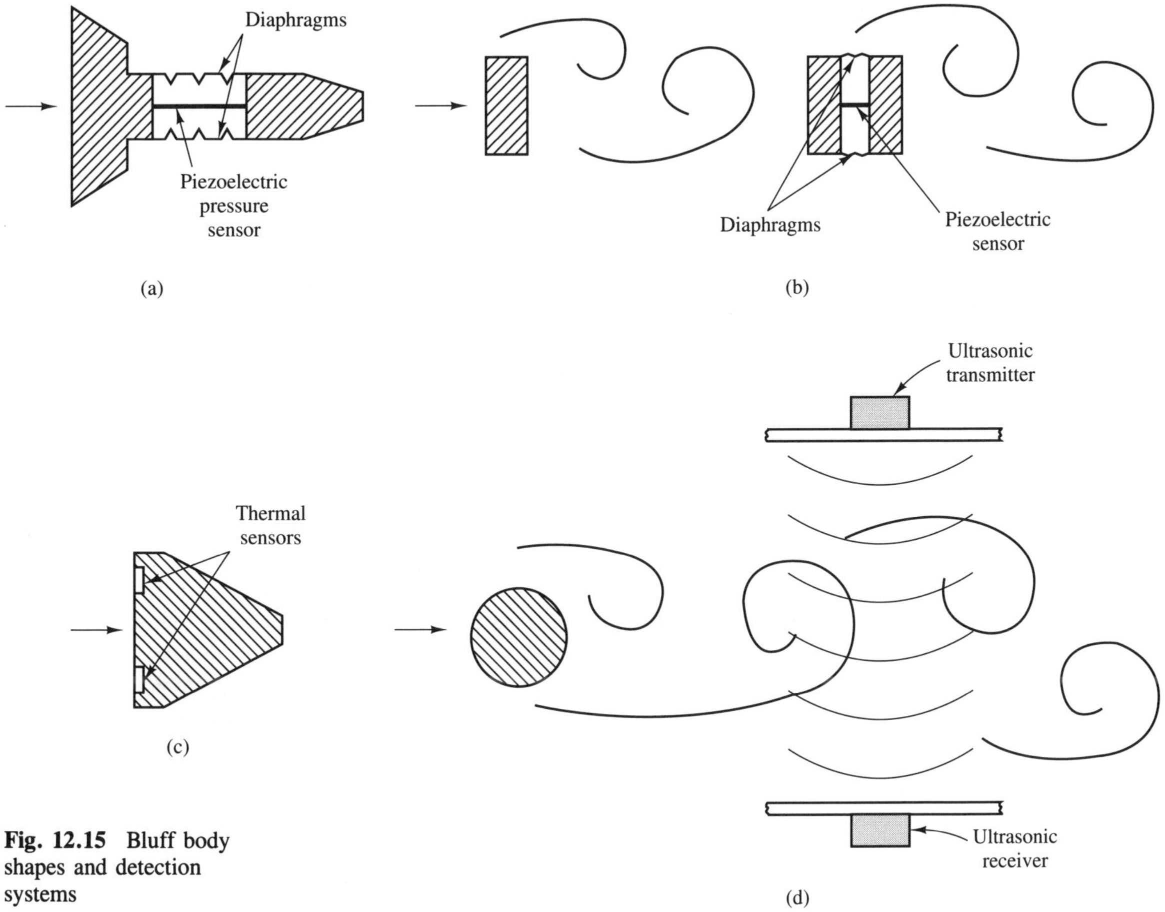

Fig. 12.15 Bluff body shapes and detection systems

process fluid detect small pressure variations due to vortex shedding. These pressure changes are transmitted to a piezoelectric differential pressure sensor which is completely sealed from the process fluid.

Figure 12.15(b) shows an arrangement of two bluff bodies in series or tandem. Vortices shed from the upstream body impinge on the downstream body causing large pressure fluctuations and enhanced vortex shedding at the downstream body. These pressure variations are again detected by a piezoelectric sensor located inside the downstream body.

(b) *Thermal* (Sections 14.2, 14.3) Figure 12.15(c) shows an approximately triangular-shaped bluff body with two semiconductor thermal sensors on the upstream face. The sensors are incorporated into constant temperature circuits which pass a heating current through each sensor, this enables small velocity fluctuations due to vortices to be detected.

(c) *Ultrasonic* (Chapter 16) Figure 12.15(d) shows a narrow circular cylinder which creates a von Karman vortex street downstream. An ultrasonic transmission link sends a beam of ultrasound through the vortex street. The vortices cause the received sound wave to be modulated both in amplitude and phase.

12.4 Measurement of mass flowrate

Liquids and gases such as crude oil, natural gas and hydrocarbon products are often transferred from one organisation to another by pipeline. Since these products are bought and sold in units of mass, it is essential to know accurately the mass M of fluid that has been transferred in a given time T. There are two main methods of measuring M; inferential and direct.

12.4.1 Inferential methods

Here mass flow rate $\dot{M}$ and total mass M are computed from volume flow rate and density measurements using $\dot{M} = \rho Q$ and $M = \rho V$. For pure liquids density ρ depends on temperature only. If temperature fluctuations are small, then ρ can be assumed to be constant and M can be calculated using only measurements of total volume V obtained from a mechanical flowmeter. If the temperature variations are significant then the density ρ must also be measured. In liquid mixtures density depends on both temperature and composition and again ρ must be measured. The same is also true of pure gases, where density depends on pressure and temperature, and gas mixtures where density depends on pressure, temperature and composition.

Figure 12.16 shows a typical system based on a turbine flowmeter and a vibrating element density transducer in conjunction with a digital micro-computer. The turbine flowmeter (Section 12.3.2) gives a pulse output signal, with frequency f proportional to volume flow rate Q, i.e.

$$f_1 = KQ \qquad [12.41]$$

where K = meter factor for turbine flowmeter. The vibrating element density transducer (Section 5.2) also gives a pulse output signal, with frequency f_2 which

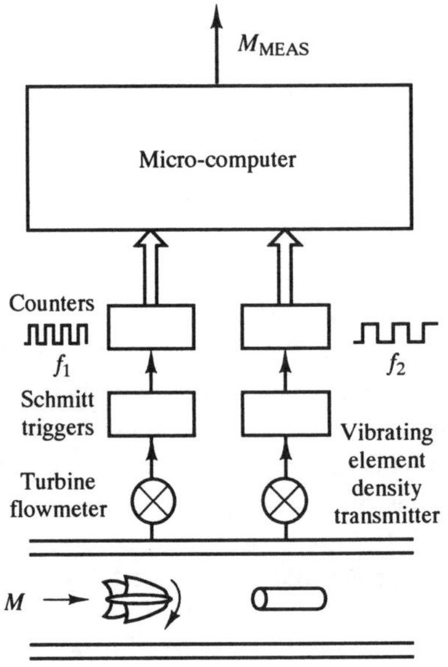

Fig. 12.16 Inferential measurement of mass flow

depends on fluid density ρ via the non-linear equation:

$$\rho = \frac{A}{f_2^2} + \frac{B}{f_2} + C \qquad [12.42]$$

Each pulse signal is input to a counter; the computer reads the state of each counter at the beginning and end of a fixed counting interval ΔT. The computer calculates frequencies f_1 and f_2 using:

$$f = \frac{N_{\text{NEW}} - N_{\text{OLD}}}{\Delta T} \qquad [12.43]$$

where N_{OLD} and N_{NEW} are the counts at the beginning and end of ΔT. Q and ρ are then calculated from [12.41] and [12.42] using the transducer constants K, A, B and C stored in memory. The total mass M transferred during time T is then:

$$M = \Delta T \sum_{i=1}^{n} \rho_i Q_i \qquad [12.44]$$

where ρ_i, Q_i are values of ρ, Q evaluated at the ith interval ΔT and $n = T/\Delta T$.

12.4.2 Direct methods

In a direct or true mass flowmeter the output of the sensing element depends on the mass flow rate $\dot{M}$ of fluid passing through the flowmeter. Such a flowmeter is potentially more accurate than the inferential type. One of the most popular and successful direct mass flowmeters in current use is based on the Coriolis effect.[10]

Coriolis flowmeter

The Coriolis effect is shown in Fig. 12.17(a). A slider of mass m is moving with velocity v along a rod, the rod itself is moving with angular velocity ω about the axis XY. The mass experiences a Coriolis force of magnitude:

$$F = 2m\omega v \qquad [12.45]$$

and direction perpendicular to both linear and angular velocity vectors.

Figure 12.17(b) shows the flowmeter. Here the fluid flows through the U-tube $ABCD$ which is rotating with an angular velocity ω about the axis XY. Here ω varies sinusoidally with time at constant frequency f, i.e. $\omega = \hat{\omega} \sin 2\pi ft$. Consider an element of fluid of length Δx travelling with velocity v along the limb AB which will have mass:

$$\Delta m = \rho A \Delta x$$

where ρ is the density of the fluid and A the internal cross-sectional area of the tube. The element experiences a Coriolis force:

$$\Delta F = 2\Delta m\omega v = 2\rho A\omega v\Delta x$$

in the direction shown. The total force on the limb AB of length l is:

$$F = 2\rho A\omega v \int_0^l dx = 2\rho A\omega v l \qquad [12.46]$$

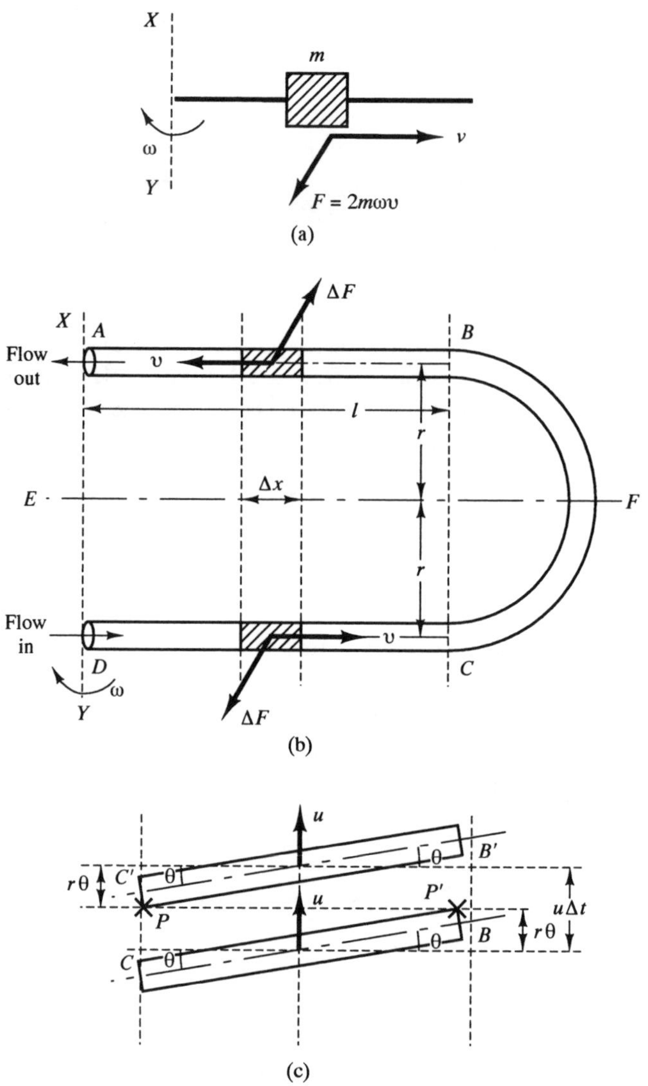

Fig. 12.17 Coriolis mass flowmeter
(a) Coriolis effect
(b) Flowmeter
(c) Measurement of twist angle

The limb *CD* experiences a force of equal magnitude *F* but in the opposite direction. In *BC* the velocity and angular velocity vectors are parallel so the Coriolis force is zero. The U-tube therefore experiences a resultant deflecting torque $T = F2r = 4lr\omega\rho Av$ about the axis *EF*. Since the mass flow rate $\dot{M}$ through the tube is equal to ρAv we have

$$T = 4lr\omega\dot{M} \qquad [12.47]$$

i.e. the deflecting torque is proportional to mass flow rate. Under the action of *T* the U-tube is twisted through an angle given by:

$$\theta = \frac{T}{c} = \frac{4lr\omega}{c}\dot{M} \qquad [12.48]$$

where c is the elastic stiffness of the U-tube. The twist angle θ varies sinusoidally with time, $\theta = \hat{\theta} \sin 2\pi f t$, in response to the sinusoidal variation in ω.

Figure 12.17(c) shows an optical method of measuring θ using optical sensors P and P'. At time t sensor P' detects the tube in position CB and emits a voltage pulse; at a later time $t + \Delta t$, sensor P detects the tube in position $C'B'$ and again emits a pulse. The time interval Δt is small compared with the period of oscillation $1/f$ of θ. The distance $BB' = CC'$ travelled by the tube in Δt is $u\Delta t$ where u is the velocity of the tube at BC. This depends on tube angular velocity ω according to:

$$u = \omega l$$

From the diagram we see that:

$$BB' = CC' = u\Delta t = 2r\theta$$

giving

$$\theta = \frac{\omega l}{2r} \Delta t \qquad [12.49]$$

Eliminating θ between [12.48] and [12.49] gives:

Equation for Coriolis flowmeter

$$\dot{M} = \frac{c}{8r^2} \Delta t \qquad [12.50]$$

Meters have not yet been successfully developed for gas measurements but they have been used on a wide variety of liquids and liquid mixtures and may be useful for liquids containing small amounts of solids. Flow rates between 0.04 and 15 kg s^{-1} in pipes with diameters between 3 and 50 mm can be measured with this type of meter. Accuracies of within $\pm 0.4\%$ of reading are claimed and a given meter has a range of between approximately 5 and 100% of maximum flow rate.

12.5 Measurement of flow rate in difficult situations

The flowmeters discussed so far will be suitable for the vast majority, perhaps 90%, of measurement problems. There will, however, be a small number of situations where they cannot be used. These 'difficult' flowmetering problems are characterised by one or more of the following features:

(a) The flow is laminar or transitional ($\text{Re} < 10^4$)
(b) The fluids involved are highly corrosive or toxic.
(c) Multiphase flows, that is mixtures of solids, liquids and gases. Important industrial examples are sand/water mixtures, oil/water/gas mixtures and air/solid mixtures in pneumatic conveyors.
(d) No obstruction or pressure drop can be tolerated (e.g. measurement of blood flow).
(e) There is also a need for a portable 'clip-on' flowmeter to give a temporary indication, usually for investigational work. This is strapped onto the outside of the pipe, thus avoiding having to shut down the plant in order to break into the pipe.

This section outlines methods which should be considered for these problem areas.

12.5.1 Electromagnetic flowmeter

The principle is based on Faraday's law of electromagnetic induction. This states that if a conductor of length l is moving with velocity v, perpendicular to a magnetic field of flux density B (Fig. 12.18), then the voltage E induced across the ends of the conductor is given by:

$$E = Blv \qquad [12.51]$$

Thus if a conducting fluid is moving with average velocity $\bar{v}$ through a cylindrical metering tube, perpendicular to an applied magnetic field B (Fig. 12.18) then the voltage appearing across the measurement electrodes is:

$$E = BD\bar{v} \qquad [12.52]$$

where D = separation of electrodes = metering tube diameter. The above equation assumes that the magnetic field is uniform across the tube. If we further assume that the fluid fills the tube then $\bar{v} = Q/(\pi D^2/4)$ giving:

Equation for electromagnetic flowmeter

$$E = \frac{4B}{\pi D} Q \qquad [12.53]$$

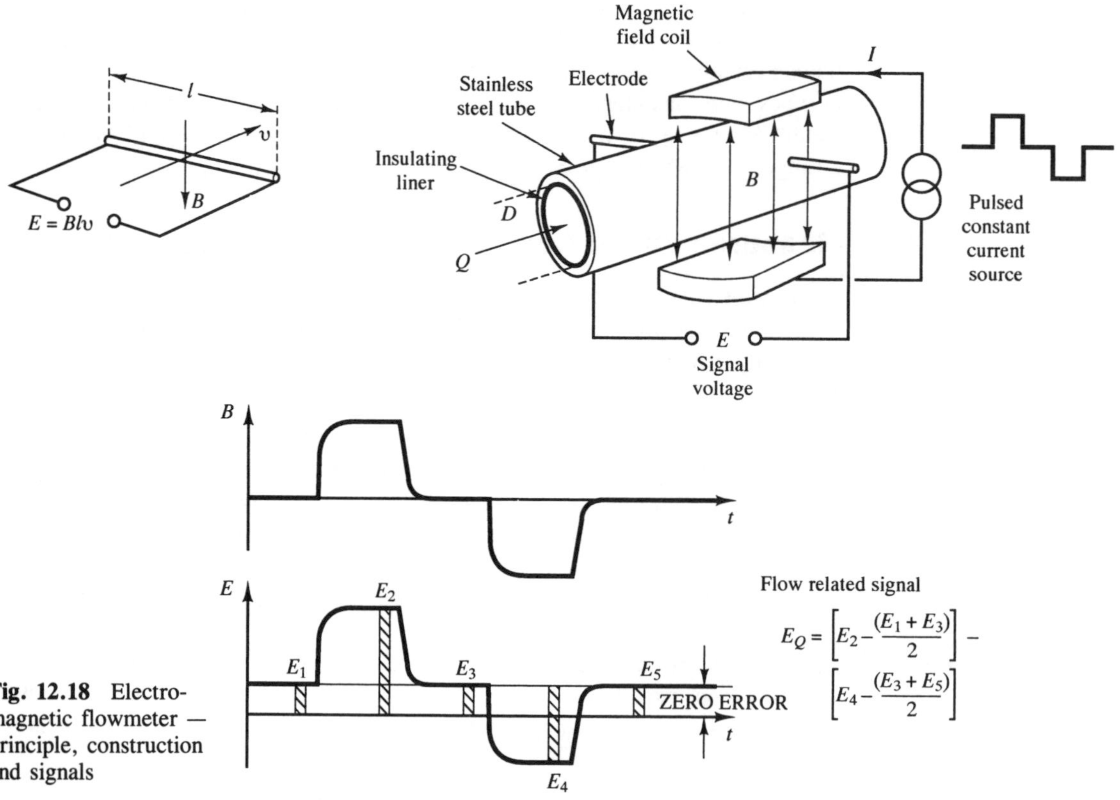

Fig. 12.18 Electromagnetic flowmeter — principle, construction and signals

If the magnetic field coils are energised by normal direct current then several problems occur: polarisation (i.e. the formation of a layer of gas around the measuring electrodes), electrochemical and thermoelectric effects all cause interfering d.c. voltages. These problems can be overcome by energising the field coils with alternating current at 50 Hz. The a.c. magnetic field induces a 50 Hz a.c. voltage across the electrodes with amplitude proportional to flow rate. However, this flow-generated voltage is subject to 50 Hz interference voltages generated by transformer action in a loop consisting of the signal leads and the fluid path. The above problems can be overcome by energising the field coils with direct current which is pulsed at a fixed period. Figure 12.19 shows typical waveforms of magnetic field B and induced signal voltage E.[11] The signal is amplified and fed to a 12-bit analogue-to-digital converter and a microcomputer. The signal is sampled 5 times during each complete period of about 400 ms. By suitable processing of the five sample values E_1, E_2, E_3, E_4, E_5 the zero error can be rejected and the flow-related signal measured.

The main features of the electromagnetic flowmeter are:

1. The electrodes are flush with the inside of the insulating liner which has the same diameter as the surrounding pipework. The meter therefore does not obstruct the flow in any way, and there is negligible pressure loss or chance of blockage.
2. The meter can only be used with fluids of electrical conductivity greater than 10 μmho cm^{-1}; this rules out all gases and liquid hydrocarbons. It is suitable for slurries provided that the liquid phase has adequate conductivity.
3. Good accuracy; $\pm$ 1% of indicated flow and the calibration is independent of changes in viscosity, pressure, density, temperature, etc.
4. Velocity range is typically between 0.5 and 10 m s^{-1} in a given flowmeter; flowmeter diameters vary between 2 mm and 1200 mm.
5. Power consumption is typically less than 30 W.

12.5.2 Ultrasonic flowmeters

Ultrasonic flowmeters use sensors which are clamped on the outside of the pipe, i.e. do not intrude into the pipe, this makes them particularly useful for multiphase flows. Doppler, transit time and cross-correlation ultrasonic flowmeters are covered in Section 16.4.

12.5.3 Cross-correlation flowmeter

A schematic diagram of the flowmeter system is shown in Fig. 12.19. This method assumes that some property of the fluid, e.g. density, temperature, velocity, conductivity is changing in a random manner. This property is detected at two positions A and B on the pipe. The corresponding detector output voltages $x(t)$ and $y(t)$ are random signals. In Section 6.2 we defined the autocorrelation function $R_{yy}(\beta)$ of a single signal $y(t)$, in terms of the average value of the product $y(t).y(t$

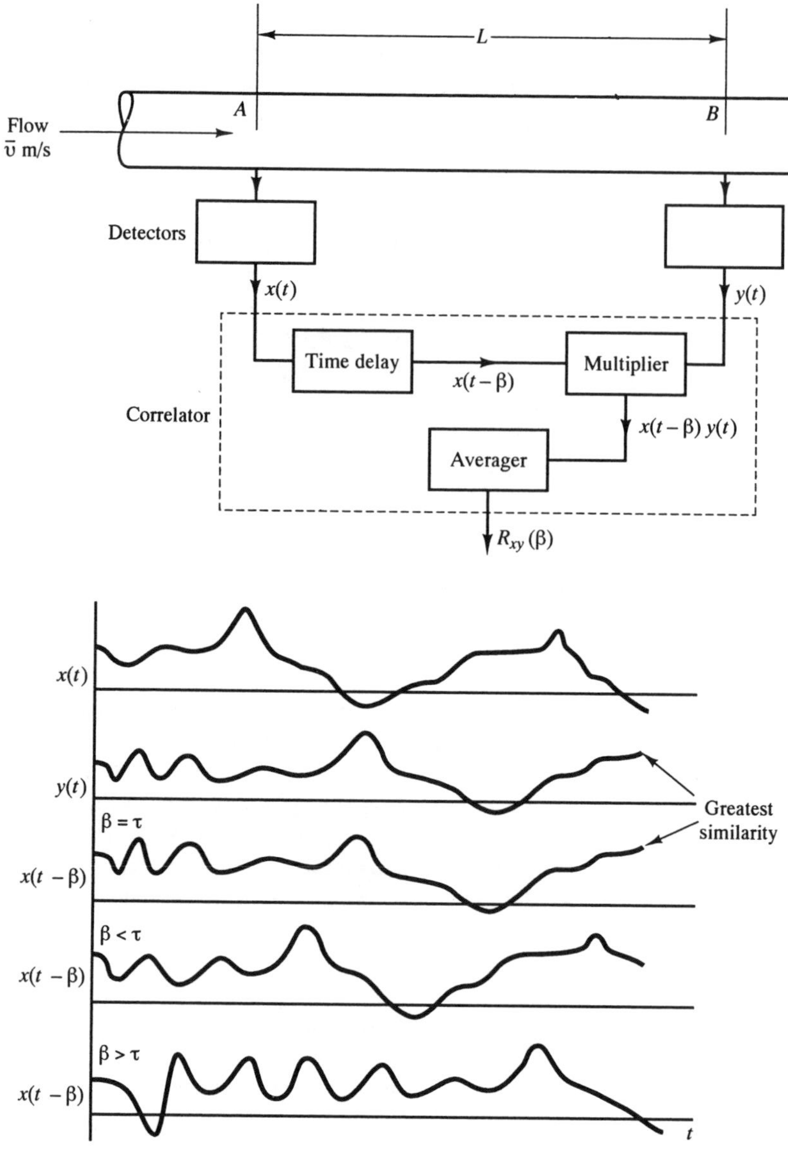

Fig. 12.19 Cross-correlation flowmeter — schematic diagrams and typical signals

− β) of the signal with a time-delayed version $y(t - \beta)$. The cross-correlation function $R_{xy}(\beta)$ between two random signals $x(t)$, y(t) is similarly defined in terms of the mean value $\overline{x(t - \beta).y(t)}$ of the product $x(t - \beta).y(t)$ of a delayed version $x(t - \beta)$ of the upstream signal with the undelayed downstream signal $y(t)$. Mathematically we have:

$$R_{xy}(\beta) = \lim_{T \to \infty} \frac{1}{T} \int_0^T x(t - \beta).y(t)\mathrm{d}t \qquad [12.54]$$

where β is the variable time delay and T the observation time. Figure 12.19 shows typical waveforms of $x(t)$, $y(t)$ and the time-delayed version $x(t - \beta)$ for three different time delays β. It can be seen that $x(t - \beta)$ is most similar to $y(t)$, when

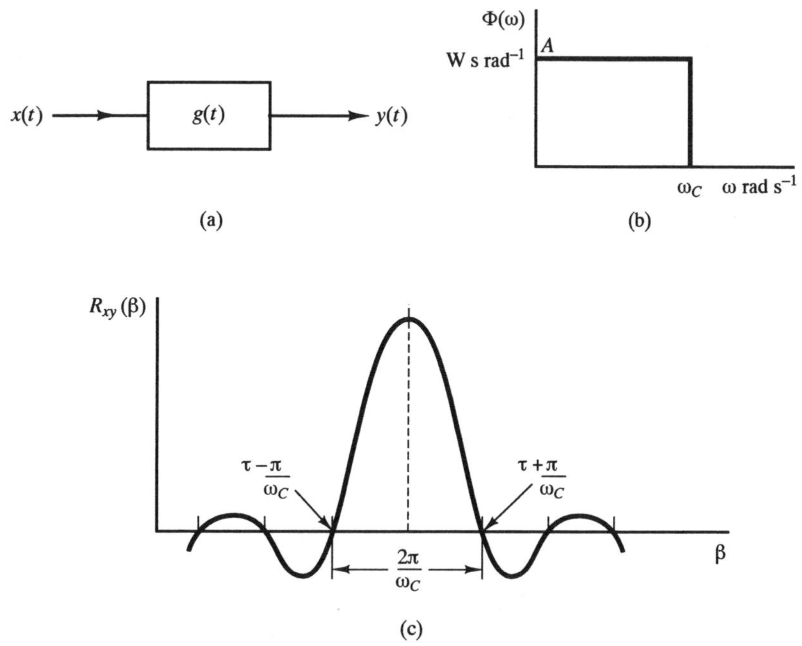

Fig. 12.20 Theoretical cross-correlation function for flowmeter

$\beta = \tau$, the mean transit time between A and B. In other words the cross-correlation function $R_{xy}(\beta)$ has a maximum when $\beta = \tau$. Thus τ can be measured by finding the value of β at which $R_{xy}(\beta)$ is maximum. Since $\tau = L/\bar{v}$, where L is the distance between A and B, the average velocity $\bar{v}$ and volume flow rate Q can then be found.

The above result can be proved more rigorously using random signal analysis. Since $x(t)$, $y(t)$ are random signals, whose time behaviour is not known explicitly, eqn [12.54] cannot be used for evaluating $R_{xy}(\beta)$. Let us regard the length of pipe between A and B as a system with input $x(t)$, output $y(t)$ and impulse response (weighting function) $g(t)$, as indicated in Fig. 12.20. We can then express $y(t)$ in terms of $x(t)$ using the convolution integral:

$$y(t) = \int_0^t g(t')x(t - t')\,\mathrm{d}t \qquad [12.55]$$

Using [12.54] and [12.55] it can be shown that:

$$R_{xy}(\beta) = \int_0^\infty g(t')\,R_{xx}(\beta - t')\,\mathrm{d}t' \qquad [12.56]$$

that is the cross correlation function can be expressed as a convolution integral involving $g(t)$ and the autocorrelation function R_{xx}. If we assume that the length of pipe can be represented by a pure time delay $\tau = L/\bar{v}$, then the corresponding system impulse response is simply a unit impulse delayed by τ, i.e.

$$g(t') = \delta(t' - \tau) \qquad [12.57]$$

This gives:

$$R_{xy}(\beta) = R_{xx}(\beta - \tau) \qquad [12.58]$$

so that the cross-correlation function is a time-shifted version of the autocorrelation function. We further assume that the signal $x(t)$ has a power spectral density $\phi(\omega)$ which is constant up to ω_c and zero for higher frequencies:

$$\begin{aligned} \phi(\omega) &= A \quad 0 \le \omega \le \omega_c \\ &= 0 \qquad \omega > \omega_c \end{aligned} \qquad [12.59]$$

The autocorrelation function is the Fourier transform of the power spectral density (Section 6.2.5), i.e.

$$R_{xx}(\beta) = \int_0^\infty \phi(\omega) \cos \omega\beta \mathrm{d}\omega \qquad [12.60]$$

Thus:

$$R_{xx}(\beta) = A \int_0^{\omega_c} \cos \omega\beta \mathrm{d}\omega = A \frac{\sin \omega_c\beta}{\beta} \qquad [12.61]$$

and

$$R_{xy}(\beta) = R_{xx}(\beta - \tau) = A \frac{\sin \omega_c(\beta - \tau)}{(\beta - \tau)} \qquad [12.62]$$

We see from Fig. 12.20(c) that $R_{xy}(\beta)$ has a maximum at $\beta = \tau$ as explained earlier. We note also that for accurate measurement of τ and $\bar{v}$ a sharp maximum is required. This means that τ should be much greater than the width of the peak, i.e. $\tau \gg 2\pi/\omega_c$ or $\tau \gg 1/f_c$.

Detection systems

The above principles are well established and much of the work in recent years has concentrated on developing detection systems suitable for a wide variety of difficult flow measurement problems. Table 12.3 summarises some of the more important work. Beck[12] gives a full review of detection systems and correlators.

The ultrasonic detection system is the most versatile and the most likely to find

Table 12.3

Detection system	Detection principle	Application
A.C. conductivity electrodes	Detects random fluctuations in electrical conductivity due to solid particles, etc.	Flow rates of liquid/solid and liquid/liquid mixtures.
Capacitance transducers	Detects random fluctuations in density	Solids velocity measurement in gas/solid systems
Radiation pyrometer with infrared detector	Detects random fluctuations in radiation intensity	Flow rate of high temperature dust laden gases
Ultrasonics	Detects modulation of ultrasonic beam by flow turbulence	Flow rate of liquids, gases, liquid/gas systems, liquid/solid systems[13]
Optical	Detects modulation of light beam due to surface ripples or turbulence	Open channel flow. Flow from chimney stacks

widespread industrial application. In this system there are two or more ultrasonic transmitter/receiver links of the type discussed in Chapter 16. Any disturbance in the flow, such as flow turbulence, solid particles or gas bubbles, modulates the ultrasonic beam in amplitude and phase. A corresponding electrical signal is obtained by demodulation at the receiver. For liquids and liquid mixtures the transmitter and receiver can be mounted on the outside of the pipe so that the method is non-invasive. A fuller account of the ultrasonic cross-correlation flowmeter is given in Section 16.4.

Conclusion

The chapter first discussed the principles of **fluid mechanics** essential to an understanding of flow measurement. Methods for the measurement of **velocity at a point** in the fluid were then studied. The next section examined systems for the measurement of **volume flow rate** and included **differential pressure**, **mechanical** and **vortex flowmeters**. The following section explained systems for the measurement of **mass flow rate** including both **inferential** and **direct flowmeters**, the **Coriolis meter** is an important example of the latter. The measurement of flow rate in difficult situations was finally discussed using **electromagnetic**, **ultrasonic** and **cross-correlation flowmeters**.

References

12.1 Bell and Howell Ltd., Electronics and Instruments Division, 1981, *Technical information on very low range pressure transmitters.*

12.2 British Standards Institution BS 1042, 1981 *Methods of measurement of fluid flow in closed conduits — Section 1.1 orifice plates nozzles and Venturi tubes in circular cross-section conduits running full.*

12.3 International Organization for Standardization ISO 5167, 1980 *Measurement of fluid flow by means of orifice plate, nozzles and Venturi tubes inserted in circular cross-section conduits running full.*

12.4 COUSINS T, FOSTER S A and JOHNSON P A *A Linear and Accurate Flowmeter using Vortex Shedding*, Kent Instruments Technical Paper.

12.5 COUSINS T 'The performance and design of vortex meters', *Proceedings of Conference on Fluid flow Measurement in the Mid 1970s*, National Engineering Laboratory, April 1975.

12.6 MILLER R W, DE CARLO J P and CULLEN J T 1977 'A vortex flow meter-calibration results and application experiences' *Institute of Measurement and Control Symposium on the Application of Flow Measuring Techniques*, Brighton.

12.7 Fisher Controls, 1983 *Technical Bulletin on DV Series vortex Flowmeter.*

12.8 Neptune Eastech, 1980 *Technical Information Vortex Shedding Flowmeters for Liquids and Gases.*

12.9 Scheme Engineering, 1985 *Technical Information on Ultrasonic Vortex Sensing.*

12.10 PLACHE K O 1980 'Measuring mass flow using the Coriolis principle', *Transducer Technology*, vol. 2, no. 3.

12.11 Flowmetering Instruments Ltd., 1985 *Technical Information on d.c. field electromagnetic flowmeters.*

12.12 BECK M S 1984 'Correlation in Instruments: cross correlation flowmeters', *Instrument Science and Technology*, vol. 2, Chapter 8, Adam Hilger.

12.13 COULTHARD J 1985 'Developments and Applications of Cross Correlation', *Institute of Measurement and Control Symposium, 'Control and Instrumentation — the changing scene*, Harrogate.

Problems

12.1 A pitot tube is used to measure the mean velocity of high pressure gas in a 0.15 m diameter pipe. At maximum flow rate the mean pitot differential pressure is 250 Pa. Use the data given below to:

(a) calculate the mean velocity of the gas at maximum flow rate;
(b) estimate the maximum mass flow rate;
(c) estimate the Reynolds number at maximum flow;
(d) explain why an orifice plate would be suitable to measure the mass flow rate of the gas.
(e) Given that a differential pressure transmitter of range 0 to 3 × 10^4 Pa is available, estimate the required diameter of the orifice plate hole (assume coefficient of discharge = 0.6, expansibility factor and velocity of approach factor = 1.0).

Data
Density of gas = 5.0 kg m^{-3}
Viscosity of gas = 5.0 × 10^{-5} Pa s

12.2 An orifice plate is to be used in conjunction with a differential pressure transmitter to measure the flow rate of water in a 0.15 m diameter pipe. The maximum flow rate is 50 $m^3\,h^{-1}$, the density of water is 10^3 kg m^{-3} and the viscosity is 10^{-3} Pa s.
(a) Explain why an orifice plate meter is suitable for this application.
(b) **Estimate** the required orifice plate hole diameter if the transmitter has an input range of 0 to 1.25 × 10^4 Pa

12.3 A Venturi is to be used to measure the flow rate of water in a pipe of diameter D = 0.20 m. The maximum flow rate of water is 1.5 × 10^3 $m^3\,h^{-1}$, density is 10^3 kg m^{-3}, and viscosity is 10^{-3} Pa s. Venturis with throat diameters of 0.10 m, 0.14 m and 0.18 m are available from the manufacturer.

(a) Choose the most suitable Venturi for the application, assuming a differential pressure at maximum flow of approximately 3 × 10^5 Pa.
(b) Calculate an accurate value for the differential pressure developed across the chosen Venturi, at maximum flow rate.
Coefficient of discharge:

$$C = 0.9900 - 0.023\left(\frac{d}{D}\right)^4 + 0.002\left(\frac{10^6}{\mathrm{Re}_D}\right)$$

where d = Venturi throat diameter,
Re_D = Reynolds number referred to pipe diameter.

12.4 Oxygen at 100 °C and 10^6 Pa is flowing down a pipe of diameter D = 0.20 m. The maximum flow rate of oxygen is 3.6 × 10^4 kg h^{-1}. It is proposed to measure this flow using an orifice plate in conjunction with a differential pressure transducer of range 0 to 5 × 10^4 Pa. Using Table 12.1, eqn [12.27], and the data given below:
(a) explain why an orifice meter is suitable for this application;
(b) make an initial estimate of the diameter d of the orifice plate hole;
(c) calculate a more accurate value of hole diameter d, using **one** iteration only.

Data
Density of oxygen = 10.0 kg m^{-3}

Viscosity of oxygen = 2.40×10^{-5} Pa s
Tappings: corner
Specific heat ratio = 1.4

12.5 A turbine flowmeter consists of an assembly of 4 ferromagnetic blades rotating at an angular velocity ω rad s^{-1} given by:

$$\omega = 4.5 \times 10^4 Q$$

where Q $m^3 s^{-1}$ is the volume flow rate of the fluid. The total flux N linked by the coil of the magnetic transducer is given by:

$$N = 3.75 + 0.88 \cos 4\theta \text{ milliwebers}$$

where θ is the angle between the blade assembly and the transducer. The range of the flowmeter is 0.15×10^{-3} to 3.15×10^{-3} $m^3 s^{-1}$. Calculate the amplitude and frequency of the transducer output signal at minimum and maximum flows.

12.6 If a circular cylinder of diameter d metres is installed as a bluff body in a pipe of diameter D metres then the frequency f Hz of vortex shedding is given by:

$$\frac{f}{Q} = \frac{4S}{\pi D^3} \frac{1}{\frac{d}{D}\left[1 - 1.4\frac{d}{D}\right]}$$

where Q = volume flow rate of fluid ($m^3 s^{-1}$)
S = Strouhal number

Use this result to calculate the correct cylinder diameter for a 0.15 m pipe carrying a water flow between 0 and 1.0 $m^3 s^{-1}$. What is the maximum vortex shedding frequency? Assume a Strouhal number of 0.2.

12.7 A turbine flowmeter has a bore of internal diameter 150 mm. The rotor consists of 8 blades, each of mean thickness 5 mm, mounted on a hub of mean diameter 35 mm. The clearance between each blade tip and the bore is 1 mm and the inlet blade angle at tip = 20°. Estimate the meter factor K in pulses/m^3.

12.8 A cross-correlation flowmeter consists of two transducers, spaced 0.15 metres apart, detecting random fluctuations in density. The velocity of flow is 1.0 m s^{-1} and the fluctuations contain frequencies up to 100 Hz. State whether the flowmeter is suitable for this application.

12.9 Steam at $P_1 = 20 \times 10^5$ Pa absolute and 250 °C is flowing down a circular pipe of diameter $D = 0.150$ m. An orifice plate with hole diameter $d = 0.080$ m and corner taps is used to measure the steam flow rate. If the measured differential pressure ΔP is 2.5×10^4 Pa, using the data given below:

(a) **Estimate** the steam mass flow rate in kg h^{-1}.
(b) Estimate the value of Reynolds number for the pipe
(c) Discuss the nature of the flow and the suitability of the orifice plate flowmeter
(d) Calculate an **accurate** value of the steam mass flow rate in kg h^{-1}.

Data
Steam density = 9.0 kg m^{-3}
Steam dynamic viscosity = 1.8×10^{-5} Pa s
Steam specific heat ratio k = 1.3
Discharge coefficient

$$C = 0.5959 + 0.0312\,\beta^2 - 0.184\,\beta^8$$

Expansibility factor

$$\epsilon = 1 - (0.41 + 0.35\beta^4)\frac{1}{k}\cdot\frac{\Delta P}{P_1}$$

where $\beta = \frac{d}{D}$

13

Intrinsically safe measurement systems

Many oil, gas and chemical plants process chemicals which are potentially explosive: common examples are hydrocarbons, i.e. compounds containing carbon and hydrogen. Any small leakage from the plant means that the atmosphere surrounding the plant will be a mixture of hydrocarbon and air; because air contains oxygen, this mixture is potentially explosive or **flammable**.

Under certain conditions, this mixture may be ignited by an electrical spark or a hot surface. Figure 13.1 shows the general relationship between the ignition energy of a hydrocarbon–air mixture and the percentage of hydrocarbon present by volume. We see that a potentially explosive mixture is one where the concentration of hydrocarbon is between the **lower explosive limit** (LEL) and **upper explosive limit**

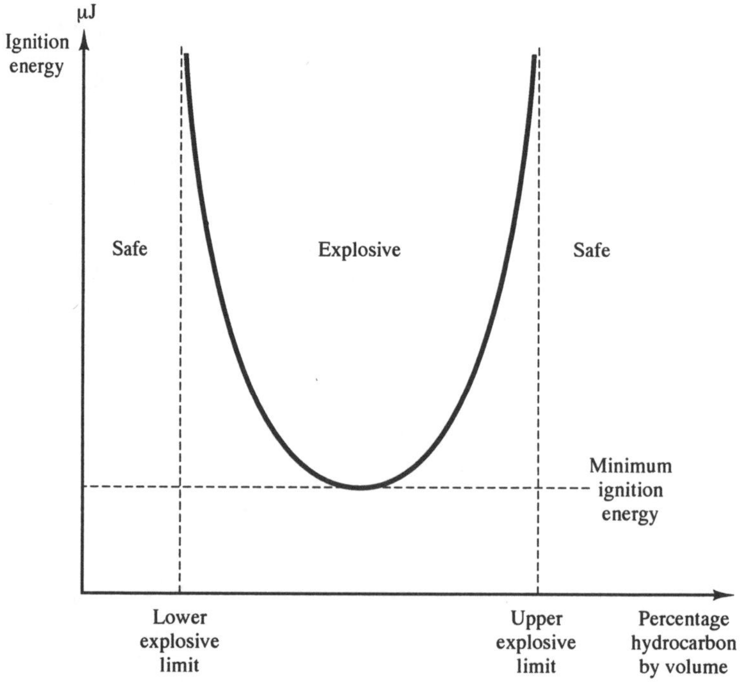

Fig. 13.1 Relationship between ignition energy and percentage of hydrocarbon for hydrocarbon–air mixture

(UEL). The graph also defines the **minimum ignition energy** (MIE) for a mixture. The values of these parameters are different for different hydrocarbons: hydrogen has LEL = 4%, UEL = 75% and MIE = 19 μJ; propane has LEL = 2.2%, UEL = 10% and MIE = 250 μJ. This shows that hydrogen–air mixture is far easier to ignite than a propane–air mixture. The ignition energy increases asymptotically as the hydrocarbon concentration is reduced towards LEL and as concentration is increased towards UEL.

There are three separate factors which determine the overall probability of an explosion which leads to three different types of hazard classification:

(1) Probability of explosive gas–air mixture being present: **area classification**
(2) The electrical spark energy required to ignite the mixture: **gas classification**
(3) The surface temperature required to ignite the mixture: **temperature classification**.

The area classification recommended by the **International Electrotechnical Commission** (IEC) and adopted by European countries is based on three zones:

Zone 0 — explosive gas–air mixture continuously present or present for long periods;
Zone 1 — explosive gas–air mixture is likely to occur in normal operation;
Zone 2 — explosive gas–air mixture is not likely to occur and if it occurs, it will exist only for a short time.

The USA and Canada use a system based on two divisions. Division 1 corresponds to Zones 0 and 1, Division 2 corresponds to Zone 2.

Since a gas–air mixture can be ignited by an electrical spark, the **maximum spark energy** under fault conditions must not exceed the **minimum ignition energy** for that mixture. Since all hydrocarbon gases have different MIEs, the IEC recommends a gas classification for equipment based on five representative (test) gases. These in order of increasing ease of ignition are Methane (Group I), Propane (Group IIA), Ethylene (Group IIB), Hydrogen/Acetylene (Group IIC). This classification means, for example, that ethylene–air cannot be ignited by Group IIB equipment since the maximum spark energy, under fault conditions, is less than the minimum ignition energy for that mixture. In the USA and Canada, the four gases propane, ethylene, hydrogen and acetylene are used in the classification and suitable equipment is designed D, C, B and A respectively.

Since gas–air mixtures can also be ignited directly by hot surfaces, the maximum surface temperature of any equipment located in a given gas–air mixture must not exceed the ignition temperature of the gas. Equipment to be installed in a potentially explosive atmosphere is therefore also classified according to the maximum surface temperature that can be produced under fault conditions. There are six classes of equipment ranging from least safe T1 (450 °C) through T2 (300 °C), T3 (200 °C), T4 (135 °C), T5 (100 °C), to the most safe T6 (85 °C). The user must ensure that the temperature class of the equipment is below the ignition temperature of any gas–air mixture that may arise.

It is therefore imperative that any measurement system which is installed in a potentially explosive gas mixture cannot ignite the gas mixture under any possible fault condition. One possible solution is to use **pneumatic measurement systems** which use compressed air as a signalling medium rather than electrical energy. Since there is no possibility of a spark at all with pneumatic equipment, they are **intrinsically safe**; the principles and characteristics of pneumatic measurement systems are discussed in the first section of this chapter. However, it is more difficult and expensive to interface pneumatic transducers to electronic signal processing and data presentation elements, so ways of making electrical transducers safe in explosive atmospheres must be found.

One method is to enclose the electrical device in a **flame-proof enclosure** which is capable of withstanding the explosion inside it and so prevent the ignition of the flammable mixture surrounding it. Other methods include pressurising or purging the enclosure with air, filling the enclosure with sand, immersing the equipment in oil, or taking special precautions to ensure that sparks do not occur. However the most universally acceptable method is to limit the electrical energy produced during any possible fault condition, so that it is below the minimum ignition energy of the gas mixture. This is referred to an **intrinsic safety**. The second section of this chapter looks at **intrinsically safe electronic** measurement systems.

13.1 Pneumatic measurement systems

Pneumatic measurement systems use compressed air as a signalling medium rather than electrical energy. The standard penumatic signal range is 0.2 to 1.0 bar (1 bar $= 10^5$ Pa), i.e. 3 to 15 lb wt in^{-2} (p.s.i.g.); these pressures are gauge pressures, i.e. pressures relative to atmospheric pressure. Thus a simple pneumatic temperature measurement system consists of a temperature transmitter giving an output of 0.2 to 1.0 bar, corresponding to an input of 0 to 100 °C, connected by copper, nylon or plastic tubing to an indicator. The indicator is a pressure gauge, incorporating a Bourdon tube elastic sensing element, with a scale marked in degrees Celsius; this means there is an indication of 0 °C for 0.2 bar input and 100 °C for 1.0 bar input. The transmitter must be supplied with clean, dry air at 1.4 bar ($\approx$ 20 p.s.i.g.). Pneumatic measurement systems are simple, robust, reliable, easy to maintain and are not affected by electrical interference.

One disadvantage of pneumatic systems is the time delay or lag in transmitting a change in pressure from transmitter to receiver. This delay can be several seconds and increases with tube length so that problems can occur with transmission distances over 500 ft (150 m). Transmission distances over 1000 ft (300 m) are not recommended, unless booster equipment is used. Another disadvantage is the possibility of condensed moisture in the pipework freezing at sub-zero ambient temperatures with open air installations. Obviously it is more difficult and expensive to interface pneumatic transmitters to digital computers and data loggers than the corresponding electrical devices.

This section begins by studying the characteristics of the flapper/nozzle displacement sensor, discusses the need for and principles of relays, explains the principles of operation and applications of torque balance transmitters and concludes by discussing pneumatic transmission and data presentation.

13.1.1 Flapper/nozzle displacement sensing element

The flapper/nozzle displacement sensor forms the basis of all pneumatic transmitters.

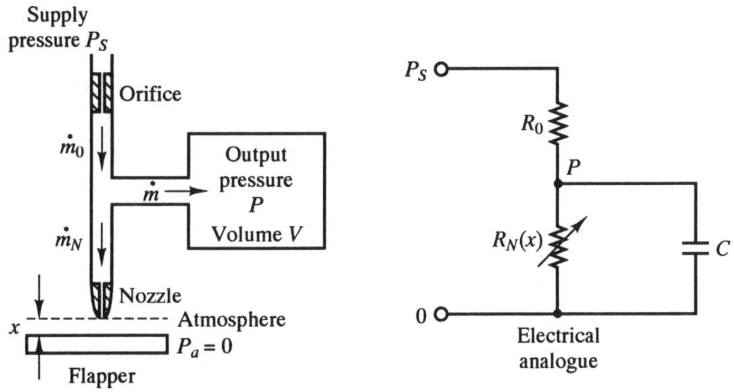

Parameters

Molecular weight of air	$w = 29$
Volume	$V = 3.2 \times 10^{-3}$ m^3 (100 m of ¼ inch (6.4 mm) OD tubing)
Ideal gas constant	$R = 8.314$ J K^{-1} mol^{-1}
Absolute ambient temperature	$\theta = 293$ K
Coefficient of discharge	$C_D = 0.6$
Density of air	$\rho = 1.0$ kg/m^{-3}
Diameter of orifice	$d_0 = 2 \times 10^{-4}$ m
Diameter of nozzle	$d_N = 8 \times 10^{-4}$ m
Supply pressure	$P_s = 1.4 \times 10^5$ Pa (gauge)
Atmospheric pressure	$P_a = 0$ (gauge)

Fig. 13.2 Principle of flapper/nozzle displacement sensor

It consists (Fig. 13.2) of a fixed restrictor (orifice) in series with a variable restrictor (flapper and nozzle). Altering the separation x of the flapper and nozzle alters the resistance to air flow and the output pressure P: an increase in x causes a reduction in resistance and a fall in pressure. The volume V represents the capacity of the transmission line connecting the sensor to an indicator. The equivalent electrical circuit (Fig. 13.2) is a potentiometer consisting of a fixed resistor in series with a variable resistor and a capacitive load across the variable resistor.

The relevant equations are:

$$\dot{m} = \dot{m}_0 - \dot{m}_N \qquad [13.1]$$

$$\dot{m} = \frac{wV}{1000R\theta}\frac{\mathrm{d}P}{\mathrm{d}t} \qquad [13.2]$$

$$\dot{m}_0 = C_D \frac{\pi}{4} d_0^2 \sqrt{[2\rho\,(P_S - P)]} \qquad [13.3]$$

$$\dot{m}_N = C_D\, \pi\, d_N\, x \sqrt{[2\rho\,(P - P_a)]} \qquad [13.4]$$

Figure 13.2 also explains the meaning and gives typical values of the element parameters. Equation [13.2] is based on the ideal gas law $PV = nR\theta$ for n moles of an ideal gas occupying volume V at absolute temperature θ. Since n = (mass in g/molecular weight) = (mass in kg $\times$ 10^3/molecular weight) = $(1000m/w)$, then

mass $m = (wV/1000R\theta)P$

mass flow rate $\dot{m} = (wV/1000R\theta)\,(\mathrm{d}P/\mathrm{d}t)$.

Equations [13.3] and [13.4] giving the mass flow rate of air through orifice and nozzle are based on eqn [12.28]. The velocity of approach factor E and expansibility factor ϵ are both assumed to be unity. The area of the orifice hole is $\pi d_0^2/4$ and the effective area of the nozzle is assumed to be the surface area of a cylinder diameter d_N and length x. This assumption is true only for small displacements x.

In the steady state dP/dt and $\dot{m}$ are zero, so that $\dot{m}_0 = \dot{m}_N$ i.e.

$$(d_0^2/4)\sqrt{(P_S - P)} = d_N x \sqrt{P} \qquad [13.5]$$

since $P_a = 0$. This gives:

Steady-state relation between pressure and displacement for flapper/nozzle

$$P = \frac{P_S}{1 + 16(d_N^2 x^2/d_0^4)} \qquad [13.6]$$

This relationship is plotted in Fig. 13.3; we see that it is characterised by non-linearity and very high sensitivity — typically $dP\,dx \approx -5 \times 10^9\ \mathrm{Pa\,m^{-1}}$. Because of these characteristics the flapper/nozzle itself is unsuitable for use as a sensing element in a measurement system. It is usually incorporated into a closed-loop torque-balance transmitter where it detects small movements due to any imbalance of torques.

The relation between dynamic changes in input displacement and output pressure can be approximately described by a first-order transfer function. The corresponding time constant τ is extremely long, typically several minutes. This is because the orifice and nozzle diameters are very small so that they present a high resistance to flow. If x decreases, air must be brought in via the orifice to increase P, the mass flow rate $\dot{m}_0$ and rate of increase in pressure dP/dt is small. Similarly if x increases, air

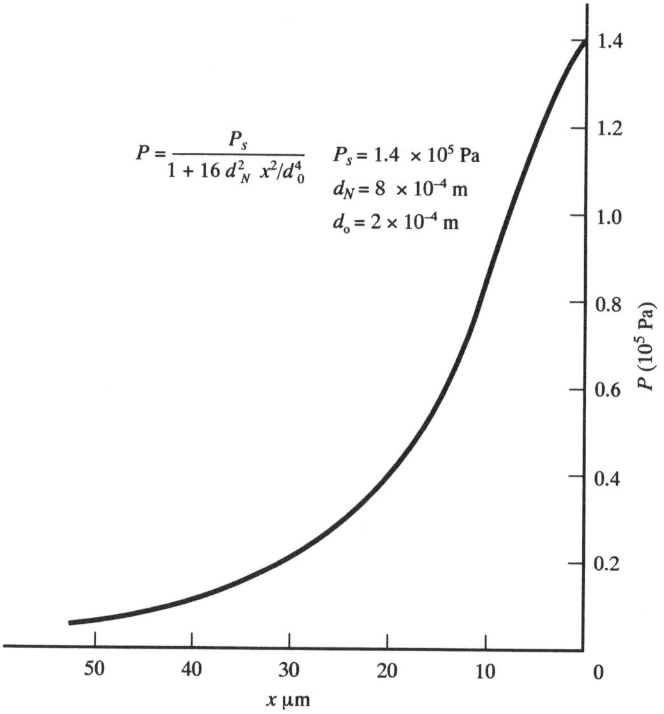

Fig. 13.3 Steady-state characteristics of flapper–nozzle displacement sensor

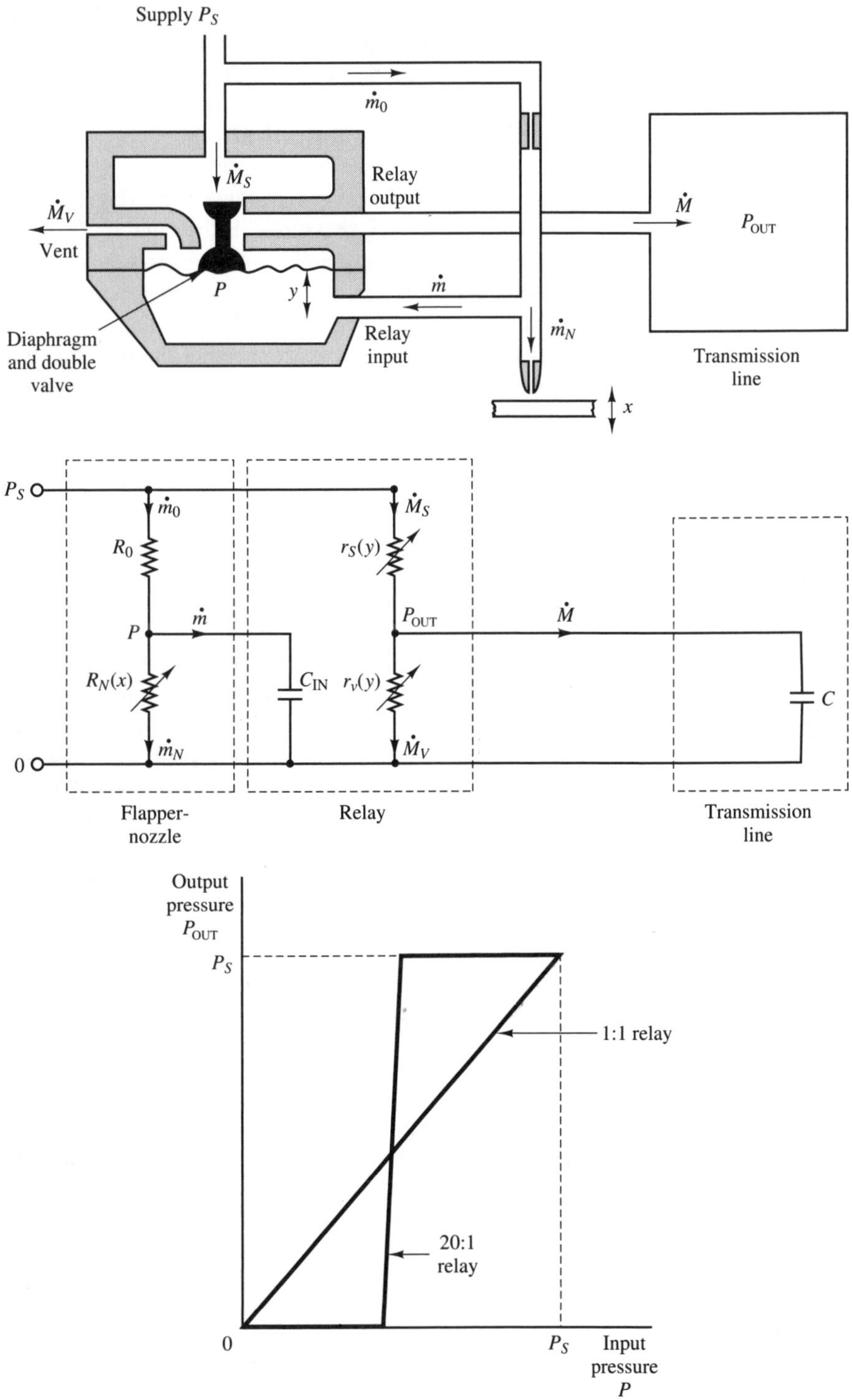

Fig. 13.4 Relay, equivalent circuit and steady-state characteristics

must be vented via the nozzle to decrease P. The mass flow rate $\dot{m}_N$ and rate of decrease in pressure is small. This means that the flapper/nozzle sensor cannot be connected directly onto a penumatic transmission line, but only via a device which gives increased air flows and therefore a reduced time constant.

13.1.2 Principle of relay amplifier

Figure 13.4 shows a relay amplifier connecting a flapper/nozzle to a transmission line. The nozzle back pressure P is the input to the relay and the transmission line is connected to the relay output. Air can now flow from supply to the transmission line via the double valve. This is a low resistance path which bypasses the orifice and allows a high flow rate of air into the line. There is also a low resistance path, again via the double valve, connecting the transmission line to a vent port. This bypasses the nozzle and allows a high flow rate of air out of the line when the line is being depressured. A change in nozzle back pressure P causes the centre of the diaphragm and the double valve to move; this adjusts the relative values of supply flow $\dot{M}_S$ and vent flow $\dot{M}_V$, thus changing the relay output pressure P_{OUT}. For example if P increases, the diaphragm and double valve move upwards; this reduces $\dot{M}_V$ but increases $\dot{M}_S$ and the netflow $\dot{M}$ into the line. The output pressure P_{OUT} rises until equilibrium is re-established.

Figure 13.4 also shows an approximate electrical equivalent circuit for the flapper/nozzle, relay and transmission line. The input capacitance C_{IN} of the relay is small and corresponds to the volume of the input chamber below the diaphragm. The resistances r_S, r_V of the supply and vent paths are small compared with the resistance R_O, R_N of orifice and nozzle. The supply and vent flows $\dot{M}_S$ and $\dot{M}_V$ are therefore large compared to the corresponding orifice and nozzle flows $\dot{m}_O$ and $\dot{m}_N$. The resistances r_S, r_V are variable and depend on the displacement y of the diaphragm and double valve. The pressure underneath the diaphragm is the nozzle back pressure P; the pressure above the diaphragm is atmospheric, i.e. a gauge pressure of zero. The resultant deflecting force on the diaphragm is $A_{RD}P$ newtons, where A_{RD} is the area of the diaphragm. If $k\,\text{N m}^{-1}$ is the effective stiffness of the diaphragm, then the spring restoring force is ky newtons, so that at equilibrium:

$$A_{RD}P = ky \quad \text{and} \quad y = \frac{A_{RD}}{k}P \qquad [13.7]$$

i.e. y is proportional to P.

The relationships between r_S, r_v and y will depend on the shape of the double valve and will in general be non-linear. An increase in P and y, causes an increase in r_V, a decrease in r_S, and increase in P_{OUT}. In the steady state, when $dP_{\text{OUT}}/dt = \dot{M} = 0$, the output pressure is determined by the potentiometer r_S, r_V. The value of P_{OUT} can be calculated by a detailed analysis based on equations similar to [13.5]. By careful design it is possible to produce an overall linear relationship between P_{OUT} and P. The steady-state sensitivity $K_R = \Delta P_{\text{OUT}}/\Delta P$ of relays in common use usually varies from unity up to 20 (Fig. 13.4). A typical relay has a steady-state air consumption $\dot{M}_S = \dot{M}_V$ of around $1\,\text{kg hr}^{-1}$. These higher air flows mean that the time constant $(r_S + r_V)C$ describing dynamic pressure variations is now only a few seconds.

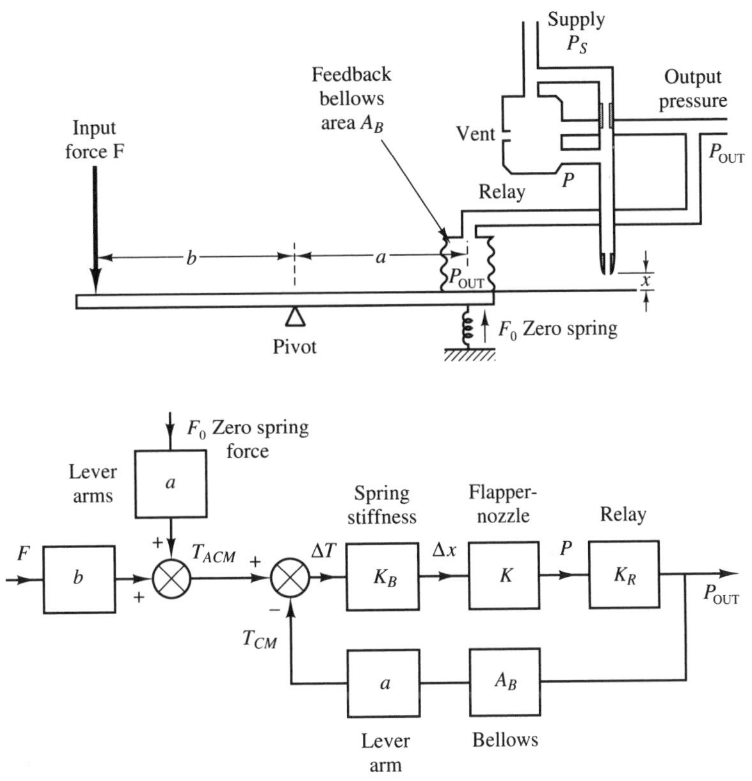

Fig. 13.5 Schematic and block diagrams of basic torque balance transmitter

13.1.3 Torque-balance transmitters

Figure 13.5 shows a flapper/nozzle and relay incorporated into a basic torque-balance transmitter. The transmitter is a closed-loop, negative feedback device and gives a pneumatic output signal, in the standard range, proportional to an input force F. An increase in F increases the anticlockwise moment on the beam, causing it to rotate in an anticlockwise direction. This reduces the flapper/nozzle separation x causing an increase in nozzle back pressure P and relay output pressure P_{OUT}. This increase in pressure is fed to the transmission line — but also to the feedback bellows, causing an increase in the feedback force $A_B P_{\text{OUT}}$. This gives an increased clockwise moment to oppose the increased anticlockwise moment due to F. The beam will rotate and the output pressure will change until clockwise and anticlockwise moments are again equal. The flapper/nozzle and relay detect small rotations of the beam due to any imbalance of torques.

From Fig. 13.5 we have:

Anticlockwise moments $T_{\text{ACM}} = Fb + F_0 a$

Clockwise moment $T_{\text{CM}} = P_{\text{OUT}} A_B a$

Assuming perfect torque balance we have:

$$P_{\text{OUT}} A_B a = Fb + F_0 a$$

i.e.

Simple model for basic torque-balance transmitter

$$P_{OUT} = \frac{b}{aA_B}F + \frac{F_0}{A_B} \qquad [13.8]$$

Thus the output pressure is proportional to the input force. According to this simple theory, the sensitivity of the transmitter is $b/(aA_B)$; i.e. sensitivity depends only on the lever arm ratio b/a and feedback bellows area A_B. Thus sensitivity is independent of flapper/nozzle, relay characteristics and supply pressure. A transmitter with $b/a = 1$, $A_B = 5 \times 10^{-4}\,\text{m}^2$ (circular bellows approximately 1 inch diameter), and zero spring force $F_0 = 10\,\text{N}$, will give an output pressure between 0.2 and 1.0 bar for an input force between 0 and 40 N. Adjusting the position of the pivot alters the ratio b/a and the sensitivity of the transmitter. This means that it is possible to adjust the input range of the transmitter while maintaining an output range of 0.2 to 1.0 bar. Thus the input range of the above transmitter can be changed from 0 to 40 N, down to 0 to 4 N, by increasing the b/a ratio from 1 to 10.

The above theory assumes that there is perfect torque balance; this is obviously an oversimplication. Figure 13.5 shows a block diagram of the system; this allows for a difference ΔT between clockwise and anticlockwise moments. K_B is the stiffness of the beam/spring arrangement, i.e. the change in flapper/nozzle separation x for unit change in torque. K is the sensitivity of the flapper/nozzle at the prevailing operating conditions, and K_R is the relay sensitivity. Using Fig. 13.5 we have:

$$\left.\begin{aligned} T_{ACM} &= bF + aF_0 \\ T_{CM} &= aA_BP_{OUT} \\ \Delta T &= T_{ACM} - T_{CM} \\ P_{OUT} &= K_BKK_R\Delta T \end{aligned}\right\} \qquad [13.9]$$

giving:

Accurate model for torque-balance transmitter

$$P_{OUT} = \frac{K_BKK_R}{(1 + K_BKK_RaA_B)}(bF + aF_0) \qquad [13.10]$$

In this more accurate model the output pressure depends on the flapper/nozzle sensitivity K and therefore on the supply pressure P_S (eqn [13.6]). Typical values are:

$K_B = 6 \times 10^{-5}\,\text{m/Nm}^{-1}$,
$K = 5 \times 10^9\,\text{Pa m}^{-1}$,
$K_R = 20$,
$a = b = 5 \times 10^{-2}\,\text{m}$,
$A_B = 5 \times 10^{-4}\,\text{m}^2$,
$F_0 = 10\,\text{N}$,
$K_BKK_RaA_B = 150$.

Since $K_BKK_RaA_B$ is large compared with 1, the effect of supply pressure variations is small: typically an increase in supply pressure of 10^4 Pa (1.5 p.s.i.) causes an increase in output pressure of only 0.2 per cent.

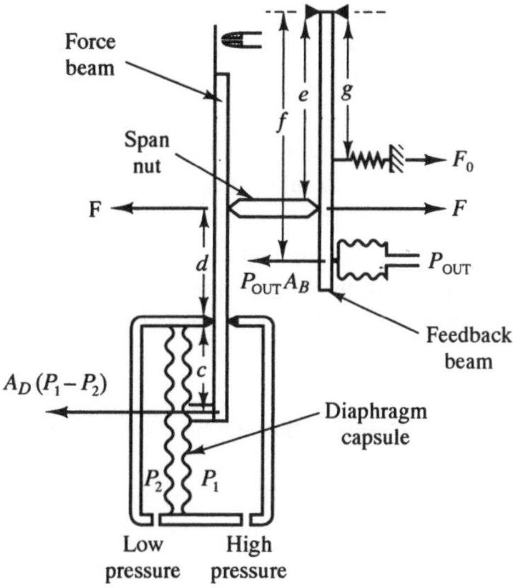

Fig. 13.6 Principle of pneumatic D/P transmitter

This characteristic is important in large pneumatic installations, where variations in compressor delivery pressure and load changes can cause fluctuations in supply pressure of up to ± 0.2 bar (± 3 p.s.i). The torque balance transmitter is an example of the use of high-gain negative feedback in reducing non-linearity and modifying input effects (Section 3.3).

A common example of the torque/balance principle is the pneumatic differential pressure (D/P) transmitter; this is shown in simplified schematic form in Fig. 13.6. It consists of two levers, the force beam and the feedback beam (or span lever). The resultant force on the diaphragm capsule is $A_D(P_1 - P_2)$, where $P_1 - P_2$ is the differential pressure and A_D is the cross sectional area of the capsule. This produces a clockwise moment $A_D(P_1 - P_2)c$ on the force beam, which is opposed by the anticlockwise moment Fd due to the action of the span nut on the force beam. Thus at equilibrium:

$$\text{Force beam:} \quad A_D(P_1 - P_2)c = Fd \qquad [13.11]$$

The span nut also produces an anticlockwise moment Fe on the feedback beam, which is supported by anticlockwise moment $F_0 g$, due to the zero spring force. These are opposed by the clockwise moment $P_{OUT}A_B f$ due to the output pressure acting on the feedback bellows. Thus at equilibrium:

$$\text{Feedback beam:} \quad Fe + F_0 g = P_{OUT}A_B f \qquad [13.12]$$

Eliminating F between [13.11] and [13.12] gives:

Simple model for pneumatic differential pressure transmitter

$$P_{OUT} = \frac{A_D}{A_B}\frac{c}{f}\frac{e}{d}(P_1 - P_2) + \frac{g}{f}\frac{F_0}{A_B} \qquad [13.13]$$

i.e.

$$\text{sensitivity} = \frac{A_D}{A_B}\frac{c}{f}\frac{e}{d} \quad \text{and} \quad \text{zero pressure} = \frac{g}{f}\frac{F_0}{A_B}$$

Therefore adjusting the position of the span nut alters the ratio e/d and the sensitivity. This means that for an output range of 0.2 to 1.0 bar, the input range of a typical transmitter can be adjusted from 0 to 1 m of water (maximum sensitivity, span nut at bottom), to 0 to 10 m of water (minimum sensitivity, span nut at top). The zero spring force F_0 is adjusted to obtain a zero pressure of 0.2 bar. Equation [13.13] can be corrected to take into account imperfect torque balance.

The pneumatic differential pressure transmitter has a similar torque/balance principle to the closed-loop electronic current transmitter discussed in Section 9.4.1.

The applications of the pneumatic transmitter are similar to those of current transmitters detailed in Section 9.4.2.

13.1.4 Pneumatic transmission and data presentation

In Sections 13.1 and 13.2 we represented the transmission line connecting transmitter and receiver as a single fluidic capacitance $C = wV/(R\theta)$. This is a considerable oversimplification, because the transmission line has both fluidic resistance and fluidic inertance (fluid analogue of electrical inductance). The resistance is due to fluid friction, and the inertance due to the mass of the air in the pipe. Moreover the transmission line cannot adequately be represented by a single fluidic L—C—R circuit. This is because the inertance, capacitance and resistance are distributed along the transmission line and therefore cannot be 'lumped' together into single values.[1]

Figure 13.7 shows the response of transmission line output pressure to a step increase in input pressure. The response is the S-shaped curve characteristic of elements or systems with multiple transfer lags. The diagram shows the response approximated by a pure time delay of length τ_D, together with a single first-order transfer lag of time constant τ_L. The corresponding approximate transfer function for the transmission line is:

$$\frac{\Delta P_R}{\Delta P_{OUT}} = \frac{e^{-\tau_D s}}{1 + \tau_L s} \qquad [13.14]$$

Both τ_D and τ_L increase with the length of the transmission line, but the ratio between the two times is constant: $\tau_D/\tau_L \approx 0.24$.[2] thus for piping of 6 mm ($\frac{1}{4}$ inch) outside diameter, a 150 m (500 ft) line has $\tau_D = 0.77$ sec, $\tau_L = 3.2$ sec, a 300 m (1000 ft) line has $\tau_D = 2.3$ sec, $\tau_L = 9.7$ sec, and a 600 m (2000 ft) line has $\tau_D = 7.4$ sec, $\tau_L = 31$ sec.

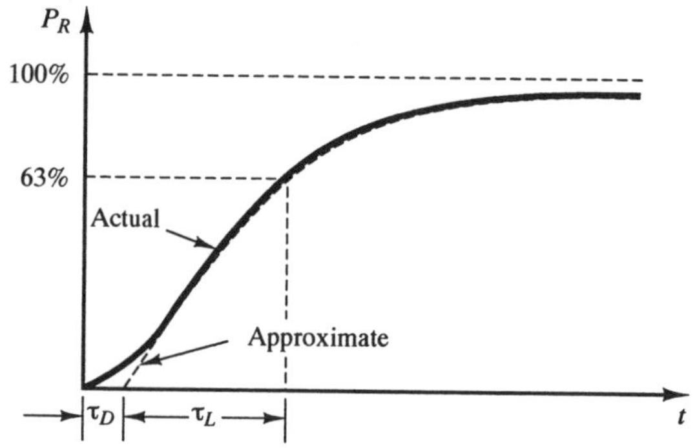

Fig. 13.7 Equivalent network and step response for pneumatic transmission line

A pressure gauge consisting of a C-shaped Bourdon tube elastic element (Figure 8.14) linked mechanically to a pointer moving over a scale is commonly used as a data presentation element in pneumatic measurement systems.

13.2 Intrinsically safe electronic systems

13.2.1 The Zener barrier

In intrinsically safe electronic systems **Zener barrier devices** are used to limit the amount of electrical energy produced in the hazardous area under any possible fault condition. Figure 13.8 shows a basic Zener barrier circuit which will be located in a safe area such as a control room. A signal conditioning, processing or data presentation element such as an amplifier, computer or recorder will be connected across terminals 1 and 2. Terminals 3 and 4 will be connected via a cable to a sensor, transducer or transmitter located in the hazardous area.

The barrier is designed so that if, under fault conditions, a high voltage up to 250 V r.m.s. is applied across terminals 1 and 2 then the resulting electrical energy in the hazardous area is limited to less than the minimum ignition energy of the mixture. The fuse F resistance R_F is present to protect the Zener diodes D_1 and D_2. The surge current rating of the fuse is significantly less than the surge current rating of the diodes, so that if the fault current is increasing rapidly, F blows before either D_1 and D_2 burn out. D_1 provides a safe path to earth for the a.c. fault current; the corresponding r.m.s. voltage across D_1 is limited to the Zener **breakdown** or **avalanche voltage** V_z of the diode. A second diode D_2 is connected in parallel with D_1 for increased realiability, so that even if D_1 fails the r.m.s. voltage across D_2, and therefore across terminals 3 and 4, is again limited to V_z. The resistor R_2 limits the fault current in the external hazardous area circuit to a maximum value of V_z/R_2. Thus under the above fault condition, the Zener barrier can be regarded as a Thévenin voltage source of V_z with series resistance R_2 across terminals 3 and 4. V_z, R_2 are the most important parameters in the barrier specification. The resistance R_1 between the diodes allows both diodes to be tested separately.

In normal operation D_1 and D_2 do not conduct and the barrier has a total or 'end-

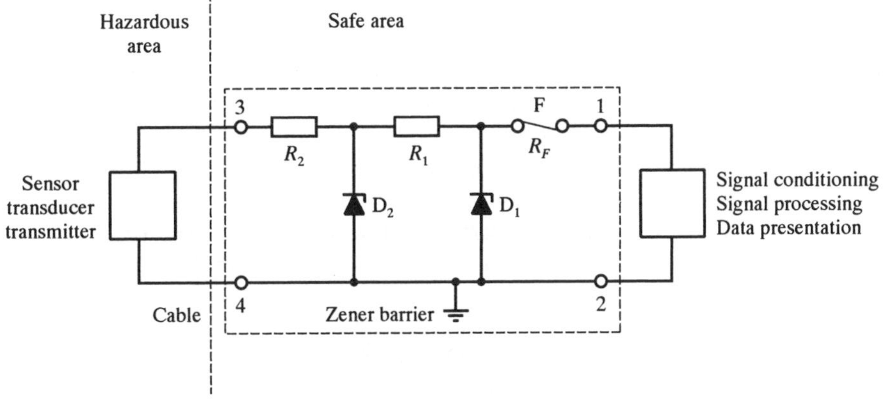

Fig. 13.8 Basic Zener barrier circuit

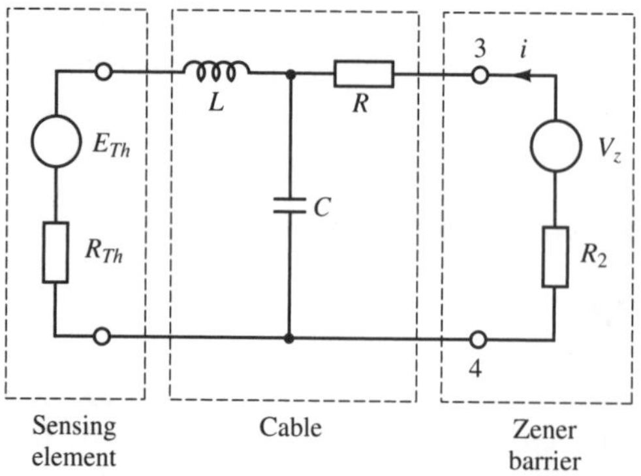

Fig. 13.9 Equivalent circuit of Zener barrier fault conditions

to-end' resistance of $R_F + R_1 + R_2$ together with the inherent capacitance of the Zener diodes. Provided that the usual conditions for maximum voltage transfer in a Thévenin signal circuit (Section 5.1.1) or maximum current transfer in a Norton signal circuit (Section 5.1.3) are obeyed, the barrier will introduce no error into the measurement system.

13.2.2 Energy storage calculation for fault conditions

The cable which connects terminals 3 and 4 of the Zener barrier to the sensor, transducer or transmitter in the hazardous area will have inductance and capacitance as well as resistance. Figure 13.9 shows a simplified approximate equivalent circuit for the cable using single lumped values L, C and R; in practice inductance, capacitance and resistance are distributed along the entire length of the cable. The diagram also shows the sensing element represented by a Thévenin equivalent circuit E_{Th}, R_{Th} (a Norton equivalent circuit i_N, R_N is also possible) and the Zener barrier under fault conditions as a Thévenin equivalent circuit V_z, R_2. The energy stored in the cable capacitance C therefore has a maximum value of $\frac{1}{2}CV_z^2$. The energy stored in the cable inductance L is $\frac{1}{2}Li^2$, where $i = V_z/R_2$, the maximum current that can flow in the circuit. The maximum total stored energy in the circuit is therefore:

Maximum stored energy in circuit

$$E_{MAX} = \tfrac{1}{2}CV_z^2 + \tfrac{1}{2}L(V_z/R_2)^2 \qquad [13.15]$$

Since E_{MAX} is the maximum energy available to create a spark it must be less than the minimum ignition energy (MIE) for the gas–air mixture.

Consider the example of the chromel–alumel thermocouple used in a hydrogen–air atmosphere with a MTL 160 barrier.[3] The barrier has $V_z = 10$ V, $R_2 = 50\,\Omega$. The maximum permissible cable capacitance and inductance for safe operation with hydrogen (group IIC) are 4 μF and 1.2 mH. The actual cable used is 100 m of standard chromel–alumel extension lead with $R = 28\,\Omega$, $C = 1.85$ nF and $L = 60\,\mu$H. This gives

$$\begin{aligned} E_{MAX} &= \tfrac{1}{2} \times 1.85 \times 10^{-9} \times 10^2 + \tfrac{1}{2} \times 60 \times 10^{-6}\,(10/50)^2\ \text{J} \\ &= 1.3\,\mu\text{J} \end{aligned}$$

This is well below 19 μJ, the minimum ignition energy for hydrogen–air, so the system is safe to use in this application.

Conclusion

The chapter begins by explaining that the atmosphere surrounding certain plants and processes may be potentially explosive, a common example being hydrocarbon–air mixtures. It then goes on to define three types of **hazard classification; area, gas** and **temperature**. Equipment can then be classified according to the hazardous environment in which it can be safely used. The chapter then describes two types of measurement systems which are **intrinsically safe**, i.e. cannot ignite the explosive mixture. The first are **pneumatic measurement systems** where there is no possibility of a spark. The second are **intrinsically safe electronic systems** where **Zener barriers** are used to limit the electrical energy available under fault conditions to a safe value.

References

13.1 ROHMANN C P and GROGAN E C 1956 'On the dynamics of pneumatic transmission lines', *Transactions of the A.S.M.E.* Cleveland, Ohio, p. 866.

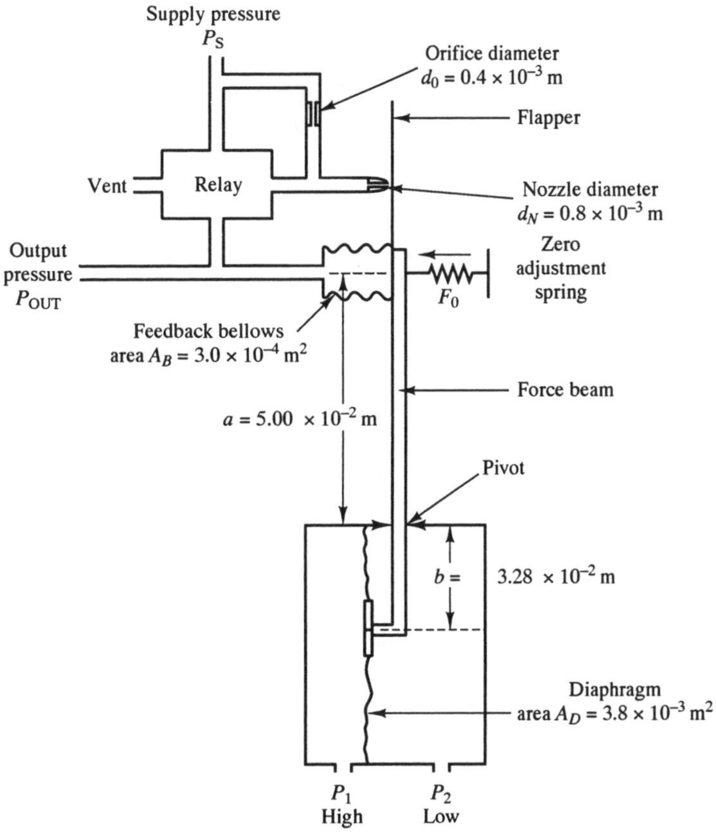

Fig. Prob. 1

13.2 SCHINSKEY F G 1979 *Process Control Systems* 2nd edition, pp. 38–9. McGraw-Hill, New York.

13.3 Measurement Technology Ltd. 1991 *Making the World Intrinsically Safe*. Technical Publication.

Problems

13.1 Figure Prob. 1 (previous page) shows a pneumatic, torque/balance transmitter to be used for measuring differential pressures in the range 0 to 10^4 Pa.

(a) Draw a labelled block diagram of the transmitter.
(b) Using the data given, calculate the range of the transmitter output signal:
  (i) when the supply pressure P_s is 1.40 bar,
  (ii) when the supply pressure P_s is 1.75 bar.

Comment briefly on the practical significance of these results.

Data

Zero spring force	$F_0 = 6.24\ \text{N}$
Flapper/nozzle sensitivity	$K = \dfrac{2.0 d_N P_S}{d_0^2}\ \text{Pa m}^{-1}$
Displacement of flapper for unit torque	$K_B = 6.0 \times 10^{-5}\ \text{m(Nm)}^{-1}$
Relay pressure gain	$K_R = 20.0\ \text{Pa}^{-1}$

13.2 A Zener barrier for use with a platinium resistance sensor has $V_z = 5$ V, $R_2 = 10\ \Omega$. The two-conductor cable connecting the sensor to the barrier has a capacitance of 0.05 μF and an inductance of 0.01 mH. Is this system safe to use with hydrogen–air atmosphere with a minimum ignition energy of 10 μJ?

14

Heat transfer effects in measurement systems

14.1 Introduction

The temperature of a sensing element at any instant of time, depends on the rate of transfer of heat both to and from the sensor. Heat transfer takes place as a result of one or more of three possible types of mechanisms – **conduction**, **convection** and **radiation**. Conduction is the main heat transfer mechanism inside solids. A solid may be regarded as a chain of interconnected atoms, each vibrating about a fixed position. An increase in temperature at one end of a solid bar causes an increase in the vibrational energy and amplitude of the atoms at that end of the chain. This energy increase is transmitted from one atom to the next along the chain, so that ultimately the temperature increase is transmitted to the other end of the bar. This chapter is concerned with **heat transfer** between a sensing element and the fluid in which it is situated. In this situation the main heat transfer mechanism is convection. Here heat is transferred to and from the sensor by the random, highly disordered, motion of molecules of fluid past the sensor. This random motion and corresponding heat transfer occur even when the average velocity of the bulk fluid past the sensor is zero. This is known as **natural convection**. If the bulk fluid is made to move so that the average velocity past the sensor is no longer zero, then there is a corresponding increase in rate of heat transfer. This is referred to as **forced convection**. Heat transfer by means of radiation involves the transmission of electromagnetic waves and will be discussed in Chapter 15.

From Newton's law of cooling the convective heat flow W watts between a sensor at T°C and fluid at T_F°C is given by:

$$W = U A(T - T_F) \qquad [14.1]$$

where U W m^{-2} °C^{-1} is the convection heat transfer coefficient and A m^2 is the heat transfer area. Heat transfer coefficients are calculated using the correlation:

$$\text{Nu} = \phi(\text{Re}, \text{Pr}) \qquad [14.2]$$

between the three dimensionless numbers:

$$\text{Nusselt} \quad \text{Nu} = \frac{Ud}{k}$$

$$\text{Reynolds Re} = \frac{vd\rho}{\eta} \qquad [14.3]$$

$$\text{Prandtl} \quad \text{Pr} = \frac{c\eta}{k}$$

where:

d m = sensor diameter
v m s^{-1} = fluid velocity
ρ kg m^{-3} = fluid density
η Pa s = fluid viscosity
c J °C^{-1} = fluid heat capacity
k W m^{-1} °C^{-1} = fluid thermal conductivity

The function ϕ is determined experimentally; its form depends on the shape of the sensor, the type of convection and the direction of fluid flow in relation to the sensor. For example a correlation for forced convection cross-flow over a cylinder is:

$$\text{Nu} = 0.48(\text{Re})^{0.5}(\text{Pr})^{0.3} \qquad [14.4]$$

From [14.3] and [14.4] we have:

$$U = 0.48\,\frac{k^{0.7}\rho^{0.5}c^{0.3}v^{0.5}}{d^{0.5}\eta^{0.2}} \qquad [14.5]$$

For two-dimensional, natural and forced convection from a cylinder, the approximate correlation is[1,2]

$$\text{Nu} = 0.24 + 0.56\text{Re}^{0.5} \qquad [14.6]$$

giving

$$U = \frac{0.24k}{d} + 0.56k\left(\frac{\rho v}{d\eta}\right)^{0.5} \qquad [14.7]$$

From [14.5] and [14.7] we see that the convection heat transfer coefficient for a given sensor depends critically on the physical properties and velocity of the surrounding fluid.

The following three sections of this chapter explain how convective heat transfer has three important applications in measurement systems. These are:

(a) in understanding and calculating the dynamic characteristics of thermal sensors;
(b) in hot wire and film systems for fluid velocity measurements;
(c) in the katharometer for gas thermal conductivity and composition measurement.

14.2 Dynamic characteristics of thermal sensors

14.2.1 Bare temperature sensor

The transfer function for a bare (unenclosed) temperature sensor has already been found to be (Section 4.1):

Transfer function for bare temperature sensor

$$\frac{\Delta\bar{T}_S}{\Delta\bar{T}_F}(s) = \frac{1}{1 + \tau_S s} \qquad [4.10a]$$

where ΔT_S, ΔT_F °C = deviations in sensor and fluid temperatures from equilibrium

τ_S = sensor time constant = MC/UA

M kg = sensor mass

C J kg^{-1} °C^{-1} = sensor specific heat

U W m^{-2} °C^{-1} = convection heat transfer coefficient between fluid and sensor

A m^2 = sensor heat transfer area

Since τ_S depends on U, the time constant of a given sensor will depend critically on the nature and velocity of the fluid surrounding the sensor (see Table 14.1).

14.2.2 Temperature sensor enclosed in a thermowell or sheath

A temperature sensor such as a thermocouple or resistance thermometer is usually enclosed in a sheath or thermowell to give chemical and mechanical protection. Figure 14.1(a) shows a typical thermocouple installation and 14.1(b) is a simplifed model where sensor and thermowell are represented by single 'lumped' masses M_S and M_W respectively.

Ignoring the thermal capacity of the space between sensor and well, the heat balance equations are:

$$\text{Sensor:}\quad M_S C_S \frac{dT_S}{dt} = U_{SW} A_S (T_W - T_S) \qquad [14.8]$$

$$\text{Well:}\quad M_W C_W \frac{dT_W}{dt} = -U_{SW} A_S (T_W - T_S) + U_{WF} A_W (T_F - T_W) \qquad [14.9]$$

which become

$$\tau_1 \frac{dT_S}{dt} = (T_W - T_S) \qquad [14.10]$$

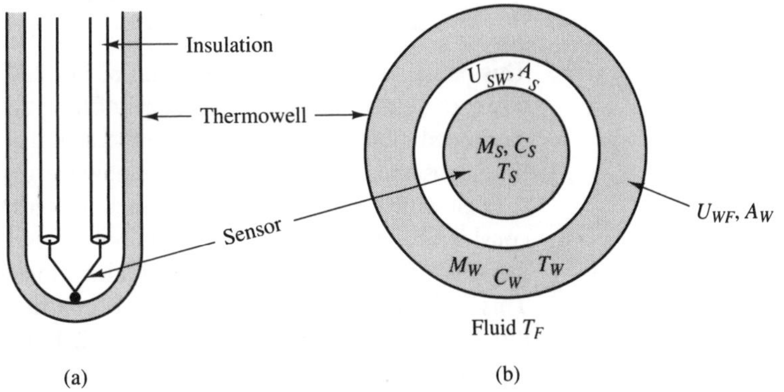

Fig. 14.1 Sensor in thermowell
(a) Typical thermowell installation
(b) Simplified model

$$\tau_2 \frac{dT_W}{dt} = -\delta(T_W - T_S) + (T_F - T_W) \qquad [14.11]$$

where $$\tau_1 = \frac{M_S C_S}{U_{SW} A_S}, \quad \tau_2 = \frac{M_W C_W}{U_{WF} A_W}, \quad \delta = \frac{U_{SW} A_S}{U_{WF} A_W} \qquad [14.12]$$

and

A_S, A_W = sensor/thermowell heat transfer areas
C_S, C_W = sensor/thermowell specific heats
U_{SW} = sensor/thermowell heat transfer coefficient
U_{WF} = fluid/thermowell heat transfer coefficient

Defining ΔT_F, ΔT_W, ΔT_S to be deviations from initial steady conditions, the Laplace transforms of [14.10], [14.11] are:

$$[1 + \tau_1 s]\Delta \bar{T}_S = \Delta \bar{T}_W \qquad [14.13]$$

$$[(1 + \delta) + \tau_2 s]\Delta \bar{T}_W = \Delta \bar{T}_F + \delta \Delta \bar{T}_S \qquad [14.14]$$

Eliminating $\bar{T}_W$ between these equations gives the overall transfer function:

Transfer function for temperature sensor in sheath

$$\frac{\Delta \bar{T}_S}{\Delta \bar{T}_F}(s) = \frac{1}{\tau_1 \tau_2 s^2 + (\tau_1 + \tau_2 + \delta \tau_1)s + 1} \qquad [14.15]$$

This is a second-order model and is a good representation of an incorrect installation where the tip of the sensing element does not touch the sheath. The effective heat transfer coefficient U_{SW} between sensor and well is now very small; this means that τ_1 is very large and the system response extremely sluggish. A correct normal installation with the sensor tip touching the sheath has a far higher U_{SW} and lower τ_1. In the limit of perfect heat transfer between sensor and well, both elements are at the same temperature T_S and the heat balance equation is now:

$$(M_S C_S + M_W C_W) \frac{dT_S}{dt} = U_{FW} A_W (T_F - T_S) \qquad [14.16]$$

giving the first-order transfer function:

$$\frac{\Delta \bar{T}_S}{\Delta \bar{T}_F}(s) = \frac{1}{1 + \tau s}, \quad \tau = \frac{M_S C_S + M_W C_W}{U_{FW} A_W} \qquad [14.17]$$

This model is a good representation of installations where special steps have been taken to improve heat transfer, namely filling the sheath with oil or mercury or using a crimped metal sleeve to increase the heat transfer area. Table 14.1 gives typical time constants for elements in thermowells under different fluid conditions.

Even with a good installation, the time constant for a sensor, in a thermowell is considerably longer than that of the sensor itself. If a fast response is required and the sensor must be protected, then a mineral insulated thermocouple would be used. This is shown in Fig. 8.11 and consists of a fine wire thermocouple inside a narrow thin-walled tube; the tube is filled with mineral material which is a good heat conductor but an electrical insulator. This device is described by the transfer function

Table 14.1

Fluid conditions	Typical U_{FW} W m^{-2} °C^{-1}	Typical τ for sensor in thermowell (min)	Typical τ for mineral insulated thermocouple (sec)
Fast liquid	625	1.0	0.7
Slow liquid	250	1.5	1.5
Fast gas	125	2	10
Med. gas	63	4	20
Slow gas	25	8	30

of eqn [14.17]; in this case M_S and M_W are small, giving time constants 100 times shorter than thermowell installations (see Table 14.1).

14.2.3 Fluid velocity sensor with self-heating current

If a current i is passed through a resistive element, like a fine metal wire or semiconductor film, then the element is heated to a temperature T which is greater than T_F, the temperature of the surrounding fluid. The element temperature T and resistance R_T depend on the balance between electrical power i^2R_T and the rate of overall convective heat transfer between element and fluid. Since convective heat transfer depends on the velocity v of the fluid, the element is used as a fluid velocity sensor (see following section). The heat balance equation is:

$$i^2R_T - U(v)A(T - T_F) = MC\frac{dT}{dt} \qquad [14.18]$$

where $U(v)$ is the convective heat transfer coefficient between sensor and fluid. If i_0, R_{T_0}, T_0, v_0 represent steady equilibrium conditions then:

$$i_0^2R_{T_0} - U(v_0)A(T_0 - T_F) = 0 \qquad [14.19]$$

Defining Δi, ΔR_T, Δv, ΔT to be small deviations from the above equilibrium values, we have:

$$\begin{aligned} i &= i_0 + \Delta i, & T &= T_0 + \Delta T \\ R_T &= R_{T_0} + \Delta R_T, & U(v) &= U(v_0) + \sigma\Delta v \end{aligned} \qquad [14.20]$$

In [14.20] $\sigma = (\partial U/\partial v)_{v_0}$; i.e. the rate of change of U with respect to v, evaluated at equilibrium velocity v_0. From [14.18] and [14.20] we have:

$$(i_0 + \Delta i)^2(R_{T_0} + \Delta R_T) - (U(v_0) + \sigma\Delta v)A(T_0 + \Delta T - T_F)$$
$$= MC\frac{d}{dt}(T_0 + \Delta T) \qquad [14.21]$$

Neglecting all terms involving the product of two small quantities gives:

$$(i_0^2 + 2i_0\Delta i)R_{T_0} + i_0^2\Delta R_T - U(v_0)A(T_0 - T_F) - U(v_0)A\Delta T$$
$$- \sigma A(T_0 - T_F)\Delta v = MC\frac{d\Delta T}{dt} \qquad [14.22]$$

Subtracting [14.19] from [14.22] gives:

$$2i_0R_{T_0}\Delta i + i_0^2\Delta R_T - U(v_0)A\Delta T - \sigma A(T_0 - T_F)\Delta v = MC\frac{\mathrm{d}\Delta T}{\mathrm{d}t} \qquad [14.23]$$

ΔT can be eliminated by setting $K_T = \Delta R_T/\Delta T$ i.e. $\Delta T = (1/K_T)\Delta R_T$, where K_T is the slope of the element resistance–temperature characteristics; thus

$$\left[\frac{U(v_0)A}{K_T} - i_0^2\right]\Delta R_T + \frac{MC}{K_T}\frac{\mathrm{d}\Delta R_T}{\mathrm{dt}} = 2i_0R_{T_0}\Delta i - \sigma A(T_0 - T_F)\Delta v \qquad [14.24]$$

which reduces to:

$$\Delta R_T + \tau_v\frac{\mathrm{d}\Delta R_T}{\mathrm{d}t} = K_I\Delta i - K_v\Delta v \qquad [14.25]$$

where

$$\tau_v = \frac{MC}{[U(v_0)A - i_0^2K_T]}, \quad K_I = \frac{2K_Ti_0R_{T_0}}{[U(v_0)A - i_0^2K_T]},$$
$$K_v = \frac{K_T\sigma A(T_0 - T_F)}{[U(v_0)A - i_0^2K_T]} \qquad [14.26]$$

Taking the Laplace transform of [14.26] gives:

$$(1 + \tau_vs)\Delta\bar{R}_T = K_I\Delta\bar{i} - K_v\Delta\bar{v}$$

i.e.

Transfer function for fluid velocity sensor

$$\Delta\bar{R}_T = \frac{K_I}{(1 + \tau_vs)}\Delta\bar{i} - \frac{K_v}{(1 + \tau_vs)}\Delta\bar{v} \qquad [14.27]$$

The corresponding block diagram for the sensing element is shown in Fig. 14.2.

From eqn [14.27] and Fig. 14.2 we note that the resistance of the element can be altered either by a change in current Δi or a change in fluid velocity Δv; the time constant for both processes is τ_v. For a metal resistance element we have $R_T \approx R_0(1 + \alpha T)$, giving $K_T = \mathrm{d}R_T/\mathrm{d}T = R_0\alpha$, where R_0 = resistance of the element at 0 °C and α °C^{-1} is the temperature coefficient of resistance. For a semiconductor resistance element (thermistor) we have $R_\theta = Ke^{\beta/\theta}$, i.e.

$$K_\theta = \frac{\mathrm{d}R_\theta}{\mathrm{d}\theta} = \frac{-\beta}{\theta^2}Ke^{\beta/\theta} = \frac{-\beta}{\theta^2}R_\theta$$

so that here K_θ is negative, and depends on thermistor temperature θ Kelvin as well as the characteristic constant β. We can now calculate τ_v in a typical situation; e.g.

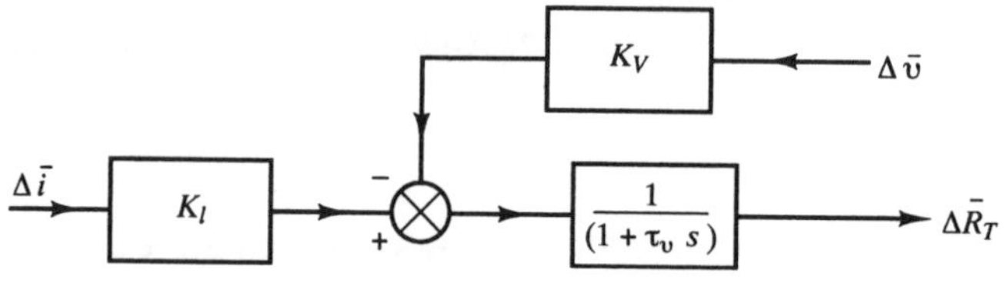

Fig. 14.2 Block diagram for thermal velocity sensor

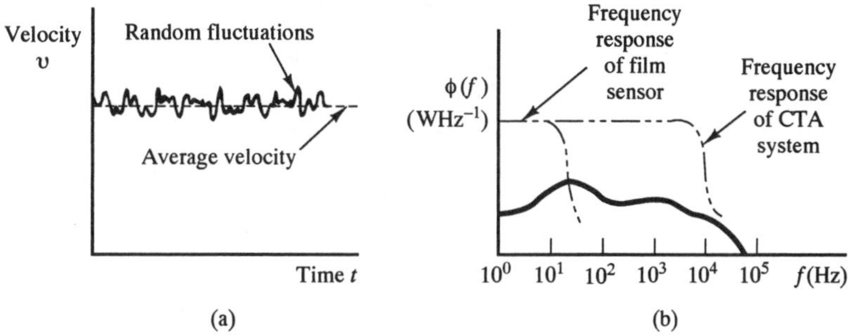

Fig. 14.3 Flow turbulence and sensor frequency response

for a thin film of semiconductor material 1 cm square, deposited on the surface of a probe inserted in a slow-moving gas stream. Typically we have:

$$\begin{aligned} \theta_0 &= 383\,\text{K}\ (110\,°\text{C}), & R_{\theta_0} &= 150\,\Omega, \\ MC &= 2.5 \times 10^{-5}\,\text{J}\,°\text{C}^{-1}, & A &= 10^{-4}\,\text{m}^2, \\ U(v_0) &= 25\,\text{W}\,\text{m}^{-2}\,°\text{C}^{-1}, & i_0 &= 25 \times 10^{-3}\,\text{A}, \\ K_\theta &= -4.0\,\Omega\,°\text{C}^{-1} \end{aligned}$$

giving

$$\tau_v = \frac{MC}{U(v_0)A - i_0^2 K_\theta} = \frac{2.5 \times 10^{-5}}{50 \times 10^{-4}} = 5.0\,\text{ms}$$

From Section 4.4 we can see that the bandwidth of the probe is between 0 and $1/(2\pi \times 5 \times 10^{-3})$ i.e. between 0 and 32 Hz. Lower time constants can be obtained with probes of a lower mass/area ratio in fluids with a higher heat transfer coefficient. It is not usually possible to reduce τ_v much below 1 ms, i.e. to increase the bandwidth much above 160 Hz.

There are, however, several potential applications of thermal velocity sensors which require a far higher bandwidth. In the testing of aircraft in wind tunnels it is important to measure the power spectral density $\phi(f)$ (Section 6.2) of the turbulence associated with air flow over the aircraft surfaces. Turbulence refers to the small random fluctuations in velocity at a point in a fluid occurring at high Reynolds numbers (Fig. 14.3(a)). The corresponding power spectral density can extend up to around 50 kHz (Fig. 14.3(b)).

Another potential application is the detection of vortices in the vortex-shedding flowmeter (Section 12.2), where vortex frequencies up to around 1 kHz are possible. This difficulty is solved by incorporating the sensor into a constant temperature anemometer system. We will see in the following Section that the CTA system time constant is considerably less than τ_v. Note that in the limit $i_0 \to 0$ i.e. negligible self heating current, $\tau_v \to MC/UA$ i.e. the time constant of a velocity sensor tends to that of a bare temperature sensor.

14.3 Constant-temperature anemometer system for fluid velocity measurements

14.3.1 Steady-state characteristics

From the previous section the steady-state equilibrium equation for a fluid velocity sensor with self-heating current is:

$$i^2R_{T_0} = U(v)A(T_0 - T_F) \qquad [14.19a]$$

In a constant-temperature anemometer system the resistance R_{T_0} and temperature T_0 of the sensor are maintained at constant values (within limits). From [14.19a] we see that if the fluid velocity v increases, causing an increase in $U(v)$, then the system must increase the current i through the sensor in order to restore equilibrium. Since sensor resistance R_{T_0} remains constant, the voltage drop iR_{T_0} across the element increases, thus giving a voltage signal dependent on fluid velocity v.

The correlation of eqn [14.6] for two-dimensional, convective heat transfer from a narrow cylinder in a incompressible fluid is the one most appropriate to fluid velocity sensors.[3] This is:

$$\mathrm{Nu} = 0.24 + 0.56\,\mathrm{Re}^{0.5} \qquad [14.6]$$

giving

$$U = \frac{0.24k}{k} + 0.56k\left(\frac{\rho v}{\mathrm{d}\eta}\right)^{0.5} \qquad [14.7]$$

i.e.

$$U = a + b\sqrt{v} \qquad [14.28]$$

where

$$a = \frac{0.24k}{d}, \quad b = 0.56k\left(\frac{\rho}{\mathrm{d}\eta}\right)^{0.5} \qquad [14.29]$$

We see that since a and b depend on the sensor dimensions d and the fluid properties k, ρ, η, they are constants only for a *given sensor* in a *given fluid*. This means that if a sensor is calibrated in a certain fluid, the calibration results will not apply if the sensor is placed in a different fluid. Figure 14.4 shows a typical hot wire velocity sensor.

Figure 14.5(a) is a schematic diagram of a constant-temperature anemometer system. This is a self-balancing bridge which maintains the resistance R_T of the sensor at a constant value R. An increase in fluid velocity v causes T and R_T to fall in the short term, thus unbalancing the bridge; this causes the amplifier output current and current through the sensor to increase thereby restoring T and R_T to their required values. Since $R_T = R$ and $R_T = R_0(1 + \alpha T)$ for a metallic sensor, then the constant temperature T of the sensor is:

$$T = \frac{1}{\alpha}\left(\frac{R}{R_0} - 1\right) \qquad [14.30]$$

From [14.19], [14.28] and [14.30] we have:

$$i^2R = A(a + b\sqrt{v})\left[\frac{1}{\alpha}\left(\frac{R}{R_0} - 1\right) - T_F\right] \qquad [14.31]$$

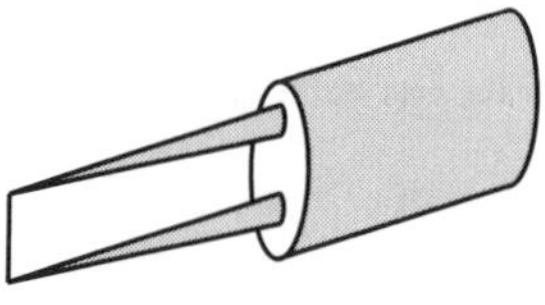

Fig. 14.4 Hot wire velocity sensor

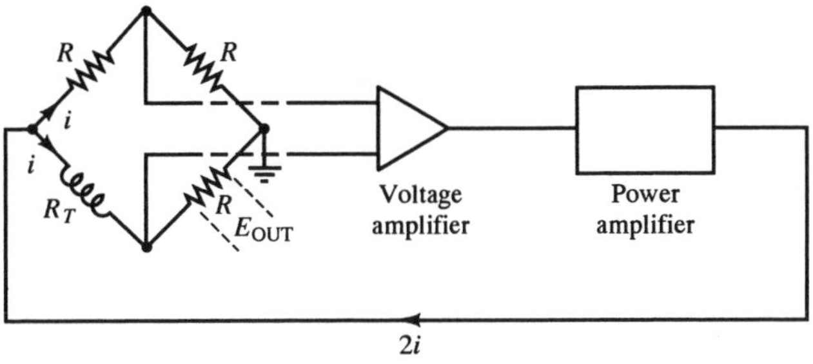

(a)

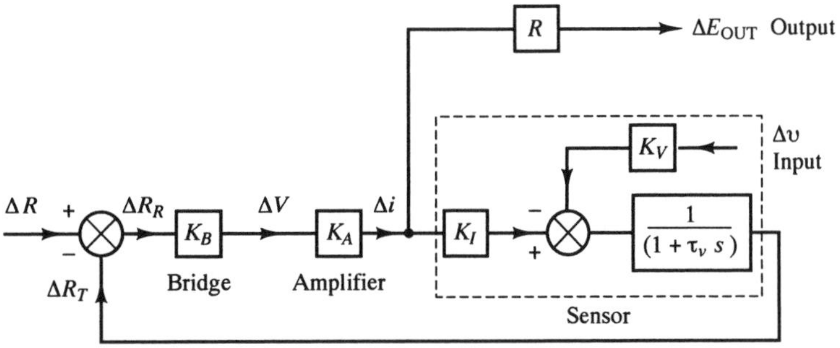

(b)

Fig. 14.5 Constant temperature anemometer system
(a) Schematic diagram
(b) Block diagram

and since $E_{OUT} = iR$:

$$E^2_{OUT} = AR(a + b\sqrt{v})\left[\frac{1}{\alpha}\left(\frac{R}{R_0} - 1\right) - T_F\right] \qquad [14.32]$$

giving

Steady state relationship between output voltage and velocity for constant temperature anemometer

$$E_{OUT} = (E_0^2 + \gamma\sqrt{v})^{1/2} \qquad [14.33]$$

where

$$E_0^2 = ARa\left[\frac{1}{\alpha}\left(\frac{R}{R_0} - 1\right) - T_F\right] \quad \text{and} \quad \gamma = ARb\left[\frac{1}{\alpha}\left(\frac{R}{R_0} - 1\right) - T_F\right] \qquad [14.34]$$

Figure 14.6 shows a typical relationship between E_{OUT} and v; these characteristics were found experimentally for a tungsten filament CTA system in air. We note that when $v = 0$, $E_{OUT} = E_0 \approx 2.0$ V (natural convection) and that the relationship is highly non-linear. The slope of this relationship, that is the sensitivity of the system, is greatest at the lowest velocities. A CTA system can therefore be used for measuring low fluid velocities of the order of a few metres/second; this is in contrast to the pitot tube (Section 12.1) which has a very low sensitivity at low fluid velocities.

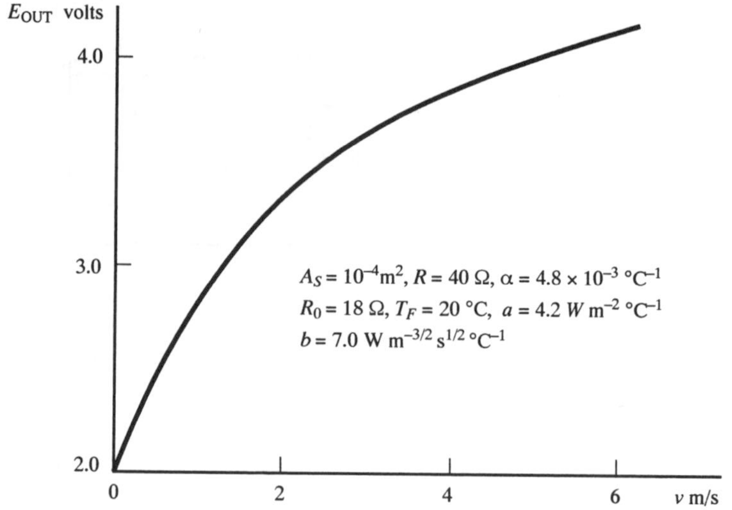

Fig. 14.6 Steady-state CTA characteristics for tungsten filament in air

A diode function generator can be used to compensate for the non-linear characteristics, but a more satisfactory method is to input E_{OUT} to a microcomputer. Experimentally determined values of E_0 and γ are stored in the computer, which then calculates v using $v = [(E_{OUT}^2 - E_0^2)/\gamma]^2$. Since E_0 and γ are dependent on fluid properties, the system must be recalibrated if the fluid is changed. From eqns [14.34] we see that if the fluid temperature T_F changes, then E_0, γ and the system calibration changes. One method of compensating for this is to incorporate a second unheated element at temperature T_F into the self-balancing bridge circuit. At higher velocities ($v \gtrsim 10$ m s) the CTA system can be calibrated using a pitot tube; at lower velocities one possibility is to use the cross-correlation method (Section 12.4) using two thermal velocity sensors.

14.3.2 Dynamic characteristics

We now calculate the transfer function of the constant-temperature anemometer system to see if the frequency response is sufficient to detect rapid velocity fluctuations due to turbulence and vortex shedding. A block diagram of the system is shown in Fig. 14.5(b); this incorporates the block diagram of a thermal velocity sensor derived in Section 14.2.3. The system equations are:

Sensor $$\Delta R_T = \frac{K_I}{(1 + \tau_v s)} \Delta i - \frac{K_v}{(1 + \tau_v s)} \Delta v \qquad [14.27]$$

Bridge $$\Delta V = K_B \Delta R_R \qquad [14.35]$$

Amplifier $$\Delta i = K_A \Delta V \qquad [14.36]$$

Output voltage $$\Delta E_{OUT} = R\Delta i \qquad [14.37]$$

Resultant change in bridge resistance $$\Delta R_R = \Delta R - \Delta R_T$$
$$= -\Delta R_T \text{ (since } \Delta R = 0) \qquad [14.38]$$

From [14.35]–[14.38]:

$$\Delta R_T = \frac{-1}{RK_AK_B}\Delta E_{\text{OUT}} \quad \text{and} \quad \Delta i = \frac{\Delta E_{\text{OUT}}}{R} \qquad [14.39]$$

Substituting [14.39] in [14.27] gives:

$$\frac{-1}{RK_AK_B}\Delta E_{\text{OUT}} = \frac{1}{(1+\tau_v s)}\left[\frac{K_I}{R}\Delta E_{\text{OUT}} - K_v\Delta v\right]$$

Rearranging we have:

$$[(1 + K_IK_AK_B) + \tau_v s]\Delta E_{\text{OUT}} = K_\theta K_AK_BR\Delta v$$

giving

Transfer function for CTA system

$$\frac{\Delta\bar{E}_{\text{OUT}}}{\Delta\bar{v}}(s) = \frac{K_{CTA}}{1+\tau_{CTA}s} \qquad [14.40]$$

where

$$K_{CTA} = \frac{K_vK_AK_BR}{1+K_IK_AK_B} \quad \text{and} \quad \tau_{CTA} = \frac{\tau_v}{1+K_IK_AK_B} \qquad [14.41]$$

We can now calculate τ_{CTA} for a typical system incorporating the semiconductor element discussed in Section 14.2.3. Using the data previously given, we have:

$$\tau_v = 5.0\,\text{ms}, \quad K_I = \frac{2K_\theta i_0 R_{\theta_0}}{[U(V_0)A - i_0^2K_\theta]} = -6 \times 10^3\,\Omega\,\text{A}^{-1}$$

For a 150 Ω bridge with supply voltage $V_S = 7.5$ V,

$$K_B = \frac{\Delta V}{\Delta R_R} = \frac{1}{4}\frac{V_S}{R} = \frac{1}{4} \times \frac{7.5}{150} = 1.25 \times 10^{-2}\,\text{V}\,\Omega^{-1}$$

In the typical case of a voltage amplifier of gain 10^3 followed by an emitter-follower power amplification stage, $K_A \approx -4\,\text{A}\,\text{V}^{-1}$. From [14.41] we have:

$$\tau_{CTA} = \frac{5.0}{1 + 6 \times 10^3 \times 1.25 \times 10^{-2} \times 4}\,\text{ms} = \frac{5.0}{301}\,\text{ms} = 17\,\mu\text{s}$$

Thus the bandwidth of the system is between 0 and $1/(2\pi \times 17 \times 10^{-6})$ Hz, i.e. between 0 and 10 kHz which is easily sufficient for vortex detection and is adequate for most turbulence measurement applications (see Fig. 14.3(b)).

14.4 Katharometer systems for gas thermal conductivity and composition measurement

The convective heat transfer coefficient U between a sensor and a moving gas depends on the thermal conductivity k of the gas as well as the average gas velocity (eqn [14.7]). In a katharometer the velocity of the gas past the element is maintained either constant or small (preferably both), so that U depends mainly on the thermal

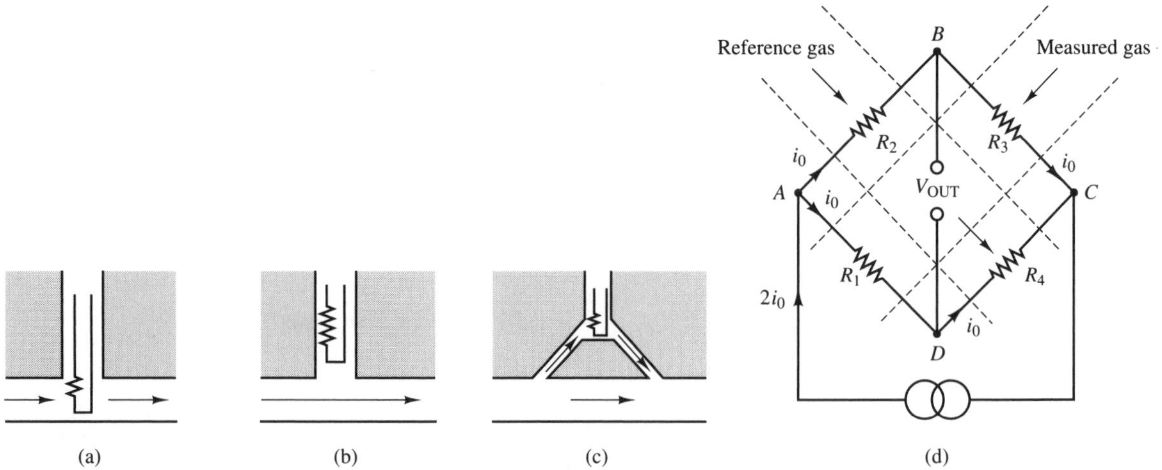

Fig. 14.7 Katharometer elements and deflection bridge

conductivity of the gas. A constant self-heating current i_0 is passed through the element. Thus from the steady-state heat balance equation,

$$i_0^2 R_T = U(k)A(\mathrm{T} - T_F) \qquad [14.19\mathrm{b}]$$

we see that if gas thermal conductivity k increases, causing $U(k)$ to increase, then the temperature T of the element falls.

Typical katharometer element configurations are shown in Figs 14.7(a)–(c) where the element is either a metal filament or a thermistor. A system normally consists of four such elements, each element being located in a separate cavity inside a metal block. The gas to be measured passes over one pair of elements and a reference gas passes over the other pair. In Fig. 14.7(a) all of the gas flow passes over the element, whereas in Fig. 14.7(b) there is only a small gas circulation around the element. Figure 14.7(d) shows the arrangement of the four elements in a deflection bridge with a constant current supply; note that measured and reference gas elements are arranged in opposite arms of the bridge.

In order to find the relationship between the resistance R_T of an element and gas thermal conductivity k, we need to eliminate T from eqn [14.19b]. For a metal filament we have:

$$R_T = R_0(1 + \alpha T) \quad \text{and} \quad R_{T_F} = R_0(1 + \alpha T_F)$$

where R_{T_F} is the resistance of the filament at the temperature T_F of the fluid ($T_F < T$). Thus:

$$\frac{R_T}{R_{T_F}} = \frac{1 + \alpha T}{1 + \alpha T_F} \approx 1 + \alpha(T - T_F) \qquad [14.42]$$

assuming that terms involving $(\alpha T_F)^2$ etc. are negligible. Rearranging gives:

$$(T - T_F) = \frac{1}{\alpha}\left(\frac{R_T}{R_{T_F}} - 1\right) \qquad [14.43]$$

Substituting [14.43] in [14.19], we have:

$$i_0^2 R_T = \frac{UA}{\alpha}\left(\frac{R_T}{R_{T_F}} - 1\right) \quad [14.44]$$

i.e.

$$BR_T = U(R_T - R_{T_F}) \quad [14.45]$$

where $B = i_0^2 \alpha R_{T_F}/A$ is a constant if gas temperature T_F is contant. Rearranging gives:

Relationship between filament resistance and fluid heat transfer coefficient for constant current

$$R_T = \frac{R_{T_F}}{\left(1 - \dfrac{B}{U}\right)} \quad [14.46]$$

In the bridge circuit of Fig. 14.7(d) we have:

$$R_1 = R_3 = \frac{R_{T_F}}{(1 - B/U_M)} \quad \text{for measured gas} \quad [14.47]$$

and

$$R_2 = R_4 = \frac{R_{T_F}}{(1 - B/U_R)} \quad \text{for reference gas.}$$

The bridge output voltage (potential difference between B and D) is thus:

$$V_{\text{OUT}} = i_0(R_1 - R_2) = i_0 R_{T_F}\left\{\frac{1}{(1 - B/U_M)} - \frac{1}{(1 - B/U_R)}\right\}$$

$$= i_0 R_{T_F} B \frac{(1/U_M - 1/U_R)}{(1 - B/U_M)(1 - B/U_R)} \quad [14.48]$$

For a typical system

$$\alpha = 4 \times 10^{-3}\,{}^\circ\text{C}^{-1}, \quad 2i_0 = 100\,\text{mA}, \quad R_{T_F} = 10\,\Omega$$
$$A = 10^{-4}\,\text{m}^2, \quad U \approx 25\,\text{W m}^{-2}\,{}^\circ\text{C}^{-1}$$

giving

$$B = \frac{i_0^2 \alpha R_{T_F}}{A} = 1.0 \quad \text{and} \quad \frac{B}{U} \approx \frac{1}{25}$$

Since B/U is small compared with 1, eqn [14.48] can be approximated by

Output voltage from katharometer system

$$V_{\text{OUT}} = \frac{i_0^3 R_{T_F}^2 \alpha}{A}\left(\frac{1}{U_M} - \frac{1}{U_R}\right) \quad [14.49]$$

From [14.49] we see that the output voltage is proportional to the cube of the sensor current and also depends non-linearly on the heat transfer coefficients U_M, U_R. Using [14.7] we have $U_M = k_M f(v)$ and $U_R = k_R f(v)$, where

$$f(v) = \frac{0.24}{d} + 0.56\left(\frac{\rho v}{d\eta}\right)^{0.5}$$

and k_M, k_R are the termal conductivities of the measured and reference gases respectively. If the flow rates of measured and reference gases are equal, then the velocity v is the same for both gases. The element of Fig. 14.7(a) is situated directly in the gas stream. This means that there is substantial forced convection i.e. the $0.56\ (\rho v/d\eta)^{0.5}$ term is large. Thus V_{OUT} depends critically on the gas flow rates: elements of this type need tight control of gas flow but respond rapidly to sudden changes in thermal conductivity. The element of Fig. 14.7(b) is situated out of the main gas stream where velocity and forced convection are much smaller. In the limit that v is negligible we have $f(v) = 0.24/d$ and:

Output voltage for katharometer with negligible gas velocity

$$V_{\mathrm{OUT}} = D\left(\frac{1}{k_M} - \frac{1}{k_R}\right) \qquad [14.50]$$

where $D = i_0^3 R_{\mathrm{T}_F}^2 \alpha d/(0.24A)$. This type of element is not sensitive to changes in gas flow rate, i.e. tight control of gas flow rates is not necessary, but responds slowly to sudden changes in thermal conductivity. Most elements in current use are of the type shown in Fig. 14.7(c); here a fraction of the gas flow is passed over the filament. These have an adequate speed of response and a low sensitivity to flow changes.

The main use of the katharometer is as a detector in a gas chromatograph system for measuring the composition of gas mixtures (Chapter 17). Here an inert gas such as nitrogen, helium or argon, referred to as the carrier, sweeps a sample of gas to be analysed through a packed column. In this application the katharometer reference gas is pure carrier: the katharometer measured gas is the gas leaving the column. Initially, while the components of the sample are inside the column, the gas leaving the column is pure carrier. Pure carrier gas flows over all four elements; therefore each arm of the bridge has the same resistance and the bridge output voltage is zero. The components then emerge from the column one at a time; as each component emerges the katharometer measured gas is a mixture of carrier gas and that component. This causes corresponding changes in thermal conductivity k_M and resistances R_1, R_3 so that the bridge output voltage is no longer zero. The thermal conductivities of reference and measured gases are now given by:

$$k_R = k_C, \quad k_M = x_i k_i + (1 - x_i)k_C \qquad [14.51]$$

where

k_C = thermal conductivity of carrier;
k_i = thermal conductivity of ith component in sample;
x_i = molar fraction of ith component in mixture of carrier and ith component.

The second equation assumes a linear relation between mixture thermal conductivity and composition. Substituting in eqn [14.50] gives:

$$V_{\mathrm{OUT}} = D\left[\frac{1}{x_i k_i + (1 - x_i)k_C} - \frac{1}{k_C}\right]$$

which simplifies to:

$$V_{\mathrm{OUT}} = D\,\frac{(k_C - k_i)x_i}{k_C[x_i k_i + (1 - x_i)k_C]} \qquad [14.52]$$

Usually x_i is small, so that this reduces to:

Approximate output voltage for katharometer detector in gas chromatograph system

$$V_{OUT} \approx D \frac{(k_C - k_i)}{k_C^2} x_i \qquad [14.53]$$

This application of the katharometer is discussed further in Chapter 17.

Conclusion

The chapter began by discussing the principles of **convective heat transfer**. It then explained three important applications of these principles to measurement systems. These are:

(a) In understanding and calculating the **dynamic characteristics** of **thermal sensors**.
(b) In **hot wire** and **film systems** for fluid velocity measurements.
(c) In the **katharometer** for **gas thermal conductivity** and **composition measurement**.

References

14.1 KING L V 1914 'On the convection of heat from small cylinders in a stream of fluid', *Philosophical Transactions of Royal Society*, Series A, Vol. 214.
14.2 COLLIS D C and WILLIAMS M J 1959 'Two-dimensional Convection from Heated wires at Low Reynolds Numbers', *Journal of Fluid Mechanics*, Vol. 6.
14.3 GALE B C *An Elementary Introduction to Hot Wire and Hot Film Anemometry*. DISA Technical Publication.
14.4 W. G. Pye and Co. *Technical Literature on Series 104 Chromatographs*.

Problems

14.1 A tungsten filament has a resistance of 18 Ω at 0 °C, a surface area of 10^{-4} m^{-2} and a temperature coefficient of resistance of 4.8×10^{-3} °C^{-1}. The heat transfer coefficient between the filament and air at 20 °C is given by $U = 4.2 + 7.0\sqrt{v}$ W m^{-2} °C^{-1}, where v m s^{-1} is the velocity of the air relative to the filament. The filament is incorporated into the constant temperature anemometer system of Fig. 14.5 which maintains the filament resistance at 40 Ω. Plot a graph of system output voltage versus air velocity in the range 0 to 10 m s^{-1}.

14.2 A heated tungsten filament is to be used for velocity measurements in a fluid of temperature 20 °C. With a constant current of 50 mA through the filament, the voltage across the filament was 2.0 V when the fluid was stationary and 1.5 V when the fluid velocity was 36 m s^{-1}. Use the data below to:

(a) identify the heat transfer characteristics of the system, and
(b) decide whether the system is suitable for measuring small velocity fluctuations, containing frequencies up to 10 kHz, about a steady velocity of 16 m s^{-1}.

Explain **briefly** why a constant temperature system would give a better performance in (b).

Filament data

Surface area	$= 10^{-4}\,m^2$
Resistance at 0 °C	$= 18\,\Omega$
Temperature coefficient of resistance	$= 4.8 \times 10^{-3}\,°C^{-1}$
Heat capacity	$= 1.3\,\mu J\,°C^{-1}$

14.3 A miniature thermistor with appreciable heating current is to be used to measure the flow turbulence in a slow moving gas stream. The frequency spectrum of the turbulence extends up to 10 Hz and the gas temperature is 90 °C. With a constant current of 23.4 mA, the steady-state voltage across the thermistor was 3.5 V. Starting from first principles, and using the thermistor data given below, show that a constant current system is unsuitable for this application. Explain briefly how the turbulence measurements can be made successfully with a modified system.

Thermistor data

Mass $= 10^{-4}$ kg.

Specific heat $= 1.64 \times 10^2\,J\,kg^{-1}\,°C^{-1}$.

Temperature T °C	88	100	105	108	110	112	118	125
Resistance $R_T\,\Omega$	240	185	170	155	150	140	130	116

14.4 A temperature sensor has a mass of 0.05 kg and a surface area of $10^{-3}\,m^2$ and is made of material of specific heat $500\,J\,kg^{-1}\,°C^{-1}$. It is placed inside a thermowell of mass 0.5 kg, surface area $10^{-2}\,m^2$ and specific heat $800\,J\,kg^{-1}\,°C^{-1}$. The heat transfer coefficient between sensor and thermowell is $25\,W\,m^{-2}\,°C^{-1}$ and between thermowell and fluid is $625\,m^{-2}\,°C^{-1}$.

(a) Calculate the system transfer function and thus decide whether it can follow temperature variations containing frequencies up to 10^{-3} Hz.

(b) What improvement is obtained if the heat transfer coefficient between sensor and thermowell is increased to $1000\,W\,m^{-2}\,°C^{-1}$?

14.5 A katharometer system is to be used to measure the percentage of hydrogen in a mixture of hydrogen and methane. The percentage of hydrogen can vary between 0 and 10 per cent. The system consists of four identical tungsten filaments arranged in the deflection bridge circuit of Fig. 14.7(d). The gas mixture, at 20 °C, is passed over elements 1 and 3; pure methane, also at 20 °C, is passed over elements 2 and 4. Assuming a linear relation between mixture thermal conductivity and composition, use the data given below to:

(a) find the range of the open circuit bridge output voltage;

(b) plot a graph of bridge output voltage versus percentage hydrogen.

Data

Temperature coefficient of resistance of tungsten	$= 5 \times 10^{-3}\,°C^{-1}$
Resistance of filament at 20 °C	$= 10\,\Omega$
Total bridge current	$= 200$ mA
Filament surface area	$= 10^{-5}\,m^2$
Heat transfer coefficient between filament and gas where k = gas thermal conductivity	$= 5 \times 10^4\,k$,
Thermal conductivity of hydrogen	$= 17 \times 10^{-2}\,W\,m^{-1}\,°C^{-1}$
Thermal conductivity of methane	$= 3 \times 10^{-2}\,W\,m^{-1}\,°C^{-1}$

15

Optical measurement systems

15.1 Introduction: types of system

Light is a general title which covers radiation in the ultraviolet (UV), visible and infrared (IR) portions of the **electromagnetic spectrum**. **Ultraviolet radiation** has wavelengths between 0.01 μm and 0.4 μm (1 micron or micrometre (μm) = 10^{-6} m), **visible radiation** wavelengths between 0.4 μm and 0.7 μm and **infrared wavelengths** between 0.7 and 100 μm. Most of the systems discussed in this chapter use radiation with wavelengths between 0.3 and 10 μm, i.e. visible and infrared radiation.

Figures 15.1 and 15.2 show two important general types of optical measurement system. Both types include a basic optical transmission system which is made up of a **source**, a **transmission medium** and a **detector**. The function $S(\lambda)$ describes how the amount of radiant power emitted by the source varies with wavelength λ. $T(\lambda)$ describes how the efficiency of the transmission medium varies with λ. The

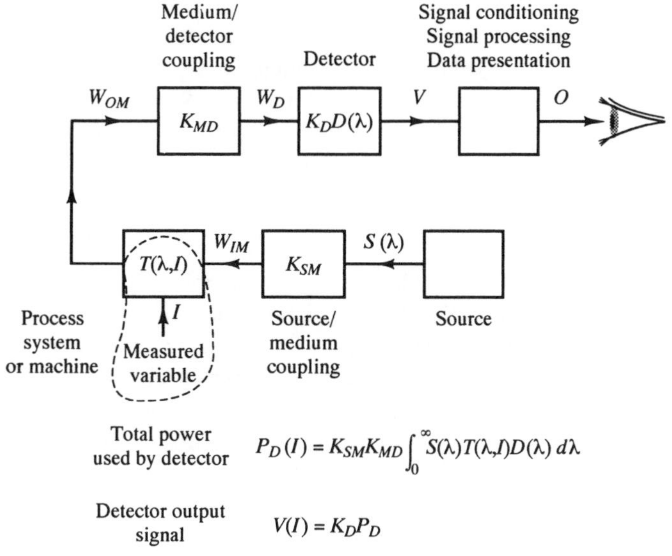

Fig. 15.1 Fixed source — variable transmission medium system

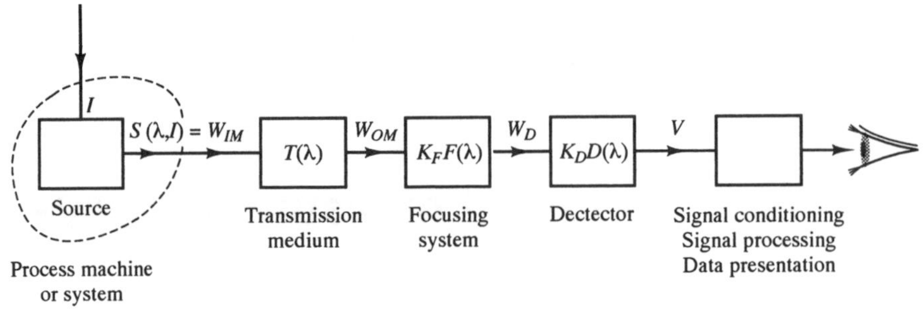

Total power used by detector $P_D(I) = K_F \int_0^\infty S(\lambda, I) T(\lambda) F(\lambda) D(\lambda)\, d\lambda$

Detector output power $V(I) = K_D P_D$

Fig. 15.2 Variable source — fixed transmission medium system

detector converts the incoming radiant power into an electrical signal. The sensitivity or responsibity K_D is the change in detector output (ohms or volts) for 1 watt change in incident power. $D(\lambda)$ describes how the detector responds to different wavelengths λ. It is important that the three functions $S(\lambda)$, $T(\lambda)$ and $D(\lambda)$ are compatible with each other. This means that the values of all three functions must be reasonably large over the wavelengths of interest, otherwise these wavelengths will not be transmitted. Both systems also include signal conditioning, signal processing and data presentation elements.

Fixed source — variable transmission medium

The system shown in Fig. 15.1 has a fixed source, i.e. constant total power; typical sources are a tungsten lamp, a light-emitting diode or a laser. However, the characteristics T of the transmission medium are not fixed but can vary according to the value of the measured variable I. In one example, the transmission medium consists of two optical fibres and the displacement to be measured changes the coupling between the fibres. In another where the transmission medium is a tube filled with gas, changes in gas composition vary the fraction of infrared power transmitted through the tube. Systems of this type normally use a **narrow band** of wavelengths. The constant K_{SM} describes the efficiency of the geometrical coupling of the source to the transmission medium (e.g. optical fibre) and the constant K_{MD} the coupling of the medium to the detector. Optical measurement systems of this type have the following advantages over equivalent electrical systems:

(a) No electromagnetic coupling to external interference voltages.
(b) No electrical interference due to multiple earths.
(c) Greater safety in the presence of explosive atmospheres. Here the source, detector, signal conditioning elements, etc., with their associated power supplies, are located in a safe area, e.g. a control room. Only the transmission fibre is located in the hazardous area, so that there is no possibility of an electrical spark causing an explosion.
(d) Greater compatibility with optical communication systems.
(e) Optical fibres may be placed close together without crosstalk.

Variable source — fixed transmission medium

The system shown in Fig. 15.2 has a variable source in that the source power S varies according to the value of the measured variable I. The most common example is a temperature measurement system where the total amount of radiant power emitted from a hot body depends on the temperature of the body. In this system the transmission medium is usually the atmosphere with fixed characteristics $T(\lambda)$. Since a hot body emits power over a wide range of wavelengths, systems of this type are usually **broad-band** but may be made **narrow-band** by the use of optical filters. Unlike optical fibres, the atmosphere cannot contain a beam of radiation and prevent it from diverging. A focusing system is usually necessary to couple the source to the detector. The constant K_F specifies the efficiency of this coupling and the function $F(\lambda)$ the wavelength characteristics of the focusing system. In temperature measurement systems of this type, the radiation receiver can be remote from the hot body or process. This means they can be used in situations where it is impossible to bring conventional sensors, e.g. thermocouples and resistance thermometers, into physical contact with the process. Examples are:

(a) high temperatures at which a normal sensor would melt or decompose;
(b) moving bodies, e.g. steel plates in a rolling mill;
(c) detailed scanning and imaging of the temperature distribution over a surface; the number of conventional sensors required would be prohibitively large.

15.2 Sources

15.2.1 Principles

All of the sources used in optical measurement systems and described in this section, emit radiation over a continuous band of wavelengths rather than at a single wavelength λ. The intensity of a source is specified by the power spectral density (PSD) function $S(\lambda)$ which is defined as follows:

The amount of energy per second emitted from 1 cm^2 of the source, into a unit solid angle, between wavelengths λ and $\lambda + \Delta\lambda$ is $S(\lambda)\Delta\lambda$. Note that:

(a) $S(\lambda)$ is the power emitted per unit wavelength at λ. $S(\lambda)\Delta\lambda$ is the power emitted between wavelengths λ and $\lambda + \Delta\lambda$. $\int_0^\infty S(\lambda)d\lambda$ is the total power emitted over all wavelengths. This is termed the **brightness** or **radiance** of the source B, i.e.

$$B = \int_0^\infty S(\lambda)d\lambda \quad \text{W cm}^{-2}\,\text{sr}^{-1} \qquad [15.1]$$

(b) For many of the sources used in optical measurement systems, the source $S(\lambda)$ is the same when viewed in all directions. Such a source is said to be Lambertian, a common example is a surface-emitting LED. Assuming that the source has uniform $S(\lambda)$ over its entire surface area A_S, then the radiant flux emitted by the source into unit solid angle is $A_S S(\lambda)$. This is termed the **luminous intensity** of the source, i.e.

$$I = A_S S(\lambda) \quad \text{W sr}^{-1}\,\mu\text{m}^{-1} \qquad [15.2]$$

This is the luminous intensity along the line $\theta = 0$ drawn normal to the radiating

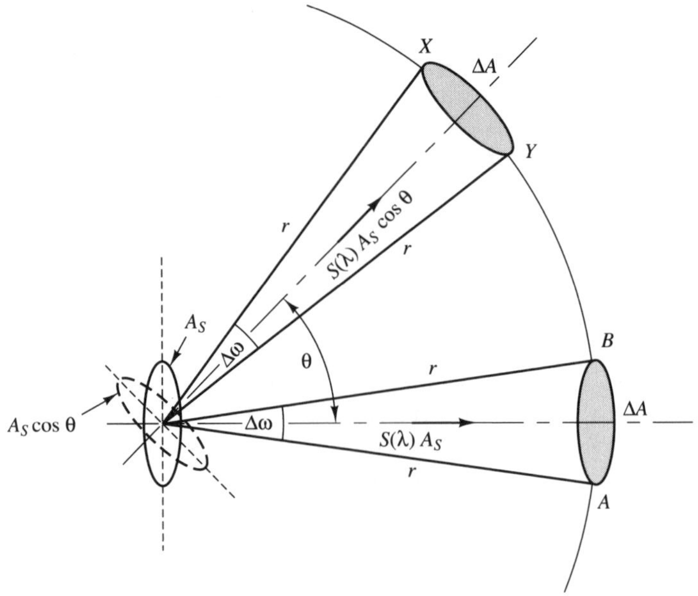

Fig. 15.3 Lambertian emitter and solid angle

surface, in Fig. 15.3. However, the projected area of the emitting surface at angle θ, relative to the normal, is $A_S\ cos\ \theta$. The corresponding luminous intensity of the source when viewed at an angle θ is:

Lambertian angular variation in intensity

$$I_\theta = S(\lambda) A_S \cos\theta = I\cos\theta \qquad [15.3]$$

(c) Figure 15.3 also explains the meaning of solid angle and its importance in radiant power calculations. *AB* and *XY* are two elements of the surface of a sphere, both elements have area ΔA and are distance r from the source. The solid angle subtended by each element as the source is:

$$\Delta\omega = \frac{\text{surface area}}{\text{distance}^2} = \frac{\Delta A}{r^2} \text{ steradians}$$

The power incident on *AB* is:

$$\Delta W = I\Delta\omega = \frac{I\Delta A}{r^2} \text{ W}\,\mu\text{m}^{-1} \qquad [15.4]$$

and the power incident on *XY* is:

$$\Delta W_\theta = I_\theta \Delta\omega = \frac{I\cos\theta\Delta A}{r^2} \text{ W}\,\mu\text{m}^{-1} \qquad [15.5]$$

15.2.2 Hot body sources

Any body at a temperature above 0 K emits radiation. The ideal emitter is termed a **black body**: from Planck's Law the power spectral density of radiation emitted by a black body at temperature TK is:

Power spectral density for black body radiator

$$W^{BB}(\lambda, T) = \frac{C_1}{\lambda^5\left[\exp\left(\frac{C_2}{\lambda T}\right) - 1\right]} \qquad [15.6]$$

where

$$\lambda = \text{wavelength in } \mu\text{m}$$
$$C_1 = 37\,413 \text{ W } \mu\text{m}^4 \text{ cm}^{-2}$$
$$C_2 = 14\,388\ \mu\text{m K}$$

Figure 15.4 is a plot of $W^{BB}(\lambda, T)$ versus λ, for various values of T. Note that:

(a) $W^{BB}(\lambda, T)$ is the power emitted per unit wavelength at λ.
$W^{BB}(\lambda, T)\Delta\lambda$ is the power emitted between wavelengths λ and $\lambda + \Delta\lambda$.
$\int_0^\infty W^{BB}(\lambda, T)d\lambda$ is the total power W_{TOT} emitted over all wavelengths, at temperature T.

W_{TOT} is therefore the total area under the $W^{BB}(\lambda, T)$ curve at a given temperature T. From Fig. 15.4 we can see quantitatively that this area increases rapidly with T. We can confirm this quantitatively by evaluating the integral to give:

Total power emitted by a black body — Stefan's Law

$$W_{TOT} = \int_0^\infty \frac{C_1 d\lambda}{\lambda^s\left[\exp\left(\frac{C_2}{\lambda T} - 1\right)\right]} = \sigma T^4 \quad \text{W cm}^{-2} \qquad [15.7]$$

where $\sigma = 5.67 \times 10^{-12}$ W cm^{-2} K^{-4} (Stefan-Boltzmann constant).

(b) The wavelength λ_P at which $W^{BB}(\lambda, T)$ has maximum value, decreases as temperature T increases according to:

$$\lambda_P = \frac{2891}{T}\ \mu\text{m} \qquad [15.8]$$

Thus, if $T = 300$ K, $\lambda_P \approx 10\ \mu$m, i.e. most of the radiant power is infrared. However, at $T = 6000$ K, $\lambda_P \approx 0.5\ \mu$m and most of the power is in the visible region.

(c) Equation [15.6] is for a source in the form of a flat surface of area 1 cm^2 radiating into a hemisphere. A hemisphere is a sold angle of 2π steradians.

A black body is a theoretical ideal which can only be approached in practice; for example by a blackened conical cavity with a cone angle of 15°. A real body emits less radiation at temperature T and wavelength λ than a black body at the same conditions. A correction factor, termed the **emissivity** $\epsilon(\lambda, T)$ of the body is introduced to allow for this. Emissivity is defined by:

$$\text{Emissivity} = \frac{\text{Actual radiation at } \lambda, T}{\text{Black body radiation at } \lambda, T}$$

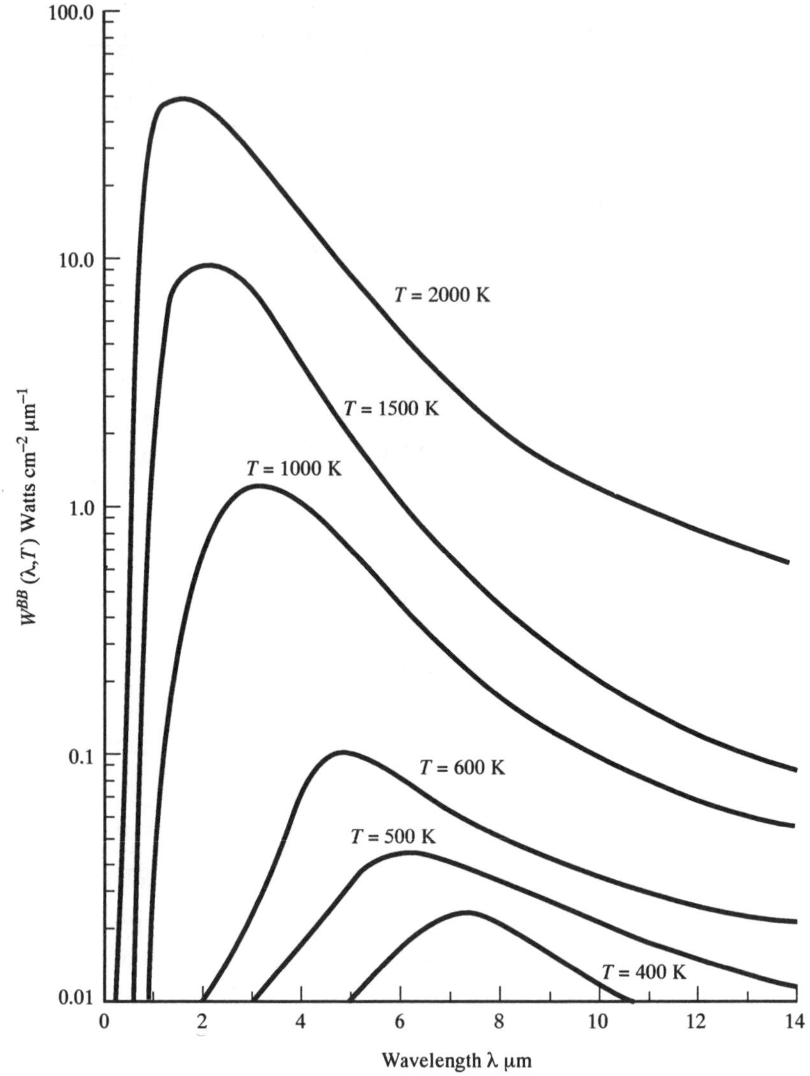

Fig. 15.4 Power spectral density for a black body radiator

Emissivity of real body

$$\epsilon(\lambda, T) = \frac{W^A(\lambda, T)}{W^{BB}(\lambda, T)} \qquad [15.9]$$

Emissivity in general depends on wavelength and temperature, although the temperature dependence is often weak. Figure 15.5 shows the emissivities of various materials. Note that a black body has $\epsilon = 1$ by definition; a grey body has an emissivity independent of wavelength, i.e. $\epsilon(\lambda) = \epsilon(< 1)$ for all λ. The emissivity of a real body must be measured experimentally by comparing the radiant power emitted by the body with that from a standard 'black body' source (typical emissivity 0.99).[1]

Summarising, a hot body source is characterised by a broadband, continuous power spectral density $S(\lambda, T)$ given by:

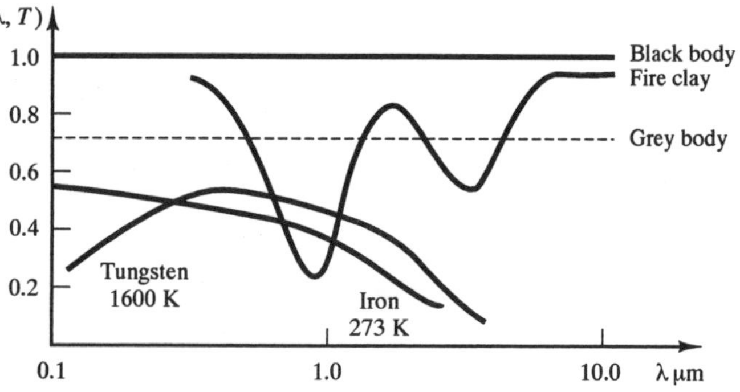

Fig. 15.5 Emissivity of various materials

$$S(\lambda, T) = \frac{1}{2\pi} \epsilon(\lambda, T)\, W^{BB}(\lambda, T) \quad \mathrm{W\,\mu m^{-1}\,cm^{-2}\,sr^{-1}} \qquad [15.10]$$

In a fixed source, variable transmission medium system (Fig. 15.1), source temperature T is held constant so that S depends on λ only and brightness B is approximately constant. A typical source is a tungsten halogen lamp operating at 3200 K and 50 W; most of the radiation emitted is between 0.4 μm and 3.0 μm with a maximum at 0.9 μm. In a variable source, fixed transmission medium (Fig. 15.2) the source is the hot body whose temperature T is varying and is to be measured. Here S depends on both λ and T and B depends on T. Uncertainty in the value of the emissivity of the hot body is the main source of error in this type of temperature measurement system.

15.2.3 Light-emitting diode (LED) sources[2,3]

Light-emitting diodes have already been discussed in Section 11.4.2. These are *PN* junctions formed from *P*-type and *N*-type semiconductors, which when forward biased emit optical radiation. LEDs emitting visible radiation are widely used in displays. Examples are gallium arsenide phosphide (GaAsP) which emits red light ($\lambda \approx 0.655$ μm) and gallium phosphide (GaP) which emits green light ($\lambda \approx 0.560$ μm). Infrared LED sources are often preferred for use with optical fibre transmission links because their wavelength characteristics $S(\lambda)$ are compatible with the fibre transmission characteristics $T(\lambda)$. LEDs based on gallium aluminium arsenide (GaAlAs) alloys emit radiation in the 0.8 to 0.9 μm wavelength region and those using indium gallium arsenide phosphide (InGaAsP) emit in the 1.3 to 1.5 μm region. Light-emitting diodes emit radiation over a narrow band of wavelengths. Figure 15.6 shows a typical power spectral density function $S(\lambda)$ for a GaAlAs LED where $S(\lambda)$ has a peak value at $\lambda_0 = 810$ nm (0.81 μm) and a bandwidth $\Delta\lambda$ of 36 nm; the area under the distribution, i.e. the brightness B, is approximately 100 W cm^{-2} sr^{-1}. The total radiant output power is typically between 1 and 10 mW.

15.2.4 Laser sources[2,3]

There are several types of lasers: the lasing medium may be gas, liquid, solid crystal or solid semiconductor. All types of laser have the same principle of operation which

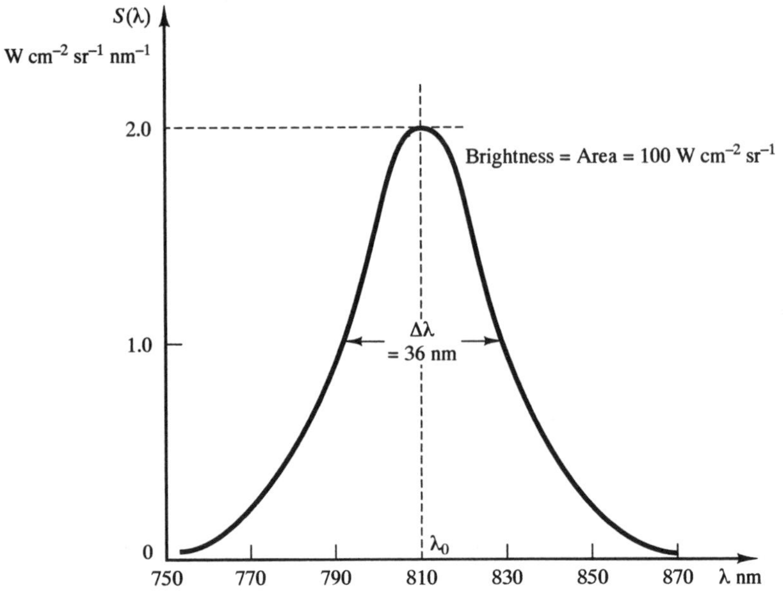

Fig. 15.6 S(λ) for GaAlAs LED

can be explained using the energy level diagram shown in Fig. 15.7(a). The ground state of the medium has energy E_1 and the excited state E_2. A transition between these two states involves the absorption or emission of a photon of energy:

$$hf = E_2 - E_1 \qquad [15.11]$$

where h is Planck's constant and f is the frequency of the radiation. If the medium is in thermal equilibrium, most of the electrons occupy the ground state E_1 and only a few have sufficient thermal energy to occupy the excited state E_2. If, however, the laser is 'pumped', i.e. energised by an external source, then a **population inversion** occurs where there are more electrons in the excited state than the ground state. Electrons then randomly return from the excited state to the ground state with emission of a photon of energy hf, this is **spontaneous emission**. If this single photon causes another electron to return to the ground state, thereby generating a second photon in phase with the first, the process is called **stimulated emission**, and is the key to laser action. The two photons then create four and so on. The process is enhanced by creating a resonant cavity. This has a mirror at either end so that the photons travel forwards and backwards through the cavity, multiplication taking place all the time. The distance between the mirrors is made equal to an integral number of wavelengths $\lambda(\lambda = c/f)$ to produce resonance. The photons leave the cavity, via a small hole in a mirror, to give a narrow, intense, **monochromatic** (almost a single wavelength), **coherent beam** of light. Coherent means that different points in the beam have the same phase.

Semiconductor injection laser diodes (ILDs) are used with optical fibre links. These use the same materials, GaAlAs and InGaAsP, as LEDs, but give a narrower, coherent beam with a much smaller spectral bandwidth $\Delta\lambda$. Figure 15.7(b) shows a typical $S(\lambda)$ for a GaAlAs ILD with $\lambda_0 = 810$ nm, $\Delta\lambda = 3.6$ nm, and brightness $B = 10^5$ W cm^{-2} sr^{-1}. The total radiant power is typically between 1 and 10 mW. The construction of a typical ILD is shown in Fig. 15.7(c).

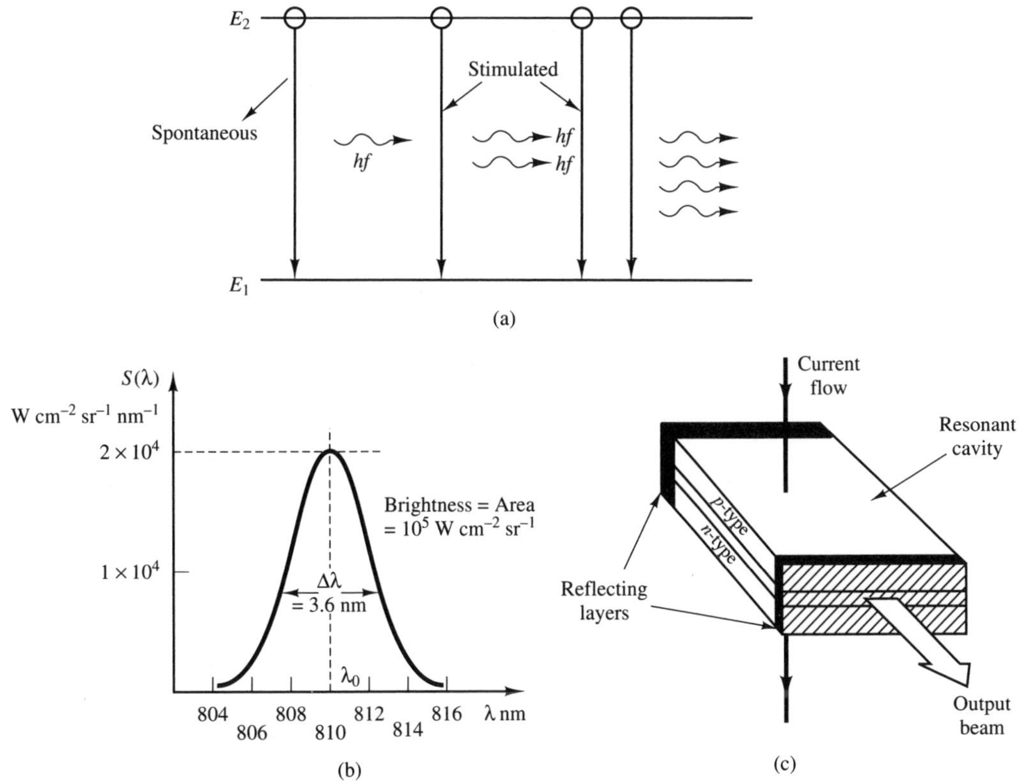

Fig. 15.7 Laser sources (a) Spontaneous and stimulated emission (b) S(λ) for GaAlAs injection laser diode (c) Construction of semiconductor laser diode

15.3 Transmission medium

15.3.1 General principles

If light is passed through any medium gas, liquid or solid, then certain wavelengths present in the radiation cause the molecules to be excited into higher energy states. These wavelengths are thus **absorbed** by the molecules; each type of molecule is characterised by a unique **absorption spectrum** which is defined by:

$$A(\lambda) = \frac{\text{Power absorbed by medium at } \lambda}{\text{Power entering medium at } \lambda} = \frac{W_{AM}(\lambda)}{W_{IM}(\lambda)} \qquad [15.12]$$

Similarly the transmission spectrum $T(\lambda)$ is defined by:

$$T(\lambda) = \frac{\text{Power leaving medium at } \lambda}{\text{Power entering medium at } \lambda} = \frac{W_{OM}(\lambda)}{W_{IM}(\lambda)} \qquad [15.13]$$

i.e. $W_{OM}(\lambda) = T(\lambda)W_{IM}(\lambda)$. From conservation of energy $A(\lambda) + T(\lambda) = 1$ so that a transmission minimum corresponds to an absorption maximum.

Figures 15.8(a) and (b) show absorption spectra for acetylene and carbon monoxide; like many hydrocarbons and other gases these show strong absorption bands in the infrared. Figure 15.8(c) shows the transmission spectrum $T(\lambda)$ for the atmosphere,

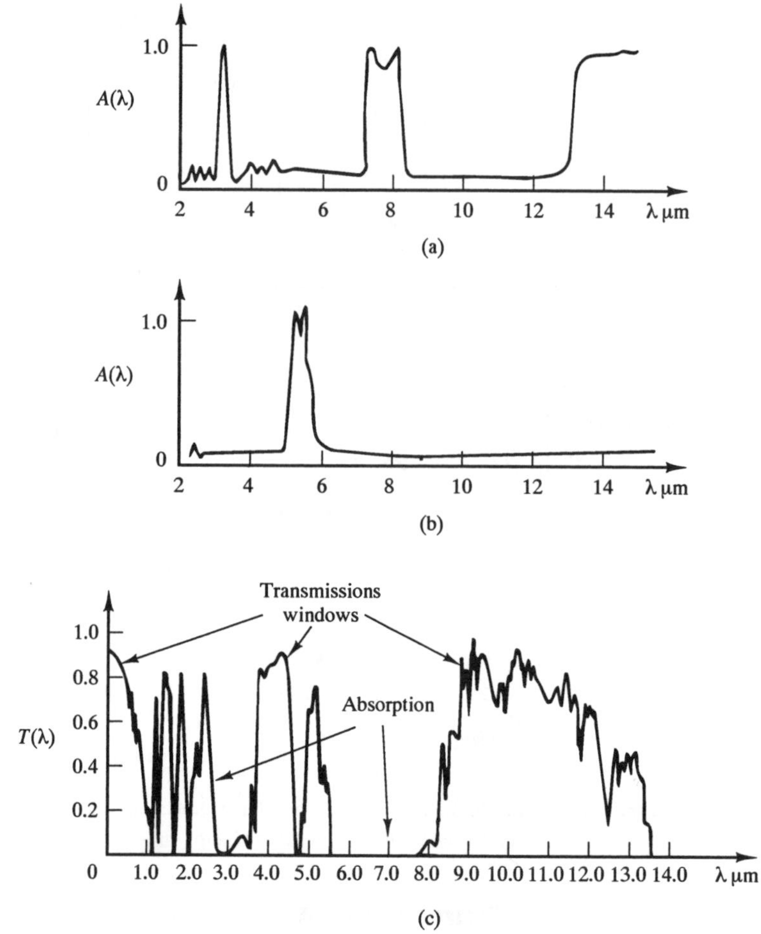

Fig. 15.8 Absorption and transmission spectra
(a) Absorption spectrum for acetylene
(b) Absorption spectrum for carbon monoxide
(c) Transmission for the atmosphere over a distance of 1 mile

this is more complicated because there are absorption bands due to water vapour, carbon dioxide and ozone.

The above effect can be used in fixed source, variable transmission medium systems to measure the composition of gas and liquid mixtures. If the percentage of the absorbing molecule in the mixture changes, then $T(\lambda)$ changes and the amount of power leaving the medium changes. Such a system is called a non-dispersive infrared analyser; it could be used, for example, to measure the percentage of an absorbing component, e.g. CO, CO_2, CH_4, C_2H_4, in a gas mixture.

The transmission characteristics of the atmosphere are also important in variable source, fixed transmission medium systems for temperature measurement. The radiation-receiving system is often designed to respond only to a narrow band of wavelengths corresponding to a **transmission window**.

15.3.2 Optical fibres

Optical fibres are widely used as transmission media in optical measurement systems.[2,3] A typical fibre (Fig. 15.9(a)) consists of two concentric dielectric

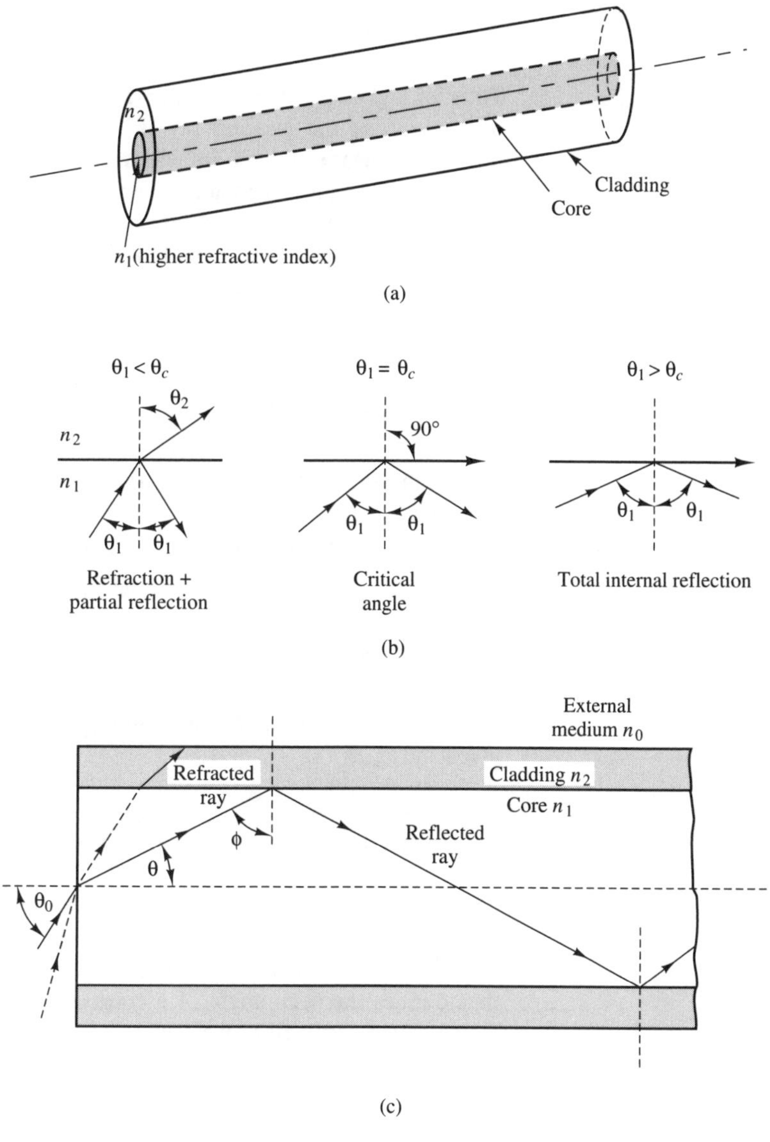

Fig. 15.9 Optical fibre principles
(a) Optical fibre construction
(b) Reflection and refraction at a boundary
(c) Total internal reflection in a fibre

cylinders. The inner cylinder, called the **core**, has a refractive index n_1: the outer cylinder called the **cladding** has a refractive index n_2 which is less than n_1.

Figure 15.9(b) shows a ray incident on the boundary between two media of refractive indices n_1 and n_2 ($n_1 > n_2$). Part of the ray is reflected back into the first medium, the remainder is refracted as it enters the second medium. From Snell's law we have:

$$n_1 \sin \theta_1 = n_2 \sin \theta_2 \qquad [15.14]$$

and since $n_1 > n_2$ then $\theta_2 > \theta_1$. When $\theta_2 = 90°$ the refracted ray travels along the boundary; the corresponding value of θ_1 is known as the **critical angle** θ_c and is given by:

$$n_1 \sin\theta_c = n_2 \text{ i.e. } \theta_c = \sin^{-1}\left(\frac{n_2}{n_1}\right) \quad [15.15]$$

Thus for a glass/air interface $n_1 = 1.5$, $n_2 = 1$ and $\theta_c = 41.8°$. When θ_1 is greater than θ_c, all of the incident ray is totally internally reflected back into the first medium. Thus the above fibre will transmit, by means of many internal reflections, all rays with angles of incidence greater than the critical angle.

Figure 15.9(c) shows total internal reflections in a **step-index fibre**; this has a core with a uniform refractive index n_1. The core is surrounded with a cladding of slightly lower refractive index n_2 and the entire fibre is surrounded by an external medium of refractive index n_0. Total internal reflection in the core occurs if $\phi \geq \theta_c$, i.e.

$$\sin\phi \geq \frac{n_2}{n_1} \text{ i.e. } \cos\phi \leq \sqrt{1 - \left(\frac{n_2}{n_1}\right)^2} \quad [15.16]$$

If θ_0 is the angle of incidence of a ray to the fibre, then the ray enters the core at an angle θ given by:

$$n_0 \sin\theta_0 = n_1 \sin\theta = n_1 \cos\phi \quad [15.17]$$

From eqns [15.16] and [15.17] total internal reflection occurs if:

$$\sin\theta_0 \leq \frac{n_1}{n_0}\sqrt{1 - \left(\frac{n_2}{n_1}\right)^2}$$

For air $n_0 = 1$, so that the **maximum angle of acceptance** θ_0^{MAX} of the fibre is given by:

Numerical aperture for step-index fibre

$$\sin\theta_0^{MAX} = \sqrt{n_1^2 - n_2^2} = \text{NA} \quad [15.18]$$

The numerical aperture (NA) of the fibre is defined to be $\sin\theta_0^{MAX}$. Thus rays with $\theta_0 \leq \theta_0^{MAX}$, have $\phi \geq \theta_c$ and are continuously internally reflected: rays with $\theta_0 > \theta_0^{MAX}$ have $\phi < \theta_c$ refract out of the core and are lost in the cladding. θ_0^{MAX} thus defines the semi-angle of a **cone of acceptance** for the fibre.

For a step-index fibre we have:

$$n_2 = n_1(1 - \Delta) \quad [15.19]$$

where Δ is the **core/cladding index difference**; Δ usually has a nominal value of 0.01. Since $\Delta \ll 1$, eqns [15.18] and [15.19] give the approximate equation for numerical aperture:

$$\text{NA} = \sin\theta_0^{MAX} \approx n_1\sqrt{2\Delta} \quad [15.20]$$

Thus if $n_1 = 1.5$, then NA = 0.21 and $\theta_0^{MAX} = 12°$.

Figure 15.10 show the three main types of fibre in current use. The **monomode step-index fibre** is characterised by a very narrow core, typically only a few μm in diameter. This type of fibre can sustain only one mode of propagation and requires a coherent laser source. The **multimode step-index fibre** has a much larger core, typically 50 μm in diameter. Many modes can be propagated in multimode fibres; because of the larger core diameter it is also much easier to launch optical power into the fibre and also to connect fibres together. Another advantage is that light

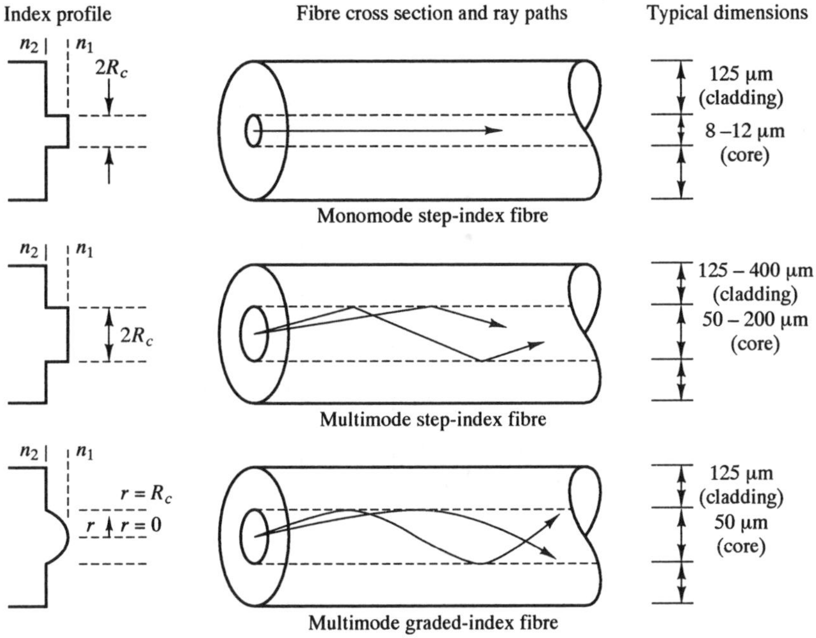

Fig. 15.10 Different types of fibre (after Keiser)[2]

can be launched into multimode fibres using LED sources, whereas single-mode fibres must be excited with more complex laser diode sources.

Multimode graded-index fibres have a core with a non-uniform refractive index; n decreases parabolically from n_1 at the core centre to n_2 at the core/cladding boundary. These fibres are characterised by curved ray paths, which offer some advantages, but are more expensive than the step-index type.

A light beam is attenuated as it propagates along a fibre, this attenuation increases with the length of the fibre. The main attenuation mechanisms are Rayleigh scattering, absorption by ions present in the fibre core and radiation. The overall attenuation loss α dB km^{-1} of a fibre length L km is defined by:

$$\alpha = \frac{10}{L} \log_{10} \left[\frac{W_{IM}}{W_{OM}} \right] = -\frac{10}{L} \log_{10} T \qquad [15.21]$$

where W_{IM} and W_{OM} are the input and output powers and T is the transmission factor. Figure 15.11 shows typical variations in α with wavelength λ for fibres made entirely of silica glass and polymer plastic. For the silica glass fibre at $\lambda = 900$ nm, $\alpha \approx 1.5$ dB km^{-1} so that $T = 0.71$ for $L = 1$ km. For the plastic fibre at $\lambda = 900$ nm, $\alpha = 600$ dB km^{-1} so that $T = 0.87$ for $L = 1$ metre.

Glass fibres must be used in telecommunication systems where long transmission distances are involved, but plastic fibres can be used in measurement systems where transmission links are much shorter.

15.4 Geometry of coupling of detector to source

The detector may be coupled to the source by an optical focusing system, an optical fibre or some combination of focusing systems and fibres. The aim of this section

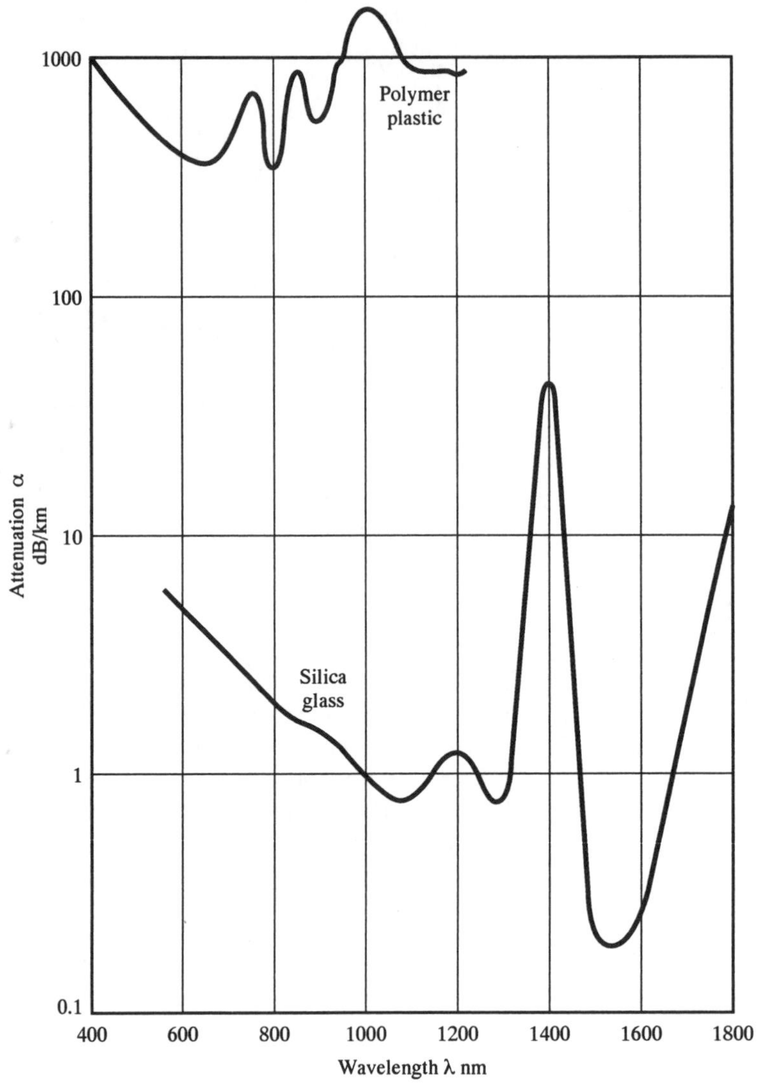

Fig. 15.11 Optical fibre attenuation characteristics

is to use the basic principles of Section 15.2.1 to study the geometry and efficiency of this coupling in some simple situations.

From eqns [15.2], [15.3], [15.5] and Fig. 15.3, the power per unit wavelength incident on an element are ΔA of a surface, due to a Lambertian source of power spectral density (PSD). $S(\lambda)$ and area A_s is:

$$\Delta W = A_s S(\lambda) \cos \theta \Delta\omega = A_s S(\lambda) \cos \theta \frac{\Delta A}{r^2} \qquad [15.22]$$

The power per unit wavelength incident on the entire surface of area A is:

$$W = A_s S(\lambda) \int_0^{\omega} \cos \theta \, d\omega = A_s S(\lambda) \int_0^{A} \frac{\cos \theta}{r^2} \, dA \qquad [15.23]$$

where ω the solid angle which the surface subtends at the source. If we make the approximations that θ is small, i.e. $\cos\theta \approx 1$ and r is a constant for all elements of the surface, then we have the approximate equations:

Approximate equations for power incident on surface

$$W(\lambda) \approx A_s \omega S(\lambda) \approx \frac{A_s A}{r^2} . S(\lambda) \qquad [15.24]$$

Equation [15.23] can also be used to evaluate the total power emitted by a circular source (such as an LED) in all directions, i.e. over a hemisphere which is a solid angle of 2π steradians. We have:

$$W(\lambda) = A_s S(\lambda)\pi = \pi^2 R_s^2 S(\lambda) \qquad [15.23(a)]$$

where R_s(cm) is the radius of the source. The total source power P_s over all wavelengths is then found by integrating [15.23(a)]:

$$P_s = \pi^2 R_s^2 \int_0^\infty S(\lambda) d\lambda = \pi^2 R_s^2 B \text{ watts} \qquad [15.23(b)]$$

where B is the brightness of the source.

15.4.1 Coupling via a focusing system

This is used in the system shown in Fig. 15.2 where the transmission medium is the atmosphere. The atmosphere cannot contain a light beam and prevent it diverging, so that a converging system is necessary to focus the beam onto a detector.

We now calculate the coupling constant K_F for the situation shown in Fig. 15.12(a), i.e. a circular detector receiving radiation via a circular aperture with a converging lens in the aperture. We assume that the image of the source just fills the detector area. From similar triangles we have:

$$\frac{R_D}{d} = \frac{R_2}{D} \quad \text{i.e. } R_2 = \frac{D}{d} R_D \qquad [15.25]$$

so that the area of the source scanned is:

$$A_s = \pi R_2^2 = \pi \frac{D^2}{d^2} R_D^2 \qquad [15.26]$$

Assuming, for the moment, that the lens is a perfect transmitter of radiation, then all of the radiation incident onto the lens from a point on the source is focused onto the detector. This means that the appropriate solid angle is that subtended by the lens at source, i.e.

$$\omega = \frac{\pi R_A^2}{D^2} \qquad [15.27]$$

In eqn [15.24] the effective source power $S(\lambda)$ is the power $W_{OM}(\lambda)$ leaving the transmission medium, thus the power coupled to the detector is:

$$W_D(\lambda) = K_F W_{OM}(\lambda) \qquad [15.28]$$

where

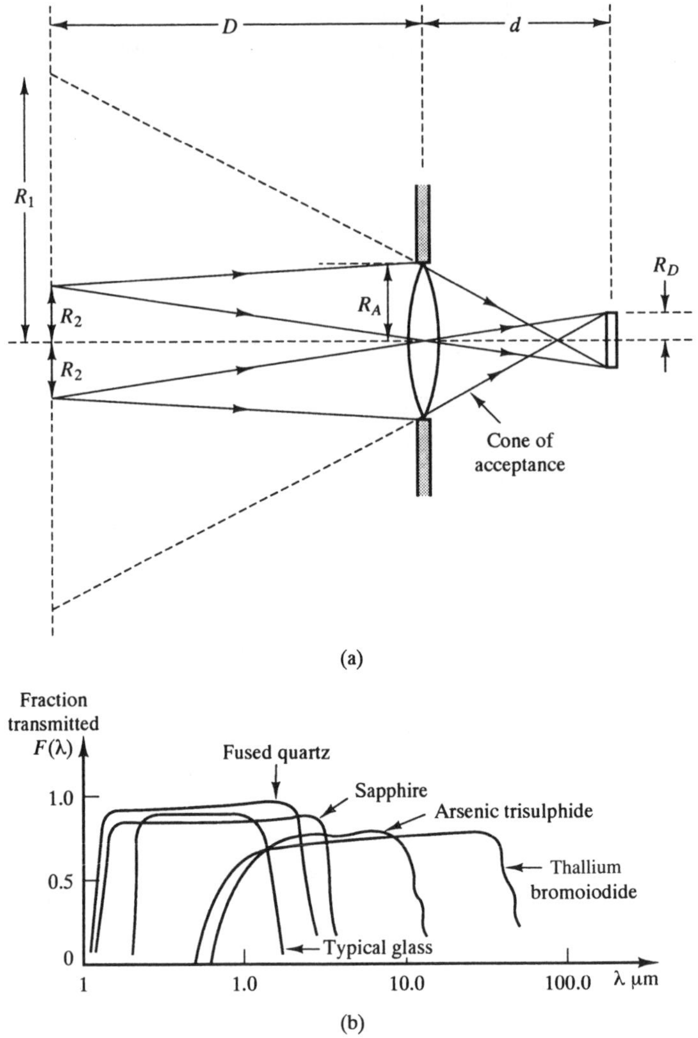

Fig. 15.12 Focusing systems
(a) Geometry of lens focusing system
(b) Transmission characteristics of lens materials (after Doebelin E.O. *Measurement Systems: Application and Design*. McGraw-Hill, New York, 1976, pp. 558–61)

$$K_F = A_s\omega = \frac{\pi^2 R_D^2 R_A^2}{d^2} \qquad [15.29]$$

We note that K_F is independent of the distance D between source and lens. This is because the amount of radiation received is limited by the 'cone of acceptance' which is defined by R_A, R_D and d but not D. Thus provided this cone is filled with radiation, the sensitivity of the radiation receiver will be independent of the distance between source and receiver. This is of great practical importance since it means that an instrument will not need re-calibration if its distance from the source is changed.

From Fig. 15.12(a) we note that if the lens were removed, then radiation from a bigger source area A_s is required to fill the cone of acceptance. However, the appropriate solid angle ω is now that subtended by the detector at the source and is therefore smaller. Since the area of source available may be limited, or we may wish to examine a small area of the source, then solid angle is a better measure of performance and a lens is preferable. For the lens we have $1/D + 1/d = 1/f$ so that if D is large, then d may be replaced by f in eqn [15.29].

In practice the material of a lens is not a perfect transmitter and the fraction of radiation transmitted depends critically on wavelength λ. These transmission characteristics are defined by $F(\lambda)$ = fraction of power transmitted by material at λ and eqn [15.29] must be modified to:

$$W_D(\lambda) = K_F F(\lambda) W_{OM}(\lambda) \qquad [15.30]$$

Figure 15.12(b) shows $F(\lambda)$ for lens and window materials in common use. We see that glass, which will transmit visible radiation, is opaque to wavelengths greater than 2 μm. For longer wavelengths either lenses of special materials, such as arsenic trisulphide, or mirrors should be used.

15.4.2 Coupling via an optical fibre

If the transmission medium is capable of containing a light beam as with an optical fibre, then it can be used to couple the detector to a source. Here there are two constants involved (Fig. 15.1): K_{SM} describes the coupling of source to medium and K_{MD} describes the coupling of the medium to the detector.

We first calculate K_{SM} for the situation shown in Fig. 15.13(a), which is a step-index optical fibre with a circular core receiving radiation from a circular source.

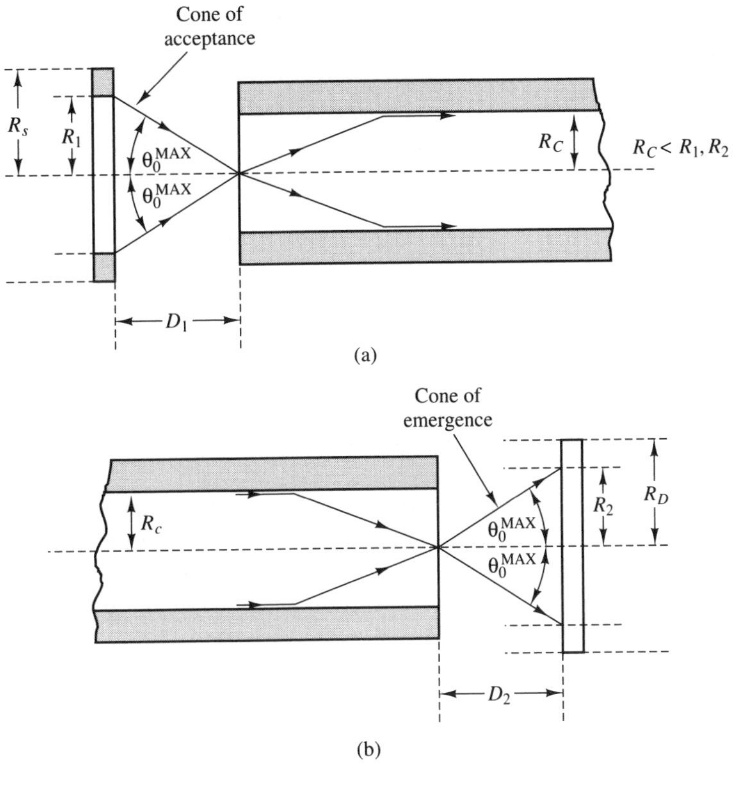

Fig. 15.13 Coupling of detector to source via an optical fibre
(a) Coupling of source to fibre
(b) Coupling of fibre to detector

The core radius R_c is less than the source radius R_s. Only rays within the cone of acceptance defined by the maximum angle of acceptance θ_0^{MAX} can be transmitted through the core. The corresponding useful source radius R_1 is defined by the cone of acceptance, i.e.

$$\frac{R_1}{D_1} = \tan \theta_0^{MAX}$$

Assuming that $R_1 \leq R_s$ where R_s is the total source radius, then the useful source area is:

$$A_s = \pi R_1^2 = \pi D_1^2 \tan^2 \theta_0^{MAX} \qquad [15.31]$$

The solid angle ω which the fibre core subtends at the source is approximately:

$$\omega \approx \frac{\pi R_c^2}{D_1^2} \qquad [15.32]$$

If the source PSD is $S(\lambda)$ then from eqn [15.24], the power per unit wavelength launched into the fibre is approximately:

$$W_{IM}(\lambda) \approx A_s \omega S(\lambda) \approx \pi^2 R_c^2 \tan^2 \theta_0^{MAX} S(\lambda) \qquad [15.33]$$

for small θ_0^{MAX}, $\tan \theta_0^{MAX} \approx \sin \theta_0^{MAX}$; also the numerical aperture NA of a fibre surrounded by air is $\sin \theta_0^{MAX}$. Thus the coupling constant K_{SM} is given approximately by:

Coupling constant source to fibre $R_c < R_s$

$$K_{SM} = \frac{W_{IM}(\lambda)}{S(\lambda)} \approx \pi^2 R_c^2 (\text{NA})^2 \qquad [15.34]$$

K_{SM} is also equal to the ratio P_{IM}/B, where P_{IM} is the input power to the fibre over all wavelengths and B the source brightness. In a situation where $R_1 > R_s$, i.e. the semi-angle θ of the inlet cone is less than θ_0^{MAX}, a converging lens can be used to increase θ to θ_0^{MAX}.

If the core radius R_c is greater than the source radius R_s then the appropriate equation for K_{SM} is:

Coupling constant source to fibre $R_c > R_s$

$$K_{SM} = \frac{W_{IM}(\lambda)}{S(\lambda)} = \frac{P_{IM}}{B} = \pi^2 R_s^2 (\text{NA})^2 \qquad [15.35]$$

We now calculate K_{MD} for the situation shown in Fig. 15.13(b), which is a circular detector receiving radiation from a step-index fibre with a circular core. The power per unit wavelength leaving the fibre is $W_{OM}(\lambda)$ which depends on the power input to the fibre $W_{IM}(\lambda)$ and the fibre transmission characteristics $T(\lambda)$. All of this power is contained within a core of emergence whose semi-angle is equal to θ_0^{MAX} the maximum angle of acceptance. The corresponding useful detector radius R_2 is defined by the cone of emergence as:

$$R_2 = D_2 \tan \theta_0^{MAX}$$

Providing $R_D \geq R_2$ where R_D is the total detector radius, then all of the power leaving the fibre is incident on the detector, i.e.

$$K_{MD} = \frac{W_D(\lambda)}{W_{OM}(\lambda)} = 1.$$

If, however, $R_D < R_2$, then only radiation within a cone of semi-angle θ, less than θ_0^{MAX}, is incident on the detector. The power incident on the detector is correspondingly reduced by a factor (ω/ω_0) where ω and ω_0 are the solid angles corresponding to θ and θ_0^{MAX} respectively, i.e.

$$W_D(\lambda) = \left(\frac{\omega}{\omega_0}\right) W_{OM}(\lambda), \text{ i.e. } K_{MD} = \left(\frac{\omega}{\omega_0}\right) \qquad [15.36)$$

Since $\omega \approx \dfrac{\pi R_D^2}{D_2^2}$, $\omega_0 \approx \dfrac{\pi R_2^2}{D_2^2}$ and NA $\approx \tan \theta_0^{MAX}$

$$K_{MD} \approx \frac{R_D^2}{R_2^2} \approx \frac{R_D^2}{D_2^2(\text{NA})^2} \qquad [15.37]$$

15.5 Detectors and signal conditioning elements

The detector converts the incident radiant power into an electrical output, that is a resistance or small voltage. A signal conditioning element, such as a bridge and/or amplifier, is usually required to provide a usable voltage signal.

The four main performance characteristics of detectors are:

(a) responsivity (sensitivity) K_D
(b) time constant τ
(c) wavelength response $D(\lambda)$
(d) Noise equivalent power (NEP) or factor of merit D^*

The sensitivity K_D is the change in detector output (ohms or volts) for a 1 W change in incident power. The wavelength response $D(\lambda)$ is the ratio between the sensitivity at wavelength λ and the maximum sensitivity; it allows for the detector not responding equally to all wavelengths, that is not *using* all wavelengths. The noise equivalent power (NEP) is the amount of incident power in watts required to produce an electrical signal of the same root-mean-square voltage as the background noise. The factor of merit D^* has been introduced in order to compare the signal-to-noise ratio of different detectors, of different sizes used with different amplifiers. It is defined by[4]:

$$D^* = \frac{(S/N)(A\Delta f)^{1/2}}{P} \text{ cm Hz}^{1/2} \text{ W}^{-1}$$

where S/N is the signal-to-noise ratio observed when P watts of power is incident on a detector of area A cm^2 used with an amplifier of bandwidth Δf. There are two main types of detector in common use, thermal and photon.

15.5.1 Thermal detectors

Here the incident power heats the detector to a temperature T_D °C which is above the surrounding temperature T_S °C. The detector is usually a resistive or thermoelectric sensor which gives a resistance or voltage output; this depends on temperature difference $T_D - T_S$ and thus incident power.

Thermal detectors respond equally to all wavelengths in the incident radiation (see Fig. 15.15), so that in effect:

$$D(\lambda) = 1 \quad 0 \le \lambda \le \infty$$

These detectors are used in radiation temperature measurement systems where the power emitted from a hot body source $S(\lambda, T)$ depends on temperature T.

Since $D(\lambda) = 1$ for all λ, the total power used by the detector is (see Fig. 15.2):

$$P_D = K_F \int_0^{\infty} S(\lambda, T)\, T(\lambda)\, F(\lambda)\, \mathrm{d}\lambda \qquad [15.38]$$

In order to examine the factors affecting both the sensitivity and time constant of a thermal detector, we now calculate the transfer function relating detector temperature T_D to power P_D. The heat balance equation for the detector is:

$$P_D - UA(T_D - T_S) = MC\frac{\mathrm{d}T_D}{\mathrm{d}t} \qquad [15.39]$$

where M = detector mass (kg)
C = detector specific heat ($\mathrm{J\,kg^{-1}\,°C^{-1}}$)
A = detector surface area ($\mathrm{m^2}$)
U = heat transfer coefficient ($\mathrm{W\,m^{-2}\,°C^{-1}}$)

Rearranging, we have:

$$T_D + \tau\frac{\mathrm{d}T_D}{\mathrm{d}t} = \frac{1}{UA}P_D + T_S \qquad [15.40]$$

where time constant $\tau = MC/UA$. The transfer function relating corresponding deviation variables ΔT_D, ΔP_D, ΔT_S is:

Transfer function for thermal detector

$$\Delta\bar{T}_D(s) = \frac{1/UA}{(1+\tau s)}\Delta\bar{P}_D(s) + \frac{\Delta\bar{T}_S(s)}{(1+\tau s)} \qquad [15.41]$$

In the steady state, $\mathrm{d}T_D/\mathrm{d}t = 0$ so that [15.40] reduces to:

$$T_D = \frac{1}{UA}P_D + T_S \qquad [15.42]$$

Commonly used thermal detectors are **thermopiles** (Fig. 15.14(a)), which consist of a large number of thermocouples in series, and **bolometers**, which are metal or semiconductor (thermistor) resistance material in the form of thin films or flakes (Fig. 15.14(b)). In both cases the surface of the detector is blackened to maximise

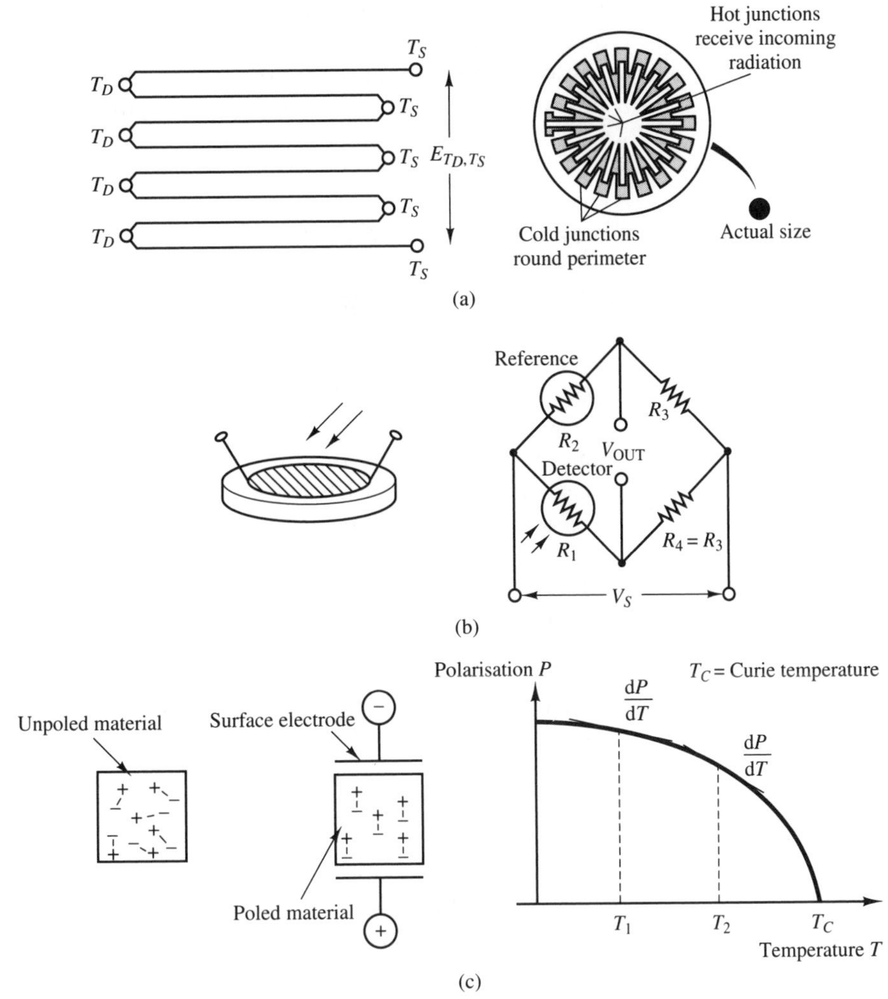

Fig. 15.14 Thermal detectors
(a) Thermopile
(b) Bolometer
(c) Pyroelectric

the absorption of incoming radiation. For a thermopile consisting of n thermocouples in series, with hot junction at T_D °C and reference junction at T_S °C (the temperature of the surroundings), the e.m.f. is given approximately by (Section 8.5)

$$E_{T_D,T_S} \approx na_1(T_D - T_S) \qquad [15.43]$$

From [15.42] the equilibrium temperature difference is $T_D - T_S = (1/UA)P_D$ giving:

$$E_{T_D,T_S} = \left(\frac{na_1}{UA}\right)P_D \qquad [15.44]$$

i.e. the detector sensitivity $K_D = \dfrac{na_1}{UA}$ mV/W.

The bridge circuit shown in Fig. 15.14(b) gives an output voltage approximately proportional to power P_D. Here R_1 is a radiation detecting bolometer at T_D °C, R_2 a reference bolometer at T_S °C and R_3, R_4 fixed equal resistors. This circuit has already been discussed in Section 9.1 (Fig. 9.4(a)). For metal resistive elements of temperature coefficient α °C^{-1} we have (from eqn [9.17]):

$$V_{OUT} \approx V_s \frac{R_0}{R_3} \alpha(T_D - T_S) \qquad [15.45]$$

provided $R_3 \gg R_0$ (resistance of bolometer at 0 °C). This gives:

$$V_{OUT} \approx \frac{V_s \, \alpha \, R_0}{UAR_3} . P_D \qquad [15.46]$$

indicating the system output voltage is approximately proportional to P_D and independent of T_S.

The time constant τ of a thermal detector is minimised by using thin flakes or films which have a large area-to-volume ratio that is a small value of M/A. However, τ cannot be reduced much below a few milliseconds because of the low heat transfer coefficient U between the detector and the surrounding air.

Pyroelectric thermal detectors are coming into wider use.[5] These are man-made ferroelectric ceramics which also show piezoelectric properties (Section 8.7). The principle of ferroelectricity is shown in Fig. 15.14(c). The ceramic is composed of a mass of minute crystallites; provided the ceramic is below the Curie temperature, then each crystallite behaves as a small electric dipole. Normally the material is **unpoled**, i.e. the electric dipoles are randomly orientated with respect to each other. The material can be poled, i.e. the dipoles lined up, by applying an electric field when the ceramic is just below the Curie temperature. After the material has cooled and the applied field has been removed, the dipoles remain lined up leaving the ceramic with a residual polarisation P. The pyroelectric effect arises because the incident radiant power causes the ceramic temperature T to increase, P decreases with T according to the non-linear relation shown in Fig. 15.14(c). This reduction in P causes a reduction in capture surface charge in the material and an excess of induced charge on the electrodes. If Δq is the excess charge caused by a temperature rise ΔT then:

$$\Delta q = \left(\frac{dP}{dT}\right) A \Delta T \qquad [15.47]$$

where A is the area of the electrodes and dP/dT the slope of the $P - T$ characteristics. The electrodes and the rectangular block of dielectric ceramic form a parallel plate capacitor C_N. The ceramic can either be regarded as a charge generator Δq in parallel with C_N or a Norton current source i_N in parallel with C_N (Section 5.1.3) where:

$$i_N = \frac{dq}{dt} = A \frac{dP}{dT} \cdot \frac{dT}{dt} \qquad [15.48]$$

A typical pyroelectric detector has a diameter of 2 mm, a responsivity or sensitivity of 250 V W^{-1} and a noise equivalent power of 2×10^{-9} W Hz^{-1} at 25 °C. When used with a silicon window the wavelength response $D(\lambda)$ is reduced to between 1 and 15 μm.

15.5.2 Photon detectors

Photon detectors are semiconductor devices in which incident photons of radiation cause electrons to be excited from the valence band to the conduction band thereby causing a measurable electrical effect. These detectors only respond to photons whose energy hc/λ is approximately equal to the energy gap E_G between valence and conduction bands. This means that photon detectors have a narrow wavelength response $D(\lambda)$ with peak wavelength $\lambda_0 \approx hc/E_G$. Photon detectors normally have higher responsivities than thermal detectors. Because they rely on atomic processes which are inherently faster than bulk heat transfer processes, photon detectors usually have shorter time constants than thermal detectors. There are two main types of photon detector, **photoconductive** and **photovoltaic**.

Photoconductive detectors

Here the presence of excited electrons in the conduction band causes an increase in electrical conductivity and a decrease in electrical resistance. The resistance R of a photoconductive detector therefore decreases as the total power P_D increases; the relationship is extremely non-linear and is best expressed in the logarithmic form (Fig. 15.15(a)).

$$\log_{10} R = a - b \log_{10} P_D \qquad [15.49]$$

Figure 5.15(b) shows the wavelength response $D(\lambda)$ for three photoconductive detectors in common use. Cadmium sulphide has $\lambda_0 \approx 0.6\ \mu\text{m}$ and is thus suitable for visible radiation. Lead sulphide with $\lambda_0 \approx 3.0\ \mu\text{m}$ and indium antimonide with $\lambda_0 \approx 5.3\ \mu\text{m}$ are suitable for infrared radiation.

It is important that the detector wavelength characteristics $D(\lambda)$, the source characteristics $S(\lambda)$ and the transmission medium characteristics $T(\lambda)$ all match each other as closely as possible. Thus a cadmium sulphide detector is ideal for measuring the temperature of a target at 4800 K ($\lambda_P \approx 0.6\ \mu\text{m}$) while indium antimonide is ideal for a target at 550 K ($\lambda_P \approx 5.3\ \mu\text{m}$). A typical indium antimonide detector of dimensions 6.0×0.5 mm has $K = 7 \times 10^3\ \text{V W}^{-1}$, $\tau = 5\ \mu\text{s}$ and NEP $= 1.6 \times 10^{-11}$ W. Photoconductive detectors must be incorporated into deflection bridge circuits to convert their output into a voltage signal.

Photovoltaic detectors[2,3,5]

These are photodiodes made by forming a junction of P-type and N-type extrinsic semiconductors. Figure 15.15(c) shows typical current/voltage characteristics for a photodiode under dark and illuminated conditions. The effect of the incident radiation is to move the whole characteristic bodily downwards by an amount equal to the **photocurrent** i_P. The photocurrent is proportional to total power P_D used by the detector:

$$i_P = K_D P_D \qquad [15.50]$$

where $K_D\ \text{A W}^{-1}$ is the sensitivity or responsivity. This type of detector can therefore by represented as a Norton current source in parallel with the diode resistance R_D and capacitance C_D (Fig. 15.15(d)). Lead–tin telluride is an example

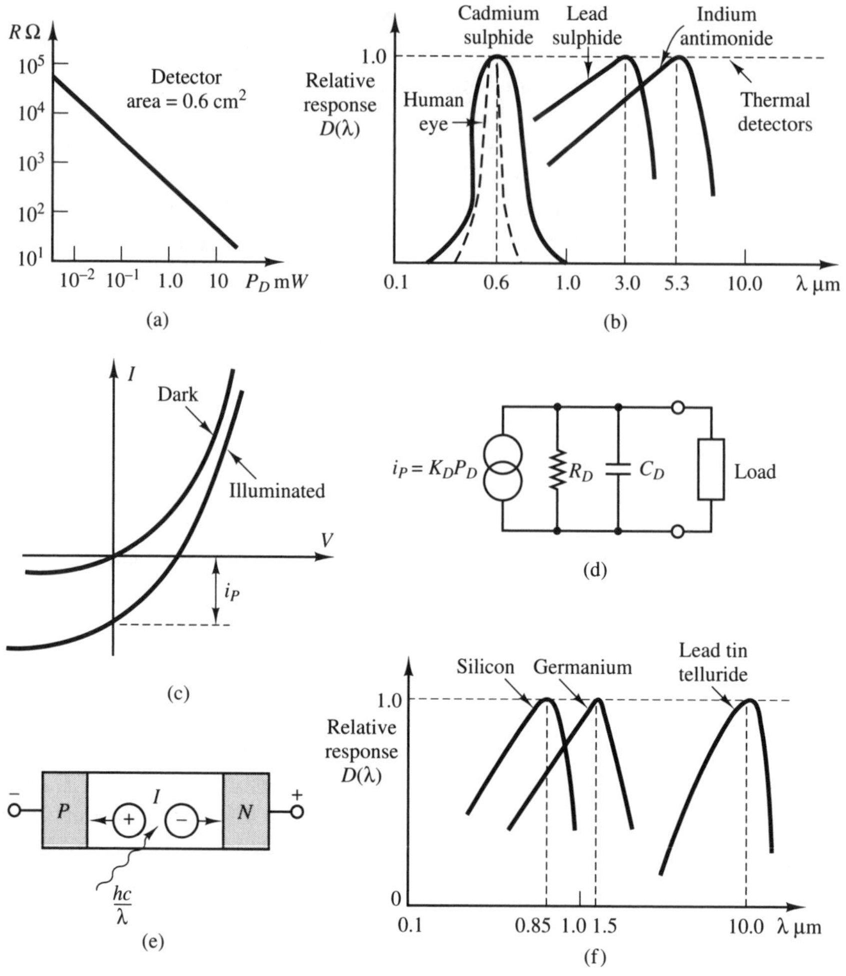

Fig. 15.15 Photon detectors
(a) Resistance-power relation for cadmium sulphide
(b) Wavelength response of photoconductive dectectors
(c) Current-voltage characteristics for photodiode
(d) Norton equivalent circuit for photodiode
(e) Construction of PIN photodiode
(f) Wavelength response of photovoltaic detectors

of an infrared photodiode. This operates at 77 K with a typical K_D of 5 A W^{-1} and wavelength response $D(\lambda)$ between 7 and 14 μm with peak wavelength $\lambda_0 \approx$ 10 μm (Fig. 15.15(f)).

Figure 15.15(e) shows the construction of a PIN diode; this has an extra layer of instrinsic (undoped) *I* material between *P* and *N* layers. The incident photons create additional electron-hole pairs in this region with a corresponding increase in photocurrent and responsivity over a comparable PN diode. A typical silicon PIN photodiode has an active area of 1 mm^2, a responsivity $K_D \approx$ 0.55 A W^{-1}, a peak wavelength $\lambda_0 \approx$ 0.85 μm and $D(\lambda)$ shown in Fig. 15.15(f). The wavelength response of this photodiode is therefore well matched to $S(\lambda)$ for GaAlAs LED and ILD sources; also their small size makes them ideal for coupling to optical fibres. Germanium PIN diodes with $K_D \approx$ 0.5 A W^{-1} and $\lambda_0 \approx$ 1.5 μm have a $D(\lambda)$ more suited to longer wavelength applications.

The principle of operation of avalanche photodiode (APD) devices differs from that of PIN diodes in one major respect. A PIN diode converts one photon to one electron, whereas in an APD multiplication takes place which results in many electrons at the output for each incident photon.

15.6 Measurement systems

Comprehensive reviews of optical measurement systems, especially involving optical fibre sensors, are given in references [6] and [7].

15.6.1 Modulation of intensity by source

Since the power emitted by a hot body source depends critically on source temperature (eqns [15.6] and [15.10]) systems of the type shown in Fig. 15.2 are commonly used for remote temperature measurement. There are two basic types of system: **broad-band** and **narrow-band**.

Broad-band

These systems use all wavelengths present in the incoming radiation and are often referred to as total radiation pyrometers. A thermal detector is therefore used and from eqns [15.38], [15.10] and [15.6] the total power used by the detector is:

$$P_D = \frac{C_1 K_F}{2\pi} \int_0^\infty \frac{\epsilon(\lambda, T)\, T(\lambda)\, F(\lambda)\, d\lambda}{\lambda^5 \left[\exp\left(\frac{C_2}{\lambda T}\right) - 1\right]} \qquad [15.51]$$

The integral can be evaluated numerically if $\epsilon(\lambda, T)$, $T(\lambda)$ and $F(\lambda)$ are known. However, if we assume that $\epsilon(\lambda, T)$, $T(\lambda)$, $F(\lambda)$ have constant values ϵ, T_M and F respectively for all λ, T then using eqn [15.7]

$$P_D = \frac{1}{2\pi}\epsilon T_M K_F F \int_0^\infty \frac{C_1 d\lambda}{\lambda^5 \left[\exp\left(\frac{C_2}{\lambda T}\right) - 1\right]} = \frac{1}{2\pi}\sigma\epsilon T_M K_F F T^4$$

The corresponding detector output signal is:

Detector output signal for broad-band temperature system

$$V_T = K_D P_D = \frac{K_D}{2\pi}\sigma\epsilon T_M K_F F T^4 \qquad [15.52]$$

The above assumptions represent a considerable over-simplification and even if they are justified there is still the problem of uncertainties in the values of ϵ, T_M, etc. For these reasons the system should be calibrated experimentally using a standard source of known emissivity. If the system is then used with a process source of different emissivity then a correction must be made.

Narrow-band

These systems use photon detectors so that only a narrow band of wavelengths is used. In some systems a narrow pass-band radiation filter $G(\lambda)$ is used to further

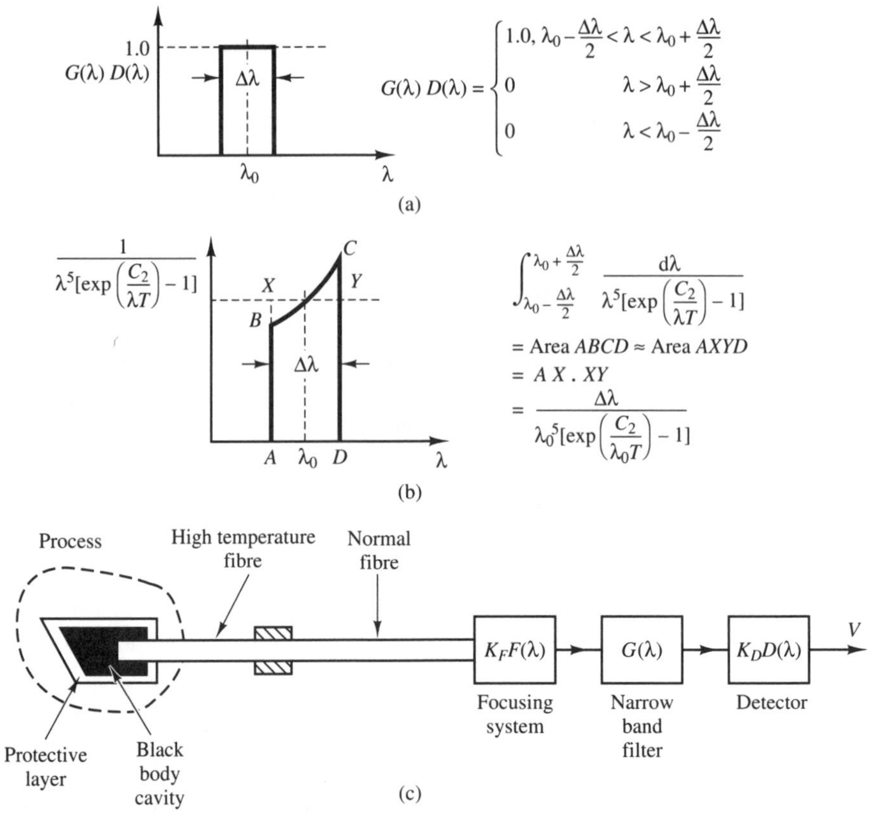

Fig. 15.16 Narrow-band radiation temperature system
(a) Wavelength characteristics
(b) Evaluation of integral
(c) Black body — optical fibre system

restrict the range of wavelengths incident on the detector. For these systems, the wavelength response $G(\lambda)D(\lambda)$ of the combined filter and detector can be approximated by the ideal band-pass filter characteristics of Fig. 15.16(a). Thus the expression for total power used is:

$$P_D = \frac{1}{2\pi} C_1 K_F \int_0^{\infty} \frac{\epsilon(\lambda, T)\, T(\lambda)\, F(\lambda)\, G(\lambda)\, D(\lambda)\, d\lambda}{\lambda^5 \left[\exp\left(\frac{C_2}{\lambda T}\right) - 1\right]} \qquad [15.53]$$

which reduces to:

$$P_D = \frac{1}{2\pi} C_1 K_F \int_{\lambda_0 - \Delta\lambda/2}^{\lambda_0 + \Delta\lambda/2} \frac{\epsilon(\lambda, T)\, T(\lambda)\, F(\lambda)\, d\lambda}{\lambda^5 \left[\exp\left(\frac{C_2}{\lambda T}\right) - 1\right]} \qquad [15.54]$$

The assumptions $\epsilon(\lambda, T) = \epsilon$, $T(\lambda) = T_M$, $F(\lambda) = F$ are far more justified for a narrow band of wavelengths and give:

$$P_D = \frac{1}{2\pi} C_1 \epsilon T_M K_F F \int_{\lambda_0 - \Delta\lambda/2}^{\lambda_0 + \Delta\lambda/2} \frac{d\lambda}{\lambda^5 \left[\exp\left(\frac{C_2}{\lambda T}\right) - 1\right]} \qquad [15.55]$$

The integral is evaluated approximately in Fig. 15.16(b) to give:

$$P_D = \frac{1}{2\pi} C_1 \epsilon T_M K_F F \frac{\Delta\lambda}{\lambda_0^5 \left[\exp\left(\frac{C_2}{\lambda_0 T}\right) - 1\right]}$$

$$\approx \frac{1}{2\pi} C_1 \epsilon T_M K_F F . \frac{\Delta\lambda}{\lambda_0^5} \exp\left(-\frac{C_2}{\lambda_0 T}\right) \qquad [15.56]$$

The approximation is valid since $\exp(C_2/\lambda_0 T) \gg 1$; for example if $\lambda_0 = 1\ \mu m$, $T = 10^3$ K, $(C_2/\lambda_0 T) = 14.4$, $\exp(C_2/\lambda_0 T) = 1.8 \times 10^6$. The detector output signal is thus:

Output signal for narrow-band radiation pyrometer

$$V(T) = K_D P_D \qquad [15.57]$$

$$= \frac{1}{2\pi} C_1 \epsilon T_M K_F F K_D \frac{\Delta\lambda}{\lambda_0^5} \exp\left(-\frac{C_2}{\lambda_0 T}\right)$$

The narrow band of wavelengths must be chosen to coincide with a 'window' in the transmission characteristics so that T_M is reasonably large. Again the system should be calibrated experimentally using a standard source of known emissivity and a correction applied if it is subsequently used with a process source of different emissivity. The above problems of variations of source emissivity and atmospheric transmission characteristics may be solved using the system shown in Fig. 15.16(c). Here a black body cavity is inserted in the process and radiation is transmitted to the radiation receiver using an optical fibre system rather than the atmosphere. The fibre system consists of a short run of special fibre, which can withstand the high process temperatures, connected to a much longer length of normal fibre.

15.6.2 Modulation of intensity by transmission medium

These are systems of the type shown in Fig. 15.1, where the measured variable alters the characteristics of the transmission medium and so modulates the intensity of the radiation reaching the detector. There are several ways in which this modulation can be achieved.

Intensity modulation using shutters and gratings

The simplest type of shutter displacement sensor is shown in Fig. 15.17(a). Figures 15.17(b), (c) and (d) show Moiré fringe gratings for the digital measurement of linear and angular position. In the linear type there are two finely ruled gratings of the same pitch P, slightly angled to each other. Relative movement of the gratings in a horizontal direction causes a series of dark fringes to move vertically across the sensor giving an approximately sinusoidal variation in light intensity and detector output signal. In the angular case, one grating has N lines, the other $N + 1$ lines, (N is typically between 50 and 1200); for every complete revolution of the gratings there are N complete revolutions of the fringe pattern and N cycles of the detector output signal. Thus in both cases the exact displacement can only be established by counting the number of detector output cycles: Moiré gratings are therefore often referred to as **incremental encoders**.

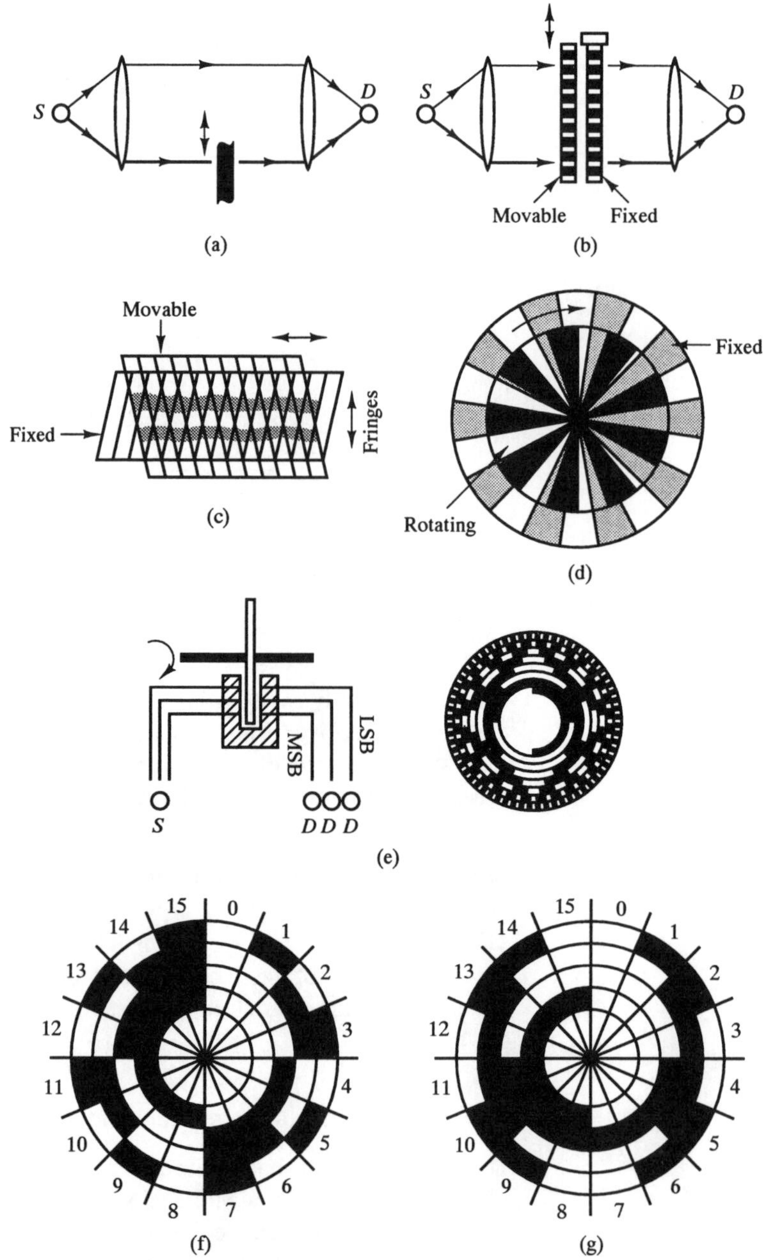

Fig. 15.17 Intensity modulation using shutters and gratings
(a) Shutter modulator
(b) Grating modulator
(c) Linear displacement moiré fringe gratings
(d) Angular displacement moiré fringe gratings
(e) Digital encoder disc
(f) Binary coded disc
(g) Gray coded disc

Figures 15.17(e), (f) and (g) show coded disc systems for the digital measurement of angular position. Here the disc is marked out in transparent and opaque segments according to the required digital code. **Gray code** (Fig. 15.17(g)) is a cyclic code in which only one digit can change between two consecutive sectors, this is in contrast to binary code (Fig. 15.17(f)) where several digits may change between consecutive sectors. If the detectors are misaligned slightly, then with a binary coded disc false codes can occur during the transition from one sector to the next. This cannot occur with Gray code; Figure 15.17(e) shows a typical 8-digit Gray coded disc. The output digital code uniquely specifies the angular position of the disc and coded discs are often referred to as **absolute encoders**.

Intensity modulation using relative displacement of fibres

Figure 15.18 shows four possible arrangements for the measurement of small linear or angular displacements. Here the intensity of the radiation incident on the detector is modulated according to the relative displacement of two optical fibres. The underlying principles have already been discussed in section 15.4.2. For example in Fig. 15.18(a), as x is decreased the solid angle ω of the cone of rays into the second fibre increases, causing the power incident on the detector to increase. This effect continues until the cone of rays into the second fibre corresponds to the cone of emergence of the first fibre. The arrangements shown in Figs 15.18(a), (b) and (c) are affected by fibre misalignment, vibration and source variations; the differential system shown in Fig. 15.18(d) should be unaffected by source variations.

Intensity modulation using reflection

Figure 15.19(a) shows a simple system for the remote measurement of angular velocity. The rotating mechanism has a reflective strip on its surface so that once in every revolution light is reflected back directly to the detector. This means that the number of detector pulses per second is equal to the revolutions of the mechanism per second.

In the liquid level measurement system shown in Fig. 15.19(b) a change in level Δh causes the reflected beam to be displaced laterally by an amount proportional to Δh. This change in lateral displacement is sensed by a two-dimensional array of detectors.

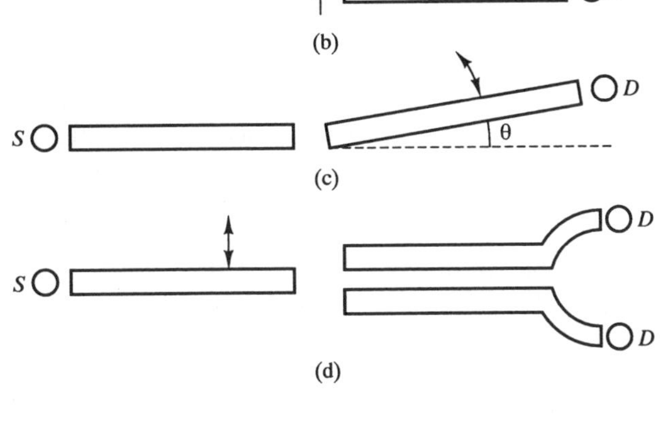

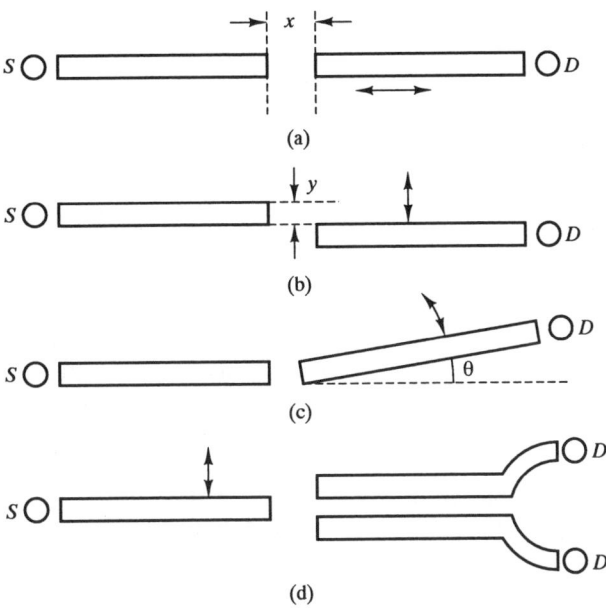

Fig. 15.18 Intensity modulation using relative displacement of fibres
(a) Longitudinal
(b) Transverse
(c) Angular
(d) Differential

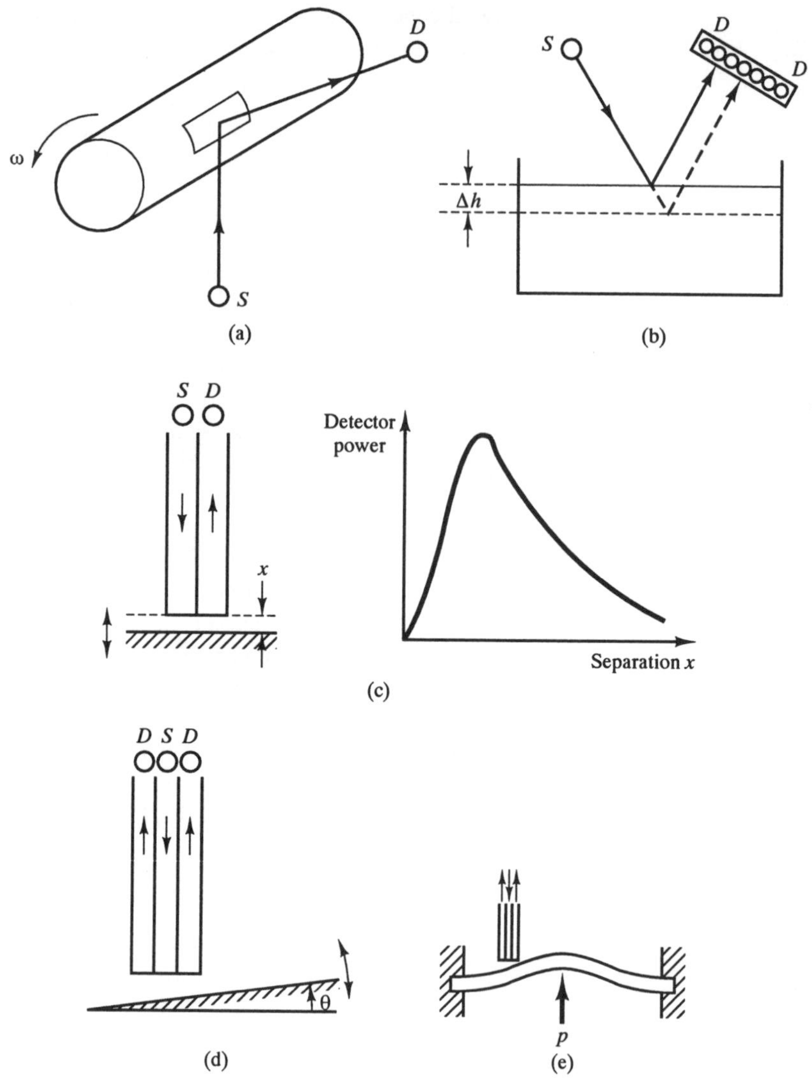

Fig. 15.19 Intensity modulation using reflection
(a) Angular velocity sensor
(b) Level sensor
(c) Two fibre linear displacement sensor and characteristics
(d) Three fibre angular displacement sensor
(e) Three fibre pressure sensor

Figure 15.19(c) shows a two-fibre linear displacement sensor and its characteristics. These characteristics show the non-linear relationship between the power incident on the detector and the separation x of fibres and reflecting surface. For small x, the detector power increases with x because the area of overlap between the cones of emergence and acceptance increases. For larger x, detector power reaches a maximum value and then decreases according to an inverse square law. This system is affected by vibration, changes in source intensity and reflection coefficient; there are also two values of separation for a given value of detector power.

A potentially more accurate sensor is the three-fibre angular displacement sensor shown in Fig. 15.19(d).[8] Here there is a single source fibre and two detector fibres; changing the angle θ alters the ratio of the powers collected by the two return fibres and the ratio of detector output signals. This ratio depends only on θ and the refractive indices of the fibre core and cladding; it is independent of source intensity, reflection coefficient and fibre-reflector separation. One application of this is the three-fibre pressure sensor shown in Fig. 15.19(e). Here the elastic element is a thin flat circular diaphragm clamped around the circumference (Fig. 8.14). The fibre system is positioned close to a point on the diaphragm where the angular deformation is maximum.

Intensity modulation using attenuation and absorption

In this type of sensor the intensity of radiation arriving at the detector is modulated either by variable attenuation in an optical fibre system or variable absorption in gases and liquids.

Figure 15.20(a) shows an optical fibre force sensor, an increase in force F causes increased microbending of the fibre. This causes rays with ϕ just greater than the critical angle θ_c (Fig. 15.9), which would normally be totally internally reflected back into the core, to be increasingly refracted into the cladding and lost from the core. The result is that the power incident on the detector decreases as F is increased.

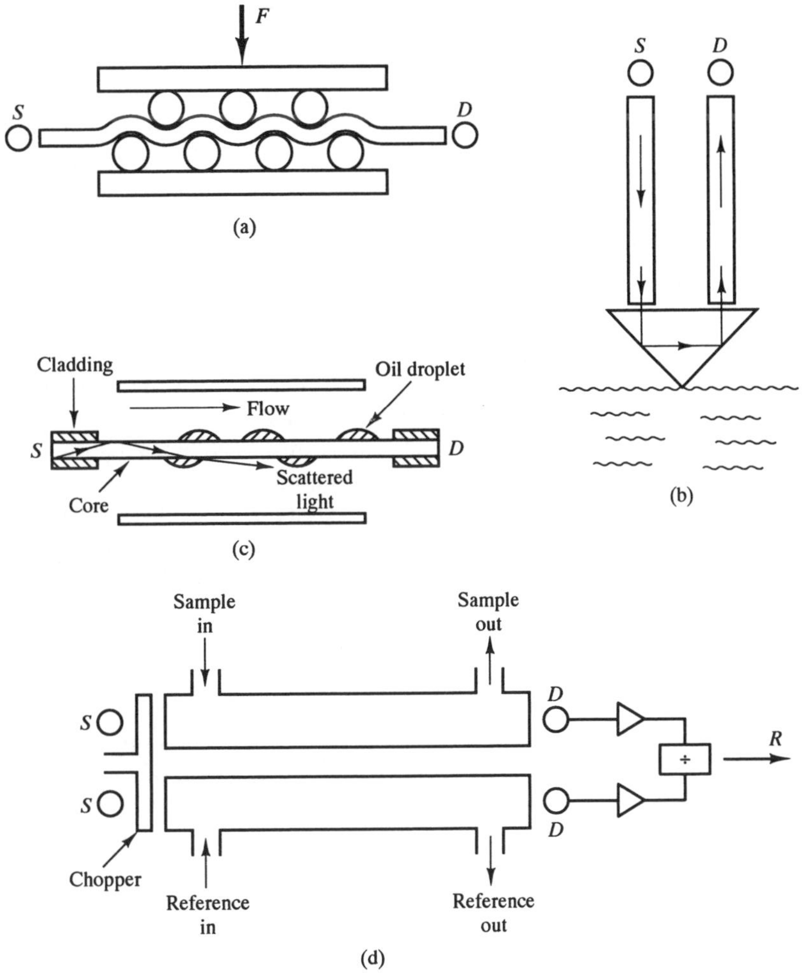

Fig. 15.20 Intensity modulation using attenuation and absorption
(a) Microbend force sensor
(b) Two fibre level detector
(c) Oil in water detector
(d) Infrared gas analyser (two-beam)

An optical level detector is shown in Fig. 15.20(b). This consists of two fibres (one inlet, one exit) cemented onto a 90° glass prism. For a low level, the prism is surrounded by air so that total internal reflection occurs at both prism faces resulting in a high detector output signal. For a high level, the prism is surrounded by liquid so that only partial internal reflection occurs resulting in a much lower detector output signal.

The system shown in Fig. 15.20(c) can be used to detect the presence of oil droplets in water. The water flows over an optical fibre core from which the cladding has been removed. The refractive index of the core is similar to that of oil, but higher than water. This means that if an oil droplet settles on the core some rays of light are refracted out of the core causing a reduction in detector output signal.

In Section 15.3 we saw that many hydrocarbon and other gases show strong absorption bands in the infrared; this means that if the percentage of the absorbing molecule in a gas mixture changes then there is a corresponding change in the transmission characteristics $T(\lambda)$. Figure 15.20(d) shows the infrared gas analyser which uses this principle. Broad-band infrared sources are used; the source radiation is chopped with a rotating mechanical chopper. This means that the detector output signals are a.c. so that electronic band-pass filters and phase-sensitive detectors can be used to reject electrical interference and noise. The system shown is a two-beam type, with the sample cell containing the gas to be analysed and the reference cell containing the pure gas of interest. For example, if we wish to measure the percentage of CO in flue gas then the reference cell will contain pure CO. The detector output signals are amplified and their ratio R computed; R depends on the percentage of the absorbing molecule in the sample gas, but should be independent of modifying effects such as source variations.

15.6.3 Two wavelength systems

The simple intensity-modulated systems discussed so far can be affected by environmental modifying and interfering inputs. For example, the source-modulated broad-band and narrow-band temperature systems are affected by changes in emissivity and transmission medium characteristics; the transmission medium-modulated systems are affected by changes in source intensity. Some of these problems can be solved using the two wavelength system shown in Fig. 15.21. Here the incoming radiation from the transmission medium is focused and split equally

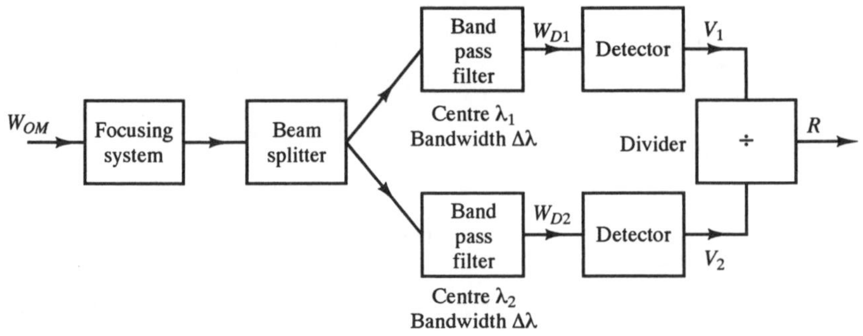

Fig. 15.21 Two wavelength system

into two beams. One beam is passed through a narrow pass-band optical filter with centre wavelength λ_1, the other through a similar filter with centre wavelength λ_2. The detectors chosen should have the same value of wavelength response $D(\lambda)$ at λ_1 and λ_2, either thermal detectors or photon detectors with a suitable wavelength response could be used. An electronic divider computes the ratio R between detector output signals V_1 and V_2.

Using eqn [15.57] for a narrow-band radiation pyrometer we can derive the equation for a two-wavelength (or two-colour) pyrometer:

$$R(T) = \frac{V_1(T)}{V_2(T)} = \frac{\frac{1}{2\pi}(C_1 \epsilon T_M K_F F K_D)_{\lambda_1} \frac{\Delta\lambda}{\lambda_1^5} \exp\left(-\frac{C_2}{\lambda_1 T}\right)}{\frac{1}{2\pi}(C_1 \epsilon T_M K_F F K_D)_{\lambda_2} \frac{\Delta\lambda}{\lambda_2^5} \exp\left(-\frac{C_2}{\lambda_2 T}\right)} \qquad [15.58]$$

If λ_1 and λ_2 are close together, so that the emissivities and transmission factors are equal, then the ratio becomes:

Equation for two-colour pyrometer system

$$R(T) = \left(\frac{\lambda_2}{\lambda_1}\right)^5 \exp \frac{C_2}{T}\left(\frac{1}{\lambda_2} - \frac{1}{\lambda_1}\right) \qquad [15.59]$$

This system does not require calibration and correction for emissivity variations.

The two-wavelength principle can also be used in the infrared gas analyser shown in Fig. 15.20(d). Instead of using a two-beam system, with reference and sample cells, a single beam-sample cell system is used. Here λ_1 coincides with the centre of the absorption band for the component of interest and λ_2 with the centre of the band for another component.

15.6.4 Modulation of wavelength by transmission medium

All of the systems discussed so far rely on the measured variable modulating the **intensity or amplitude** of light waves. These systems are inherently susceptible to unwanted variations in source intensity and transmission characteristics. Two channel/wavelength systems overcome some of these problems, but wavelength modulation systems, where the **wavelength** λ of the light is determined by the measured variable, are completely independent of these unwanted effects. Two such systems have been developed at the National Physical Laboratory[9] and will now be described.

Figure 15.22(a) shows a 'Littrow' diffraction grating used as a sensor to convert angular displacement θ into changes in wavelength. Broad-band radiation emerging from the input fibre is first collimated into a parallel beam, this is incident at an angle θ onto a reflection grating. the grating consists of a regular array of grooves whose spacing d is comparable with the wavelength of the incident beam. The resulting diffraction pattern is complex but only light with wavelengths λ satisfying the equation

$$2d \sin\theta = m\lambda \qquad [15.60]$$

where $m = 1, 2, 3 \ldots$ etc., is reflected back along its own path to the output fibre.

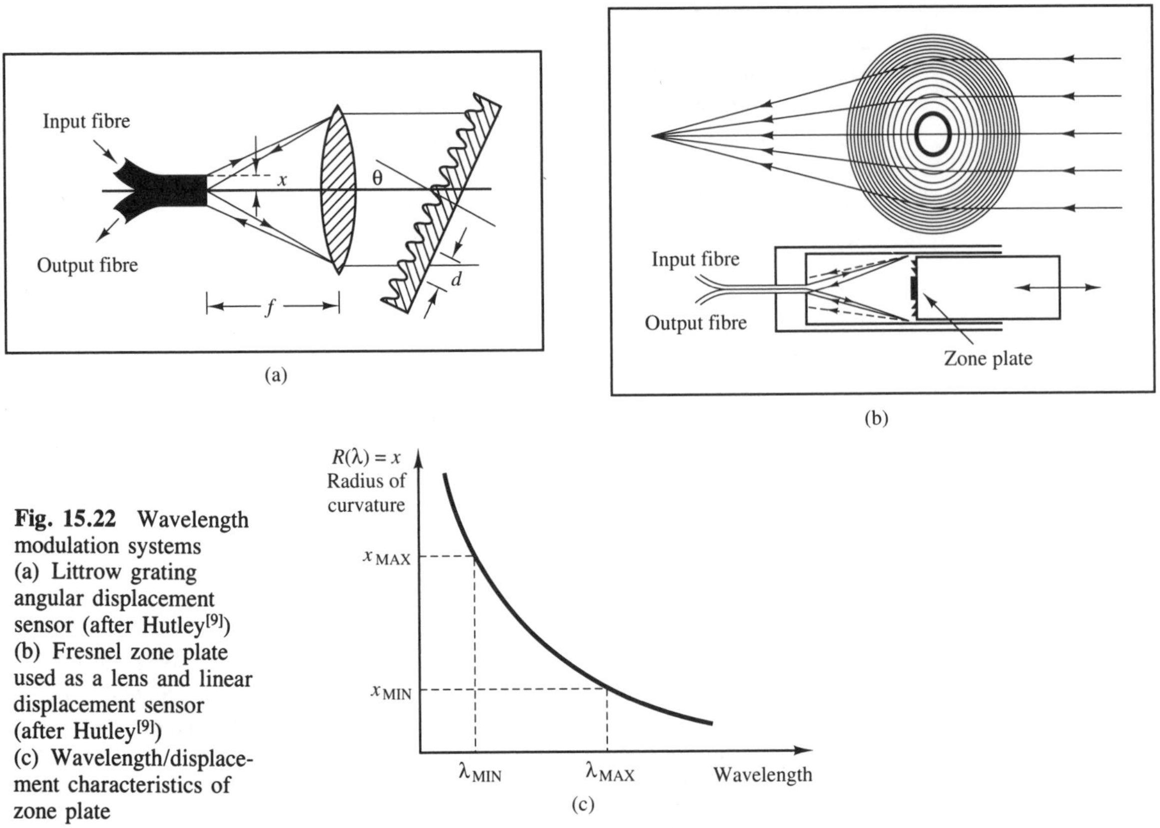

Fig. 15.22 Wavelength modulation systems
(a) Littrow grating angular displacement sensor (after Hutley[9])
(b) Fresnel zone plate used as a lens and linear displacement sensor (after Hutley[9])
(c) Wavelength/displacement characteristics of zone plate

By limiting the range of incident wavelengths, only first-order effects with $m = 1$ need to be considered, giving the following unique relation between reflected wavelength λ and grating angle θ:

Wavelength-encoded angular displacement sensor

$$\lambda = 2d \sin \theta \qquad [15.61]$$

The angular range of this device is limited to between 20° and 60°.

Figure 15.22(b) shows a **Fresnel zone plate**; it is a circular grating consisting of a number of concentric lines, the spacing between the lines decreasing with distance from the centre. The top diagram shows a parallel beam of monochromatic light incident on the zone plate. All rays are brought to a focus at a single point so that the plate acts as a lens of focal length f. However, the deviation of each ray depends on wavelength, so that different wavelengths are brought to a focus at different points; the focal length f is inversely proportional to wavelength λ.

The bottom diagram shows a reflecting zone plate, used as a linear displacement sensor. The plate acts as a spherical concave mirror with a radius of curvature R inversely proportional to wavelength (Fig. 15.22(c)). Broad-band radiation from the input fibre is incident on the zone plate a distance x away; the radiation reflected back to the output fibre is narrow-band, with centre wavelength λ such that:

Wavelength-encoded linear displacement sensor

$$R(\lambda) = x \qquad [15.62]$$

Thus if the range of displacement to be measured is x_{MIN} to x_{MAX}, the required range of wavelengths in the incoming radiation is λ_{MIN} to λ_{MAX} where

$$R(\lambda_{MAX}) = x_{MIN} \text{ and } R(\lambda_{MIN}) = x_{MAX}$$

Because of the need to exclude overlapping higher orders, λ_{MAX} cannot usually exceed $2\lambda_{MIN}$ with corresponding limits on the displacement range.

A wavelength decoder is necessary with both of these devices to convert wavelength variations into a usable voltage signal. One possibility is to split the beam from the output fibre equally between two detectors, with very different wavelength responses $D(\lambda)$ over the wavelength range of interest. The ratio of detector output voltages should vary monotonically with wavelength.

15.6.5 Interferometers

We firstly need to discuss the phenomenon of **interference** which is exhibited by any type of wave motion. Consider a monochromatic light source which is emitting light at a single angular frequency ω rad/s and corresponding wavelength λ. The light beam is then split into two beams of equal amplitude a. These beams then travel along different optical paths of lengths x_1 and x_2 before both arriving at a detector. The beams can be described by the wave equations (Section 16.3.1):

$$\phi_1 = a \sin\left(\omega t - \frac{2\pi}{\lambda} x_1\right), \; \phi_2 = a \sin\left(\omega t - \frac{2\pi}{\lambda}\tfrac{1}{2} x_2\right) \qquad [15.63]$$

so that the resultant light beam incident on the detector is given by:

$$\phi = \phi_1 + \phi_2 = a\left\{\sin\left(\omega t - \frac{2\pi}{\lambda} x_1\right) + \sin\left(\omega t - \frac{2\pi}{\lambda} x_2\right)\right\} \qquad [15.64]$$

$$= 2a \cos \frac{2\pi}{\lambda} \frac{(x_2 - x_1)}{2} \sin\left[\omega t - \frac{2\pi}{\lambda} \frac{(x_1 + x_2)}{2}\right]$$

This has amplitude:

$$a_R = 2a \cos \frac{\pi}{\lambda} x \qquad [15.65]$$

where x = optical path difference $x_2 - x_1$, and intensity:

Variation in resultant intensity with path difference

$$I_R = a_R^2 = 4a^2 \cos^2 \frac{\pi}{\lambda} x \qquad [15.66]$$

Figure 15.23(a) shows the form of eqn [15.66]. We see that when $x = 0, \lambda, 2\lambda, 3\lambda$, etc., $I_R = 4a^2$, i.e. the intensity of the resultant beam corresponds to the sum $(2a)$ of the amplitudes of the individual beams, then this is **constructive interference**.

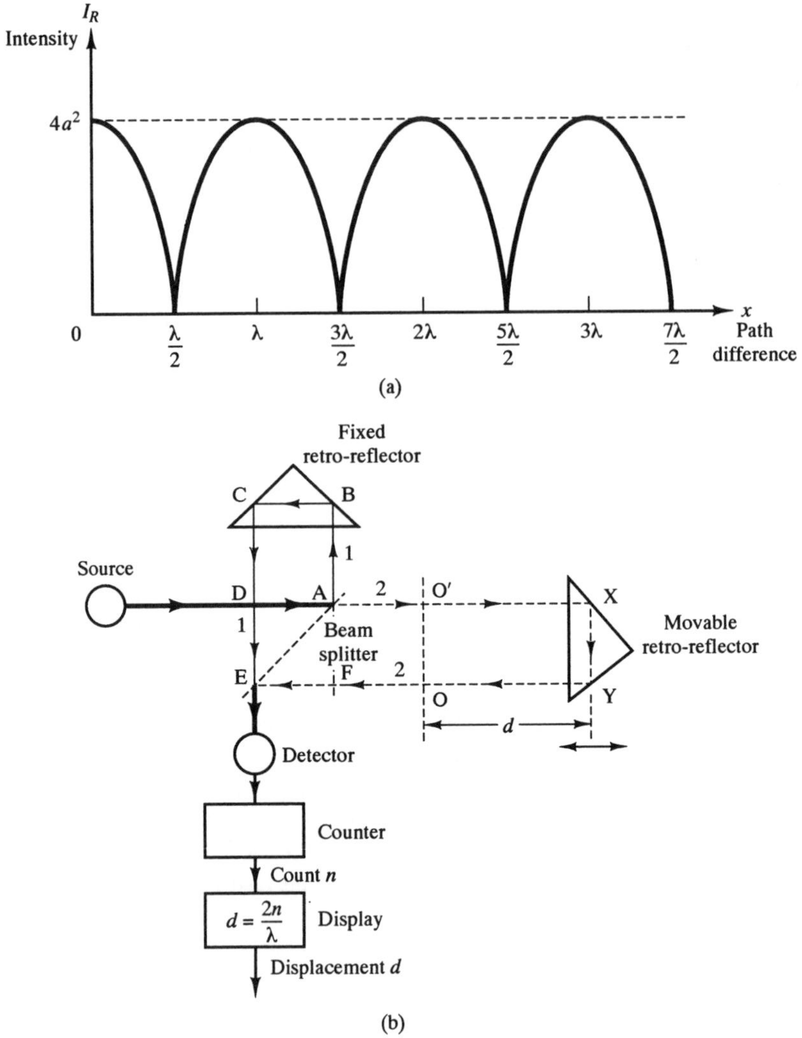

Fig. 15.23
Interferometers
(a) Variation in resultant intensity with path difference
(b) Michelson interferometer

When $x = \lambda/2, 3\lambda/2, 5\lambda/2$, etc., i.e. the intensity of the resultant beam corresponds to $I_R = 0$, the **difference** (i.e. 0) of the amplitudes of the individual beams, then this is **destructive interference**. Thus if path difference x is initially zero, I_R is maximum and a bright fringe is observed. If x is increased to $\lambda/2$, I_R is zero and a dark fringe is observed. If x is further increased to λ, I_R returns to maximum and a bright fringe is again observed. Thus the variation in the intensity of the light incident on the detector, and therefore the detector output signal, with path difference x, has a period equal to λ.

Figure 15.23(b) shows a **Michelson interferometer** which is used in the accurate measurement of displacement. Here a monochromatic beam of light from a laser source (Section 15.2.4) is split into two equal beams 1 and 2 at A using a beam splitter (in its simplest form this is a half-silvered mirror). Beam 1 travels to a fixed retro-reflector and is reflected back to the point E, so that the total optical path length x_1 = ABCDE. Beam 2 travels to a movable retro-reflector and is again reflected back to the point E, so that here the total optical path length x_2 = AXYFE. The optical path difference x is therefore:

$$x = x_2 - x_1 = \mathrm{AXYFE} - \mathrm{ABCDE}$$
$$= (\mathrm{AX}+\mathrm{XY}+\mathrm{YF}+\mathrm{FE}) - (\mathrm{AB}+\mathrm{BC}+\mathrm{CD}+\mathrm{DE}) \quad [15.67]$$

Since XY = DE, FE = BC then

$$x = (\mathrm{AX} + \mathrm{YF}) - (\mathrm{AB} + \mathrm{CD}) = 2\mathrm{AX} - 2\mathrm{AB} \quad [15.68]$$

If displacement d is measured from the line OO′, where AO′ = AB then AX = AB + d and

$$x = 2(\mathrm{AB} + d) - 2\mathrm{AB} = 2d \quad [15.69]$$

This means that a change δd in d results in a change δx in x of $2\delta d$, so that if $\delta d = \lambda/2$, i.e. half a wavelength, then $\delta x = \lambda$, i.e. a full wavelength.

Thus with a Michelson interferometer the detector output signal varies periodically with displacement d, the period being equal to $\lambda/2$, one half wavelength of the laser radiation. A Michelson interferometer used as a standard for length (Section 2.4.1) would typically use a helium–neon laser source with wavelength λ = 633 nm. Thus a displacement d = 1 mm would produce:

$$n = \frac{d}{\lambda/2} = \frac{2 \times 10^{-3}}{633 \times 10^{-9}} = \frac{2 \times 10^{6}}{633} = 3159.56$$

cycles of detector output signal. This displacement can therefore be measured to within a resolution of $\pm\lambda/4$ by counting the number of whole cycles, 3160 (rounded up) in this case.

We see therefore that a basic interferometer has a very high sensitivity when used for displacement measurement. For all displacements greater than λ or $\lambda/2$ it is necessary to count the total number of cycles of intensity variation. The sensitivity can be reduced by using a source of much longer effective wavelength λ' obtained by mixing two laser beams with much shorter wavelengths λ_1 and λ_2. This is called **synthetic wave** or **multisource heterodyne interferometry**. Thus, if we add two waves with the same amplitude a but with different frequencies ω_1, ω_2 and wavelengths λ_1, λ_2, the resultant wave is:

$$\phi = a \sin\left(\omega_1 t - \frac{2\pi}{\lambda_1}x\right) + a \sin\left(\omega_2 t - \frac{2\pi}{\lambda_2}x\right)$$
$$= 2a \cos\left[\frac{(\omega_1-\omega_2)}{2}t - 2\pi x.\frac{1}{2}\left(\frac{1}{\lambda_1} - \frac{1}{\lambda_2}\right)\right] \times$$
$$\sin\left[\frac{(\omega_1+\omega_2)}{2}t - 2\pi x.\frac{1}{2}\left(\frac{1}{\lambda_1} + \frac{1}{\lambda_2}\right)\right] \quad [15.70]$$

This is an **amplitude modulated wave**, with a carrier frequency of $(\omega_1 + \omega_2)/2$ and a much lower amplitude modulating frequency of $(\omega_1 - \omega_2)/2$. The amplitude or envelope function:

$$2a \cos\left[\frac{(\omega_1-\omega_2)}{2}t - 2\pi x.\frac{1}{2}\left(\frac{1}{\lambda_1} - \frac{1}{\lambda_2}\right)\right]$$

is a progressive wave of the form:

$$2a \cos\left[\omega' t - \frac{2\pi}{\lambda'} x\right]$$

with frequency $\omega' = (\omega_1 - \omega_2)/2$ and wavelength:

$$\frac{1}{\lambda'} = \frac{1}{2}\left(\frac{1}{\lambda_1} - \frac{1}{\lambda_2}\right) \text{ i.e. } \lambda' = \frac{2\lambda_1\lambda_2}{(\lambda_2 - \lambda_1)} \qquad [15.71]$$

Thus if $\lambda_2 = 1001$ nm and $\lambda_1 = 1000$ nm, $\lambda' = 2$ mm. This means that if the envelope wave with wavelength λ' is used as a source in an interferometer, displacements of up to one millimetre can be detected within one cycle of the intensity variation. The carrier wave with wavelength $(\lambda_1+\lambda_2)/2$ can still be used as an alternative source if high precision measurements are required.

Conclusion

This chapter has discussed:

(1) The two general types of optical measurement systems: fixed source, variable transmission medium and variable source, fixed transmission medium.
(2) The principles and types of sources: hot body, light-emitting diodes and lasers.
(3) The principles and characteristics of optical transmission media.
(4) The geometry of coupling of detector to source and the use of focusing systems and optical fibres.
(5) The principles and characteristics of thermal and photon detectors.
(6) Measurement systems using the modulation of intensity by source and transmission medium, two-wavelength systems, wavelength modulation and interferometry.

References

15.1 Polarizers Technical Products, 1988, *Technical Information on Temperature Standards and Black Body Sources*.
15.2 KEISER G 1983, *Optical Fibre Communications*, McGraw-Hill Book Company, Japan Ltd., Tokyo, pp. 12–44, 8–115, 145–67.
15.3 CHERIN A H 1983, *An Introduction to Optical Fibres*, McGraw-Hill Book Company, Japan Ltd., Tokyo, pp. 270–4.
15.4 HOUGHTON J and SMITH S D 1966, *Infrared Physics*, Oxford University Press, London, pp. 220–3.
15.5 Plessey Optoelectronics and Microwave Ltd, *Application Notes on Pyroelectric and Lead Tin Telluride Infra-red Detectors*.
15.6 JONES B E 1985, 'Optical Fibre sensors and systems for Industry', *J. Phys. E.: Scientific Instruments*, vol 18, 770–82.
15.7 MEDLOCK R S 1986, 'Review of Modulating Techniques for fibre optic sensors', *Measurement and Control*, vol. 19, no. 1, pp. 4–17.
15.8 HILL S *Optical Fibre Reflectance Transducer*, Teesside Polytechnic project report, unpublished.

15.9 HUTLEY M C 1985, 'Wavelength Encoded Optical Fibre Sensors', *N.P.L. News*, No. 363.

Problems

15.1 A broad-band radiation pyrometer has a thermistor bolometer detector. The radiant power incident on the thermistor in the steady state is given by:

$$P_D = 10T^4 \text{ pW}$$

where T is the temperature of the distant target in degrees Kelvin. The beam of radiation is chopped by a semicircular disc rotating at 60 r.p.m. The thermistor is incorporated into the deflection bridge circuit shown in Fig. Prob. 15.1 together with an identical thermistor at 290 K, the temperature of the detector surroundings. Using the detector data given below, sketch the form of the bridge output signal when the target temperature is 1000 K. Hint: assume a square wave variation in detector power which can be regarded as a series of steps.

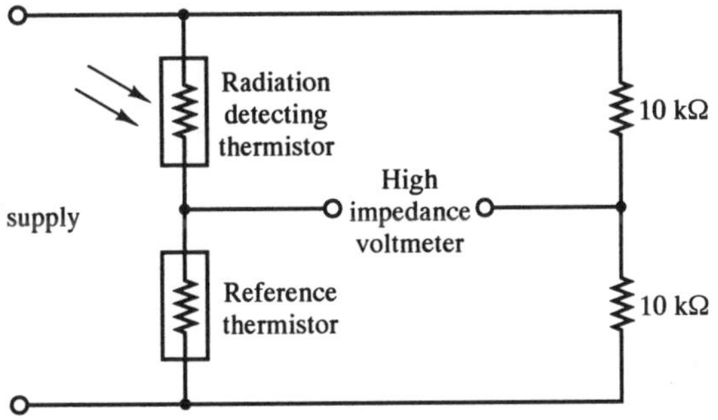

Fig. Prob. 1

Thermistor data

Mass = 1 g
Surface area = 10^{-4} m^2
Specific heat = 10^2 J kg^{-1} K^{-1}
Heat transfer coefficient = 10^4 W m^{-2} K^{-1}

$$\text{Resistance} = 1.0 \exp\left(\frac{3000}{\theta}\right) \Omega$$

where θ = thermistor temperature in degress Kelvin.

15.2 A thermocouple has an e.m.f. of 5 mV when the hot junction is at 100 °C and the cold junction is at 0 °C. A thermopile consisting of 25 such thermocouples in series is used as a detector in a chopped, broad-band pyrometer. When unchopped the power in the beam of radiation incident on the hot junction of the thermopile is given by $7.5T^4$ pW, where TK is the temperature of the distant target. The beam is chopped by a semicircular disc rotating at 6000 r.p.m. The reference junction of the thermopile is at the temperature of the pryometer enclosure. Assuming the thermocouples are linear and the detector data given below, sketch the form of the output signal when T = 2000 K.

Detector data

heat transfer coefficient = 2×10^4 W m^{-2} °C^{-1}.
Surface area = 1.6×10^{-4} m^2
Time constant = 1 ms

15.3 A narrow-band optical pyrometer is to be used to measure the temperature of the radiant section of a furnace in the range 1500 to 2000 K. The emissivity of the furnace wall is 0.5 at a wavelength of 0.5 μm. The transmission factor of the atmosphere between the furnace and the pyrometer is 0.90 at 0.5 μm.

Calculate the range of the pyrometer output signal given the following data:

Lens
Diameter of aperture = 3.0 cm
Focal length at 0.5 μm = 10.0 cm
Transmission factor at 0.5 μm = 0.95

Photoconductive detector
Diameter of detector disc = 0.5 cm
Detector sensitivity = 10 Ω W^{-1}
Centre wavelength = 0.5 μm
Bandwidth = 0.05 μm

Deflection bridge and amplifier
Overall sensitivity = 10 V Ω^{-1}
Temperature for bridge balance = 1500 K

The Planck–Wien law for the hemispherical spectral intensity of radiation from a black body is:

$$W(\lambda, T) = \frac{C_1}{\lambda^5\left[\exp\left(\frac{C_2}{\lambda T}\right) - 1\right]}$$

where
T = absolute temperature of body (K)
λ = wavelength of radiation (μm)
C_1 = 37 400 W μm^4 cm^{-2}
C_2 = 14 400 μm K

15.4 A two-colour radiation thermometer is used to measure the temperature of a body at 800 °C. The system incorporates two detectors, one measuring radiation at 0.5 μm and the other at 1.0 μm. What is the ratio of the detector output signals? State clearly all the assumptions made in the calculation. Explain the advantage of this system over broad-band and narrow-band systems.

15.5 An optical fibre transmission system consists of a circular LED source, a 2-metre length of optical fibre and a circular PIN diode detector. Both the source and detector are positioned 100 μm from the ends of the fibre. Detailed data for the source, a glass fibre, a polymer fibre and the detector are given below. Use this data to perform the following calculations.

(a) The total power P_s emitted by the source in all directions (use eqn [15.23(b)]).
(b) The numerical aperture and maximum angle of acceptance of the glass fibre.
(c) The numerical aperture and maximum angle of acceptance of the polymer fibre.
(d) The source and detector are linked by the glass fibre; calculate the power input to the fibre, fibre transmission factor, and power output from the fibre. (Hint: use $P_{IM}/B = K_{SM}$ and $P_{OM}/P_{IM} = T$)
(e) The source and detector are now linked by the polymer fibre; calculate the power input to the fibre, fibre transmission factor, and power output from the fibre.
(f) For both the glass system of (d) and the polymer system of (e), calculate the power incident on the detector and the detector output current. (Hint: use $P_D/P_{OM} = K_{MD}$)

LED source
Brightness = 10 W cm^{-2} sr^{-1}
Diameter = 200 μm
Centre wavelength = 810 nm

Glass fibre
Multimode step index
Core diameter = 100 μm
Core refractive index = 1.5
Core-cladding index difference = 0.015
Attenuation at 810 nm = 5 dB km^{-1}
Fibre length = 2 m

Polymer fibre
Multimode step index
Core diameter = 1.0 mm
Core refractive index = 1.65
Core cladding index difference = 0.04
Attenuation at 810 nm = 500 dB Km^{-1}
Fibre length = 2 m

PIN diode detector
Diameter = 2.0 mm
Centre wavelength = 810 nm
Responsibity at 810 nm = 0.55 A W^{-1}

15.6 A system for the measurement of linear displacement consists of a reflecting Fresnel zone plate (Fig. 15.22(b)), acting as a wavelength encoder, and a PIN diode photon detector, acting as a wavelength decoder. The distance between input/output fibres and the zone plate (input displacement range) can be varied between 12.5 and 25 mm. The radius of curvature $R(\lambda)$ of the zone plate is given by $R(\lambda) = 1.25 \times 10^4/\lambda$ where R is in mm and λ in nm. The detector has $D(\lambda) = 9 \times 10^{-4}\lambda - 0.35$. Calculate the wavelength output range for the encoder and the corresponding ratio of decoder output signals.

16

Ultrasonic measurement systems

Ultrasound refers to sound waves at frequencies higher than the range of the human ear, i.e. at frequencies greater than about 18 kHz. Ultrasonic waves obey the same basic laws of wave motion as lower frequency sound waves; they have, however, the following advantages:

(a) Higher frequency waves have shorter wavelengths; this means that diffraction or bending around an obstacle of given dimensions is correspondingly reduced. It is therefore easier to direct and focus a beam of ultrasound.
(b) Ultrasonic waves can pass easily through the metal walls of pipes and vessels. This means that the entire measurement system can be mounted completely external to the fluid, i.e. is **non-invasive**. This is extremely important with hostile fluids, such as those with corrosive, radioactive, explosive or flammable properties. There is also no possibility of blockage occurring with dirty fluids or slurries.
(c) Ultrasound can be launched into and propagated through biological tissue making it useful for medical applications.
(d) The silence of ultrasound means that it has important military applications.

This chapter studies ultrasonic transmitters and receivers, and the principles of transmission, and examines a range of ultrasonic measurement systems.

16.1 Basic ultrasonic transmission link

This forms the basis of any ultrasonic measurement system and is shown in Fig. 16.1. It consists of an ultrasonic transmitter, the transmission medium and an ultrasonic receiver. The most commonly used devices for ultrasonic transmitters and receivers are piezoelectric sensing elements (Section 8.7). The piezoelectric effect is reversible, i.e. mechanical energy can be converted into electrical energy and electrical energy into mechanical. The ultrasonic transmitter uses the inverse piezoelectric effect; if a sinusoidal voltage $\hat{V}_s \sin \omega t$ is applied to the transmitting crystal, then the crystal undergoes a corresponding sinusoidal deformation x (eqn (8.67)). This vibration of the crystal is transmitted to the particles at the beginning of the medium, and these are set in sinusoidal motion, causing other particles to

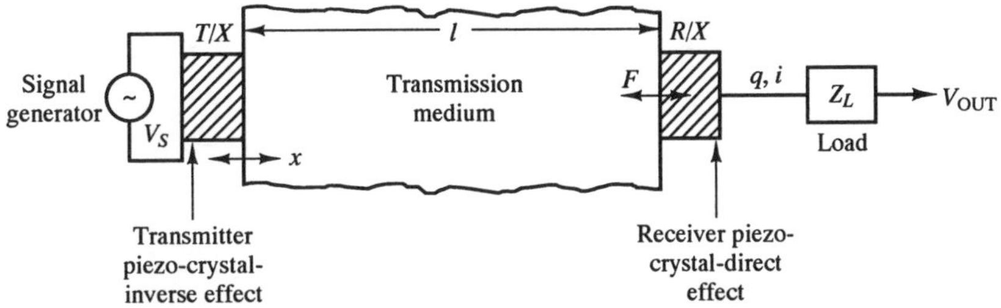

Fig. 16.1 Basic ultrasonic transmission link

vibrate, until eventually the disturbance is transmitted to the other end of the medium. These sinusoidal particle displacements set up an accompanying sinusoidal pressure or stress in the medium. This is detected by the ultrasonic receiver, which is simply a force sensor using the direct piezoelectric effect. The fluctuating pressure causes a sinusoidal force F over the area of the crystal, thus producing a corresponding time-varying charge q and current i. This current produces a corresponding output voltage V_{OUT} across a load Z_L.

16.2 Piezoelectric ultrasonic transmitters and receivers

16.2.1 Basic principles

In Chapter 5 we introduced the concept of **effort** and **flow** variables; voltage and force are examples of effort variables, current and velocity are examples of flow variables. The interconnection of any two elements in a system is then represented by two terminals and the appropriate effort/flow pair, e.g. voltage/current at an electrical connection and force/velocity at a mechanical connection. This leads to the concept of an element being represented by a four-terminal or **two-port network**. Thus in Fig. 5.17 a force sensor is represented as two-port network with a mechanical input port and an electrical output port. Two-port networks provide a good representation of piezoelectric transmitters and receivers; because the piezoelectric effect is reversible the same crystal can act as a transmitter (electrical input port/mechanical output port) or as a receiver (mechanical input port/electrical output port) and is said to be **bilateral**. Figure 16.2(a) shows transmitter and receiver two-port networks in block diagram form.

In order to derive the detailed form of these circuits we need to revise the properties of piezoelectric elements (Section 8.7). Starting with the basic equations for the direct and inverse effects we have:

Direct effect $q = dF$ [8.66]

Inverse effect $x = dV$ [8.67]

We also have the equation $F = kx$ relating applied force F and deformation x of a crystal of stiffness k. Thus $q = dkx$ and:

Relationship between flow variables for piezo-crystal

$$i = \frac{dq}{dt} = dk\dot{x} \qquad [16.1]$$

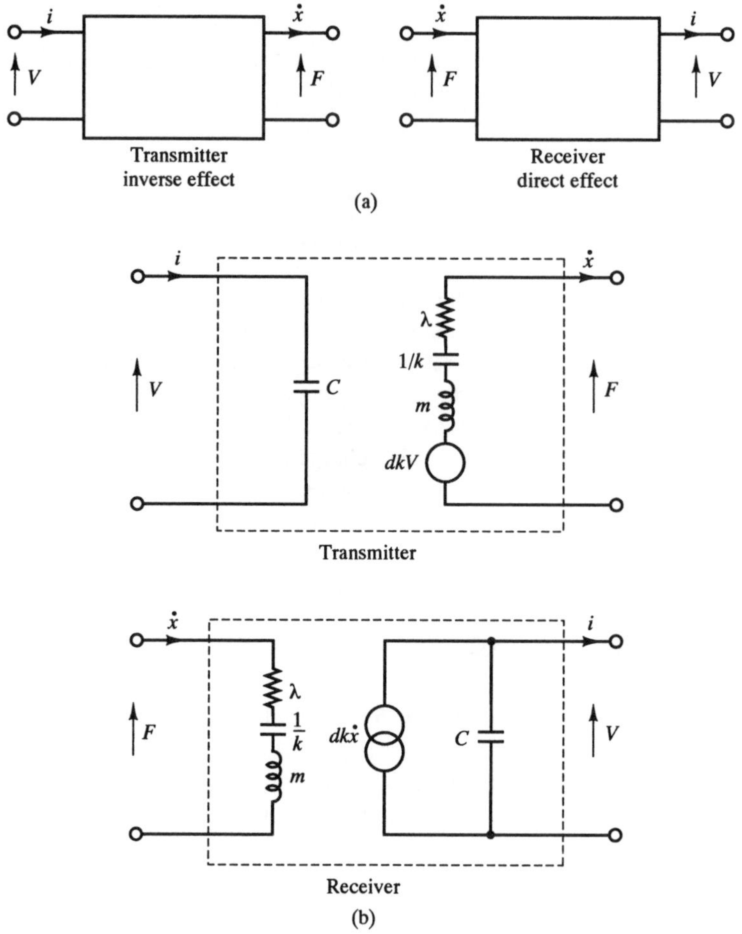

Fig. 16.2 Two port networks for piezoelectric transmitters and receivers
(a) Block diagrams
(b) Detailed form

Similarly $F/k = dV$ giving:

Relationship between effort variables for piezo-crystal

$$F = dkV \qquad [16.2]$$

The crystal has electrical capacitance C and the corresponding electrical impedance is purely capacitive:

$$Z_E(s) = \frac{\Delta\bar{V}(s)}{\Delta\bar{i}(s)} = \frac{1}{Cs} \qquad [16.3]$$

The mechanics of the crystal can be represented most simply using single mass (inertance m), spring (stiffness k) and damping (resistance λ) elements. The corresponding mechanical circuit consists of m, $1/k$ and λ in series and has mechanical impedance (Section 5.2).

$$Z_M(s) = \frac{\Delta\bar{F}(s)}{\Delta\bar{\dot{x}}(s)} = ms + \lambda + \frac{k}{s} \qquad [16.4]$$

Figure 16.2(b) shows the detailed form of the transmitter and receiver two-port networks. In the transmitter network, a voltage V at the input electrical port produces a force dkV which drives a velocity $\dot{x}$ through the output mechanical impedance. In the receiver network, a force F at the input mechanical port drives a velocity $\dot{x}$ through the input mechanical impedance, this produces a current $dk\dot{x}$ at the output electrical port.

Figure 16.3(a) shows the input electrical port of the transmitter connected to an

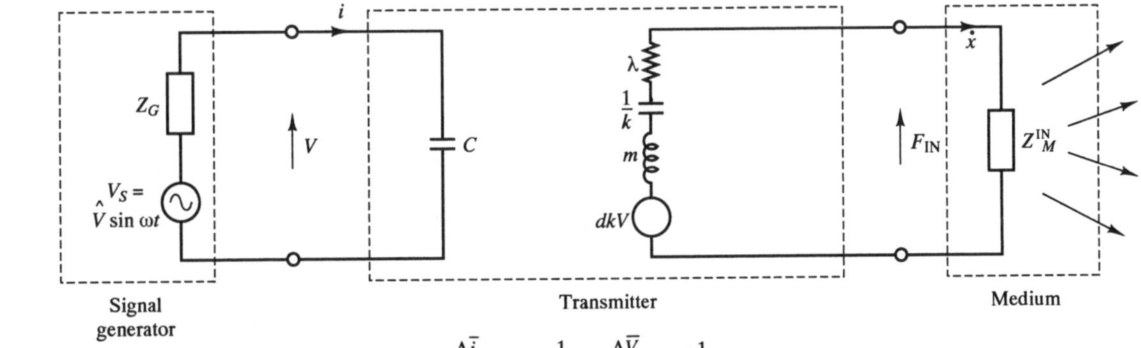

$$\frac{\Delta\bar{i}}{\Delta\bar{V}}(s) = \frac{1}{Z_G + \frac{1}{Cs}}, \quad \frac{\Delta\bar{V}}{\Delta\bar{i}}(s) = \frac{1}{Cs}$$

$$\frac{\Delta\bar{\dot{x}}}{\Delta\bar{V}}(s) = \frac{dk}{(ms+\lambda+k/s)+Z_M^{IN}}, \quad \frac{\Delta\bar{x}}{\Delta\bar{\dot{x}}}(s) = \frac{1}{s}$$

$$\frac{\Delta\bar{x}}{\Delta\bar{V}}(s) = \frac{dk}{s[(ms+\lambda+k/s)+Z_M^{IN}]}\,\frac{1/Cs}{\left(Z_G+\frac{1}{Cs}\right)}$$

(a)

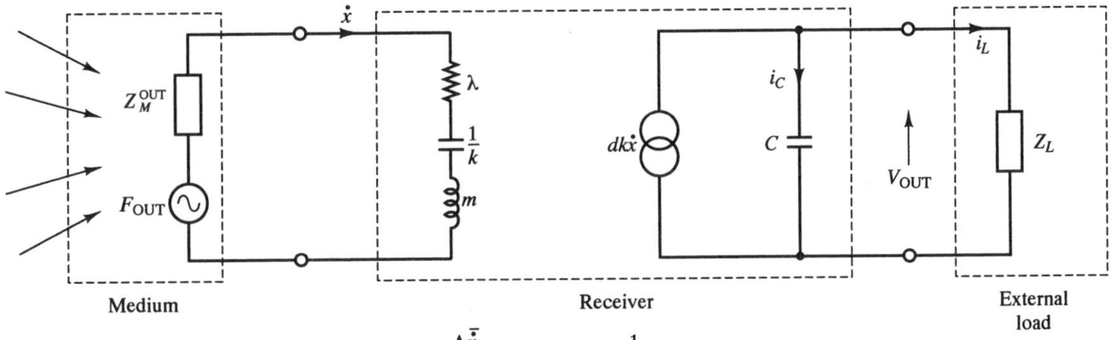

$$\frac{\Delta\bar{\dot{x}}}{\Delta\bar{F}_{OUT}}(s) = \frac{1}{(ms+\lambda+k/s)+Z_M^{OUT}}$$

$$\Delta\bar{V}_{OUT}(s) = \frac{\frac{1}{Cs}\cdot Z_L}{\frac{1}{Cs}+Z_L}\cdot dk\Delta\bar{\dot{x}}(s), \quad \frac{\Delta\bar{V}_{OUT}}{\Delta\bar{\dot{x}}}(s) = dk\,\frac{Z_L}{(1+Z_LCs)}$$

$$\frac{\Delta\bar{V}_{OUT}}{\Delta\bar{F}_{OUT}}(s) = \frac{dk}{[(ms+\lambda+k/s)+Z_M^{OUT}]}\cdot\frac{Z_L}{(1+Z_LCs)}$$

(b)

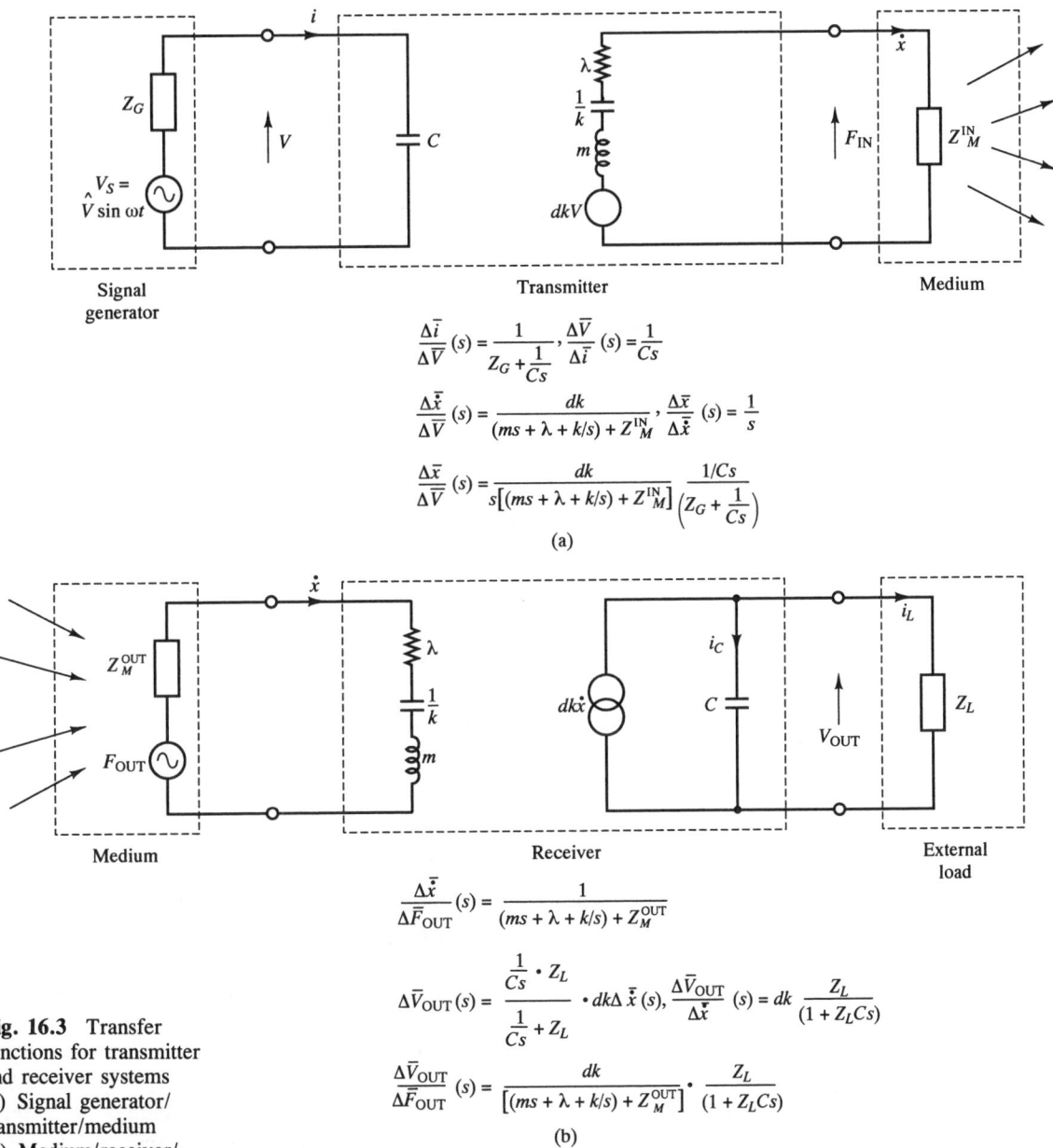

Fig. 16.3 Transfer functions for transmitter and receiver systems (a) Signal generator/transmitter/medium (b) Medium/receiver/external load

external signal generator and the output mechanical port to the beginning of the transmission medium. The figure derives the overall transfer function relating crystal displacement and generator voltage. In the special case that the output impedance of the generator and the input impedance of the medium are negligible, i.e. $Z_G \approx 0$, $Z_M^{\text{IN}} \approx 0$, the overall transfer function simplifies to the standard second-order form:

Transmitter system transfer function

$$\frac{\Delta \bar{x}}{\Delta \bar{V}_s}(s) = \frac{d}{\dfrac{1}{\omega_n^2}s^2 + \dfrac{2\xi}{\omega_n}s + 1} \qquad [16.5]$$

where ω_n and ξ are the mechanical natural frequency and damping ratio for the crystal (see Section 8.7).

Figure 16.3(b) shows the input mechanical port of the receiver connected to the end of the transmission medium and the output electrical port connected to an external load such as an oscilloscope or amplifier. The figure derives the overall transfer function relating voltage across the load V_{OUT} to force F_{OUT} (due to medium output pressure acting over the area of the receiver crystal). In the special case that the output impedance of the medium is negligible and the load is purely resistive, i.e. $Z_M^{\text{OUT}} \approx 0$ and $Z_L = R_L$ the overall transfer function simplifies to:

Receiver system transfer function

$$\frac{\Delta \bar{V}_{\text{OUT}}}{\Delta \bar{F}_{\text{OUT}}}(s) = \frac{d}{C}\,\frac{\tau s}{(1 + \tau s)}\,\frac{1}{\left(\dfrac{1}{\omega_n^2}s^2 + \dfrac{2\xi}{\omega_n}s + 1\right)} \qquad [16.6]$$

This is identical to the transfer function for the basic piezoelectric force measurement system discussed in Section 8.7: an ultrasonic receiver is simply a piezoelectric force sensor applied to the specific problem of measuring high frequency sinusoidal pressure variations.

16.2.2 Crystal oscillators and resonators

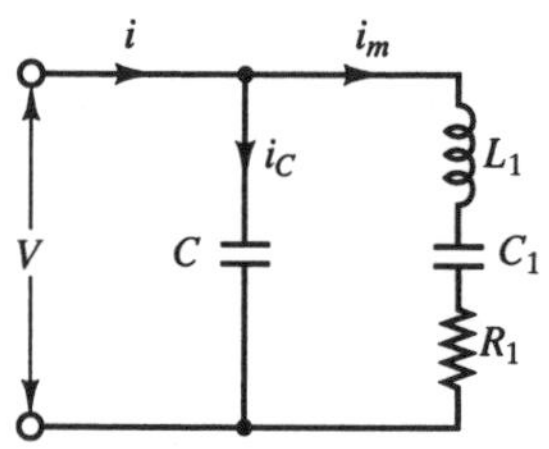

Fig. 16.4 Equivalent electrical circuit for crystal

Rather than energise the crystal with an external signal generator, the crystal is usually incorporated into a closed-loop system which oscillates at the mechanical natural frequency ω_n.

In order to discuss the principles of crystal oscillators we need to find the overall electrical impedance of the crystal. Figure 16.4 shows an approximate electrical equivalent circuit for the crystal. We see that the total current i drawn by the crystal is the sum of the capacitance current i_C together with the current i_m required to drive the mechanical mass/spring/damper system. C is therefore shunted by a series electrical circuit L_1, C_1, R_1 representing these mechanical losses.

The impedance Z_E of the series circuit is given by:

$$Z_E(s) = \frac{\Delta\bar{V}}{\Delta\bar{i}_m}(s) = \frac{\Delta\bar{V}}{\Delta\bar{F}}\,\frac{\Delta\bar{\dot{x}}}{\Delta\bar{i}_m}\,\frac{\Delta\bar{F}}{\Delta\bar{\dot{x}}} \qquad [16.7]$$

where the ratios $\Delta\bar{V}/\Delta\bar{F}$, $\Delta\bar{\dot{x}}/\Delta\bar{i}_m$ are both equal to $1/(dk)$ (eqns [16.1] and [16.2]), and $\Delta\bar{F}/\Delta\bar{\dot{x}}$ is the crystal mechanical impedance $Z_m(s)$ (eqn [16.4]). Thus:

$$Z_E(s) = \frac{1}{(dk)^2}\left(ms + \lambda + \frac{k}{s}\right)$$

i.e.

Equivalent electrical impedance of mechanical system

$$Z_E(s) = L_1 s + R_1 + \frac{1}{C_1 s}$$

$$\text{where } L_1 = \frac{m}{(dk)^2},\quad R_1 = \frac{\lambda}{(dk)^2},\quad C_1 = d^2k \qquad [16.8]$$

The overall electrical impedance transfer function is thus given by:

$$\frac{1}{H(s)} = \frac{1}{1/(Cs)} + \frac{1}{Z_E(s)} = Cs + \frac{1}{R_1 + L_1 s + 1/(C_1 s)} \qquad [16.9]$$

For sinusoidal signals we replace s by $j\omega$ to give:

$$\frac{1}{H(j\omega)} = j\omega C + \frac{1}{R_1 + j[\omega L_1 - 1/(\omega C_1)]} \qquad [16.10]$$

and

Overall electrical impedance of piezoelectric crystal

$$H(j\omega) = \frac{\omega R_1 C_1 - j(1-\omega^2 L_1 C_1)}{\omega[(C+C_1)-\omega^2 L_1 C C_1] + j\omega^2 C C_1 R_1} \qquad [16.11]$$

Figure 16.5 gives values of all the above parameters for a typical crystal, and shows the corresponding frequency variations of impedance magnitude $|H(j\omega)|$ and phase $\arg H(j\omega)$.

From the figure we see that there are two important frequencies. The first is the natural frequency $\omega_n = \sqrt{(k/m)} = 1/\sqrt{(L_1 C_1)}$ of the mechanical system. Here we have $1 - \omega_n^2 L_1 C_1 = 0$, so that the corresponding impedance magnitude $|H(j\omega_n)|$ is almost minimum; ω_n is termed the series resonant frequency. At the second frequency:

$$\omega_1 = \sqrt{\frac{C + C_1}{L_1 C_1 C}}$$

i.e. $(C + C_1) - \omega_1^2 L_1 C C_1 = 0$, so that the corresponding impedance $|H(j\omega_1)|$ is almost maximum. ω_1 is termed the parallel resonant frequency. Below ω_n and above ω_1, $\arg H(j\omega) = -90°$, i.e. the crystal behaves as a capacitance. Between ω_n and ω_1, $\arg H(j\omega) = +90°$, i.e. the crystal is inductive.

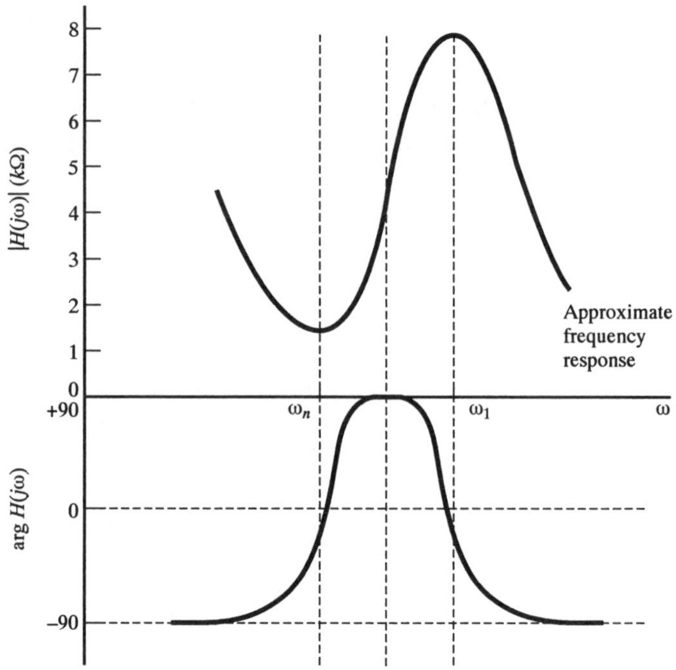

<u>Crystal data</u>

$f_n = 31.8$ kHz $k = 2 \times 10^9$ N/m $m = 5 \times 10^{-2}$ kg

$\xi = 0.01$ $\lambda = 200$ Ns/m $d = 2 \times 10^{-10}$ C/N

$C = 1600$ pF $dk = 0.4$

$$L_1 = \frac{m}{(dk)^2} = 312.5 \text{ mH}, R_1 = \frac{\lambda}{(dk)^2} = 1250\ \Omega, C_1 = d^2k = 80 \text{ pF}$$

$$\omega_n = \sqrt{\frac{k}{m}} = \sqrt{\frac{1}{(L_1C_1)}} = 2.00 \times 10^5 \text{ rad s}^{-1}\ Q = \frac{\sqrt{mk}}{\lambda} = \frac{1}{2\xi} = 50$$

$$\omega_1 = \sqrt{\frac{C + C_1}{L_1 + C_1C}} = 2.05 \times 10^5 \text{ rad s}^{-1}$$

$$H(j\omega_n) = \frac{1250}{1 + 0.4j} \begin{cases} |H(j\omega_n)| = 1161\ \Omega \\ \arg H(j\omega_n) = -22° \end{cases}$$

$$H(j\omega_1) = \frac{3049 + 7512j}{j} \begin{cases} |H(j\omega_1)| = 8105\ \Omega \\ \arg H(j\omega_1) = -22° \end{cases}$$

Fig. 16.5 Electrical impedance of typical piezoelectric crystal

A schematic diagram of a crystal oscillator is shown in Fig. 16.6. It is a closed-loop system, with a 'maintaining amplifier' of impedance $1/G(j\omega)$ in the forward path, and a piezoelectric element of impedance $H(j\omega)$ in the feedback path. From Fig. 16.6 we have:

$$\frac{i}{V_{REF} - V} = G(j\omega) \quad \text{and} \quad \frac{V}{i} = H(j\omega)$$

giving

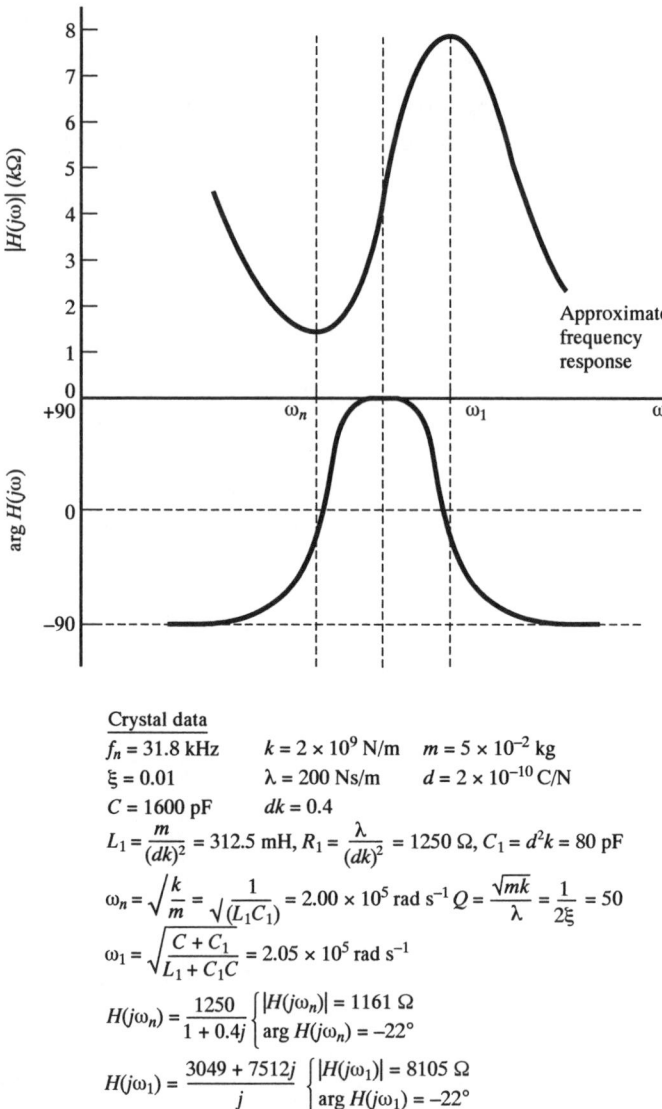

$$\frac{i}{V_{REF}} = \frac{G(j\omega)}{1 + H(j\omega)G(j\omega)} \qquad [16.12]$$

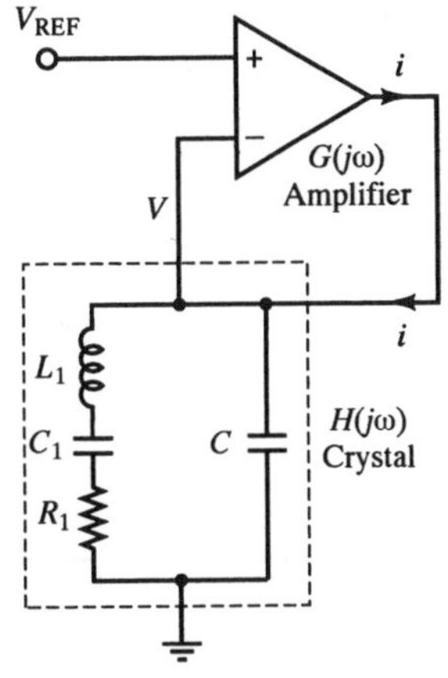

Fig. 16.6 Schematic diagram of crystal oscillator

Using the arguments of Section 9.5, we can show that continuous oscillations at frequency ω can be maintained, with $V_{REF} = 0$ if:

$$|G(j\omega)|\,|H(j\omega)| = 1 \qquad [9.46]$$

and

$$\arg G(j\omega) + \arg H(j\omega) = -180°$$

Thus to maintain oscillations at ω_n we require:

$$|G(j\omega_n)| = \frac{1}{|H(j\omega_n)|} \qquad [16.13]$$

and

$$\arg G(j\omega_n) = -180° - \arg H(j\omega_n)$$

For the crystal specified by Fig. 16.4 we require the maintaining amplifier to have:

$$|G(j\omega_n)| = \frac{1}{1161} = 8.61 \times 10^{-4} \quad \text{and} \quad \arg G(j\omega_n) = -158° \qquad [16.14]$$

It is possible to adjust the frequency of oscillation slightly, by adjusting $G(j\omega)$ for the maintaining amplifier.

16.2.3 Characteristics and applications

Table 16.1 gives physical data for commonly used piezoelectric materials.[1] **Characteristic acoustic impedance** will be discussed in Section 16.3, **mechanical quality factor** Q is related to damping coefficient ξ by the equation $Q = 1/2\xi$. Ultrasonic transmitters and receivers are normally in the form of a thin disc of piezoelectric material with metal electrodes deposited on each face. A disc of diameter 10 mm, thickness 1 mm would typically have a natural frequency f_n of 5 MHz, the amplitude of the crystal deformation is around a few microns. The two usual modes of vibration are shown in Fig. 16.7(a); the **thickness expander** mode is used in

Table 16.1 Physical properties of piezoelectric materials (adapted in part from O'Donnell *et al.*[1])

Parameter / Material	Density ρ $\times 10^3$ kg m^{-3}	Sensitivity d pCN^{-1}	Dielectric constant ϵ	Longitudinal velocity of sound c_L $\times 10^3$ m s^{-1}	Characteristic acoustic impedance R_A $\times 10^7$ kg m^{-2} s^{-1}	Mechanical quality factor $Q = \frac{\sqrt{mk}}{\lambda}$
Quartz	2.65	2.3	4.5	5.65	1.5	>25 000
$BaTiO_3$	5.7	149	1700	4.39	2.5	<400
PZT−5H	7.5	593	3400	4.0	3.0	<65
$PbNb_2O_6$	5.9	75	240	2.7	1.6	<5
PVDF	1.8	≈10	13	2.2	0.4	14

the generation and reception of longitudinal waves and the **thickness shear** mode in the generation of transverse waves.

The performance characteristics of piezoelectric transmitters and receivers can be summarised by four graphs: Fig. 16.7(b) shows the form of the graphs in a typical case.[2]

Graph 1 shows a decibel plot of the transmitter frequency response. This is based on the pressure/voltage transfer function $(\Delta\bar{P}_{IN}/\Delta\bar{V}_s)\,(s)$, which can be derived from Fig. 16.3(a) using:

$$\frac{\Delta\bar{P}_{IN}}{\Delta\bar{V}_s}(s) = \frac{Z^M_{IN}}{A}\, s\, \frac{\Delta\bar{x}}{\Delta\bar{V}_s}(s) \qquad [16.15]$$

where A is the area of the circular face of the disc. The decibel plot is relative to 1 μbar i.e. dB = 20 $\log_{10}(\hat{P}_{IN}/1\ \mu\text{bar})$, and for a supply voltage V_s of 1 V_{rms}. The graph shows a sharp resonance peak at f_n which is characteristic of a second order mechanical system with low ξ or high Q.

Graph 2 shows a decibel plot of the receiver frequency response. This is based on the voltage/pressure transfer function $\Delta\bar{V}_{OUT}/\Delta\bar{P}_{OUT}$ which can be derived from Fig. 16.3(b) using

$$\frac{\Delta\bar{V}_{OUT}}{\Delta\bar{P}_{OUT}}(s) = A\,\frac{\Delta\bar{V}_{OUT}}{\Delta\bar{F}_{OUT}}(s) \qquad [16.16]$$

The decibel plot is relative to 1 volt, i.e. dB = 20 $\log_{10}(\hat{V}_{OUT}/1\text{ V})$, and for a pressure of 1 μbar. The graph again shows a sharp resonance peak at f_n.

Graph 3 shows how the magnitude $|H(j2\pi f)|$ of the transducer electrical impedance varies with frequency (eqn [16.11] and Fig. 16.5).

Graph 4 shows the directional response of the transmitter, i.e. how the pressure of the sound wave (at frequency f_n), launched from the transmitter, varies with angle.

Ultrasonic transmission links are usually either **narrow-band** or **broad-band**. In narrow-band systems a narrow band of sound frequencies, close to the transducer natural frequency, is transmitted over the link. Here both transmitter and receiver should have a high Q to give a sharp resonant peak. Quartz is therefore an ideal material; a composite resonator is usually set up consisting of transmitter, transmission medium and receiver. In broad-band systems either a broad band of frequencies or a pulse of sound is transmitted over the link. Here both transmitter and receiver

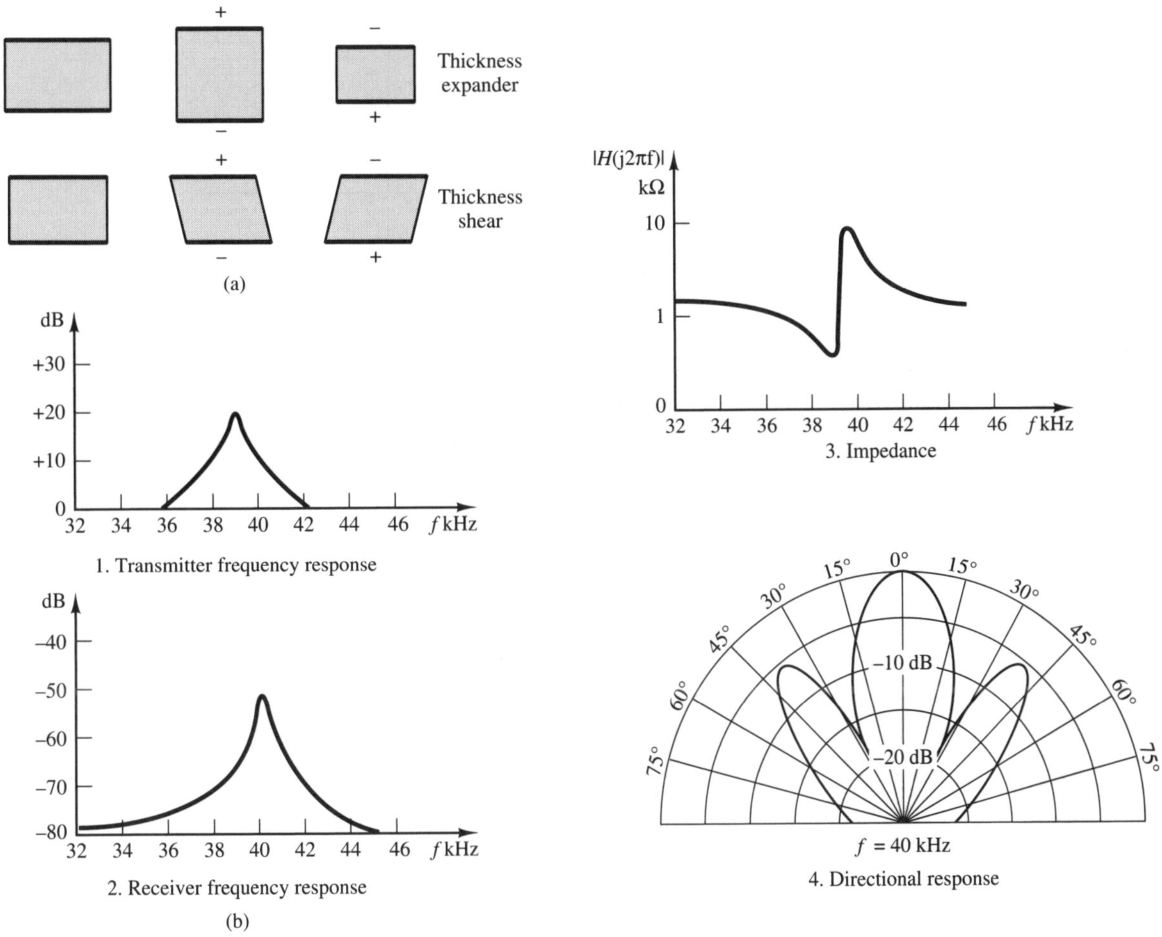

Fig. 16.7 Piezoelectric transmitters and receivers (a) Vibration displacement modes (b) Performance characteristics

should have a low Q to give a fairly flat frequency response. In pulse systems, Q should be ideally around 0.7, i.e. $\xi = 0.7$, give an optimum step response and minimum pulse distortion. Here low Q materials such as $PbNb_2O_6$ or PVDF should be used with a backing layer of absorbing material to further reduce Q (Section 16.3.5).

16.3 Principles of ultrasonic transmission

As mentioned earlier, ultrasonic waves obey the same transmission principles as lower frequency acoustic waves.

16.3.1 Propagation of plane waves

Figure 16.8 shows the transmission medium represented as a series of layers of particles. If an acoustic plane wave is passing through the medium, then each particle

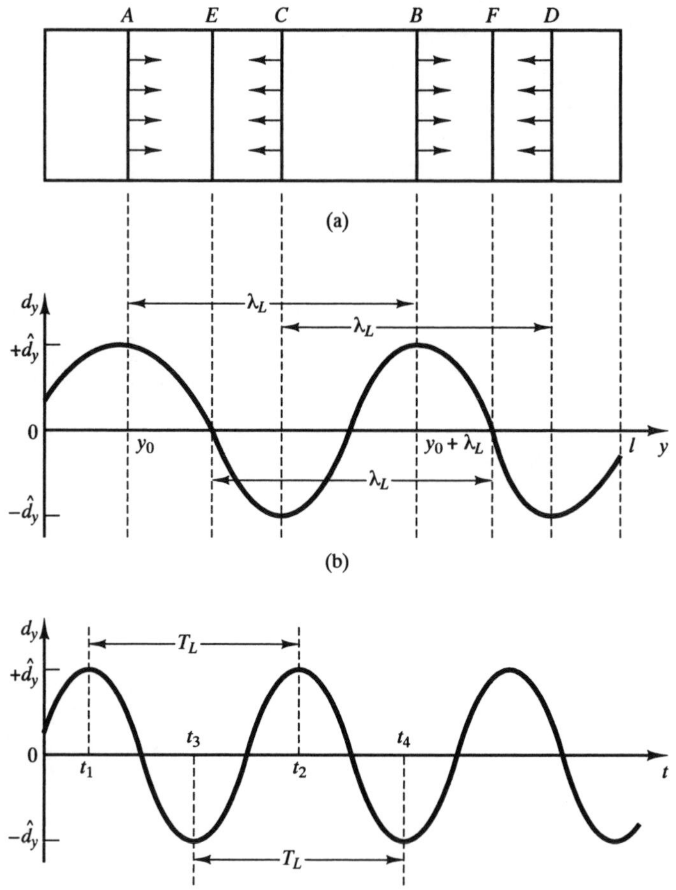

Fig. 16.8 Longitudinal plane waves
(a) Wave fronts
(b) Variation of d_y with y at $t = t_0$, $d_y = \hat{d}_y \sin \omega_L (t_0 + y/c_L)$
(c) Variation of d_y with t for layer A with $y = y_0$, $d_y = \hat{d}_y \sin \omega_L (t + y_0/c_L)$

in a given layer, at a given time, is displaced by the same amount. For a longitudinal plane wave, the displacement d_y, of a layer at position y, is parallel to the direction of propagation of the wave, i.e. parallel to the Oy axis. Longitudinal plane waves are described by the partial differential equation:

$$\frac{\partial^2 d_y}{\partial y^2} = \frac{1}{c_L^2} \frac{\partial^2 dy}{\partial t^2} \qquad [16.17]$$

where dy is the displacement of layer y at time t and c_L is the velocity of propagation of the longitudinal wave. For a progressive wave moving along the positive y direction, the solution to [16.17] is:

Longitudinal plane wave along positive Oy direction

$$dy = \hat{d}y \sin \omega_L \left(t - \frac{y}{c_L} \right) \qquad [16.18]$$

where $\hat{d}y$ is the amplitude of the layer displacement and $\omega_L = 2\pi f_L$ is the angular frequency. For a progressive wave moving along the negative y direction the solution to [16.17] is:

Longitudinal plane wave along negative Oy direction

$$dy = \hat{d}y \sin \omega_L \left(t + \frac{y}{c_L}\right) \qquad [16.19]$$

In each case, dy is a sinusoidal function of both time t and layer position y. Figures 16.8(a) and (b) represent 'still' pictures of the displacements of each layer, taken at time $t = t_0$. Since time is now constant, dy becomes a sinusoidal function of position y only, i.e.

$$dy = \hat{d}y \sin \left(\omega_L t_0 + \frac{\omega_L y}{c_L}\right) \qquad [16.20]$$

The distance between layers such as A and B, C and D, E and F, which have equal displacements, is defined to be the wavelength λ_L. At A and B, $dy = \hat{d}y$, so that:

$$\text{at } A, \quad \omega_L t_0 + \frac{\omega_L y_0}{c_L} = \frac{\pi}{2} \qquad [16.21]$$

$$\text{at } B, \quad \omega_L t_0 + \frac{\omega_L}{c_L}(y_0 + \lambda_L) = \frac{5\pi}{2}$$

Subtracting gives $\omega_L \lambda_L / c_L = 2\pi$, and since $\omega_L = 2\pi f_L$, we have

Relationship between velocity, wavelength and frequency for a longitudinal plane wave

$$c_L = f_L \lambda_L \qquad [16.22]$$

Thus for a 1 MHz wave travelling with longitudinal velocity $6 \times 10^3 \text{ m s}^{-1}$ in steel, the corresponding wavelength is 6 mm. Figure 16.8(c) shows how the displacement of a given layer A, at $y = y_0$, varies sinusoidally with time t according to:

$$d_y = \hat{d}_y \sin \omega_L \left(t + \frac{y_0}{c_L}\right) \qquad (16.23)$$

The period T_L of the wave is the interval between time instants t_1 and t_2, t_3 and t_4, where the layer displacement is the same: as usual $T_L = 1/f_L = 2\pi/\omega_L$.

For a transverse plane wave, the displacement dz of a layer at position y, is perpendicular to the direction Oy of propagation of the wave, i.e parallel to the Oz axis. The medium is therefore subject to time-varying shear strains. The corresponding partial differential equation is:

$$\frac{\partial^2 d_z}{\partial y^2} = \frac{1}{c_T^2} \frac{\partial^2 d_z}{\partial t^2} \qquad [16.24]$$

where d_z is the transverse displacement of layer y at time t, and c_T is the transverse wave velocity. For a progressive wave travelling in the positive y direction, the solution to [16.24] is similar to [16.18], i.e.

Transverse plane wave along positive Oy direction

$$d_z = \hat{d}_z \sin \omega_T \left(t - \frac{y}{c_T}\right) \qquad [16.25]$$

where $\hat{d}_z$ is displacement amplitude and $\omega_T = 2\pi f_T$ is the angular frequency. Using arguments similar to above, we can show that the relationship between velocity, wavelength and frequency of a transverse wave is similar to [16.22], i.e.

$$c_T = f_T \lambda_T \qquad [16.26]$$

The velocity of sound in a medium depends on the appropriate elastic modulus and density ρ, according to the general equation:

$$c = \sqrt{\frac{\text{elastic modulus}}{\rho}} \qquad [16.27]$$

Gases will support compressive and tensile stresses but not shear stresses, so that only longitudinal acoustic waves can be passed through gases. The corresponding velocity is given by:

Velocity of longitudinal sound waves in a gas

$$c_L = \sqrt{\left(\frac{\gamma P}{\rho}\right)} \qquad [16.28]$$

where P is the pressure of the gas and γ the specific heat ratio. In general, liquids collapse under the action of shear stresses, so that usually transverse waves cannot be transmitted through liquids. The velocity of longitudinal waves in liquids is given by:

Velocity of longitudinal waves in a liquid

$$c_L = \sqrt{\left(\frac{B}{\rho}\right)} \qquad [16.29]$$

where B is the bulk modulus of elasticity of the liquid. Since solids will support compressive/tensile and shear stresses, both longitudinal and transverse waves can be transmitted through solids. The corresponding velocities are given by[3]

Velocity of longitudinal and transverse waves in a solid

$$c_L = \sqrt{\left[\frac{E(1 - \nu)}{(1 + \nu)(1 - 2\nu)\rho}\right]} \qquad [16.30]$$

$$c_T = \sqrt{\left(\frac{s}{\rho}\right)}$$

where E is Young's modulus, s is the shear modulus and ν is Poisson's ratio for the solid.

16.3.2 Acoustic impedance and power

In Chapter 5 we saw that mechanical force F and velocity $\dot{x}$ were an effort/flow pair, so that the product $F\dot{x}$ represents mechanical power and the ratio $F/\dot{x}$ represents mechanical impedance. In acoustic work the particle velocity $u = \partial d/\partial t$ is the flow variable, and the accompanying pressure or stress P is the effort variable. Specific acoustic impedance is thus defined by:

Specific acoustic impedance

$$Z_A = \frac{P}{u} \qquad [16.31]$$

As with electrical impedance; Z_A is, in general, a complex quantity of the form $Z_A = R_A + jX_A$. However, for plane progressive waves, the imaginary component X_A is zero, leaving $Z_A = R_A$. It can be shown that R_A is the product of density ρ and velocity of sound c for the material, i.e.

Characteristic impedance of a medium

$$R_A = \rho c \qquad [16.32]$$

Table 16.2 (adapted from Blitz[3]) gives values of longitudinal velocity, density and characteristic impedance for commonly used materials.

The power or intensity at any point in an acoustic field is the product Pu of through and across variables. For a plane progressive wave, we can use eqn [16.31] to give three equations for acoustic intensity:

Acoustic intensity or power

$$W = Pu = \frac{P^2}{R_A} = u^2 R_A \qquad [16.33]$$

W is the rate of flow of acoustic energy, through unit area, at right angles to the direction of propagation of the wave, and has the units of watts metre^{-2}.

16.3.3 Attenuation of a plane wave in a medium

So far we have assumed that as a plane acoustic wave travels through a material, the amplitudes $\hat{d}$, $\hat{P}$ of particle displacement and pressure variation respectively remain constant. In practice the wave becomes attenuated as it passes through the medium, so that the amplitudes decrease exponentially with distance y travelled.

Table 16.2 Characteristic impedance of some common materials (adapted from Blitz[3])

Material	Longitudinal velocity of sound c_L m s^{-1}	Density ρ kg m^{-3}	Characteristic impedance $R_A = \rho c_L$ kg m^{-2} s^{-1}
Steel	6.0×10^3	7.80×10^3	4.7×10^7
Iron	5.90×10^3	7.90×10^3	4.7×10^7
Brass	3.50×10^3	8.60×10^3	3.0×10^7
Aluminium	6.40×10^3	2.70×10^3	1.7×10^7
Bone	$\approx 5.3 \times 10^3$	$\approx 1.5 \times 10^3$	$\approx 0.8 \times 10^7$
Glass	5.50×10^3	2.50×10^3	1.3×10^7
Biol. tissue	$\approx 1.5 \times 10^3$	$\approx 1.0 \times 10^3$	$\approx 0.15 \times 10^7$
Polystyrene	2.35×10^3	1.10×10^3	0.25×10^7
Oil	1.40×10^3	0.90×10^3	0.13×10^7
Water	1.50×10^3	1.0×10^3	0.15×10^7
Air	0.34×10^3	1.3	430

Thus for displacement amplitudes we have:

Displacement attenuation for a plane wave

$$\hat{d}(y) = \hat{d}(0)\exp(-\alpha_d y) \qquad [16.34]$$

where $\hat{d}(0)$ is the amplitude at the beginning of the medium, $\hat{d}(y)$ is the amplitude at position y, and α_d metre^{-1} is the displacement attenuation coefficient. A similar equation relates the acoustic intensity $W(y)$ at position y, to the intensity $W(0)$ at $y = 0$, i.e.

Power attenuation for a plane wave

$$W(y) = W(0)\exp(-\alpha_W y) \qquad [16.35]$$

where α_W metre^{-1} is the power attenuation coefficient. Various mechanisms are responsible for this power loss. In liquids viscous friction effects are important; in solids scattering at grain boundaries and absorption at crystal lattice defects are the main loss mechanisms.

16.3.4 Reflection and refraction at a boundary between two materials

Figure 16.9 shows a plane wave incident on a plane boundary separating two materials of different characteristic impedances. From the figure, we see that part of the incident wave I is reflected (R), and part is transmitted (T). The laws relating the angles θ_I, θ_R, θ_T of incidence, reflection and transmission are those for any wave motion.

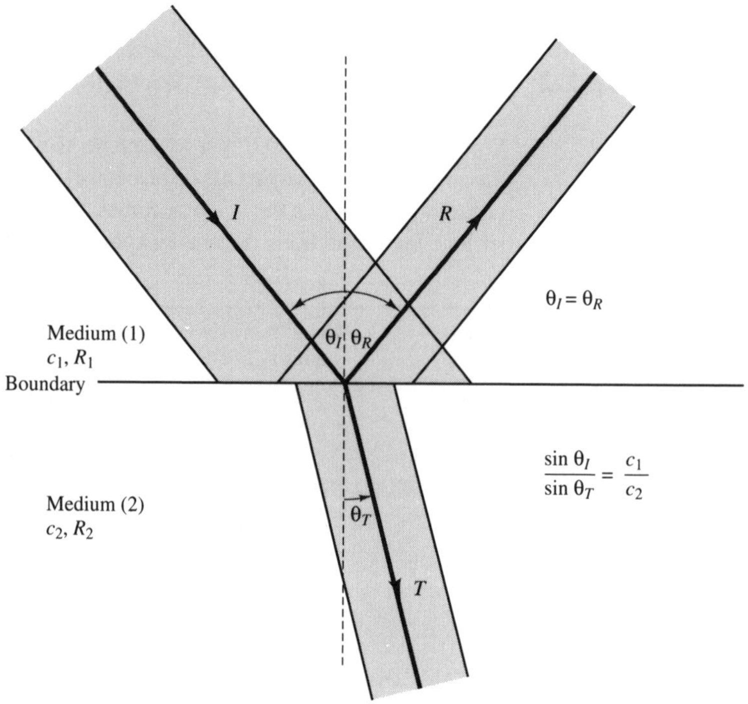

Fig. 16.9 Reflection and refraction of sound at a boundary between two materials

Figure 16.9 takes no account of possible conversion of longitudinal waves into transverse, and vice versa, at the boundary. The ratio between intensity W_R and incident intensity W_I is termed the reflection coefficient α_R; the ratio between transmitted intensity W_T and incident intensity W_I is the transmission coefficient α_T, i.e.

$$\alpha_R = \frac{W_R}{W_I}, \ \alpha_T = \frac{W_T}{W_I} \qquad [16.36]$$

For normal or near-normal angles of incidence, i.e. $\theta_I \approx \theta_R \approx \theta_T \approx 0$; it can be shown that[4]

Reflection and transmission coefficients for near-normal incidence

$$\alpha_R = \frac{(R_2 - R_1)^2}{(R_2 + R_1)^2}, \quad \alpha_T = \frac{4R_1R_2}{(R_2 + R_1)^2} \qquad [16.37]$$

We note that $\alpha_R + \alpha_T = 1$; this is a consequence of energy conservation, i.e. $W_R + W_T = W_I$. Table 16.3 uses R values from Table 16.2 to give α_R, α_T values for boundaries of five important materials. We see that the α_T values for quartz/steel and steel/water are reasonably large. This is a great practical value, because it means that sound waves can be launched from quartz crystal transmitters into solid metal bars, and also into liquids contained in metal pipes and vessels. The α_R values for steel/air and water/air are very close to unity (α_T very small); this means there is almost perfect reflection of sound waves at these boundaries. Detection of cracks and flaws in metal samples and at the liquid/gas interface in level measurement is therefore possible.

We can now examine the efficiency of the simple transmission link of Section 16.1 and Fig. 16.1. As an example, consider a quartz transmitter and receiver attached to opposite sides of a steel billet 10 cm thick. The ratio $W_{R/X}/W_{T/X}$ of received power to transmitted power depends on attenuation losses in the steel, together with interface losses at the quartz/steel and steel/quartz boundaries. Using eqns [16.35] and [16.36] we have:

$$\frac{W_{R/X}}{W_{T/X}} = (\alpha_T)_{Q/S} e^{-\alpha_W l} (\alpha_T)_{S/Q} \qquad [16.38]$$

Since $(\alpha_T)_{Q/S} = (\alpha_T)_{S/Q} = 0.73$, $l = 0.1$ m, and assuming $\alpha_W = 2.0$ for steel,

$$\frac{W_{R/X}}{W_{T/X}} = (0.73)^2 e^{-0.2} = 0.53 \times 0.82 = 0.44$$

i.e. the efficiency of energy transmission is 44%.

Table 16.3 Values of α_R, α_T at the boundaries of five common materials

	Steel	Quartz	Polystyrene	Water	Air
Steel	0.0, 1.0	0.27, 0.73	0.81, 0.19	0.88, 0.12	≈ 1.0, 3.7×10^{-5}
Quartz		0.0, 1.0	0.51, 0.49	0.67, 0.33	≈ 1.0, 1.1×10^{-4}
Polystyrene			0.0, 1.0	0.06, 0.94	≈ 1.0, 6.9×10^{-4}
Water				0.0, 1.0	≈ 1.0, 1.1×10^{-3}
Air					0.0, 1.0

From Table 16.3 we see that the α_T values for quartz/air and steel/air are very small. This means that it is very difficult to launch a sound wave from a piezoelectric transmitter into a gas, either directly, or via a steel pipe. One way of overcoming this problem is to place a **matching layer** between the transmitter and the gas. If R_1, R and R_2 are the characteristic impedances of the transmitter, matching layer and gas respectively then:

$$\alpha_{T,M} = \frac{4R_1R}{(R_1 + R)^2} \quad \text{and} \quad \alpha_{M,G} = \frac{4RR_2}{(R + R_2)^2} \qquad [16.39]$$

and the overall transmission coefficient between transmitter and gas is:

$$\alpha_{T,G} = \alpha_{T,M} \times \alpha_{MG} = \frac{16R_1R_2R^2}{(R_1 + R)^2\,(R + R_2)^2} \qquad [16.40]$$

It can be shown (Problem 16.9) that this overall transmission coefficient is a maximum when:

Optimum impedance of matching layer

$$R = \sqrt{R_1R_2} \qquad [16.41]$$

Thus in order to maximise power transfer between quartz ($R_1 = 1.5 \times 10^7$) and air ($R_2 = 430$) a matching layer with $R \approx 8 \times 10^4$ is required. From Table 16.1, we see that $R_A \approx 0.4 \times 10^7$ for PVDF and from Table 16.2, $R_A \approx 0.15 \times 10^7$ for biological tissue. Acoustic power can therefore be efficiently launched from PVDF transducers directly into the human body without the need for matching layers.

16.3.5 Stationary waves and resonance

Consider a progressive plane wave travelling along the positive y direction through a medium with zero attenuation coefficient. If it meets the boundary with a medium of very different characteristic impedance then almost perfect reflection occurs. This means that there is now a reflected wave travelling along the negative y direction with equal amplitude to the incident wave. For longitudinal waves the incident wave is:

$$dy = \hat{d}_y \sin \omega_L \left(t - \frac{y}{c_L}\right)$$

and the reflected wave is:

$$d_y = \hat{d}_y \sin \omega_L \left(t + \frac{y}{c_L}\right)$$

The resultant particle displacement is therefore:

$$d_y^R = \hat{d}_y\left\{\sin \omega_L \left(t - \frac{y}{c_L}\right) + \sin \omega_L \left(t + \frac{y}{c_L}\right)\right\}$$

Equation for stationary wave

$$d_y^R = 2\hat{d}_y \cos \frac{\omega_L y}{c_L} \sin \omega_L t \qquad [16.42]$$

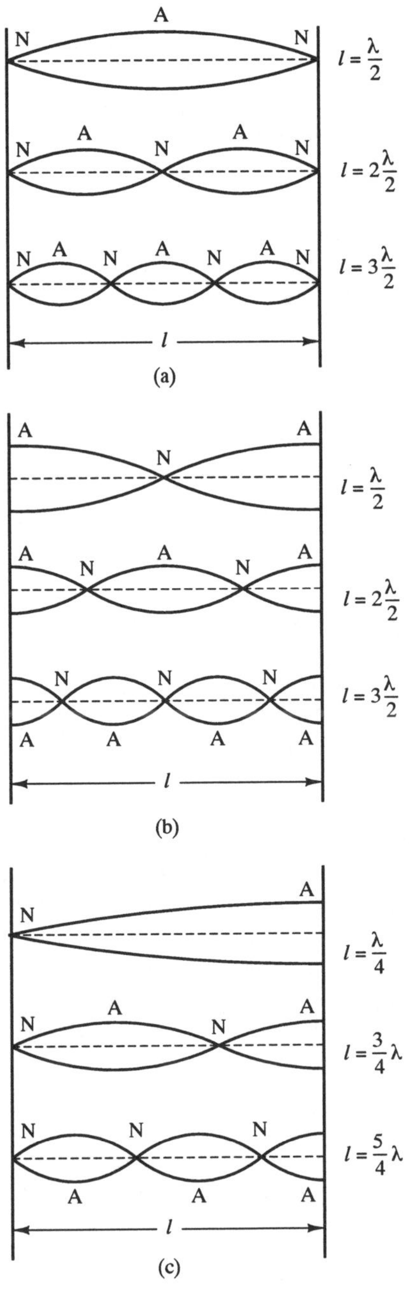

Fig. 16.10 Resonant stationary wave patterns
(a) $\frac{1}{2}$ wave resonance — node at each end
(b) $\frac{1}{2}$ wave resonance — antinode at each end
(c) $\frac{1}{4}$ wave resonance — node at one end, antinode at other end

The function $2\hat{d}_y \cos(\omega_L y/c_L)$ defines the wave amplitude or envelope. Since this is independent of t, i.e. does not move forward with time, the wave is **stationary**. At values of y, where $\cos(\omega_L y/c_L) = 0$, the envelope has a minimum value of zero; these are **nodes**. At values of y where $\cos(\omega_L y/c_L) = 1$, the envelope has a maximum value of $2\hat{d}_y$, these are **antinodes**.

In practice the medium will have finite length l and the type of stationary or **standing wave** set up will depend on the boundary conditions at each end of the

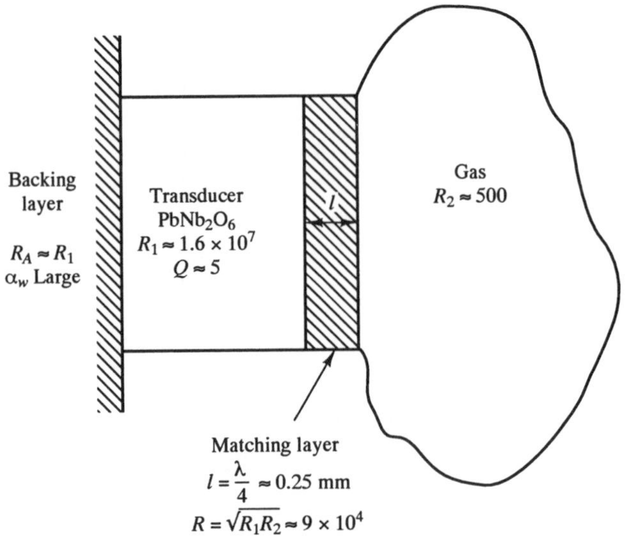

Fig. 16.11 Transmitter for launching sound pulses into a gas

medium, i.e. whether there are nodes or antinodes. In the special cases that the frequency of the sound wave is such that l is a whole number of either half or quarter wavelengths, then **resonance** is produced where $\hat{d}_y$ has a large resonant value.

Two types of resonance are possible. In **half wave resonance** (Figs 16.10(a) and (b)) there are identical boundary conditions at each end of the medium. Figure 16.10(a) shows possible resonant modes for a node at each end; these modes could occur if a medium of low characteristic impedance is bounded on both sides by a medium of high impedance, for example a gas inside a metal pipe. Figure 16.10(b) shows possible resonant modes for an antinode at each end; these modes could occur if a medium of high impedance is bounded on both sides by a medium of low impedance, for example a metal bar surrounded by gas. The other type of resonance is **quarter wave resonance**; here there are different boundary conditions, i.e. a node and antinode, at each end of the medium. Figure 16.10(c) shows the possible resonant modes; these modes could occur if a medium of intermediate impedance is bounded on one side by a material of high impedance and on the other by a material of low impedance.

The principles discussed in this and previous sections are important in the design of efficient transmitters. As an example, consider the problem of launching a pulse of ultrasound from a piezoelectric transducer into a gas. There are three essential requirements:

(1) the sound wave launched from the rear face of the transducer should not enter the gas, otherwise pulse transit time measurements will be confused;
(2) the overall Q should be ideally around 0.7 to minimise pulse distortion;
(3) maximum power should be transmitted from the front face of the transducer into the gas.

The transmitter shown in Fig. 16.11 fulfils these requirements. The backing layer, with characteristic impedance similar to that of the transducer and high attenuation coefficient, accepts and absorbs the wave from the rear face. This layer also reduces the Q of the overall transmitter to a value well below that of the transducer itself.

The matching layer has length $l = \lambda/4$, so that a quarter wave resonant mode is present, and impedance $R = \sqrt{R_1 R_2}$. This ensures maximum power transfer from transducer front face to the gas.

16.3.6 Doppler effect

When a source (or transmitter) and observer (or receiver) of sound waves are in relative motion, then the frequency of the received signal differs from the frequency of the transmitted signal by an amount that depends on the relative velocities of source and observer. This shift in apparent frequency is called the Doppler effect and occurs in all types of wave motion. There are two cases to consider.[5] One where the source is stationary and the observer is moving; the other where the source is moving and the observer stationary.

Figure 16.12(a) shows a stationary source of sound S, frequency f and a stationary observer O. If the disturbance takes time Δt to travel from S to O, then O receives $f\,\Delta t$ cycles in this time and the distance $OS = f\lambda\Delta t$. If O moves towards S with velocity v, then in time Δt it travels a distance $v\Delta t$ to O' (Fig. 16.12(b)), and receives an additional $v\Delta t/\lambda$ cycles of sound. The total number of cycles received in Δt is thus $(f + v/\lambda)\Delta t$ and the apparent frequency is:

$$f' = f + \frac{v}{\lambda} = f + \frac{f}{c}v = f\frac{(c + v)}{c} \qquad [16.43]$$

since $1/\lambda = f/c$. If O moves away from S with velocity v, then $(f - v/\lambda)\Delta t$ cycles are received in time Δt, so that the apparent frequency is:

$$f' = f\frac{(c - v)}{c} \qquad [16.44]$$

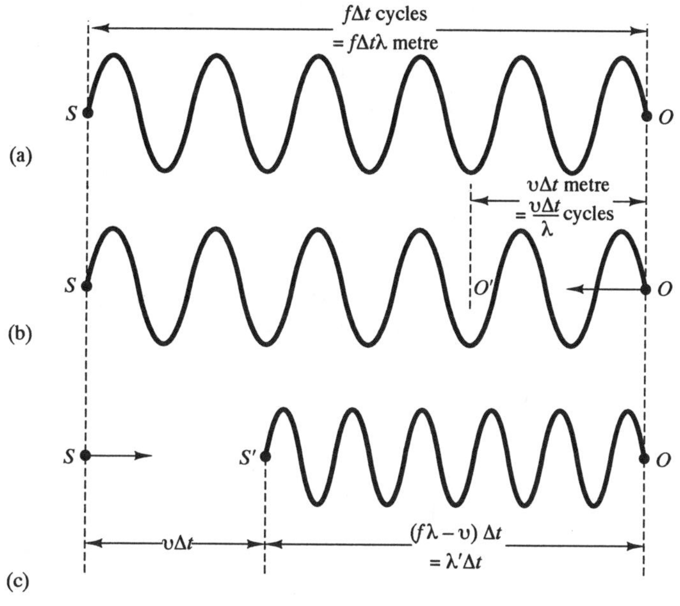

Fig. 16.12 Doppler effect

We can generalise these two results by the single equation:

Doppler shift: fixed source, moving observer

$$\frac{f'}{f} = \frac{\text{velocity of waves relative to observer}}{\text{normal wave velocity}} \quad [16.45]$$

If S now moves towards O (stationary) with velocity v, then in time Δt it travels a distance $v\Delta t$ to S' (Fig. 16.12(c)). In this case $f\Delta t$ cycles of sound occupy the distance $S'O = (f\lambda - v)\Delta t$. The apparent wavelength is thus:

$$\lambda' = \frac{\text{total distance}}{\text{total number of cycles}} = \frac{(f\lambda - v)\Delta t}{f\Delta t} = \lambda - \frac{v}{f} \quad [16.46]$$

Since

$$\frac{1}{f} = \frac{\lambda}{c}, \lambda' = \lambda - \lambda\frac{v}{c} = \lambda\frac{(c - v)}{c} \quad [16.47]$$

If S moves away from O we have:

$$\lambda' = \lambda\frac{(c + v)}{c} \quad [16.48]$$

and we can generalise these two results by the single equation:

Doppler shift: moving source, fixed observer

$$\frac{\lambda'}{\lambda} = \frac{\text{velocity of waves relative to source}}{\text{normal wave velocity}} \quad [16.49]$$

16.4 Examples of ultrasonic measurement systems

16.4.1 Pulse reflection or pulse echo systems

Figure 16.13(a) shows a typical system. A piezoelectric crystal acting as a transmitter/receiver is attached to medium 1, the characteristic impedances of media 1 and 2 are substantially different. When the crystal acts as a transmitter, an oscillator giving a sinusoidal voltage at radio frequency f is connected to the crystal for a time T_W using switch A. Thus a pulse of ultrasound of width T_W enters medium 1 and most of the pulse energy is reflected at the boundary of medium 1 with medium 2. The reflected pulse returns to the crystal at a time T_T after the outgoing pulse. Since T_T is the time for the 'round trip' of distance $2l$ then:

Transit time for pulse echo system

$$T_T = \frac{2l}{c} \quad [16.50]$$

where l is the distance of the interface from the crystal and c is the velocity of sound in medium 1.

The crystal now acts as a receiver and converts the reflected pulse into a voltage pulse. Switch B is now closed so that the pulse passes to the echo signal conditioning

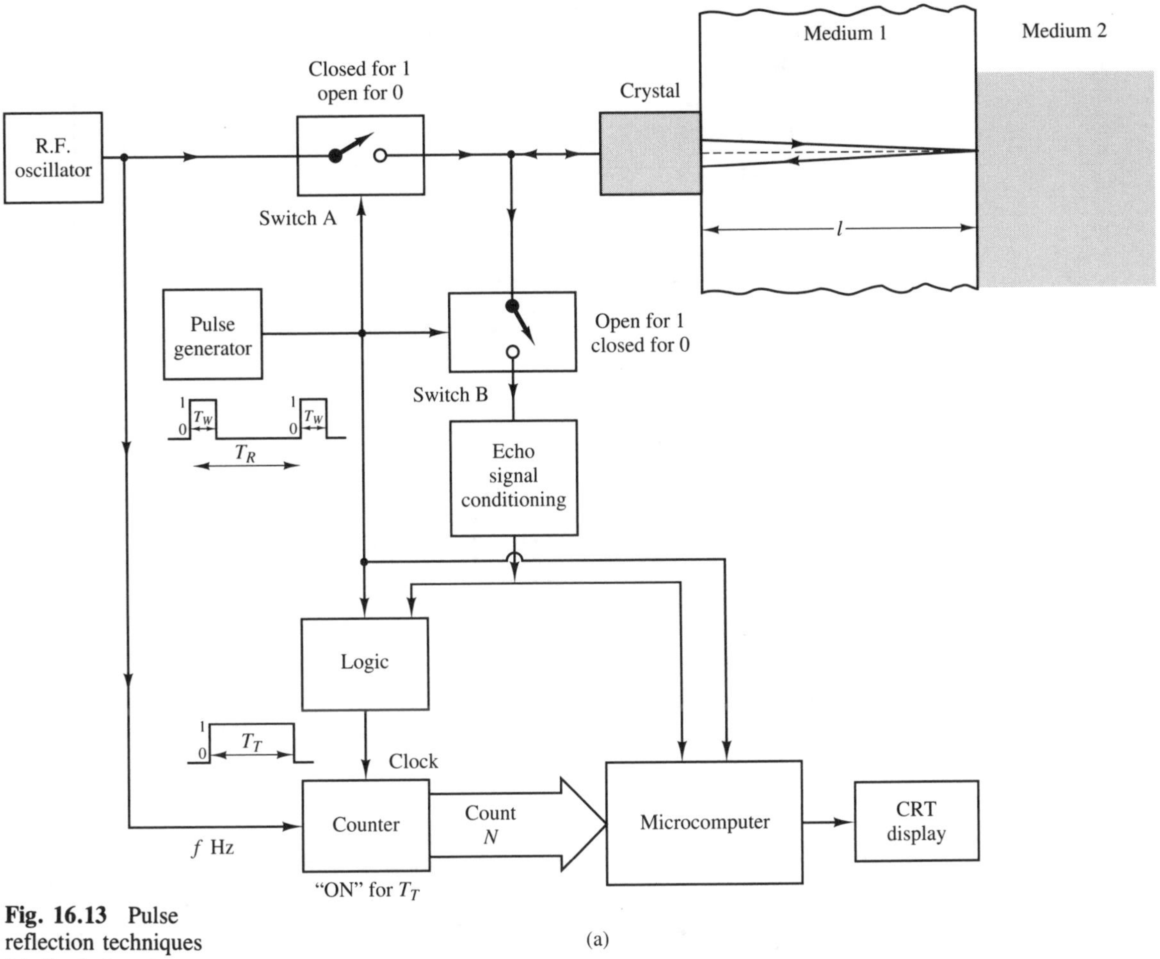

Fig. 16.13 Pulse reflection techniques (a) Typical system

circuit where it is amplified, rectified and 'squared up' using a Schmitt Trigger. Figure 16.13(b) shows the idealised outgoing and reflected pulse waveforms. These pulse waveforms pass to a logic circuit which detects the leading edge of the outgoing pulse and that of the first reflected pulse, the resulting output is a pulse signal which is in the '1' state during the transit time T_T. This is used to control a counter which also receives clock pulses at frequency f from the signal generator. The total count N during the 'ON' time is therefore $N = fT_T$. This is transferred to a microcomputer using a parallel digital signal. The computer calculates T_T from N and uses equation [16.50] with a known value of c to find l.

The measurement is complicated by the creation of multiple reflections or 'echoes'. Part of the first reflected pulse is reflected at the boundary of medium 1 and the crystal, and reflected again at the boundary of media 1 and 2 to give a second reflected pulse. The process is repeated many times, the amplitude of the reflected pulses dying away due to attenuation losses in medium 1 and reflection losses at the boundaries. Figure 16.13(b) shows these multiple reflected pulses. The following conditions should be obeyed by the pulse signal

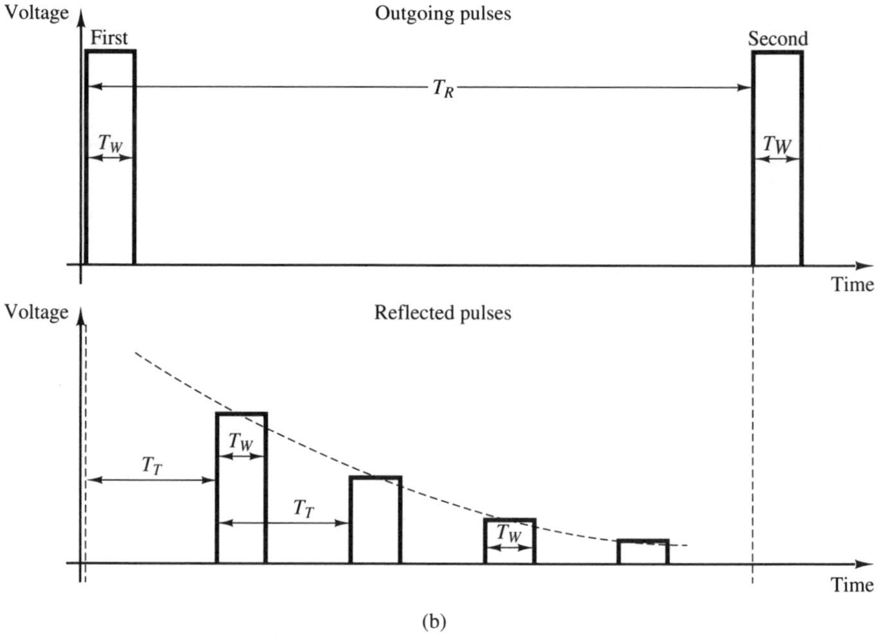

Fig. 16.13 (continued) (b) Idealised outgoing and reflected pulse waveforms

(a) The pulse width T_W should be large compared with the period $1/f$ of the sound wave; this ensures that there are many cycles, i.e. sufficient energy, in each pulse:

$$T_W \gg \frac{1}{f} \qquad [16.51]$$

(b) The transit time T_T should be large compared with the pulse width T_W to avoid interference between outgoing and reflected pulses:

$$T_T \gg T_W \qquad [16.52]$$

(c) The repetition time T_R between successive outgoing pulses should be large compared with transit time T_T; this ensures that all reflections, following one outgoing pulse, are attenuated before the next enters the material:

$$T_R \gg T_T \qquad [16.53]$$

Thus for a metal specimen with $l = 0.2$ m and $c = 5 \times 10^3\,\text{ms}^{-1}$ $T_T = 80\,\mu\text{s}$. If $f = 1$ MHz, then the above conditions are satisfied with $T_W = 15\,\mu\text{s}$, $T_R = 1$ ms.

Because of the large difference in characteristic impedance between most solids and air, this method can be used for thickness measurement. Pulse reflection techniques are also commonly used for the detection of flaws in metals.[6] Here frequency f is chosen so that the sound wavelength is small compared with the size of defects it is desired to detect.

The large difference in characteristic impedance between gases and liquids means that almost perfect reflection occurs at a liquid/gas interface and that these techniques are applicable to level measurement. Analysis of the interface losses (Problem 16.2) would suggest it is better to mount the crystal at the base of the vessel, directing the waves upwards through the liquid, rather than to mount the crystal at the top of the vessel, directing the waves downwards through the gas. However, most

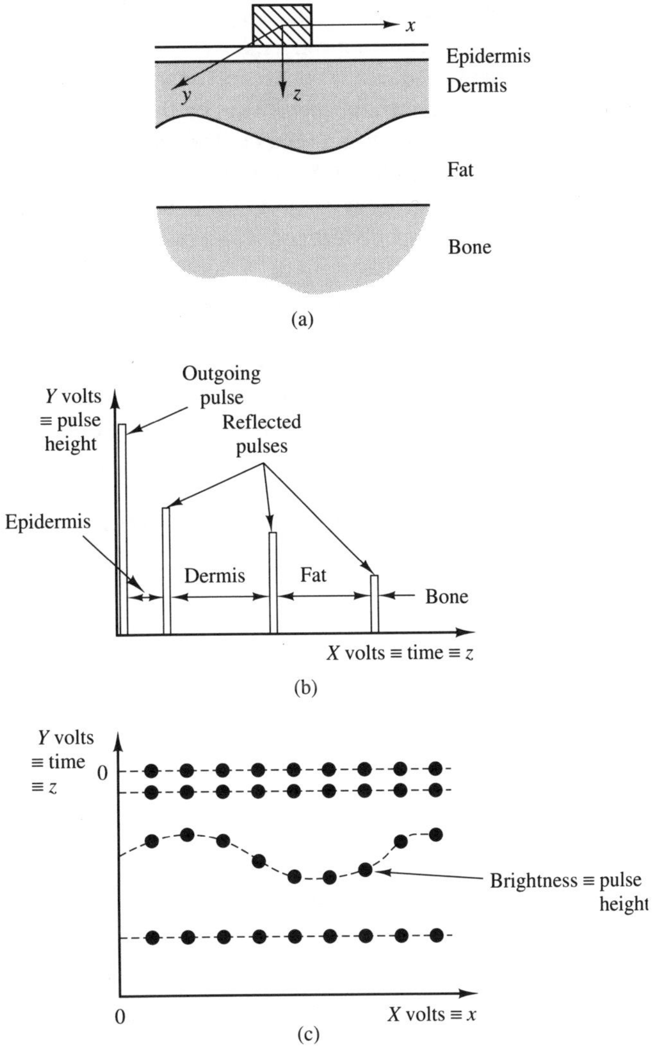

Fig. 16.14 Ultrasonic imaging of human body (adapted from Payne[12]) (a) Layers and coordinate system (b) A — scan display (c) B — scan display

commercial ultrasonic level measurement systems[7–10] use the latter method, because it offers greater ease of installation and maintenance.

Another important application of pulse reflection techniques is in the 'imaging' of areas of the human body.[11] Figure 16.14(a) shows, in simplified form, the various layers of tissue.[12] The characteristic impedance of these layers will be different, for example the impedance of bone is around 0.8×10^7 whereas that of soft biological tissue is around 0.15×10^7 (Table 16.2). A piezoelectric transducer is placed on the epidermis layer, which has either a characteristic impedance close to that of soft tissue (e.g. PVDF) or a matching layer. This minimises internal reflections at the transducer/epidermis boundary and the problem of multiple echoes.

Figure 16.14(b) shows the CRT trace obtained when the system of Fig. 16.13 is used with the layer system of Fig. 16.14(a). The three reflected pulses correspond to reflections at the epidermis/dermis, dermis/fat and fat/bone boundaries respectively; the time intervals between successive reflected pulses are proportional to the thickness of each layer. This trace, referred to as an **A-scan display** is rather difficult to interpret; a more realistic image is obtained using a **B-scan display**. Here the transducer is connected to two displacement sensors which measure transducer x and y position coordinates on the body surface. The output voltage of the x sensor is applied to the X plates of the CRT and a voltage proportional to time, i.e. distance z travelled through the body, is applied to the Y plates. The brightness of the image on the screen is proportional to the transducer output voltage (Z modulation) so that a bright spot corresponds to a reflected pulse. By keeping the transducer y coordinate fixed and adjusting the x coordinate, an image of the body in the $x - z$ plane is built up and stored (Fig. 16.14(c)). Thus the B-scan images a 'slice' through the body, normal to the surface. Another alternative is the **C-scan display**; this is an image of the body in the $x - y$ plane, i.e. a slice parallel to the surface of the body. This is obtained by applying x sensor output voltage to CRO X plates, y sensor output voltage to the Y plates and using Z modulation.

16.4.2 Ultrasonic Doppler flowmeter

This flowmeter uses the Doppler effect discussed in Section 16.3.6 and is shown in Fig. 16.15. We see that the flowmeter is completely external to the pipe; it is

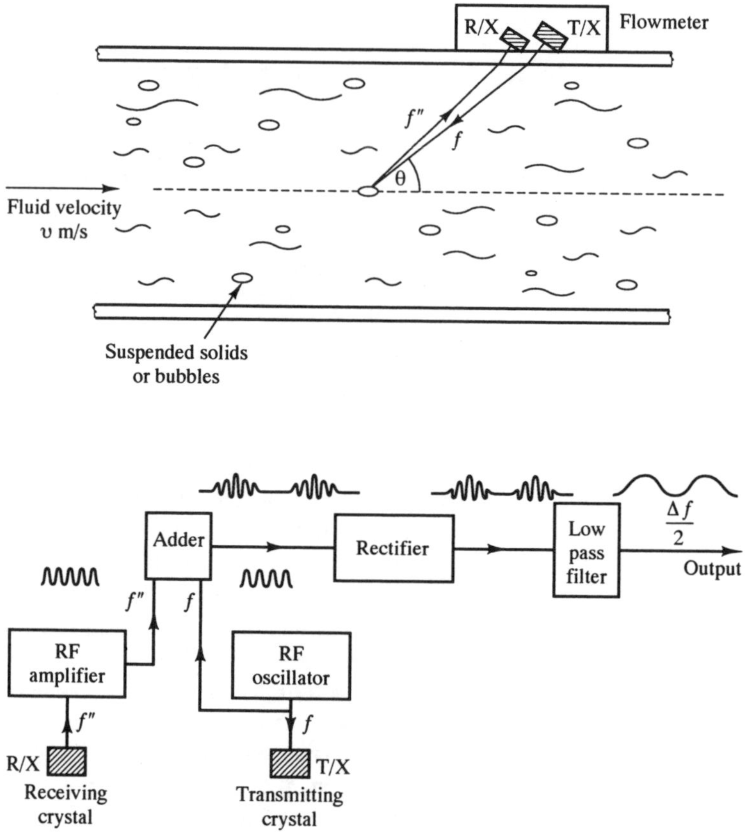

Fig. 16.15 Doppler flowmeter — layout and signal processing

thus suitable for the 'difficult' flowmetering problems discussed in Section 12.5, (i.e. corrosive fluids, slurries) and as a 'clip-on' flowmeter.[13] The transmitting crystal sends an ultrasonic wave of frequency f and velocity c into the fluid, at an angle θ relative to the direction of flow. Bubbles, solid particles, or eddies in the flow stream can be regarded as 'observers', moving with velocity v relative to the fixed transmitter.

The velocity of sound waves relative to the observer is $c + v\cos\theta$, so that from eqn [16.45] the apparent frequency f' seen by the observer is given by:

$$\frac{f'}{f} = \frac{c + v\cos\theta}{c} \qquad [16.54]$$

The particle scatters the incident sound wave in all directions, but a small proportion is back-scattered in the direction of the receiving crystal. The particle now acts as a 'source' moving with velocity v relative to the receiving crystal, acting as a fixed observer. The velocity of the sound waves relative to the source is $c - v\cos\theta$, so that from eqn [16.49], the apparent wavelength seen by the receiving crystal is given by:

$$\frac{\lambda''}{\lambda'} = \frac{c - v\cos\theta}{c} \qquad [16.55]$$

since $\lambda''f'' = \lambda'f' = c$, the corresponding received frequency f'' is:

$$\frac{f''}{f'} = \frac{c}{c - v\cos\theta} \qquad [16.56]$$

Eliminating f' between [16.54] and [16.56] gives:

$$f'' = f\frac{(c + v\cos\theta)}{(c - v\cos\theta)} = f\frac{\left(1 + \dfrac{v}{c}\cos\theta\right)}{\left(1 - \dfrac{v}{c}\cos\theta\right)} \qquad [16.57]$$

Typically we have $c \approx 10^3\ \text{m s}^{-1}$, $v \approx 10\ \text{m s}^{-1}$, i.e. $v/c \approx 10^{-2}$. Thus we can ignore terms involving second and higher powers of v/c. With this approximation [16.57] reduces to:

Frequency shift in Doppler flowmeter

$$\Delta f = f'' - f = \frac{2f}{c}(\cos\theta)v \qquad [16.58]$$

Thus the frequency difference Δf is proportional to fluid velocity v, and thus volume flowrate. If $f = 1$ MHz, $\theta = 30°$, then $\Delta f = 17.3$ kHz for the above values of c and v.

Figure 16.15 shows a possible system for processing the electrical signal from the receiver crystal. Due to attenuation, interface and scattering losses the receiver signal is at a low level, and is first amplified to the amplitude $\hat{V}$ of the transmitter drive signal. Both signals are input to an adder. The adder output signal is:

$$\begin{aligned} V_{ADD} &= \hat{V}\sin 2\pi f''t + \hat{V}\sin 2\pi ft \\ &= 2\hat{V}\cos\frac{2\pi(f'' - f)t}{2}\sin\frac{2\pi(f + f'')t}{2} \qquad [16.59] \end{aligned}$$

This is a sine wave of frequency $(f + f'')/2$ and amplitude $2\hat{V}\cos(2\pi\Delta ft/2)$ which is an amplitude modulated signal with a carrier frequency of $(f + f'')/2$ and a modulating frequency of $\Delta f/2$. The signal is demodulated to give a final output sinusoidal signal whose frequency $\Delta f/2$ is proportional to fluid velocity v.

16.4.3 Ultrasonic cross-correlation flowmeter

The principle of the cross-correlation flowmeter has already been explained in Section 12.5. In general it consists of two sensors, distance L apart, which detect random fluctuations in some property of the fluid. The sensor output signals $x(t)$ (upstream) and $y(t)$ (downsteam) are input to a correlator which evaluates the cross-correlation function:

$$R_{xy}(\beta) = \lim_{T\to\infty} \frac{1}{T}\int_0^T x(t - \beta)y(t)\,dt$$

This has a maximum at $\beta = L/v$, when there is maximum similarity between the delayed upstream signal $x(t - \beta)$ and the downstream signal $y(t)$.

Figure 16.16 shows an ultrasonic cross-correlation flowmeter[14,15] consisting of two transmission links $T/X(1)-R/X(1)$ and $T/X(2)-R/X(2)$. The meter is completely external to the pipe and is especially suitable for two-phase flow measurements, that is liquids with suspended gas bubbles or solid particles. A bubble or particle has a different acoustic impedance from that of the liquid. This means that if one is present

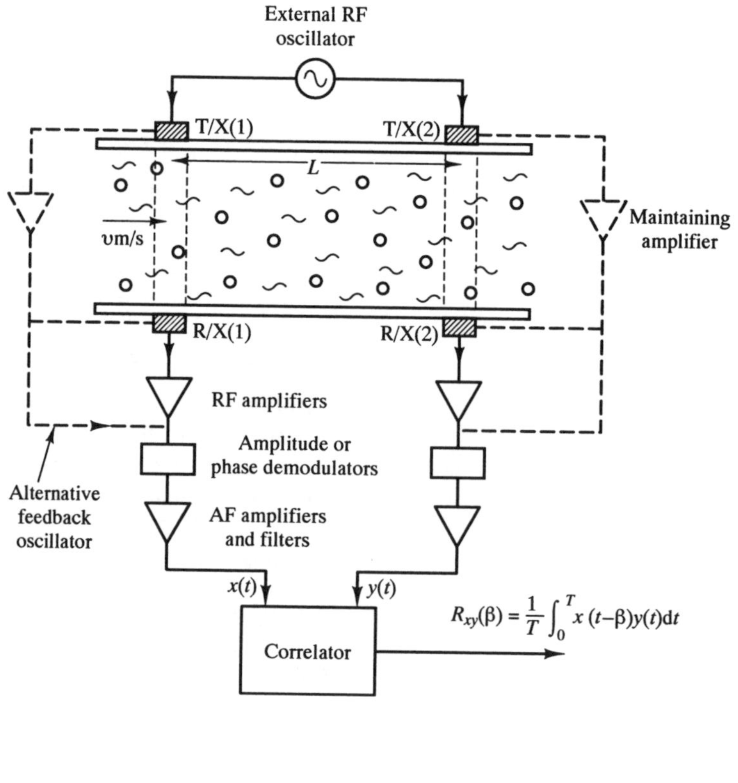

Fig. 16.16 Ultrasonic cross-correlation flowmeter

in either of the transmission links, there are two extra interfaces and corresponding reflection losses. The ultrasonic wave reaching the receiver is reduced in amplitude and changed in phase relative to the received wave with no particles present. The number of particles in either link varies randomly with time, causing the received ultrasonic waves to be randomly modulated, both in amplitude and phase. The receiver signals are first amplified and then demodulated to remove the high-frequency carrier signal. The demodulated signals, representing the random fluctuations in flow, are then amplified and input to the correlator. Each transmission link can be either energised by an external r.f. oscillator (solid line), or incorporated into a feedback oscillator system of the type described in Section 16.2 (dotted line).

16.4.4 Ultrasonic transit time flowmeter

Figure 16.17 shows an ultrasonic transit time flowmeter of the 'wetted sensor' type where the piezoelectric transducers are in contact with the fluid. This is suitable for clean, single-phase liquids and gases. For dirty, hostile or multiphase flows, a 'clamp on' type, with the transducers mounted on the outside of the pipe, is preferred. Firstly, transducer B acts as a transmitter and sends a pulse of ultrasound to transducer A, acting as a receiver. The corresponding transit time from B to A is:

$$T_{BA} = \frac{L}{c - v\cos\theta} \qquad [16.60]$$

where path length $L = D/\sin\theta$ and c is the velocity of sound in the fluid. If then A acts as a transmitter and sends a pulse to B acting as a receiver, the corresponding transit time from A to B is:

$$T_{AB} = \frac{L}{c + v\cos\theta} \qquad [16.61]$$

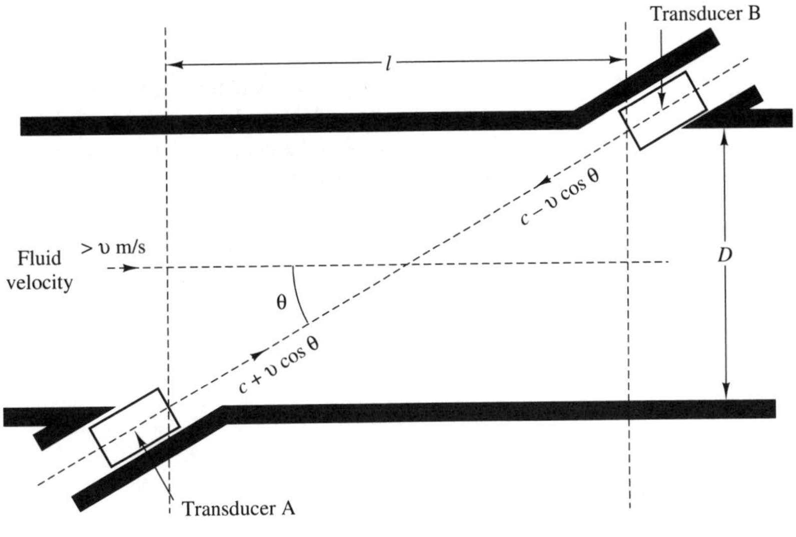

Fig. 16.17 Ultrasonic transit time flowmeter

The **differential transit time** ΔT is therefore given by:

$$\Delta T = T_{BA} - T_{AB} = \frac{D}{\sin\theta}\left[\frac{1}{c - v\cos\theta} - \frac{1}{c + v\cos\theta}\right]$$

$$= \frac{2D\cot\theta v}{c^2\left(1 - \frac{v^2}{c^2}\cos^2\theta\right)} \qquad [16.62]$$

As with the Doppler flowmeter the ratio v/c is typically 10^{-2} so that $(v^2/c^2)\cos^2\theta \ll 1$, eqn [16.62] reduces to:

Differential transit time

$$\Delta T = \frac{2D\cot\theta}{c^2}v \qquad [16.63]$$

Thus differential transit time ΔT is proportional to fluid velocity v but is normally very small. For water in a 0.1 m diameter pipe, the transit time at $\theta = 45°$, in the stationary fluid, is 95 μs. If the water velocity is 1 m s^{-1} the corresponding differential transit time is 89 ns. Thus to obtain a measurement error within 1% of reading, the maximum error in ΔT should be 0.9 ns. This high degree of precision can be obtained using a **sing-around system** where transducers A and B are continuously switched between transmitter and receiver modes.[16] Here the differential transit time is converted into a frequency difference $\Delta f = (\sin 2\theta/D)v$ which is independent of sound velocity c.

16.4.5 Ultrasonic vortex flowmeter

This is described in Section 12.3.3.

Conclusion

This chapter begins by explaining the basic ultrasonic transmission link which consists of a **transmitter**, **transmission medium** and a **receiver**. The next section examines the detailed characteristics of ultrasonic transmitters and receivers based on **piezoelectric** crystals. The following section studies the principles of ultrasonic transmission including **plane wave propagation**, **acoustic impedance and power**, **attenuation**, **reflection**, **refraction** and **stationary** waves. The final section looks at examples of ultrasonic measurement systems including **pulse reflection**, **Doppler**, **cross-correlation** and **transit time** systems.

References

16.1 O'DONNELL M, BUSSE L J and MILLER J G 1981 'Piezoelectric Transducers', *Methods of Experimental Physics*, vol. 19, Academic Press.

16.2 Massa Products Corporation 1983 *Technical Information on Model Tr-89/B series transducers*.

16.3 BLITZ J 1971 *Ultrasonics: Methods and Applications*. Butterworths, London.

16.4 BLITZ J 1967 *Fundamentals of Ultrasonics*, 2nd Ed., Butterworths, London.

16.5 LONGHURST R S 1963 *Geometrical and Physical Optics*, Longman, London, pp. 112–14.

16.6 WHITTLE M J 1979 'Advances in ultrasonic flaw detection' *C.E.G.B. Research*, No. 10, November.

16.7 BENSON F W 1967 'Ultrasonics for liquid level control', *Measurement and Control*, Vol. 9, No. 11, November.

16.8 Robertshaw Controls Co. 1979 Industrial Instrumentation Division, *Publicity Material for Model 165 Ultrasonic Level System*.

16.9 Bestobell Mobrey Ltd 1979 *Instruction Manual for 'Sensall' Continuous Level Monitoring System*.

16.10 Endress and Hauser (UK) Ltd, *Technical Information, DU 210, DU 211 Ultrasonic Level Sensors*.

16.11 CRECRAFT D I 1983 'Ultrasonic instrumentation: principles, methods and applications', *J. Phys. E. Scientific Instruments*, Vol 16, pp. 181–9.

16.12 PAYNE, P A 1985 'Medical and industrial applications of high resolution ultrasound', *J. Phys. E. Scientific Instruments*, Vol 18, pp. 465–73.

16.13 Bestobell Mobrey Ltd, *Publicity material for Doppler Flowmeter*.

16.14 MEDLOCK R S 1985 'Cross Correlation Flow Measurement', *Measurement and Control*, Vol 18, No 8, Oct. pp. 293–7.

16.15 KEECH R P 1982 'The KPC multichannel correlation signal processor for velocity measurement', *Transaction of the Institute of Measurement and Control*, Vol 4, No 1, Jan–Mar.

16.16 SANDERSON M L 1982 'Electromagnetic and Ultrasonic flowmeters: their present states and future possibilities', *Electronics and Power*, vol 28, No 2, February, pp. 161–4.

Problems

16.1 A piezoelectric crystal has an effective mass of 10^{-2} kg, stiffness of 10^{10} N m^{-1} and damping constant 200 Ns m^{-1}. The electrical capacitance of the crystal is 1000 pF and the charge sensitivity is 2×10^{-10} C N^{-1}

(a) Calculate the series and parallel resonant frequencies of the crystal.
(b) Calculate the magnitude and phase of the overall electrical impedance of the crystal at the above frequencies.
(c) The crystal is incorporated into a closed-loop oscillator system which is to oscillate at the crystal series resonant frequency. Calculate the required gain and phase of the maintaining amplifier at this frequency.

16.2 Fig. Prob. 2 shows a cross-section through a steel pipe through which sewage is flowing. The level of sewage in the pipe can also vary between the limits shown. Quartz crystals are attached to the outside of the pipe to form the four transmission links: $T_1 - R_1$, $T_2 - R_2$, $T_3 - R_3$ and $T_4 - R_4$.

(a) Assuming that only interface losses take place, calculate the fraction of received power to transmitted power for each link. Use the data given in Tables 16.1 and 16.2, and assume that sewage has the same transmission characteristics as water.
(b) Explain how these links can form the basis of systems to measure:

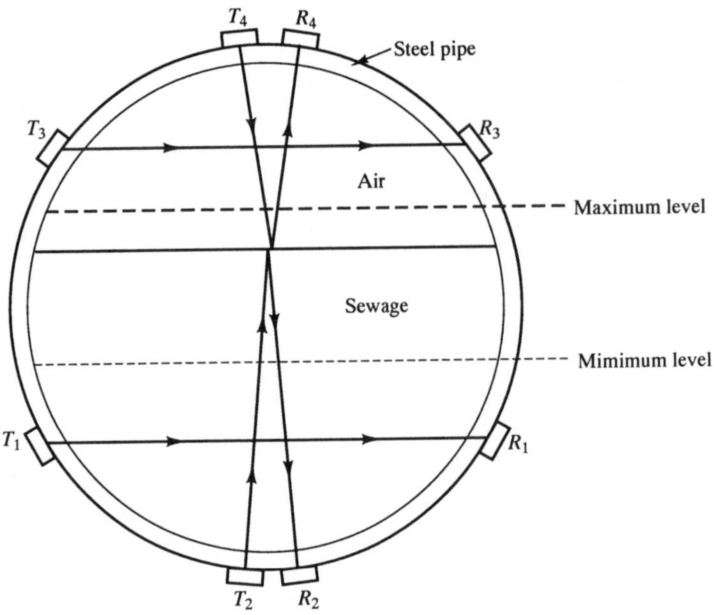

Fig. Prob. 2

(i) the level of sewage;
(ii) the velocity of sewage.

16.3 An open steel vessel contains liquid metal to a depth of about 0.75 m. It is proposed to measure the depth of liquid using ultrasonic pulse reflection techniques. A quartz crystal attached to the base of the vessel is to act alternatively as a transmitter and receiver. Using the data given below and in Tables 16.1 and 16.2

(a) Calculate the 'round trip' time T_T and the fraction of received power to transmitted power.
(b) Choose suitable values for pulse width and repetition times.

Data: Velocity of sound in liquid metal $= 1.5 \times 10^3\ \text{ms}^{-1}$
Density of liquid metal $= 5 \times 10^3\ \text{kg m}^{-3}$
Power attenuation coefficient $= 0.1\ \text{m}^{-1}$
Natural frequency of quartz crystal $= 1$ MHz

16.4 An ultrasonic Doppler flowmeter is to be used to measure the volume flowrate of a slurry in a steel pipe of diameter 0.2 m. Two piezoelectric crystals, each having a natural frequency of 1 MHz, are positioned, a few millimetres apart, on the outside of the pipe to form an ultrasonic transmission link. The transmitting crystal directs an ultrasonic beam into the pipe so that the beam is moving in an opposite direction to the flowstream. The angle between the ultrasonic beam and the direction of flow is 60°. On average 10 per cent of the ultrasonic power reaching each solid particle is scattered back in the direction of the receiving crystal.

Using the data given below

(a) Find the difference between the frequencies of the transmitted and received beams when the flowrate is $1.13 \times 10^3\ \text{m}^3\,\text{hr}^{-1}$.
(b) Estimate the ultrasonic power incident on the receiving crystal for each watt of ultrasonic power leaving the transmitting crystal. State any assumptions made in your calculation.

Assume that the slurry has the same density and sound velocity as water (Table 16.2) and a power attenuation coefficient of $1.0\ \text{m}^{-1}$.

16.5 An ultrasonic transmitter is in the form of a piezoelectric disc of diameter 2.5 cm and thickness 1.0 cm. The front face of the disc is placed directly onto biological tissue, the rear face is in contact with air. A pulse launched from the centre of the disc divides into two equal pulses, each of power 1 W and width 0.5 μs, one travelling towards the front face and one towards the rear face.

(a) Use the data given below to derive the form of the signal entering the tissue.
(b) Explain in detail what modifications should be made to the transducer so that the signal entering the tissue is a close approximation to a single pulse.

Data

Material	Velocity of sound m s^{-1}	Density kg m^{-3}
Piezoelectric	5×10^3	2.6×10^3
Tissue	1.3×10^3	1.0×10^3
Air	0.3×10^3	1.33

16.6 Figure Prob. 6 shows transmitting and receiving quartz crystals attached to a composite metal slab. The composite slab consists of a slab of metal *A* bonded to a slab of metal *B*; metal *B* contains an air-filled crack of negligible thickness in the position shown. An ultrasonic pulse of mean power 1 W and duration 5 μs is launched from the transmitting crystal into the slab. Use the data given below to discuss the main features of the signal at the receiving crystal. Estimates of times and power levels should be made.

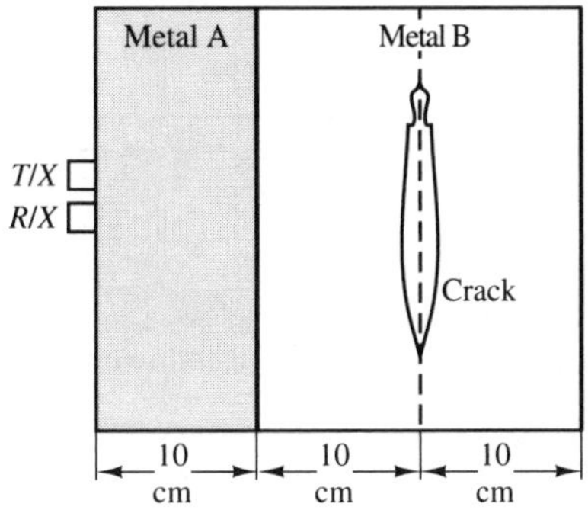

Fig. Prob. 6

Data

Material	Velocity of sound m s^{-1}	Density kg m^{-3}	Power attenuation coefficient m^{-1}
Metal A	6.0×10^3	3.0×10^3	2.0
Metal B	6.0×10^3	8.0×10^3	4.0
Quartz	6.0×10^3	2.5×10^3	—
Air	0.30×10^3	1.33	—

16.7 A differential transit time ultrasonic flowmeter is to measure gas velocities between 0.3 and 24 m s^{-1} to an accuracy within $\pm 1\%$. Use the data given below to critically assess the feasibility of the flowmeter according to the following criteria.

(a) Ratio of received pulse power to transmitted pulse power. Explain how this ratio may be improved.

(b) Accuracy of pulse transit time measurement necessary to meet the above specification.

Data

Material	Velocity of sound $m\,s^{-1}$	Density $kg\,m^{-3}$
Quartz	5700	2600
Natural gas	500	50

Pipe Diameter $D = 300$ mm, Flowmeter length $l = 600$ mm.

16.8 An organic liquid containing suspended plastic pellets is flowing in a steel pipe of internal diameter 0.2 m and thickness 0.01 m. An ultrasonic transmission link is set up across the pipe; this consists of identical transmitting and receiving quartz discs as shown in Fig. Prob. 8. The plastic pellets can be regarded as discs of thickness 0.01 m, diameter 0.02 m, moving parallel to the direction of flow.

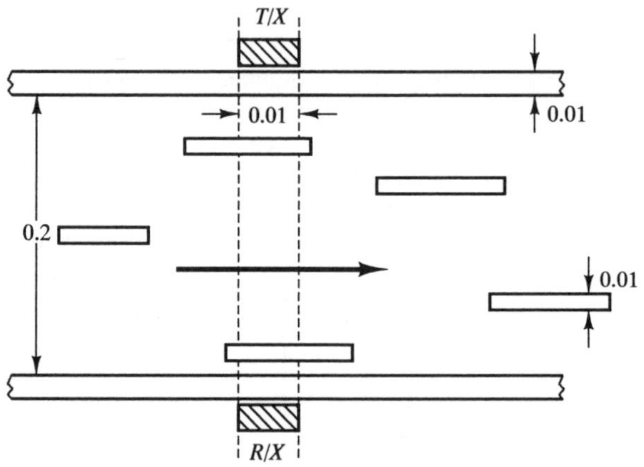

Fig. Prob. 8

Using the data given in the table below, calculate the power incident on the receiver for each watt of transmitter power in the following cases:

(a) no pellets present in the transmission link;
(b) a single pellet completely intersecting the transmission link;
(c) *n* pellets completely intersecting the transmission link, all pellets being separated by the liquid phase.

Material	Velocity of sound ($m\,s^{-1}$)	Density ($kg\,m^{-3}$)	Power attenuation coefficient (m^{-1})
Steel	6.0×10^3	7.8×10^3	4.0
Quartz	5.7×10^3	2.6×10^3	—
Plastic	2.4×10^3	1.1×10^3	2.0
Liquid	1.5×10^3	1.0×10^3	1.0

16.9 The overall transmission coefficient $\alpha_{T,G}$ between a transmitter of characteristic impedance R_1, and a gas of impedance R_2 separated by a matching layer of impedance R is given by eqn [16.40]. By differentiating [16.40] and setting the derivative equal to zero (or any other method), show that $\alpha_{T,G}$ is a maximum when $R = \sqrt{R_1 R_2}$.

17

Gas chromatography

In many industrial processes, the materials passing between different stages of the process are mixtures of compounds or elements. These mixtures are often a single phase — i.e. gas, liquid or solid — but may be multiphase — e.g. gas and liquid. In order to operate many processes at maximum throughput or efficiency, it is essential to have some information on the **composition** of important process streams. In some cases a detailed analysis giving the fraction of each component in the mixture may be required; in others it is sufficient to know the percentage of a single key component.

Important industrial examples are:

complete analysis of the hydrocarbon gas stream leaving a naphtha cracking furnace;
the percentage of carbon monoxide in boiler flue gas;
the acidity/alkalinity of boiler feed water;
the concentration of dissolved metal ions in industrial effluent discharged into rivers.

The measurement of composition is usually referred to as **analysis** and the corresponding systems as **analytical measurement systems**. The study of analytical techniques and systems is a wide subject which warrants a complete book in itself; the interested reader is referred to Refs. [1–3]. This chapter discusses the widely used technique of **gas chromatography** and the systems based on this technique. The use of katharometer and non-dispersive infrared systems for analysing gases has already been discussed in Chapters 14 and 15. The use of electrochemical sensors for measuring the concentration of ions in solution has been discussed in Section 8.9.

17.1 Principles and basic theory

Gas chromatography is a technique for measuring the composition of a gas or volatile liquid sample by separating the sample into its constituent components. The separation is performed using two phases: a moving gas phase, and a stationary phase which is either liquid or solid. In **gas–liquid chromatography** the stationary phase is a thin layer of non-volatile liquid, coated onto solid particles acting only as a support.

In **gas–solid chromatograpy** the solid particles themselves provide the stationary phase.

A typical column for **gas–liquid chromatography** (GLC) is shown a Fig. 17.1. The moving gas phase is called the carrier gas; this is an inert gas such as helium, argon or nitrogen which does not react with the sample and is not dissolved by the liquid phase. A known volume of sample gas is injected into the carrier at time t_0. This causes a pulse increase in the flow rate of gas into the column. The column is a long coiled tube, typically 2–3 m long and of 3 mm internal diameter, packed with small particles of solid ceramic material such as ground fire brick. The particles are coated with the liquid phase to give a large surface area for dissolving the gas. The liquid should not react with any of the components, and should have negligible vapour pressure at the column temperature. Commonly used liquids are silicones and polyglycols.

The carrier gas 'sweeps' the sample through the packing and brings it into intimate contact with the liquid phase. A dynamic equilibrium is set up with molecules passing

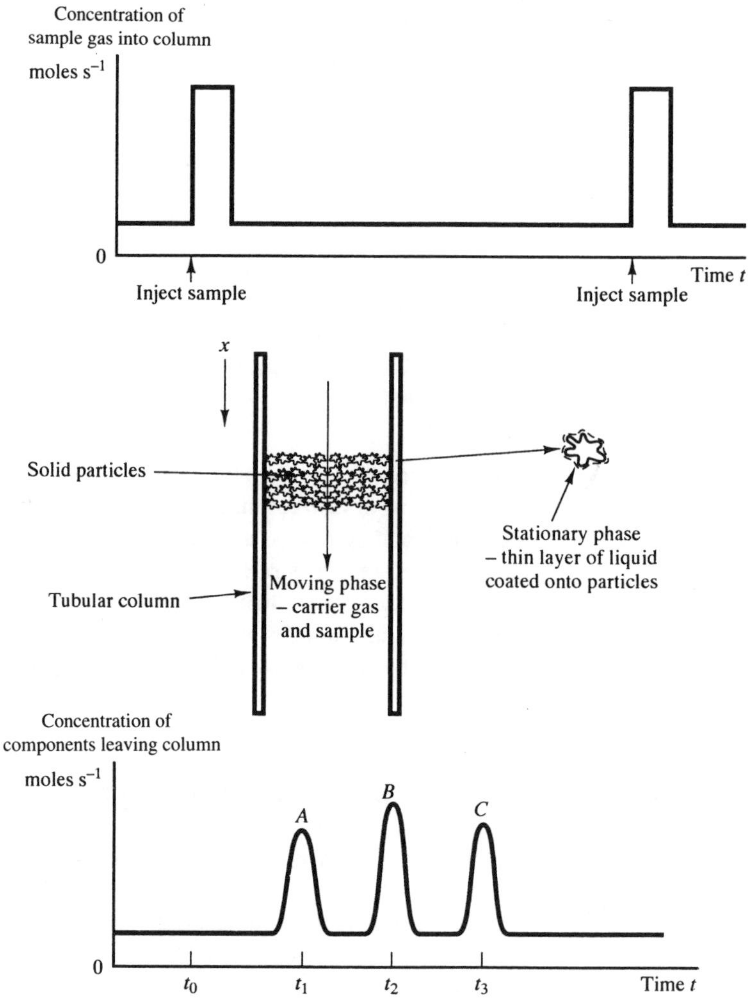

Fig. 17.1 Typical GLC column and gas flow rates in gas–liquid chromatography

from gas to liquid and liquid to gas at equal rates. However, molecules of different components spend, on average, different amounts of time in the liquid phase. When molecules are dissolved in the liquid phase they have zero velocity down the column. The overall result is that the different components of the sample emerge or 'elute' from the end of the column at different times.

Figure 17.1 shows the variation in the flow rate of gas leaving the column with time; this is called a **chromatogram**. Component A spends very little time in the liquid phase and elutes from the column first (at time t_1) whereas component C spends a lot longer in the liquid phase and is last to emerge from the column (time t_3).

The average time for the ith component of the sample mixture to travel through the column is termed the **retention time** T_i. This is the sum of the time T_{Mi} spent in the moving gas phase and the time T_{Si} spent in the stationary phase, i.e.

$$T_i = T_{Mi} + T_{Si} \qquad [17.1]$$

The time T_{Mi} depends only on the length L of the column and the average carrier gas velocity $\bar{u}$; this is the same for all components, i.e.

$$T_{Mi} = \frac{L}{\bar{u}} \quad \text{for all } i \qquad [17.2]$$

Time T_{Si} is usually different for each component and depends on the distribution ratio K_i of the component between the two phases, i.e.

$$K_i = \frac{T_{Si}}{T_{Mi}} \qquad [17.3]$$

From [17.1]–[17.3] we have

Retention time for ith component

$$T_i = \frac{L}{\bar{u}}(K_i + 1) \qquad [17.4]$$

Figure 17.2 is an idealised distance/time graph for four components, with $K_i = 0$, 1, 2, 3, travelling down the column. The component with $K_i = 0$ spends no time in the liquid phase and the graph is a straight line of slope $\bar{u}$. The components with $K_i = 1, 2, 3$ spend increasing time in the liquid phase and the corresponding graphs are straight lines of slope $\frac{1}{2}\bar{u}$, $\frac{1}{3}\bar{u}$, $\frac{1}{4}\bar{u}$ respectively. The average carrier gas velocity $\bar{u}$ can be found using[4]

$$\bar{u} = \frac{3(P^2 - 1)}{2(P^3 - 1)}\bar{u}_0 \qquad [17.5]$$

where $\bar{u}_0$ is the velocity at the column outlet, and $P = P_I/P_O$ is the ratio of column inlet and outlet pressures.

The molecules of the ith component do not all elute at the same time; there is a statistical distribution of times (usually Gaussian) about the mean value T_i. The width of the distribution depends on the width of the original injection pulse, as well as molecular diffusion effects inside the column, and can be described by a standard deviation σ_i. An alternative measure is the base width Δt_i; defined by the points of

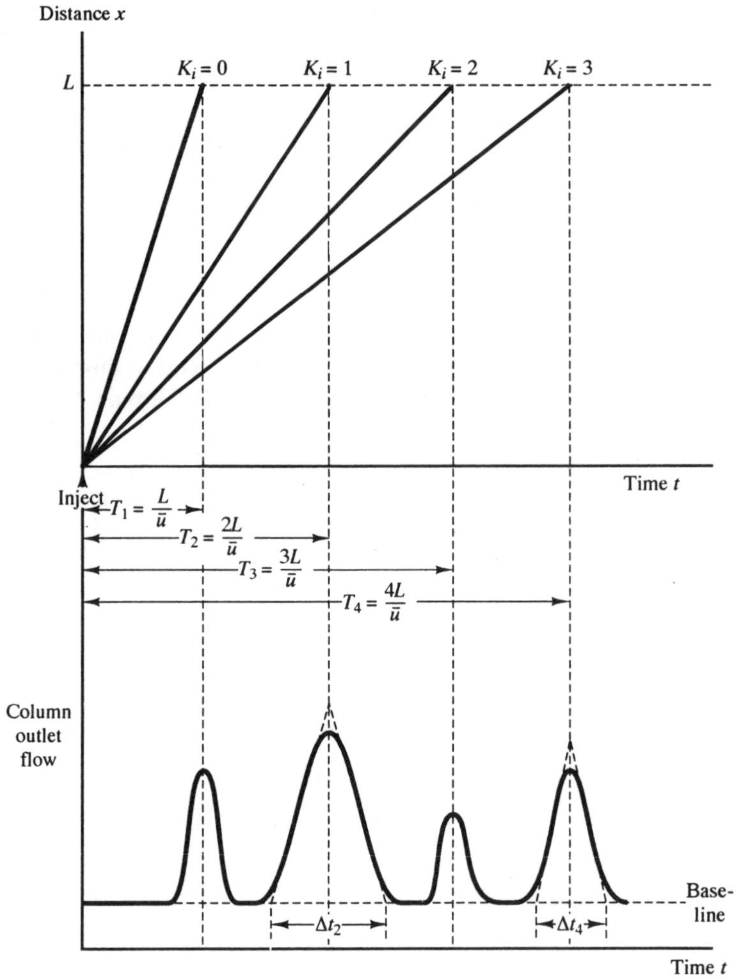

Fig. 17.2 Retention times and peak widths

intersection of the base line, and tangents drawn to the distribution at the half-peak height values (Fig. 17.2). For a Gaussian distribution $\Delta t_i = 4\sigma_i$.

The resolution R_{ij} of two adjoining components i, j is defined by:

Resolution of adjacent components

$$R_{ij} = \frac{2(T_j - T_i)}{\Delta t_i + \Delta t_j} \qquad [17.6]$$

A resolution of 1.5 for two identical Gaussian peaks means that $T_j - T_i$ is equal to six standard deviations and that the peaks are almost completely separated. A commonly used measure of the separating efficiency of a column is **height equivalent to a theoretical plate** (HETP). This quantity arises because of the similarity between packed chromatograph columns and packed distillation columns, and the need to find the height of a packed column equivalent to one theoretical stage in a plate column. To calculate HETP for a given column of length L, we first calculate the number of theoretical plates N in the column which is defined by

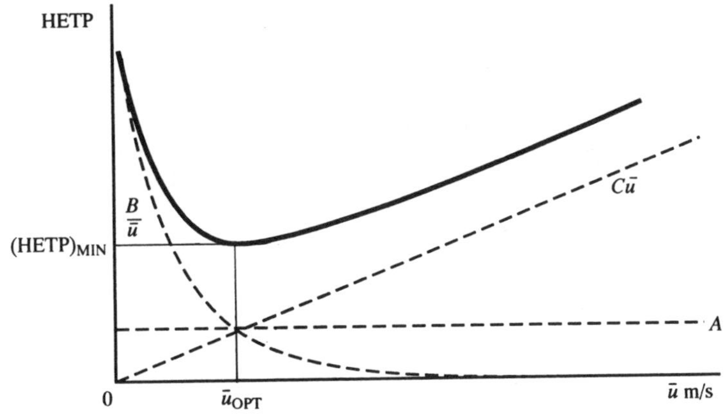

Fig. 17.3 Relation between HETP and carrier gas velocity $\bar{u}$

$$N = 16\left(\frac{T_i}{\Delta t_i}\right)^2 \qquad [17.7]$$

and then

$$\text{HETP} = \frac{L}{N} \qquad [17.8]$$

For a typical column we may have $L = 1$ m, $T_i = 120$ sec, $\Delta t_i = 10$ sec, giving $N = 2304$ and HETP ≈ 0.44 mm. HETP is not a constant for a given column, but depends on the mean carrier gas velocity $\bar{u}$ and column temperature T°C. The relationship between HETP and $\bar{u}$ is given by the Van Deemter equation:

$$\text{HETP} = A + \frac{B}{\bar{u}} + C\bar{u} \qquad [17.9]$$

where A, B, C are constants for a given system at a given temperature. The individual terms represent diffusion and mass transfer effects and are plotted in Fig. 17.3. We see that there is an optimum carrier gas velocity $\bar{u}_{OPT}$ at which HETP is a minimum, i.e. the best separation of components is obtained.

17.2 Typical gas chromatograph

Figure 17.4 shows a typical gas chromatograph system for composition measurement. Because column HETP depends on carrier gas velocity and column temperature, it is essential to control the flow rate of carrier gas and to place the sample injection valve, column and detector in a temperature-controlled enclosure. The sample injection valve is used to inject the sample into the carrier gas stream.

Figure 17.5 shows the principle of operation of a sliding plate valve used with a gas sample. The valve consists of a plastic plate with grooved channels, sandwiched between two metal blocks. Prior to injection the valve is in position A such that the carrier gas passes directly to the column and the sample gas passes through the sample loop before being vented. This ensures that the sample loop is filled with sample gas; typically the sample loop has a volume of a few millilitres. On receiving an 'injection' signal from the computer the plate moves across to position B. The carrier

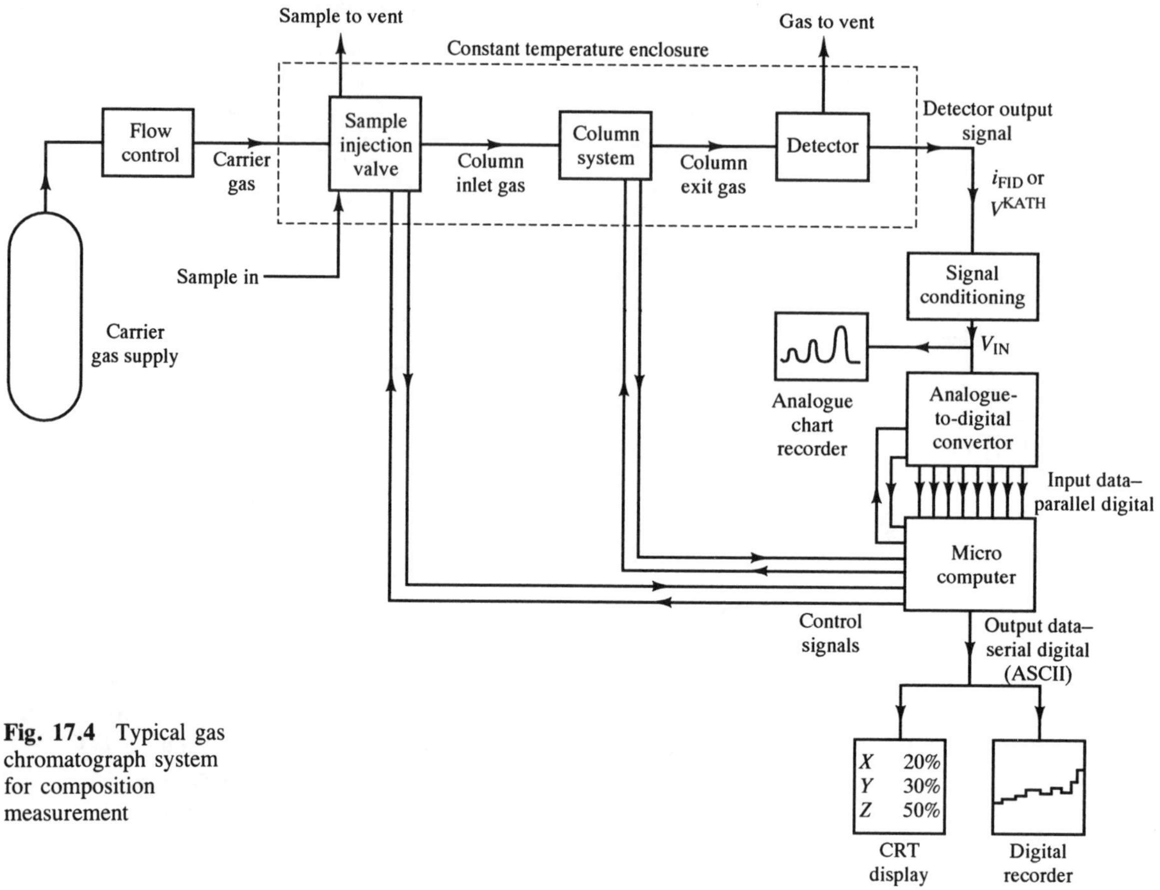

Fig. 17.4 Typical gas chromatograph system for composition measurement

Fig. 17.5 Injection of a gas sample with a sliding plate valve (after Pine[5])

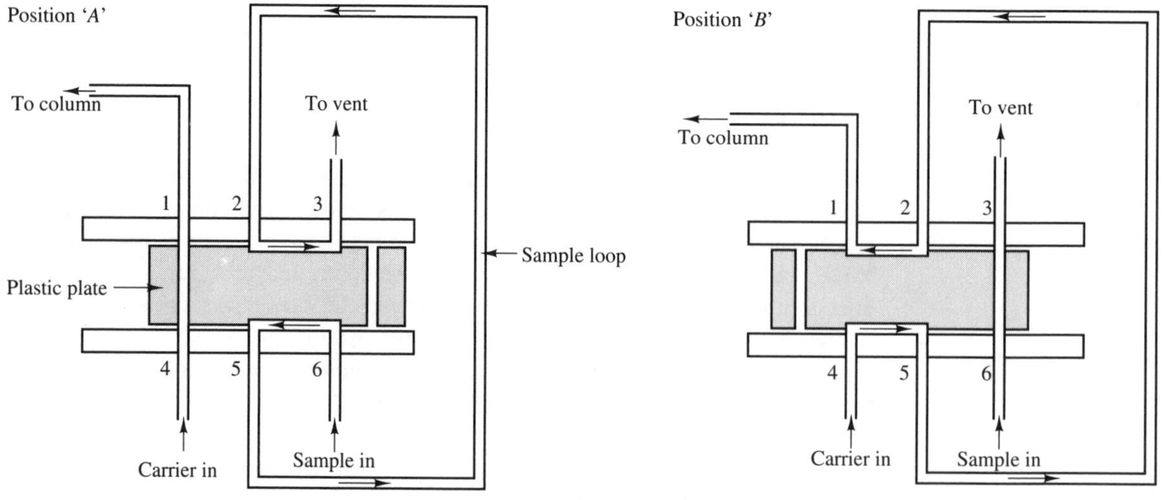

gas now passes through the sample loop and flushes the sample volume into the column. The sample gas stream passes directly to vent. The valve is then returned to position A ready for the next injection. The operation is similar for a liquid sample, except that the volume of the sample is much smaller; this vaporises on entering the carrier gas stream.

The construction of a typical GLC column has already been described. A column for **Gas–Solid Chromatography** (GSC) is similar, except that the solid particles themselves provide the active stationary phase rather than just acting as a support. Materials such as charcoal, silica gel, activated alumina and molecular sieve can adsorb gas molecules onto their surface. Many chromatograph systems incorporate more than one column, enabling sample and gas streams to be switched between columns.

The purpose of the detector is to give a voltage or current output, depending on either concentration or flow rate, for each of the component gases eluting from the column. An ideal detector will give an output proportional to either component concentration or component flow rate.[1] The detector sensitivity (e.g. current/flow rate) should be high and the same for all components. The detector should also have good repeatability and fast speed of response.

The **katharometer** (Section 14.4) has medium sensitivity, and is widely used as a detector in gas chromatograph systems for measuring compositions from several per cent down to around 1000 p.p.m. It consists of four heated, matched metal filaments arranged in a deflection bridge energised by a constant current $2i_0$. Pure carrier gas is passed over two of the filaments (Fig. 14.7), and the gas eluting from the column over the other two. The bridge is balanced when pure carrier gas leaves the column, but for each eluting component there is an output voltage given approximately by:

$$V_i^{\mathrm{KATH}} = D\,\frac{(k_c - k_i)x_i}{k_c[x_i k_i + (1 - x_i)k_c]}, \qquad D = \frac{i_0^3 R_{T_F}^2 \alpha d}{0.24\,A} \qquad [14.52],\ [14.50]$$

where k_c = thermal conductivity of carrier, k_i = thermal conductivity of the eluting component and x_i = molar concentration of component i in the carrier + component mixture. From [14.52] we see that the carrier thermal conductivity must be substantially different from all the component conductivities. For small x_i,

$$V_i^{\mathrm{KATH}} \approx D\left(\frac{k_c - k_i}{k_c^2}\right)x_i;$$

i.e. an approximate linear relation between V_i^{KATH} and x_i. However, sensitivity depends on k_i, i.e. sensitivity is different for different components. The **flame ionisation detector** has a high sensitivity for organic molecules and is used in the measurement of low compositions down to a few p.p.m. Here the gas eluting from the column is passed to a controlled oxy-hydrogen flame. The flame causes the organic molecules to be split up into charged ions, which flow to the collector electrode under the action of an applied d.c. voltage. This ionisation current is proportional to component flow rate, i.e. the number of molecules per second of component eluting from the column.

The detector output signal is at a low level (see Problem 14.5 for example) and requires amplification before being input to the A/D converter. This typically (Chapter 10) accepts an input signal of 0 to +5.0 V and converts it into an 8-, 12- or 16-bit binary parallel digital signal, which is passed to the microcomputer for further processing. The microcomputer performs two main functions which are explained more fully in the following section. These are as follows:

Control of sequence of operations. Several operations must be performed on the system of Fig. 17.4 including sample injection, column switching, control of A/D converter, signal processing and control of data presentation elements. These must be performed in the correct order and at the correct times.

Signal Processing. The time variation of converter input voltage is closely related to the flow rate of gas leaving the column as a series of peaks. The computer calculates the area under each peak which is approximately proportional to the total mass of the corresponding component. The composition of the sample mixture can then be found. These data can then be presented to the observer, possibly as a complete analysis of the sample on the CRT display or as a record showing the trend of a key component.

17.3 Signal processing and operations sequencing

The input voltage V_{IN} to the A/D converter will be proportional to the detector output signal. For a flame ionisation detector, the detector output current is proportional to the number of molecules $\dot{n}_i$ of the ith component, eluting from the column per second. In this case the converter input voltage is:

$$V_{\mathrm{i}}^{\mathrm{IN}} = G\dot{n}_i \qquad [17.10]$$

where G is a constant, ideally the same for all components.

Integrating [17.10], we have:

$$n_i = \frac{1}{G}\int V_{\mathrm{i}}^{\mathrm{IN}}\,\mathrm{d}t = \frac{1}{G}A_i \qquad [17.11]$$

which shows the total number of molecules n_i of the ith component in the sample is proportional to the integral of converter input voltage, i.e. to the **area** A_i under the ith peak. Thus the total number M of molecules present in a sample of m components is given by:

$$M = \sum_{i=1}^{m} n_i = \frac{1}{G}\sum_{i=1}^{m} A_i \qquad [17.12]$$

i.e M is proportional to the sum of the areas under all peaks. The percentage molar concentration c_i of the ith component in the sample is therefore given by:

Calculation of concentrations in ideal case

$$c_i = \frac{n_i}{M} \times 100 \text{ per cent} = \frac{A_i}{\sum_{i=1}^{m} A_i} \times 100 \text{ per cent} \qquad [17.13]$$

Thus in the ideal case, the composition of the sample can be found by evaluating the area under each component peak. The output voltage of the katharometer, however, is different for different components, and is approximately proportional to the concentration x_i of the component in the component/carrier mixture. If $\dot{n}_c$ is the number of molecules of carrier leaving the column per second, then:

$$x_i = \frac{\dot{n}_i}{\dot{n}_c + \dot{n}_i} \approx \frac{\dot{n}_i}{\dot{n}_c} \quad \text{for small } x_i \qquad [17.14]$$

Thus for dilute mixtures, x_i is proportional to $\dot{n}_i$ provided that $\dot{n}_c$ is a constant, so that there is tight control of carrier gas flow rate. The A/D converter input voltage for a katharometer detector is therefore given approximately by:

$$V_i^{\text{IN}} \approx G_i \dot{n}_i \qquad [17.15]$$

where G_i is the sensitivity for the ith component, which depends on thermal conductivity k_i. In this case the corresponding expression for the concentration of the ith component in the original sample is:

Calculation of concentrations in practical case

$$c_i = \frac{(1/G_i)A_i}{\sum_{i=1}^{m} \frac{1}{G_i} A_i} \times 100 \text{ per cent} = \frac{A_i^*}{\sum_{i=1}^{m} A_i^*} \times 100 \text{ per cent} \qquad [17.16]$$

where the A_i^* are the corrected peak areas $A_i^* = (1/G_i)A_i$.

The input to the computer is a sampled, coded version of the A/D converter input signal V_{IN}. The converter has, typically, a conversion time of a few milliseconds, so that samples can be read by the computer at intervals of a few milliseconds. These sampled values may be affected by high-frequency interference or noise and should be filtered. One simple filtering method is to calculate the average value of a group of several samples. Thus if the A/D conversion time is 5 ms an average of 40 samples is available every 0.2 sec. The computer uses these filtered values to estimate the area of the peak by dividing it into a series of rectangular elements, as shown in Fig. 17.6.

Bentley 1000/17-06

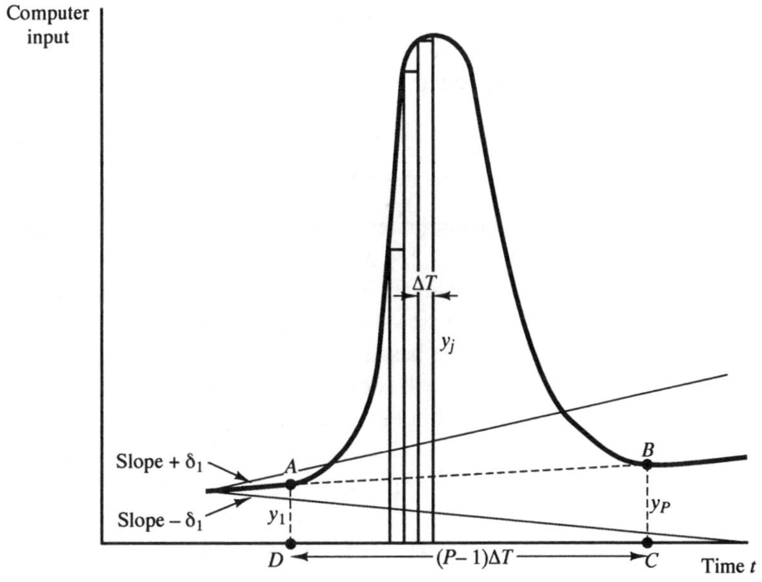

Fig. 17.6 Calculation of peak area

If y_j is the jth filtered value and ΔT the interval between values, then the area of the jth rectangle is $y_j\Delta T$, giving:

Approximate expression for peak area

$$A \approx \sum_{j=1}^{P} y_j\Delta T = \Delta T \sum_{j=1}^{P} y_j \qquad [17.17]$$

where P is the number of filtered values used in the calculation. The above expression is accurate if ΔT is small compared with the peak standard deviation σ. Thus if σ = 2.5 sec, a ΔT of 0.2 s is suitable, giving $P = 75$ for the complete peak of width 6σ.

Figure 17.7 shows a flowsheet of a simple program for evaluating the area of a given peak. After initialisation, the computer reads the current filtered value y_j and then evaluates the corresponding slope or gradient s_j. This is given by:

$$s_j = \frac{y_j - y_{j-1}}{\Delta T} \qquad [17.18]$$

where y_{j-1} is the previous filtered value and ΔT the interval between values. This slope value is used to decide whether or not a peak is eluting from the column. The baseline will not always be constant with time but may drift either upwards or downwards. If the maximum slope of the baseline is $+\delta_1$ or $-\delta_1$, then the condition for a peak is:

Peak: $\quad s_j > +\delta_1$ or $s_j < -\delta_1$ $\qquad$ [17.19]

and the condition for baseline is:

Baseline: $\quad s_j \leq +\delta_1$ or $s_j \geq -\delta_1$ $\qquad$ [17.20]

The integration, i.e. summation of y_j values, proceeds only while the peak condition is satisfied. However, integration must also be carried out around the top of the peak where the slope is small, which means that the baseline condition rather than the peak condition is obeyed. At the baseline there is only a small difference between current and previous slopes s_j and s_{j-1}, whereas at the top of a peak the difference between these slopes is large. Thus we can differentiate between baseline and peak top using the conditions:

Baseline $\quad s_j \leq +\delta_1$ or $s_j \geq -\delta_1$ and $|s_j - s_{j-1}| \leq \delta_2$ $\quad$ [17.21]

Peak top $\quad s_j \leq +\delta_1$ or $s_j \geq -\delta_1$ and $|s_j - s_{j-1}| > \delta_2$ $\quad$ [17.22]

The computer counts the total number of times $I + J + K$ that conditions [17.19] and [17.22] are satisfied; this gives the total number of values P used in the calculation. The computer finds the first value to satisfy the condition $s_j > \delta_1$ which is the first value used in the integration and is designated y_1. This gives the position A of the baseline at the start of the peak. The position B of the baseline at the end at the peak is the last value to satisfy the condition $s_j < -\delta_1$, that is the last value y_P used in the integration. The baseline is assumed to follow the straight line AB, and the peak area A_i is calculated by subtracting the area of trapezium $ABCD$ from eqn [17.17], i.e.

$$A_i = \Delta T \sum_{j=1}^{P} y_j - \Delta T \frac{(P-1)}{2}(y_1 + y_P) \qquad [17.23]$$

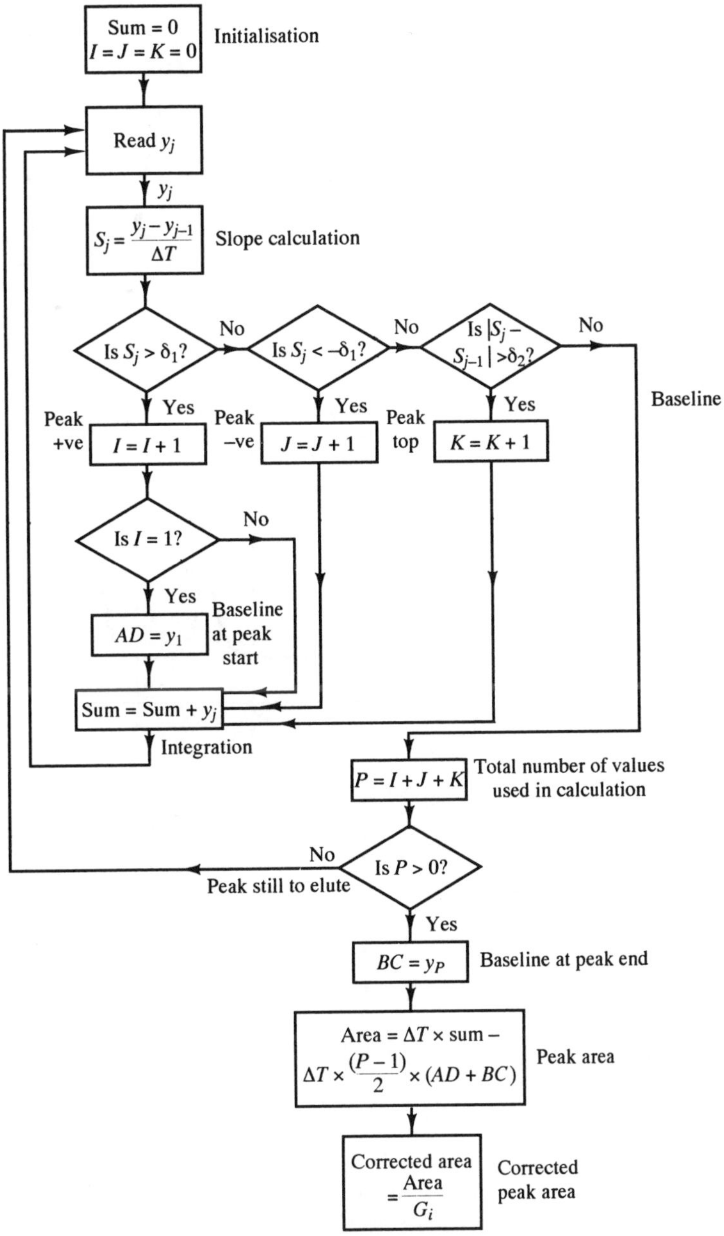

Fig. 17.7 Flowsheet of program for evaluating corrected peak area

As mentioned earlier the detector sensitivity may be different for different components i. The corrected area $A_i^*$is evaluated using $A_i^* = A_i/G_i$, where G_i is the sensitivity for the ith component. The G_i values can be found experimentally by injecting a standard mixture of known composition and comparing the areas of the corresponding voltage peaks. The computer uses eqn [17.16] to calculate component concentrations c_i from the A_i^*.

Figure 17.8(a) shows the sequence of operations for a simple system involving one column and one carrier stream. There are, however, many analyses which require

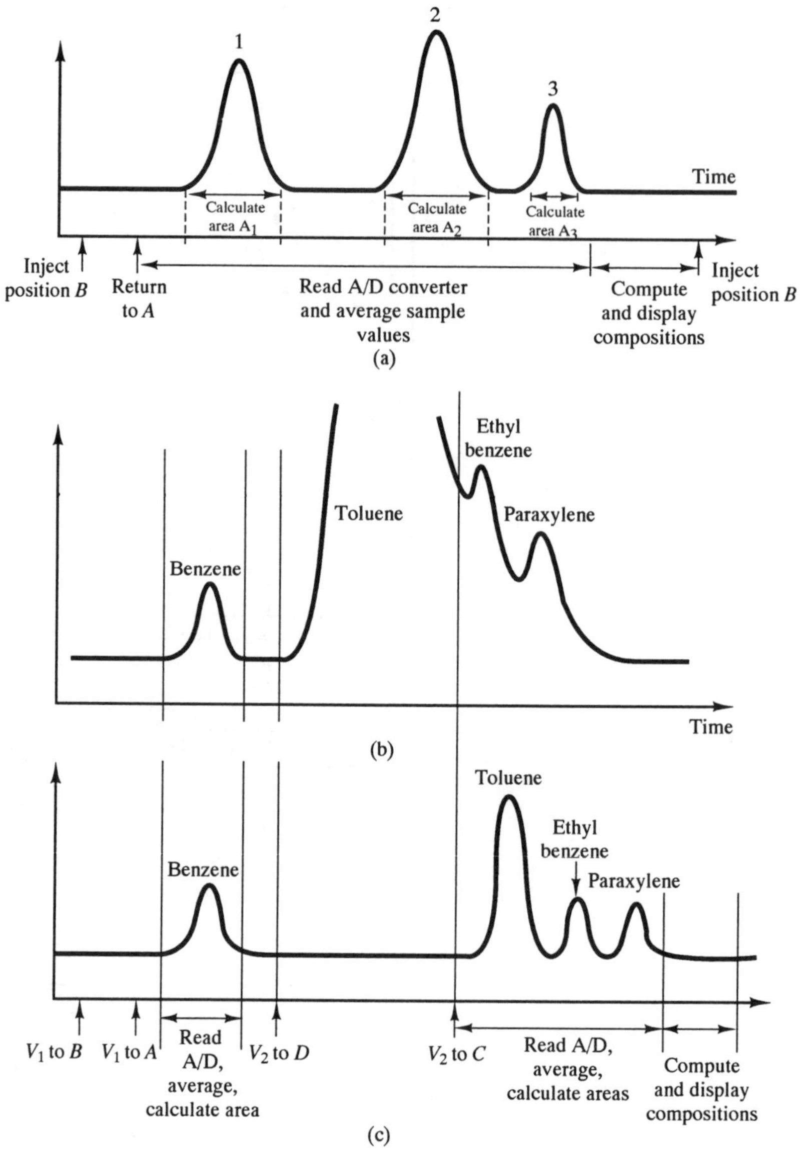

Fig. 17.8 Sequences of operations in gas chromatography
(a) Normal sequence
(b) Chromatogram without 'heartcutting'
(c) Chromatogram and sequence of operations with 'heartcutting'

multiple column and carrier operation.[5] One example is the measurement of small concentrations of benzene, ethyl benzene and paraxylene in a mixture which is mainly toluene. The resulting chromatogram, for a single column, is shown in Fig. 17.8(b). We see that the peaks are not clearly resolved; the small ethyl benzene and paraxylene peaks sit high on the 'tail' of the large toluene peak. The problem is solved using the technique of **heartcutting** which involves switching gas streams between two columns, as shown in Fig. 17.9. Two sliding plate valves are used, one for sample injection (V_1) and one for column switching (V_2). The sample is injected into column 1 using V_1, in the normal way. Valve V_2 is initially in position C, which enables the benzene peak to pass from column 1 to column 2. As the toluene peak is about to elute from column1, V_2 is switched to position D, so that the gas leaving

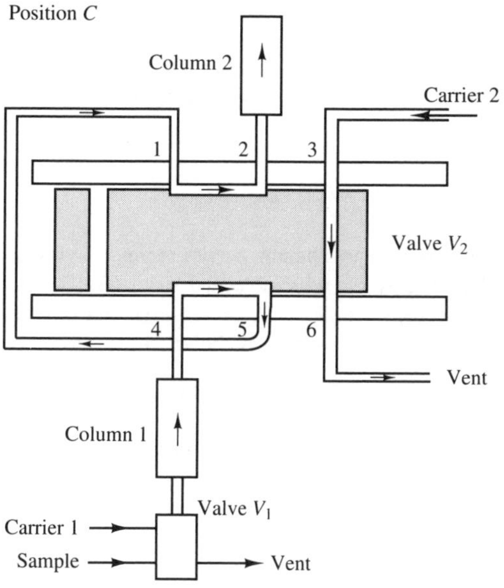

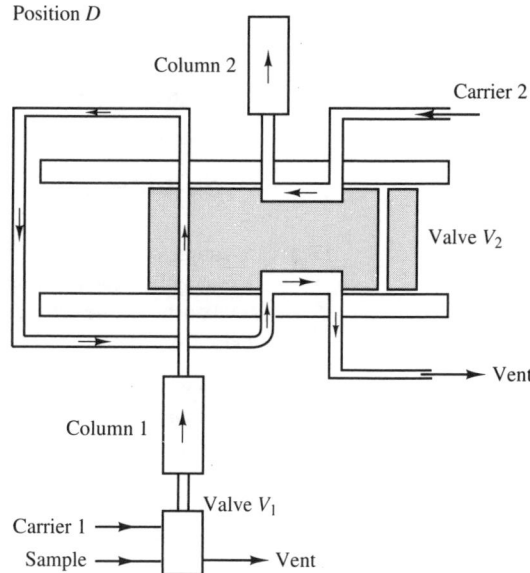

Fig. 17.9 Heartcutting using two sliding plate valves

column 1 is vented and carrier gas 2 is passed through column 2. When most of the toluene has been vented, V_2 is returned to position *C*. Thus the remainder of the toluene peak, together with the associated ethyl benzene and paraxylene peaks, enter column 2. These peaks can then be effectively separated in column 2 without having to use very long retention times. The resulting chromatogram is shown in Fig. 17.8(c), together with the operations sequence.

Another switching technique involving two columns and two carrier supplies is called **backflushing**. This technique allows components of interest to pass through columns 1 and 2 to the detector. Components not required for the analysis, or which may harm the detector, are prevented from entering column 2 and 'flushed' to vent by carrier 2 flowing through column 1 in a reverse direction.

Conclusions

The chapter first discussed the principles of **gas chromatography** and a typical *chromatograph* for measuring the compositions of gas mixtures. Associated **signal processing** and **sequencing operations** were then studied.

References

17.1 EWING G W 1975 *Instrumental Methods of Chemical Analysis*. 4th edn, McGraw-Hill, New York.

17.2 STROBEL H A 1973 *Chemical Instrumentation; A Systematic Approach*, Addison Wesley, New York, 2nd edn.

17.3 WILLARD H H, MERRITT L L and DEAN J A 1974 *Instrumental Methods of Analysis*, 5th edn, Van Nostrand, New York.

17.4 AYERS B O, LLOYD R J and DEFORD D D 1961 'Principles of high speed gas chromatography with packed columns, *Analytical Chemistry*, Vol 33, No 8, July, pp. 986–91.

17.5 PINE C S F 'Process gas chromatography' *Talanta 1967*, vol 14, pp. 269–97.

Problems

17.1 A sample containing oxygen and nitrogen is injected into a helium carrier at time $t = 0$ sec. The sample is swept through a column 1.0 m long packed with molecular sieve. The eluting components are detected by a katharometer detector which has an equal sensitivity for oxygen and nitrogen. The time variation in katharometer output voltage is shown in Fig. Prob. 1.

(a) Assuming that the distribution ratio K for oxygen is 2.0, estimate the mean carrier velocity $\bar{u}$ and K for nitrogen.
(b) Estimate base width Δt for both peaks and hence find the resolution R.
(c) Estimate the number of theoretical plates N and HETP.
(d) Estimate the percentage composition of the sample (assume peaks are approximately triangular).

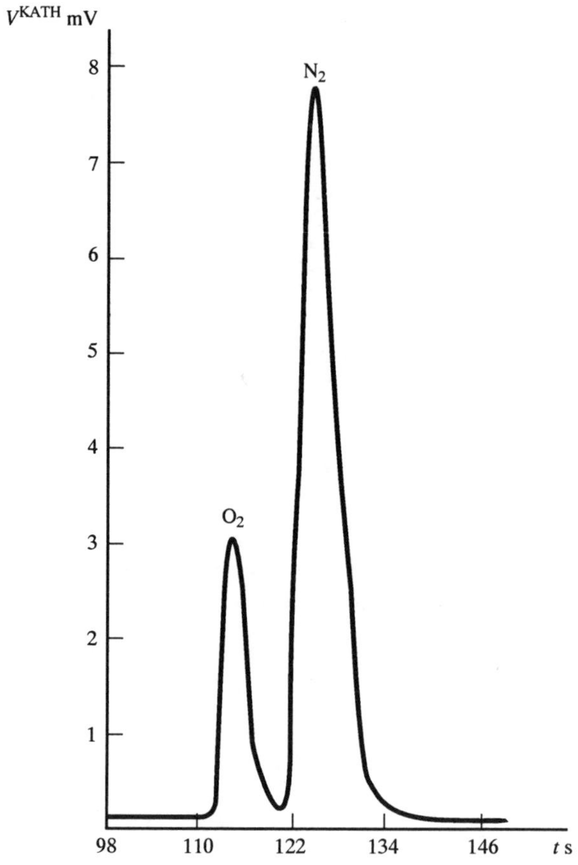

Fig. Prob. 1

18

Data acquisition and communication systems

All of the measurement systems discussed so far have presented the measured value of a single variable to an observer; i.e. the systems were single input/single output. However, there are many applications where it is necessary to know, simultaneously, the measured values of several variables associated with a particular process, machine or situation. Examples are measurements of flow rates, levels, pressures and compositions in a distillation column; temperature measurements at different points in a nuclear reactor core; components of velocity and acceleration for an aircraft. It would be extremely uneconomic to have several completely independent systems and a single multi-input/multi-output **data acquisition system** is used. Here several elements are 'time shared' amongst the different measured variable inputs. This technique of **time division multiplexing** is discussed in the first section of this chapter and a typical data acquisition system is described in the following section.

The oil, water and gas industries are characterised by complex distribution systems involving the transfer of fluids by long pipelines from producing to consuming areas. Similarly, an electricity distribution system involves the transfer of electrical power from power stations to consumers, via a network of high voltage cables. These systems also include several items of equipment or plant, e.g. pumping stations, compressors, storage tanks and transformers, each with associated measured variables. These plant items are often located several miles from each other, in remote areas. It is essential for the effective supervision of these distribution systems that all relevant network measurement data are transmitted to a central control point.

To do this a complex **communications system** is required. This usually consists of a **master station** (at the central control point) and several **outstations** (at the plant items). The system must be capable of transmitting large amounts of information in two directions (M/S to O/S and O/S to M/S), over long distances, in the presence of interference and noise. This chapter discusses the principles of **parallel digital signalling**, **serial digital signalling**, **error detection/correction**, and **frequency shift keying** which are used in communications systems, and concludes by describing the implementation of communications systems for measurement data with special regard to the **Fieldbus standard**.

18.1 Time division multiplexing

Figure 18.1 shows a simple schematic diagram of a time division multiplexer, with four channels, 0, 1, 2, 3. The input signal at each channel is a continuous voltage corresponding to a measured variable. The multiplexer also requires a two-bit parallel channel address signal to specify which input signal is connected to the output line. Thus if the binary address signal is 10, the switch in channel 2 is closed and input 2 is connected momentarily to the output line. The multiplexer output signal is thus a series of samples (Chapter 10), the samples being taken from different measurement signals at different times. In **sequential** addressing the channels are addressed in order, i.e. first 0, followed by 1, then 2 and 3 returning to channel 0 and repeating, so that the pattern of samples for the multiplexed signal is as shown in the diagram. **Random** addressing, whereby an observer selects a channel of interest at random, is also possible.

If ΔT is the sampling interval, i.e. the time interval between samples of a given input e.g. 0 or 1; then the corresponding sampling frequency $f_s = 1/\Delta T$ must satisfy the conditions for the Nyquist sampling theorem (eqn (10.1]). These require that f_S be greater than or equal to $2f_{MAX}$ ... where f_{MAX} is the highest significant frequency present in the power spectral density of the measurement signal.

In Fig. 18.1 four samples occur during ΔT, so that the number of samples per second for the *multiplexed* signal is $4f_S$. In general, for m signals, each sampled f_S times per second, the number of samples per second for the multiplexed signal is:

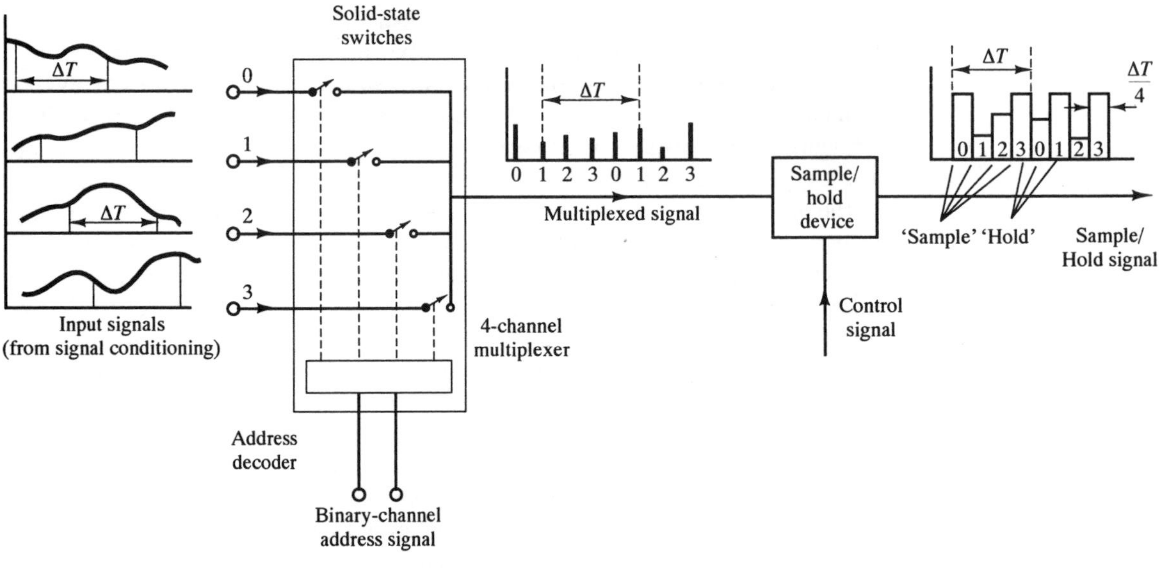

Binary address	Channel selected
00	0
01	1
10	2
11	3

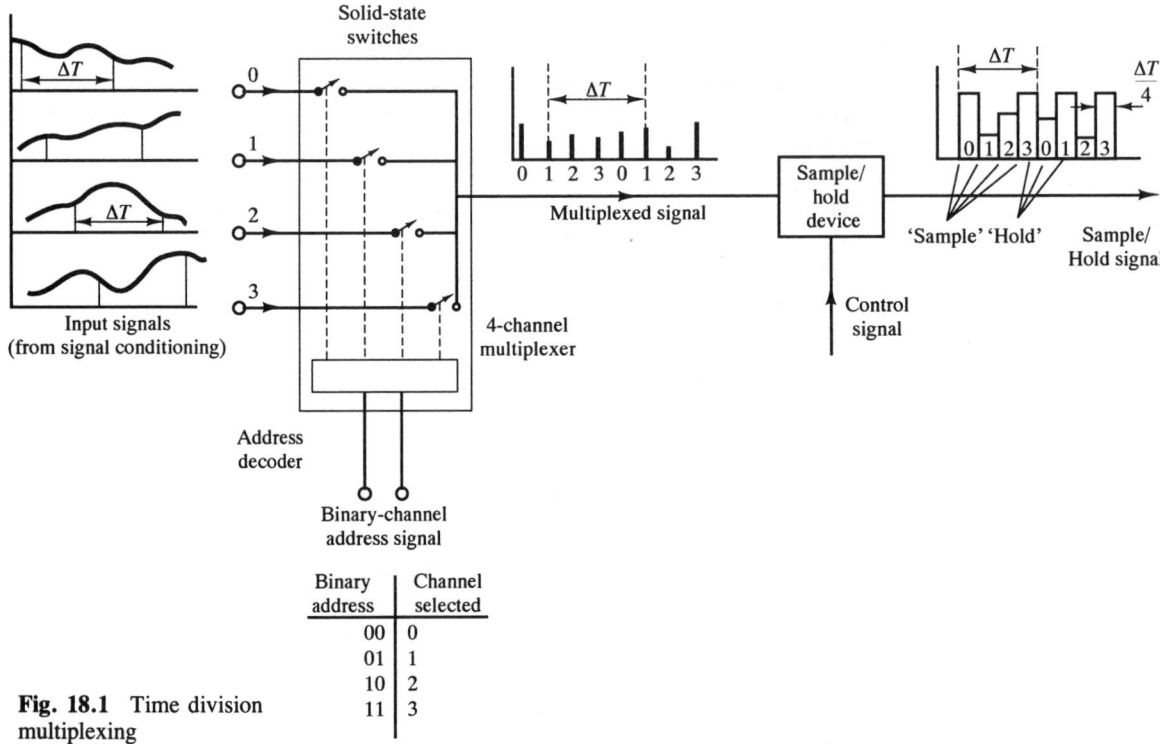

Fig. 18.1 Time division multiplexing

Sample rate for m multiplexed signals

$$f_S^M = mf_S \quad [18.1]$$

Different measured variables may have frequency spectra with different maximum frequencies: thus the power spectrum of a flow measurement may extend up to 1 Hz, but that of a temperature measurement only up to 0.01 Hz. The sampling frequency of the flow measurement must therefore be 100 times that of the temperature measurement. In the multiplexed signal there will be 100 samples of the flow measurement between each temperature sample. The multiplexed signal is normally fed to a sample-and-hold device (Section 10.1). Figure 18.1 shows the sample-and-hold waveform.

18.2 Typical data acquisition system

Figure 18.2 shows a typical microcomputer-based data acquisition system.[1,2] The signal conditioning elements are necessary to convert sensor outputs to a common signal range, typically 0 to 5V; Table 18.1 gives sensing and signal conditioning elements for different measured variables. The voltage signals are input to a 16-channel time division multiplexer and the multiplexed signal is passed to a single sample/hold device and analogue-to-digital converter (Section 10.1). In cases where all the sensors are of an identical type, for example 16 thermocouples, it is more economical to multiplex the sensor output signals. Here the multiplexed sensor signal is input to a single signal conditioning element, such as the reference junction circuit and instrumentation amplifier, before passing to the sample/hold and ADC.

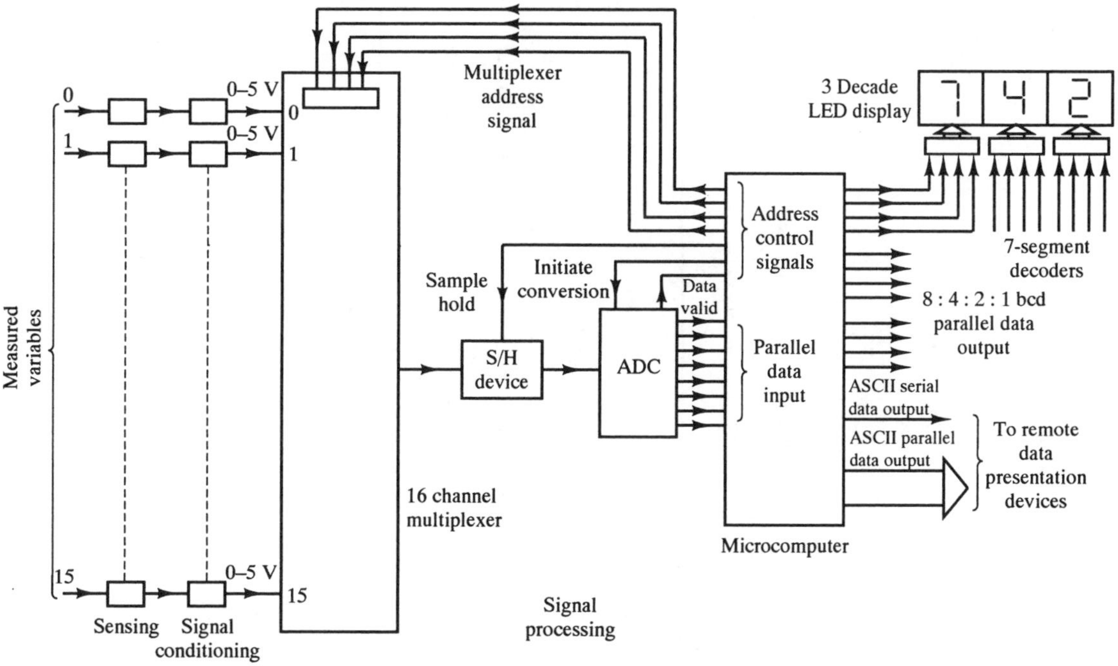

Fig. 18.2 Typical microcomputer based data acquisition system

Table 18.1 Typical measured variables, sensing and signal conditioning elements

Measured variable	Sensing element(s)	Signal conditioning elements
Temperature	Thermocouple	Reference junction circuit + inst. amplifier
Temperature	Platinum resistance thermometer	Deflection bridge + instrumentation amplifier
Flow rate	Orifice plate	Electronic D/P transmitter (4–20 mA) + current-to-voltage converter
Weight	Strain gauge load cell	Deflection bridge + instrument amplifier
Level	Electronic D/P transmitter (4–20 mA)	Current (e.g. 4 to 20 mA) to voltage (e.g. 0 to 5 V) converter
Angular velocity	Variable reluctance tachogenerator	Frequency-to-voltage converter
Linear displacement	Linear variable differential transformer (LVDT)	A.C. amplifier + phase-sensitive demodulator + low pass filter
Pressure	Diaphragm + capacitance displacement sensor	A.C. bridge + a.c. amplifier + phase sensitive demodulator + LPF.
Acceleration	Piezoelectric crystal	Charge amplifier

The ADC gives a parallel digital output signal which passes to one of the parallel input interfaces of the microcomputer. Another parallel input/output I/O interface provides the address and control signals necessary for the control of multiplexer, sample/hold and ADC.

These are a four-bit multiplexer address signal, **sample/hold** control signal, **initiate conversion** signal to the ADC and **data valid** signal from the ADC. The microcomputer performs whatever calculations (on the input data) are necessary to establish the measured value of the variable. A common example is the solution of the non-linear equation relating thermocouple e.m.f. and temperature (Section 10.4). The computer converts the measured value from hexadecimal into binary coded decimal form (Section 10.3). This b.c.d. data is written into a computer parallel output interface. Each decade is then separately converted into 7-segment code and presented to the observer using a 7-segment LED display (Section 11.4). The computer also converts each decade of the b.c.d. to ASCII form (Section 10.4). The resulting ASCII code is then written into a serial and/or parallel output interface. These can transmit ASCII data in serial and/or parallel form to remote data representation devices such as a monitor, printer or host microcomputer.

18.3 Parallel digital signals[3]

Parallel digital signals were introduced in Section 10.1; one path is required for each data bit and all of the bits are transmitted at the same time. Therefore, if 8 data bits (1 byte) are to be transmitted there are 8 paths in parallel, the voltage on each path being typically 5 V for a 1 and 0 V for a 0. The total collection of parallel paths is called a **data bus** or **data highway** and is similar to the internal data bus in a microcomputer. Since, however, an internal computer bus can only handle low power levels, it must be connected to an external data highway via a buffered interface. One commonly used parallel data highway conforms to the IEE 488/IEC 625 standard. This is a bit parallel, byte serial transmission system capable of a maximum transmission rate of 1 M bytes sec^{-1} up to a maximum transmission distance of 15 m. The standard is intended for high speed, short distance communication in a laboratory type environment, where there is relatively low electrical interference. The bus comprises 16 lines; 8 lines are used for data (usually 7-bit ASCII + parity check bit), 3 for 'handshaking' (see following section) and 5 for bus activity control. Up to 15 devices can be connected onto the bus. Each device must be able to perform at least one of the following three functions:

Listener	— a device capable of receiving data from other devices: e.g. printer, monitor.
Talker	— a device capable of transmitting data to other devices; e.g. counter, the data acquisition system of Fig. 18.2.
Controller	— a device capable of managing communications on the bus by sending addresses and commands; e.g. a computer.

If the transmission distance for parallel signals is increased beyond a few metres, imperfections in the transmission line result in some of the bits in a given byte arriving out of synchronisation with the rest. Similarly the presence of external interference again results in loss of synchronisation or corruption of data. In conclusion, parallel digital signals are suitable for high speed, short distance communication in laboratory environments. They are not suitable for long distance communication in industrial environments where significant external interference may be present.

18.4 Serial digital signals

18.4.1 Introduction

Serial digital signals can be used to transmit data over much longer distances (typically up to around 1 km) and are therefore commonly used in telemetry systems.[3] Here all of the data bits are transmitted one bit at a time in a chain along a single path. A serial digital signal is therefore a time sequence of two voltage levels; for example 0 V for a 0, 5 V for a 1 (**unipolar**), or −2.5 V for a 0, +2.5 V for a 1 (**bipolar**). The transmission path can vary from a standard twisted pair cable to a low loss coaxial cable or an optical fibre cable. Serial digital signalling is often referred to as **pulse code modulation**.

Figure 18.3(a) shows the use of an 8-stage shift register to convert an 8-bit parallel digital signal into serial form. The parallel signal $b_7 \ldots b_0$ is first loaded into the register and a clock signal applied. On receipt of the first clock pulse the contents of the register are shifted one place to the right, causing the least significant bit b_0 to appear at the register output, i.e. the least significant bit is transmitted first. The second clock pulse causes the register contents again to be shifted one place to the right, causing the next bit b_1 to be transmitted. The process is repeated until the register is empty: the most significant bit b_7 is the last to be transmitted. Figure 18.3(b) shows the register used to convert a serial signal into parallel form. On receipt of the first clock pulse the least significant bit is loaded into the storage element on the extreme left of the register. The second clock pulse causes the register contents to move one place to the right, allowing the next significant bit to enter the register. After eight clock pulses the entire signal is loaded into the register, with the LSB b_0 on the extreme right and the MSB b_7 on the extreme left.

Digital transmission links may be divided into three categories, depending on whether the communication is one-way or two-way. These categories are:

Simplex One way communication from A to B where B is not capable of transmitting back to A. This may be sufficient where a remote outstation is merely sending data to a master station. However, the master station cannot acknowledge receipt of data or request retransmission of corrupted data.

Half duplex Transmission from A to B and B to A but not simultaneously. With this system, after an outstation has transmitted data to a master station, the master station can then send an acknowledgement and if necessary request retransmission.

Full duplex Simultaneous transmission from A to B and B to A. This can be done using two paths; however, using modulation techniques, it is possible to transmit in two directions along a single path.

In each of the above systems it is important that the receiver is ready to receive and identify each set of data from the transmitter. There are two ways in which this can be achieved; **asynchronous transmission** and **synchronous transmission**. In

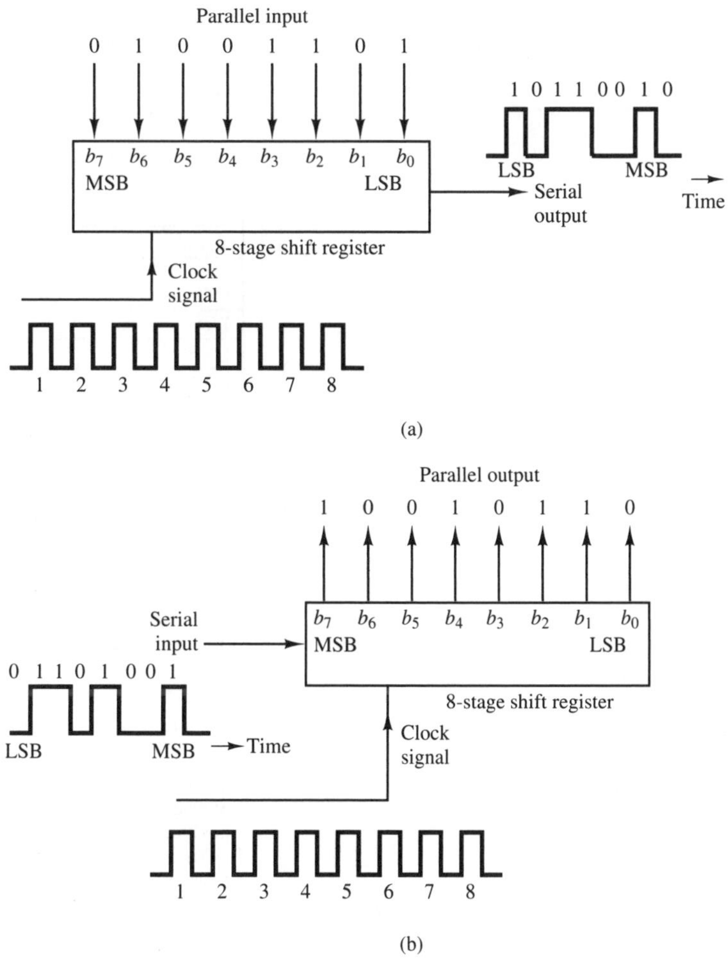

Fig. 18.3 Parallel and serial digital signals
(a) Parallel to serial conversion
(b) Serial to parallel conversion

asynchronous transmission each byte or **frame** of data is preceded by a start bit and concluded by a stop bit (Fig. 18.4(a)), so that the receiver knows exactly where the data starts and finishes. The transmission rate of serial digital signals is specified using **bit rate** R; this is the number of bits transferred in unit time, usually expressed in bits per second. Because of the need to check start and stop bits, the maximum transmission rate possible with asynchronous transmission is around 1200 bits s^{-1}. This method is therefore more suited to slower transmission systems.

For rates greater than 1200 bit s^{-1} synchronous transmission is used. Here a regular clocking signal is used to keep the receiver exactly in step with the transmitter. The transmitted data is preceded by a synchronising character which acts as a clocking pulse at the receiver. The receiver will then 'clock in' each bit of data.

There are several standard methods of **serial digital communication** available. The choice of method depends on several criteria including the following:

Transmission distance
Bit rate R
Resistance to external interference and noise
Number of multiplexed signals over a single link.

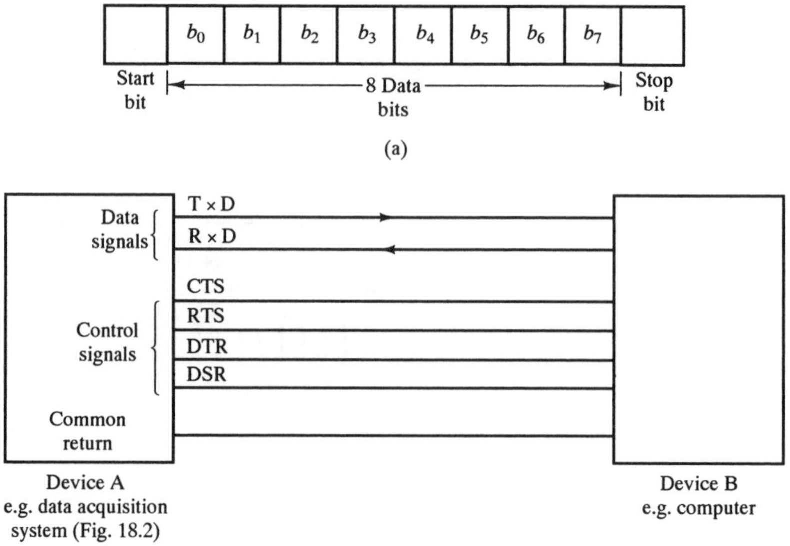

T × D ≡ Transmitted data
R × D ≡ Received data

CTS ≡ Clear to send
RTS ≡ Ready to send
DTR ≡ Data terminal ready
DSR ≡ Data set ready

(b)

A sets DTR to ON and awaits a response from B
B, if ready, sets DSR to ON
A now sets RTS to ON and awaits a response from B
B, when ready for data, sets CTS to ON
A transmits its data on T × D

(c)

Fig. 18.4 Asynchronous transmission and 'handshaking'
(a) Asynchronous data framing
(b) Connection for asynchronous transmission using RS 232
(c) 'Handshaking' sequence — A transmitting to B

These criteria are often conflicting; for example a high bit rate is incompatible with a long transmission distance and high noise immunity. From Section 18.4.2, we see that the bandwidth required for transmission of PCM is proportional to the bit rate R, i.e. the greater R the higher the required bandwidth. However, the available bandwidth of a given electrical cable decreases with length as the effects of resistance and capacitance increase. From Section 18.4.3, we see that for PCM affected by 'white' noise, the standard deviation of noise present at the PCM receiver is proportional to $\sqrt{R}$, i.e. the greater R the greater the noise present. However, since the receiver has simply to decide whether a 1 or a 0 has been transmitted, this decision can be made correctly even if the pulses are severely distorted by noise. Any errors that do occur, as a result of noise and interference, can be detected by adding check bits to the serial data signal (Section 18.5).

From Section 18.4.2, the bit rate R for m multiplexed signals is m times greater than for a single signal, so that high m is again incompatible with long transmission distance and high noise immunity. Finally the amount of external noise and interference will generally increase with transmission distance.

One commonly used standard for serial digital signals is the **RS 232 C/V 24 interface**, this specifies a 25-line connector and can be used for asynchronous and synchronous communication. In asynchronous communication only seven lines are used, 2 for data (transmitted and received), 4 for control signals and 1 for a common return (Fig. 18.4(b)). Figure 18.4(c) shows how the control signals are used in a '**handshaking**' sequence; this is necessary to ensure that the receiver is ready to receive data from the transmitter. RS 232 is capable of a bit rate of up to 20 k bits s^{-1} over short distances, i.e. up to 15 metres; longer transmission distances can be used at lower bit rates. However, RS 232 is vulnerable to external interference and cable resistance/capacitance effects and is best used over short transmission distances.

For higher bit rates over longer distances, RS 232 is gradually being replaced by a new standard **RS 449**, which is capable of a bit rate of 10 k bits s^{-1} over a distance of 1 km. The reason for this longer transmission distance is that, in RS 449, the transmitting and receiving data lines each have their own separate return lines (rather than sharing a common return). Each data line and return line can then form a twisted pair to give shielding from inductively-coupled interference (Section 6.5.2). However, in industrial environments, where there is high external interference, RS 449 cannot be used successfully even at low bit rates. The solution here is to use **current loop transmission**: as explained in Section 6.3, a current transmission system has far greater immunity to series mode interference than an equivalent voltage transmission system. Here the current in the loop is serially switched between 0 to 20 mA; 0 mA corresponds to a 0 and 20 mA to a 1. With current loop transmission the bit rate is normally limited to 4800 bits s^{-1}, also in order to transmit data to and from a computer a converter is necessary to convert the current serial signal into computer-compatible RS 232.

The serial digital techniques discussed above are suitable for transmission distances up to around 1 km. Successful transmission over longer distances in the presence of high interference can be obtained by frequency modulating the serial digital signal onto a carrier (Section 18.6).

18.4.2 Transmission bandwidth

Figure 18.5 shows a simple PCM transmission system. A transmitter consisting of a sample/hold device, ADC and parallel-to-serial converter, converts an input analogue voltage into a serial digital signal, which is sent over a transmission link to a receiver. The transmission link may be cable, radio link or optical fibre.

In order to estimate the bandwidth required for the transmission link it is necessary to find the extent of the frequency spectrum of the PCM signal. We first need to find the **bit rate** of the PCM signal; this is the number of bits per second or **baud** (1 baud = 1 bit s^{-1}).

Consider a single signal, sampled f_S times per second, each sample being encoded into an n-bit code. There are f_S samples per second and n bits per sample, so that the bit rate is:

Bit rate for a single signal

$$R = nf_S \qquad [18.2]$$

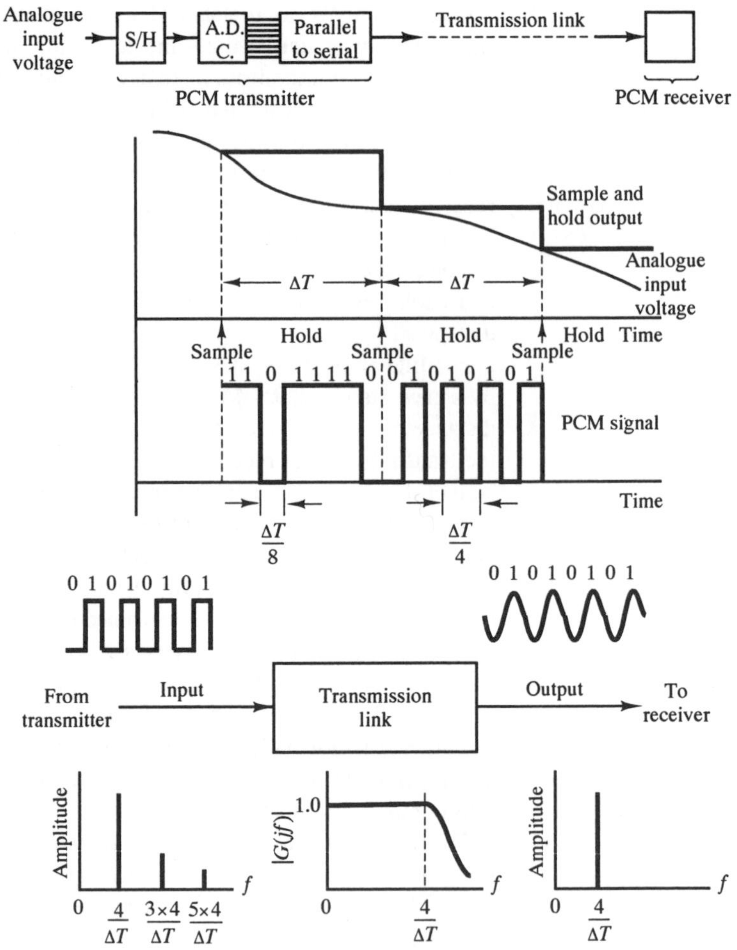

Fig. 18.5 Calculation of PCM transmission bandwidth

For m multiplexed signals, each sampled f_S times per second, there are mf_S samples per second, so that in this case the bit rate is:

Bit rate for m multiplexed signals

$$R = nmf_S \qquad [18.3]$$

Figure 18.5 shows corresponding time variations in input analogue voltage, sample-and-hold output signal and PCM signal. The graphs assume that the sample/hold device is in the SAMPLE state for an infinitely short time. This means that, for a single signal, the time in the HOLD state is equal to the sampling interval ΔT. If the ADC has an 8-bit encoder, i.e. $n = 8$, then 8 bits (either 0 or 1) of information must be transmitted during this time interval ΔT. Thus the width of each bit of information in the PCM signal is $\Delta T/8$. There are 256 possible pulse patterns during each sampling interval, but the pulse pattern corresponding to 01010101 has the shortest period and the highest frequency components.

From Fig. 18.5 we see that this pulse pattern is a square wave of period $\Delta T/4$. The frequency spectrum of this square wave (Section 4.3) consists of a fundamental of frequency $4/\Delta T$ Hz, together with harmonics at frequencies $3 \times 4/\Delta T$, $5 \times 4/\Delta T$,

$7 \times 4/\Delta T$, etc. If this square wave signal is transmitted over a link with bandwidth between 0 and a little over $4/\Delta T$ (Fig. 18.5), then the received signal contains only the fundamental frequency $4/\Delta T$, i.e. it is a sine wave of frequency $4/\Delta T$ Hz. The receiver can still decide correctly that the transmitted message was 01010101, so that the minimum bandwidth required for transmission of the square wave is 0 to $4/\Delta T$, i.e. 0 to $4f_S$ Hz (since $f_S = 1/\Delta T$). Since this square wave has the highest frequency components of all possible pulse patterns, then the minimum bandwidth required for transmission of the PCM signal is 0 to $4f_S$. Thus in the general case of a single signal, sampled at f_S, and encoded into an n-bit code, we have:

Minimum PCM bandwidth for a single signal

$$\text{PCM bandwidth} = 0 \text{ to } \tfrac{1}{2}nf_S \qquad [18.4]$$

For m multiplexed signals, each sampled as $f_S = 1/\Delta T$, the time in the HOLD state is $\Delta T/m$. This means that n bits of information must be transmitted during time $\Delta T/m$; i.e. the width of each bit of information is now $\Delta T/mn$. The corresponding PCM bandwidth in this case is:

Minimum PCM bandwidth for m multiplexed signals

$$\text{PCM bandwidth} = 0 \text{ to } \tfrac{1}{2}mnf_S \qquad [18.5]$$

From [18.2]–[18.5] we see that a single general expression for minimum PCM bandwidth is:

$$\text{PCM bandwidth} = 0 \text{ to } \tfrac{1}{2}R \qquad [18.6]$$

Thus a PCM signal, derived from 16 multiplexed signals, each sampled once per second and encoded into 12 bits, has a bit rate of $16 \times 12 = 192$ bauds and requires a transmission link with a minimum bandwidth of 0 to 96 Hz.

18.4.3 Effect of noise of PCM signal

The transmission link connecting PCM transmitter and receiver may be affected by external interference and noise as shown in Fig. 18.6(a). Figure 18.6(b) shows the power spectral density ϕ_s of the PCM signal, extending effectively from 0 to $\tfrac{1}{2}R$ Hz. The figure also shows the power spectral density ϕ_N of 'white' noise; here ϕ_N has a constant value of A W Hz^{-1} over an infinite range of frequencies. The first stage of the receiver is a low-pass filter of bandwidth between 0 and $\tfrac{1}{2}R$ Hz. This rejects noise frequencies greater than $\tfrac{1}{2}R$, but noise frequencies inside the signal bandwidth, i.e. between 0 and $\tfrac{1}{2}R$, are allowed to pass to the comparator. The total power W_N of the noise in the comparator input signal is given by the area *PQRS* under the noise power spectral density curve (eqn [6.23]), i.e.

$$W_N \approx \tfrac{1}{2}AR \text{ watts} \qquad [18.7]$$

Assuming the noise signal has zero mean $\bar{y}$, then the standard deviation σ is equal to the root mean square value y_{RMS}. From eqn [6.33] $y_{RMS}^2 = W_N$, so that the standard deviation of the noise present in the comparator input signal is given by:

$$\sigma = \sqrt{W_N} = \sqrt{\left(\frac{AR}{2}\right)} \text{ volts} \qquad [18.8]$$

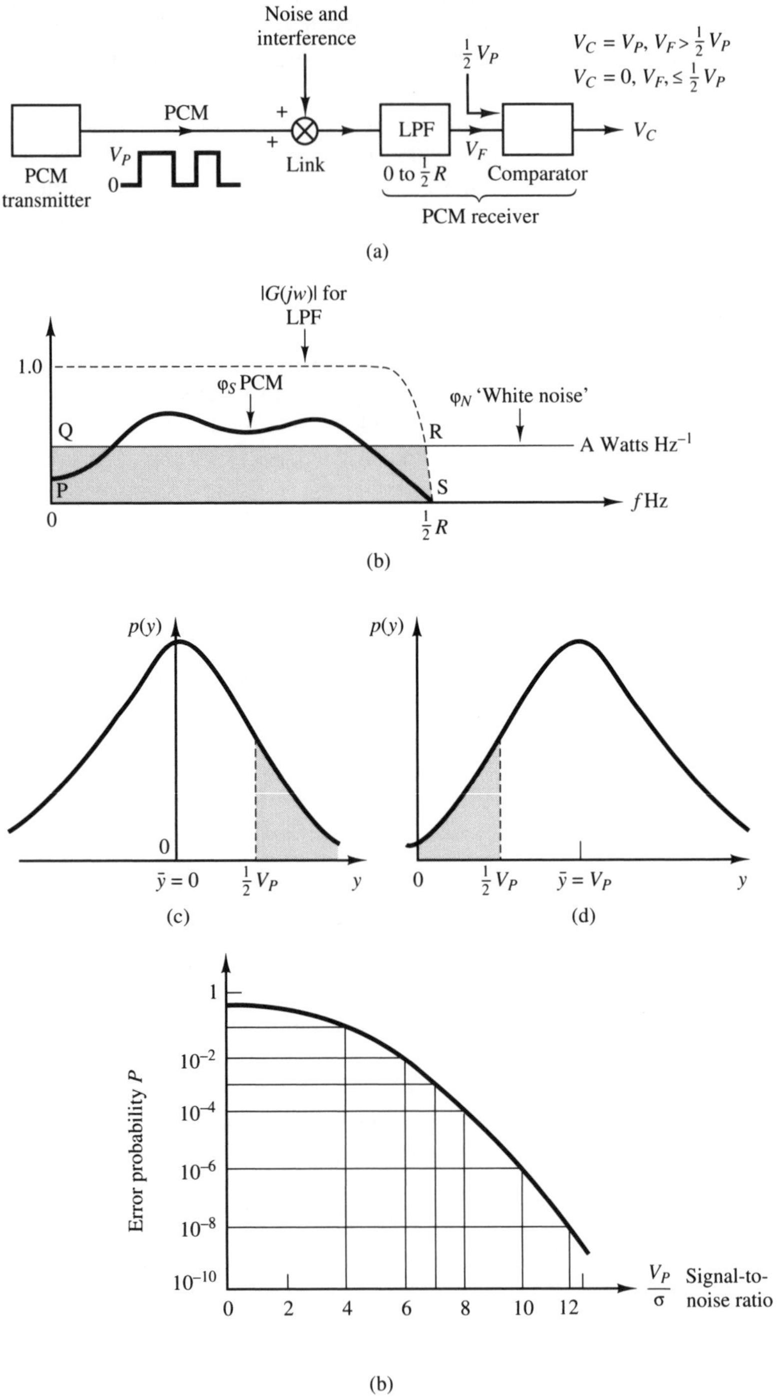

Fig. 18.6 Effect of noise of PCM signal

The PCM signal leaving the transmitter has a value V_P volts for a 1 and 0 V for a 0. The second stage of the receiver is a comparator which compares the filter outputs V_F with $\frac{1}{2}V_P$, i.e. one half of the original pulse amplitude. The comparator output voltage is given by:

$$\begin{aligned} V_C &= V_P, \quad \text{if} \quad V_F > \tfrac{1}{2}V_P \\ V_C &= 0, \quad \text{if} \quad V_F \leq \tfrac{1}{2}V_P \end{aligned} \qquad [18.9]$$

so the receiver decides that a 1 has been transmitted if $V_F > \frac{1}{2}V_P$, and a 0 has been transmitted if $V_F \leq \frac{1}{2}V_P$. This decision is often made correctly, even if the received pulses are distorted by noise. The presence of noise, however, does mean that some decisions are made incorrectly, i.e. the receiver decides that a transmitted 1 is a 0 or a transmitted 0 is a 1. The probability of these errors occurring can be evaluated using the probability density function $p(y)$ for the noise.

Suppose that a 0 is transmitted: we can assume, then, that the comparator input signal is noise alone with standard deviation σ and zero mean value ($\bar{y} = 0$). The probability of a wrong decision here is the probability of the noise being greater than $\frac{1}{2}V_P$, i.e. $P_{y>\frac{1}{2}V_P}$. This probability is equal to the shaded area under the $p(y)$ curve in Fig. 18.6(c), i.e.

Probability of a 0 being received as a 1

$$P_{y>(1/2)V_P} = \int_{(1/2)V_P}^{\infty} p(y)\mathrm{d}y \qquad [18.10]$$

If a 1 is transmitted, then the comparator input signal is noise superimposed on a d.c. voltage of V_P, i.e. noise with a mean value of V_P. The probability of a wrong decision in this case is the probability that the noise will be less than $\frac{1}{2}V_P$. This probability is equal to the shaded area in Fig. 18.6(d), i.e.

Probability of a 1 being received as a 0

$$P_{y<(1/2)V_P} = \int_{-\infty}^{(1/2)V_P} p(y - V_P)\mathrm{d}y \qquad [18.11]$$

In the special case of noise with a Guassian probability density function:

$$p(y) = \frac{1}{\sigma\sqrt{(2\pi)}} \exp\left[-\frac{(y-\bar{y})^2}{2\sigma^2}\right] \qquad [6.14]$$

it can be shown[3] that both the above probabilities are equal and given by

PCM error probability (error rate) for Gaussian noise

$$P\left(\frac{V_P}{\sigma}\right) = \frac{1}{2} - \frac{1}{\sqrt{\pi}} \int_0^{V_P/2\sqrt{2}\sigma} \exp(-x^2)\,\mathrm{d}x \qquad [18.12]$$

Assuming that a given digit is equally likely to be 0 or 1, then [18.12] gives the probability of an error in the decoding of any digit. Figure 18.6(e) shows how error probability varies with V_P/σ (signal-to-noise ratio). We see that for V_P/σ greater than around 8, a small increase in V_P/σ causes a very large reduction in error probability. Thus increasing V_P/σ from 8 to 12 reduces the error probability from approximately 10^{-4} to 10^{-8}. At $V_P/\sigma = 7$, the probability of error in a single bit is ≈ 10^{-3}. This means that the probability of error in a 16-bit signal is 0.016. There is an equal probability that the error will occur in any of the digits in the PCM signal. If the error occurs in the least significant digit (LSB) then the resulting measurement error will be small; if the error occurs in the most significant digit (MSB) then the measurement error will be 50% of full scale.

18.5 Error detection and correction

For the reasons given above, it is important that any errors occurring during the decoding of a noise-affected PCM signal are detected, and in some cases corrected. This is achieved by the use of **redundancy**. Redundancy here means the addition of extra **check** digits to the **information** digits containing the measurement data. Thus each complete code word consists of n digits made up of k information (measurement) digits and $r = n - k$ check digits as shown below. Such a code word is referred to as an (n, k) code and has a redundancy of $(r/n) \times 100\%$.

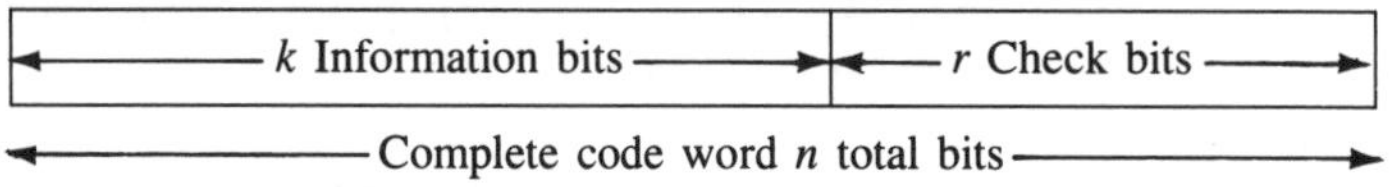

18.5.1 Single parity check bit system

The simplest error detection system uses a single check bit, i.e. $r = 1$. The check bit is chosen using the concept of **parity**.[4] A complete code word has **even parity** if the total number of 1s is even, and **odd parity** if the total number of 1s is odd. Thus in an even parity check system the check digit is set so that the total number of 1s in the complete code word is even: in an odd parity check system the check digit is set so that the total number of 1s is odd. Examples are given below:

Information bits	Even parity code word	Odd parity code word
1011	10111	10110
1000	10001	10000
0101	01010	01011
1111	11110	11111

The parity check bit is added to the information bits at the PCM transmitter using **modulo 2** addition. This process is characterised by the rules:

$$0 \oplus 0 = 0, \quad 0 \oplus 1 = 1, \quad 1 \oplus 0 = 1, \quad 1 \oplus 1 = 0$$

and can be implemented either by an **exclusive-or** logic gate or a read only memory. The transmitter performs modulo 2 addition on the information bits. Thus in the above example:

$$1 \oplus 0 \oplus 1 \oplus 1 = 1$$
$$1 \oplus 0 \oplus 0 \oplus 0 = 1$$
$$0 \oplus 1 \oplus 0 \oplus 1 = 0$$
$$1 \oplus 1 \oplus 1 \oplus 1 = 0$$

In an even parity system the check bit is the result of modulo 2 addition of the information bits; in an odd parity system the check bit is the inverse of this result.

The PCM receiver checks the parity of the complete received code word for correctness. Thus in an even parity system, a received code word with even parity is deemed to be correct, one with odd parity incorrect. This checking is performed by modulo 2 addition of *all* the digits in the code word. For an even parity system the result of addition is zero if there is no error.

This simple system has several limitations. It only detects the presence of an odd number of errors, e.g. 1 or 3 in the above example. An even number of errors, e.g. 2 and 4, gives the correct code word parity and goes undetected. Even if an error is detected, this system cannot decide which bit or bits are in error and therefore cannot correct the code word.

18.5.2 Practical error detecting systems

In industrial telemetry systems the amount of random noise present is often small; i.e. there is usually a high signal-to-noise ratio V_P/σ and consequently a low probabililty of errors. Occasionally, however, large interference voltages lasting a short time occur; these voltage transients are often caused by switching electrical equipment on or off. In this situation it is obviously important to detect as many error combinations as possible, but it is not worthwhile attempting to correct errors. If the receiver detects an error it simply requests a retransmission of the code word. Since substantial interference and corresponding 'single burst' errors occur infrequently, interruptions to normal operation due to requests for retransmission also occur infrequently.

By using several check bits, each checking the parity of a different combination of information bits, it is possible to detect practically all error combinations. A typical arrangement, used in an industrial telemetry system[5] is shown below:

Information bits												Check bits			
b_{11}	b_{10}	b_9	b_8	b_7	b_6	b_5	b_4	b_3	b_2	b_1	b_0	c_3	c_2	c_1	c_0
x		x		x		x		x		x		x			
x	x			x	x			x	x				x		
x	x	x	x					x	x	x	x			x	
x	x	x	x	x	x	x	x	x	x	x	x				x

Here there are 12 information bits and 4 parity check bits, i.e. $k = 12$, $r = 4$ and $n = 16$. For example the transmitter sets c_3 so that the combination $b_{11}\ b_9\ b_7\ b_5\ b_3\ b_1\ c_3$ has even parity. The complete scheme is represented by:

$$\left.\begin{aligned}
c_3 &= b_{11} \oplus b_9 \oplus b_7 \oplus b_5 \oplus b_3 \oplus b_1 \\
c_2 &= b_{11} \oplus b_{10} \oplus b_7 \oplus b_6 \oplus b_3 \oplus b_2 \\
c_1 &= b_{11} \oplus b_{10} \oplus b_9 \oplus b_8 \oplus b_3 \oplus b_2 \oplus b_1 \oplus b_0 \\
c_0 &= b_{11} \oplus b_{10} \oplus b_9 \oplus b_8 \oplus b_7 \oplus b_6 \oplus b_5 \oplus b_4 \oplus b_3 \oplus b_2 \oplus b_1 \oplus b_0
\end{aligned}\right\} \quad [18.13]$$

A similar system with 16 information bits and 5 check bits, generated using the polynomial $1 + x^2 + x^5$, is described in Ref. [6].

18.5.3 Error correction systems

Suppose we require to correct a single incorrect digit in a complete code word of n digits. The check digits must contain the following information: either that the complete code word is correct or the position of the incorrect digit. This means that the check digits must be able to signal $(n + 1)$ possible situations; the number of check bits r required is therefore given by:

$$2^r = n + 1$$

i.e.

$$r = \log_2(n + 1) = 3.33 \log_{10}(n + 1) \qquad [18.14]$$

Thus 3 check digits will correct a 7-digit code word, 4 check digits a 15-digit code word and 5 check digits a 31-digit code word. However, it is possible to have more than one error in a code word, so that the check bits must signal more than $(n + 1)$ possible situations. For example, in order to correct both single errors and double errors in adjacent digits $(2n + 1)$ situations must be signalled, $r = \log_2(2n + 1)$.

One practical error-correcting method is to send the information digits in groups or blocks, i.e. n groups, each containing m digits. The information digits are arranged in a $n \times m$ matrix form and parity check bits added to each row and column at the transmitter. A 4×4 system is shown below:

Transmitted

Odd parity check on rows				
0	0	0	0	1
1	0	0	1	1
1	0	1	1	0
1	1	0	0	1
0	0	0	1	0

Odd parity check on columns (bottom row)

Received

	1	2	3	4	
0	1*	0	0	1	1
1	0	0	1	1	2
1	0	1	1	0	3
1	1	0	1*	1	4
0	0	0	1	0	

Each column and row in the received code is then checked for parity. In the examples shown, rows 1 and 4 and columns 1 and 3 have even parity, so that the incorrect digits are in the positions marked. This method is useful for information in b.c.d. code, since here the digits are arranged in groups of four, one group for each decade.

18.6 Frequency shift keying

In the previous section we saw that the frequency spectrum of the PCM signal extends, effectively, from 0 to $\frac{1}{2}R$ Hz. This means that part of the PCM spectrum may coincide with the spectrum of interference voltages, usually at 50 Hz, due to nearby power circuits (Fig. 6.13(b)). Also the bandwidths of practical transmission links do not normally extend down to 0 Hz: for example a British Telecom landline may have a bandwidth between 300 and 3300 Hz, and a VHF radio link bandwidth between 107.9 and 108.1 MHz. These two problems can be solved by modulating the PCM signal onto a carrier signal, whose frequency lies within the bandwidth of the transmission link. The spectrum of the signal is now shifted up to the carrier frequency, away from the interference spectrum, so that the latter can be rejected by a band pass filter (Fig. 6.13(d)).

Two types of modulation were discussed in Chapter 9. In amplitude modulation (AM) the modulating signal alters the amplitude of a sinusoidal carrier: in frequency modulation (FM) the modulating signal alters the frequency of the carrier. It will be shown later in this section that FM requires a greater bandwidth than AM. However, providing the change in carrier frequency is sufficiently large, an FM receiver is better at improving a given signal-to-noise ratio than an AM receiver.

For this reason, in most telemetry systems, the PCM signal is frequency modulated onto a carrier; this is called *frequency shift keying* (FSK).

18.6.1 FSK transmitters and receivers

The **voltage controlled oscillator** (VCO) is the basis of both FSK transmitters and receivers. The principle of feedback oscillators was discussed in Section 9.5; we saw that the frequency of an electrical oscillator depends on the inductance and capacitance of an L—C—R circuit. The frequency of oscillation of a VCO is determined by the magnitude of the input voltage. Thus if V and f are corresponding values of input voltage and oscillation frequency, we have:

Frequency of voltage controlled oscillator

$$f = f_C + kV \qquad [18.15]$$

where f_C is the frequency at zero voltage (unmodulated carrier frequency), and $k\,\mathrm{HzV}^{-1}$ is the VCO sensitivity. The corresponding VCO output signal is:

$$V_{VCO} = \hat{V} \sin 2\pi(f_C + kV)t \qquad [18.16]$$

If a PCM signal is input to a VCO (Fig. 18.7(a)), the VCO output is an FSK signal; this has two frequencies, f_1 for a 1 and f_0 for a 0. For bipolar PCM with $V = +V_P$ for a 1 and $V = -V_P$ for a 0; the FSK frequencies are:

$$f_1 = f_c + kV_P = f_c + D, \quad \text{for a } 1$$
$$f_0 = f_c - kV_P = f_c - \mathrm{D}, \quad \text{for a } 0 \qquad [18.17]$$

where $D = kV_P$ is the maximum frequency deviation of the carrier. A typical FSK transmitter, suitable for a British Telecom land line, has $f_c = 1080$, $f_0 = 960$, $f_1 = 1200$, i.e. $D = 120\,\mathrm{Hz}$.[7]

Figure 18.7(b) shows an FSK receiver which converts an incoming FSK signal back to PCM. The first stage of the receiver is a band pass filter which rejects all noise and interference outside the FSK bandwidth. The second stage is a Phase Locked Loop (PLL), which consists of a VCO, multiplier and low pass filter in a closed loop system. The multiplier detects any difference in phase between the input signal and the VCO signal.

Suppose that the input signal is $\hat{V} \sin 2\pi(f_C + D)t$ (corresponding to a 1) and initially $V_{OUT} = 0$, so that the VCO output signal is $\hat{V} \sin 2\pi f_C t$. The multiplier output signal contains sum and difference frequencies, i.e. $2f_C + D$ and D; the LPF removes the $2f_C + D$ component so that now V_{OUT} is a low amplitude signal of

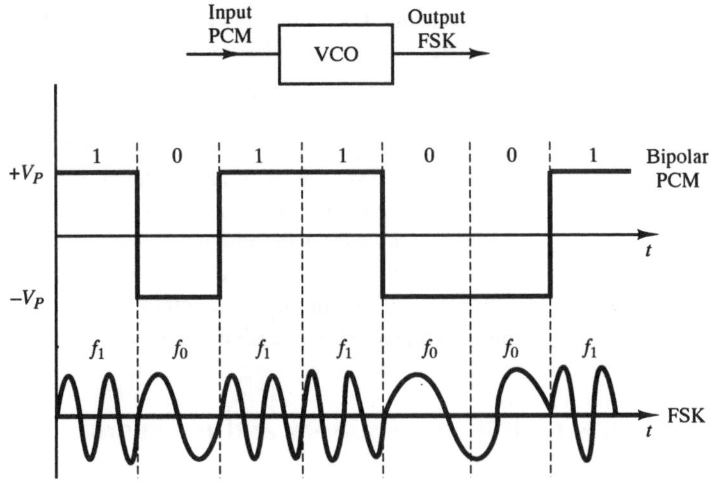

(a)

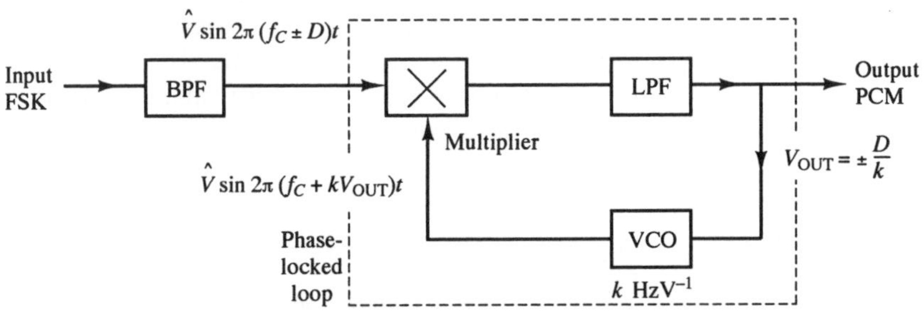

(b)

Fig. 18.7 FSK transmitter and receiver
(a) Transmitter
(b) Receiver

frequency D. This causes the frequency of the VCO output signal to increase until it is equal to that of the input signal, i.e. until $f_C + kV_{OUT} = f_C + D$. At the same time the frequency of V_{OUT} falls from D to zero as the amplitude of V_{OUT} increases. When the system settles out, V_{OUT} is equal to a d.c. voltage of magnitude $+ D/k$ (for a 1). Similarly, when the input frequency is $f_C - D$, the system settles out with $f_C - D = f_C + kV_{OUT}$ and V_{OUT} a d.c. voltage of magnitude $-D/k$ (for a 0). In both cases the VCO frequency is said to 'lock' onto the input frequency.

18.6.2 Bandwidth of FSK signal

In Section 9.3 we saw that if a single sine wave of frequency f_i is amplitude modulated onto a sinusoidal carrier of frequency f_S, then the spectrum of the AM signal consists of two lines at frequencies $f_S - f_i$, $f_S + f_i$. This means that if a random signal, with spectrum between 0 and f_M, is amplitude modulated onto f_S, the spectrum of the AM signal lies between $f_S - f_M$ and $f_S + f_M$. If a single sine wave $\hat{V}_i \sin 2\pi f_i t$ is frequency modulated onto a sinusoidal carrier using a VCO, then from eqn [18.15] the instantaneous frequency of the FM signal is:

$$\begin{aligned} f &= f_C + k\hat{V}_i \sin 2\pi f_i t \\ &= f_C + D \sin 2\pi f_i t \end{aligned} \qquad [18.18]$$

Here f_C is the unmodulated carrier frequency and $D = k\hat{V}_i$ is the maximum deviation of f from f_C. The resulting FM signal is given by:

$$V_{FM} = \hat{V} \sin 2\pi(f_C + D \sin 2\pi f_i t)t \qquad [18.19]$$

From Fig. 18.8 we see that the spectrum of this FM signal is wider and more complex than the corresponding AM spectrum. In FM there are several lines, symmetrically arranged about f_C, the number and relative amplitudes of the lines depending on the **modulation index** D/f_i.[8]

For a random modulating signal, containing frequencies between 0 and f_M, the FM spectrum consists of a large number of lines. In this case, the number of lines and width of the spectrum depends on the appropriate modulation index D/f_M. For very small and very large values of D/f_M, i.e. $D/f_M \ll 1$ and $D/f_M \gg 1$, the frequency spectrum of the FM signal extends approximately from $f_C - (D + f_M)$ to $f_C + (D + f_M)$.[8] For $D/f_M \approx 1$, the FM spectrum extends approximately from $f_C - (D + 2f_M)$ to $f_C + (D + 2f_M)$.

Summarising these results we have:

$$\begin{aligned} \text{Approx FM Bandwidth} &= f_C - (D + f_M) \text{ to } f_C + (D + f_M), \\ &\quad \frac{D}{f_M} \ll 1, \frac{D}{f_M} \gg 1 \qquad [18.20] \\ &= f_C - (D + 2f_M) \text{ to } f_C + (D + 2f_M), \\ &\quad \frac{D}{f_M} \approx 1 \end{aligned}$$

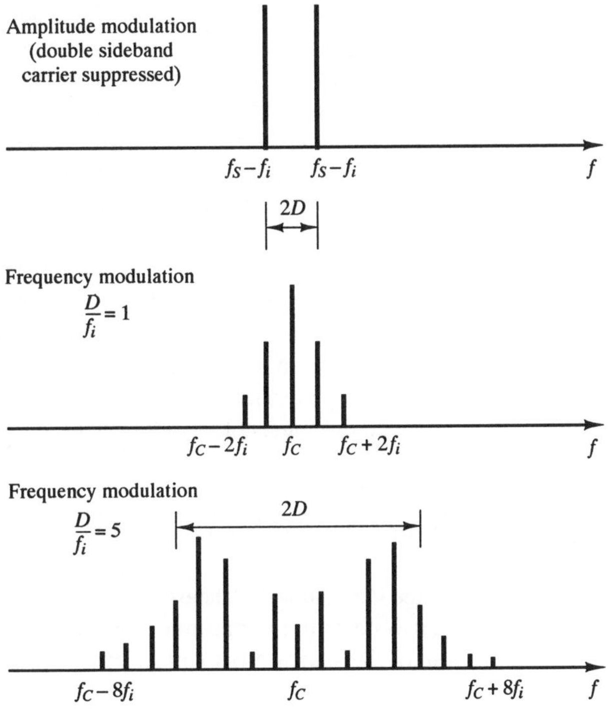

Fig. 18.8 Comparison of AM and FM spectra

In FSK the modulating signal is PCM. The PCM signal has a frequency spectrum effectively between 0 and $\frac{1}{2}R$ (Section 18.4.2); here $f_M = \frac{1}{2}R$, where R is the PCM bit rate. Equations [18.20] can therefore be used to give corresponding expressions for the bandwidth necessary to transmit an FSK signal, i.e.

$$\begin{aligned}\text{Approx FSK Bandwidth} &= f_C - (D + \tfrac{1}{2}R) \quad \text{to} \quad f_C + (D + \tfrac{1}{2}R), \\ &\quad \frac{2D}{R} \ll 1, \frac{2D}{R} \gg 1 \qquad [18.21] \\ &= f_C - (D + R) \quad \text{to} \quad f_C + (D + R), \\ &\quad \frac{2D}{R} \approx 1\end{aligned}$$

Thus an FSK signal with $f_C = 1080$ Hz, $D = 120$ Hz and $R = 200$ bauds, has $2D/R = 1.2$, and requires a bandwidth approximately between 760 and 1400 Hz.

18.7 Communication systems for measurement

18.7.1 Introduction

In the introduction to this chapter, we saw that in many industrial situations, it will be necessary to transmit measurement data from transducers/transmitters (outstations, O/S), located at different items of plant equipment, to signal processing and data presentation elements (master stations, M/S), located in a central control room. The distances between individual plant items and between the items and the control room may be up to a few kilometres. A communications system is therefore required which is capable of transmitting large amounts of information in two directions (M/S to O/S and O/S to M/S), over long distances in the presence of external interference and noise. Such systems are often referred to as **telemetry systems**.

In Section 9.4.3 we discussed **intelligent** or **smart transmitters**. These transmitters incorporate a microcomputer which is not only used to calculate the measured value of the variable but also to control a **digital communications module**. Since this module can both transmit and receive digital information an intelligent transmitter can therefore act as an outstation in a telemetry system. The main problem here is that there are a large number of digital communication standards in current use. This means that transmitting/receiving equipment made by one manufacturer may not be compatible with that made by another manufacturer. There is a need therefore for an agreed signal standard for the digital communication of measurement data, just as a 4–20 mA current loop has been used as an analogue transmission standard.

18.7.2 Reference Model for Open Systems Interconnection (OSI)

In 1983 the International Standards Organisation (ISO) approved the Reference Model for Open Systems Interconnection (OSI) as an international communication standard. The aim of the standard is to allow open communication between equipment from

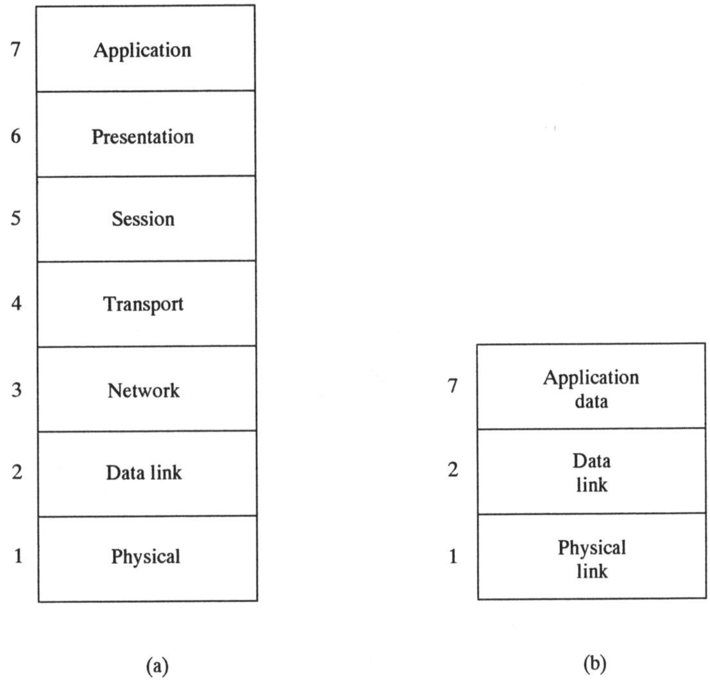

Fig. 18.9 Layered protocol models
(a) ISO OSI model
(b) Fieldbus model

different vendors. It will be many years before this aim is achieved but the OSI model will provide the basis of the aim. In the meantime the Model can act as a bridge between two different proprietory systems which previously could not communicate.

The OSI model is an abstract concept consisting of seven layers (Figure 18.9(a)). Each layer represents a group of related functions or tasks. **OSI protocols** are used to define these functions but do not define how these functions are implemented. The main purpose of the model is to provide a structure whereby vendor independent systems can be implemented. The functions of the seven layers can be summarised as follows:

Application Layer (7) This is the highest order layer in the Model and its purpose is to ensure that a user application program can both receive and transmit data.

Presentation Layer (6) This layer ensures that an application correctly interprets the data being communicated by translating any differences in representation of data.

Session Layer (5) This layer is responsible for controlling communication sessions between applications programs.

Transport Layer (4) This layer provides those layers above it with a reliable data-transfer mechanism which will be independent of any particular network implementation.

Network Layer (3) This layer provides the actual communication service to the transport layer; it controls communications functions such as routing, relaying and data link connection.

Data-Link Layer (2) This layer controls the transfer of data between two physical layers and provides for error-free (ideally) sequential transmission of data over a link in a network.

Physical Layer (1) This layer defines the actual signalling method employed, together with the type of transmission medium e.g. copper wire, fibre optic, radio waves, etc.

18.7.3 Fieldbus

Work is currently in progress to produce a single standard for two-way digital communication of measurement and control data between intelligent sensors/transmitters located in the field and a computer-based master station located in a control room. This standard is referred to as 'Fieldbus' and is particularly, but not exclusively, applicable to the process industries. The standard is based on a simplified three layer version of the seven layer OSI model (Figure 18.9(b)), the **physical layer (1)**, the **data-link layer (2)** and the **application layer (7)**. Detailed discussion has taken place over many years with the aim of producing an agreed standard for each of the three layers. These standards are now in the process of being finalised. We now discuss how each of the layers can be implemented.

Physical layer[9]

The first decision to be made is the **network topology** i.e. the geometry of the interconnection between the out stations and the master station. Figure 18.10(a) shows three classic network topologies the **star**, **ring** and **bus**. In any given link in the star arrangement there is only communication between the master station and a single outstation. This means that less multiplexing but more wiring is required, also if the master station fails the whole network fails. The ring arrangement also has limited reliability. The network will only function if all the stations are working; if any station fails then all stations beyond the failed station are also inaccessible. In the bus, or multi-drop arrangement several outstations and master stations share a common transmission path. Some method of multiplexing is therefore required to separate the signals from each outstation. In **time division multiplexing** (TDM, Section 18.1) the signals are separated in time so that for example O/S 1 transmits to M/S during a given time slot, O/S 2 transmits to M/S during the next time slot and so on. The

Fig. 18.10 Network topologies
(a) Star
(b) Ring
(c) Bus

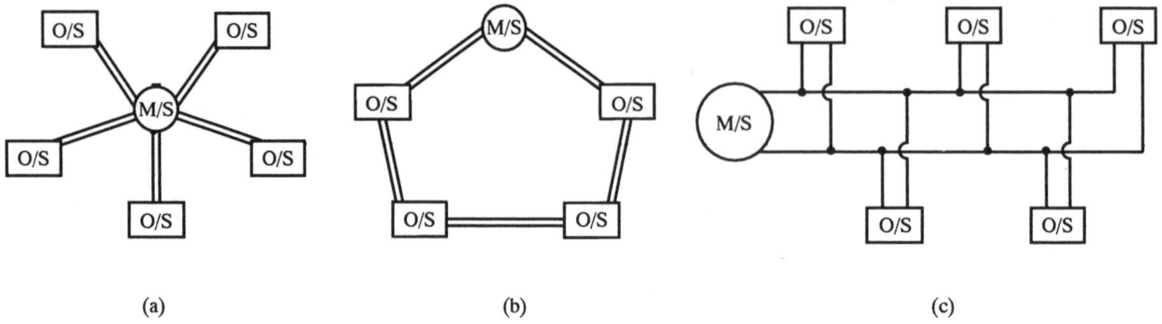

alternative is **frequency division multiplexing** (FDM) where the signals are separated in frequency by modulating them on to different carrier frequencies (Sections 9.3 and 18.6). However, FDM requires more complex hardware so that TDM is usually preferred. A protocol will be necessary to arbitrate between the different outstations and avoid data collisions. The bus arrangement has high reliability: the network will still function if any outstation fails and if a suitable protocol is used will still function, to a limited extent, if the master station fails. The bus arrangement is therefore the preferred topology for Fieldbus.

The next decision to be made concerns the **interconnection medium** to be used to implement the bus arrangement. This could be copper wire, optical fibre or a radio link. It is generally extremely difficult to implement the bus arrangement using optical fibres and there are problems in making reliable fibre/fibre connections in hostile environments. If the distances between the outstations and master station are many kilometres then radio links may be required. However, in many industrial situations these distances are only a few kilometres at most; this means that copper wire is feasible. Since it is also the cheapest practical way of implementing the bus arrangement it is preferred for Fieldbus. In order to minimise the effects of inductive and capacitive coupling to interference sources (Section 6.5) screened twisted pair wires should be used.

The final decision in the physical layer is the **signalling method** to be used. There are two methods of transmitting digital data; either in **parallel** form (Section 18.3) or in **serial** form (Section 18.4). We saw in Section 18.3 that parallel digital signals are suitable for high speed, short distance (up to 15 m) communication in laboratory environments where there is low level electrical interference. Serial digital signals (pulse code modulation) are more suited to longer transmission links where significant electrical interference is present. If the serial signal is not modulated onto a carrier so that the signal bandwidth is determined entirely by the bit rate R (eqn [18.6]), this is referred to as **base-band** transmission.

Figures 18.11(a), (b) show two base-band transmission methods **non-return to zero** (NRZ) and **bi-phase Manchester**. NRZ is characterised by one voltage level

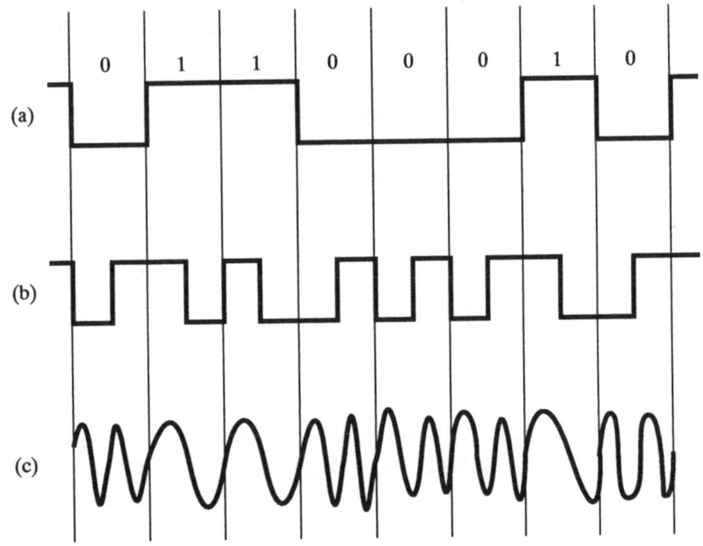

Fig. 18.11 Serial digital signalling
(a) Non-return to zero (NRZ)
(b) Bi-phase Manchester
(c) Frequency shift keying (FSK)

(e.g. 0 V) corresponding to binary 0 and another (e.g. 5 V) corresponding to binary 1. The major disadvantage of this method is that a long sequence of 1s or 0s results in the voltage on the line remaining constant. This problem is overcome in the bi-phase method where one phase e.g.  corresponds to a 0 and another phase e.g.  corresponds to a 1, this guarantees a change in voltage at least once per bit period. The noise immunity of the serial signal can be increased by modulating it onto a carrier signal, this is **carrier band** transmission. Figure 18.11(c) shows **frequency shift keying** (FSK, Section 18.6). Here there are two distinct frequencies: the lower frequency corresponds to binary one and the higher frequency to binary zero.

One possible implementation of the physical layer is provided by the HART protocol which is marketed by Rosemount.[10] This is compatible with the Rosemount range of smart transmitters (Section 9.4.3). A multidrop bus topology is used which can accommodate up to 15 smart devices with one power source. The interconnection medium is a single shielded twisted pair (maximum length 3 km) or multiple twisted pairs with an overall shield (maximum length 1.5 km). The signalling method is FSK based on the Bell 202 communications standard which uses 1200 Hz to represent binary 1 and 2200 Hz to represent 0; the bit rate is 1200 bits/s. This FSK signal is superimposed on the normal 4–20 mA analogue signal in a current loop (Figure 9.17). The amplitude of the FSK signal is 0.5 mA, so that if the normal analogue current is 12 mA, the minimum and maximum instantaneous currents are 11.5 and 12.5 mA. The average value of loop current is therefore unchanged at 12 mA. The Bell 202 communications system also allows smart devices to be directly connected to leased telephone lines. This allows the device to communicate with a central control point many kilometres away. Since the transmitter power is also isolated from the communications in leased line applications, any number of devices can be networked.

Data Link Layer[9]

As a result of the above discussions, the physical layer will be a single multidrop bus which is shared between several intelligent devices using time division multiplexing. The data link layer is concerned with the transfer of data between these devices. The first requirement is for a method of controlling the bus which specifies the time interval during which a given device will transmit and receive information. The simplest method is to have a single fixed central master station which controls the times at which each of the outstations transmits and receives. This method has however limited reliability, if the master station fails then the whole network fails even though all of the outstations may be working correctly. The reliability is increased if two or more stations are designated to be masters. Another method is decentralised mastership; here each of the stations connected to the bus takes it in turns to be the master for a given period of time. This can be achieved using **token passing**. The station holding the token is responsible for initiating all communications on the network for a given period of time. At the end of the period it must pass on the token to the next station.

In order that data is correctly routed between stations addressing information must be added to the data. Thus if a given station is transmitting data to another station, then the data should be preceded by an **address code** which uniquely defines the

Start flag	Address field	Command field	Information field	Parity check field	Stop flag

Fig. 18.12 Typical frame format

source and destination addresses. An 8-bit address code could specify up to 16 source addresses and up to 16 destination addresses. There is also a requirement for **control** or **command instruction codes** where, for example, a given device is required either to receive (read) incoming data or to transmit (write) data to another device. Finally there is a need for any errors present in the transmitted information to be detected and ideally corrected. This is done by adding additional **parity check bits** (Section 18.5).

The total information transmitted from one outstation to another now consists of different types, one type following another in time. It is essential therefore that the data is arranged in **packets or frames**. Each frame should have **format** which clearly separates and distinguishes between different types of data, for example addressing codes and check codes. Figure 18.12 shows a typical frame format; it commences with a beginning of message or START **flag code**, followed by address field, command field, information field, parity check field and concludes with an end-of-message or STOP flag code.

Application layer

The role of the application layer can be summarised as follows:

(1) It enables the master station to obtain measurement data from the data link layer in a form suitable for further processing in a user applications program. An example is a master station obtaining volume flow rate data Q and density data ρ from smart field transmitters in a form suitable for an applications program to calculate mass flow rate $\dot{M} = \rho Q$.

(2) It enables the master station to output data from an applications program to the data link layer in a form suitable for transmission to outstations. An example is an applications program in the master station changing the input range of a smart field differential pressure transmitter from 0 to 2.5 m of water to 0 to 3.0 m of water.

The application layer therefore specifies a set of control or command instructions which occur as a corresponding code in the data frame (Fig. 18.12). Three necessary basic instructions are:[10]

READ e.g. Master station reads measured value;
Transmitter reads input range data.

WRITE e.g. Master station specifies identification number and input range of transmitter;
Transmitter outputs measured value.

CHANGE e.g. Master station changes transmitter input range.

Conclusion

This chapter commenced by discussing the principles of **time division multiplexing** and its use in **data acquisition systems**. This was followed by a discussion of the principles of a range of techniques which form the basis of **communications systems**. These are **parallel digital signalling**, **serial digital signalling**, **error detection/correction** and **frequency shift keying**. The chapter concluded by studying how communication systems for measurement data can be implemented making special reference to the emerging **Fieldbus** standard.

References

18.1 Mowlem Microsystems, 1985. *Technical Information on ADU Autonomous Data Acquisition Unit.*

18.2 RDP Electronics, 1986. *Technical Information on Translog 500 Data Acquisition System.*

18.3 MATTHEWS P R 1983. 'Communications in Process Control' *Measurement and Control*, November and December 1982, January.

18.4 CARLSON A B 1975. 'Communication Systems'. McGraw-Hill–Kogakusha International Student Edition, 2nd ed. pp. 410–17.

18.5 Kent Process Control, 1985. *Technical Information on Kent P4000 Telemetry Systems.*

18.6 Serck Controls, 1977. *Product Data Sheet on a Telemetry Drive Module PDS 10/9.*

18.7 A.T.S. (Telemetry) Ltd. 1987. *Technical Information on Type 1100 F.S.K. Data Modules.*

18.8 CARLSON A B 1975. 'Communication Systems' McGraw-Hill–Kogakusha International Student Edition, 2nd ed. pp. 225–37.

18.9 ATKINSON J K 1987. 'Communications protocols in instrumentation', *J. Phys. E: Scientific Instrumentation*, Vol 20, pp. 484–91.

18.10 Rosemount, 1991. Product Data Sheet PDS 2000, *HART field communications protocol.*

Problems

18.1 Sixteen analogue input voltages, each with a frequency spectrum between 0 and 5 Hz, are input to a time division multiplexer. The multiplexed signal passes to a serial digital (PCM) transmitter consisting of a sample/hold device, 12-bit binary ADC and a parallel-to-serial converter. The PCM signal is transmitted to a distant receiver over a link affected by 'white' noise with a power spectral density of $0.2\ \mathrm{mW\ Hz^{-1}}$.

(a) Suggest a suitable sampling frequency for each input signal.
(b) What is the corresponding number of samples per second for the multiplexed signal?
(c) What is the maximum length of time the sample hold device can spend in the HOLD state?
(d) What is the maximum percentage quantisation error for the ADC?
(e) Find the bit rate and minimum transmission bandwidth for the PCM signal.
(f) The first stage of the PCM receiver is a low pass filter with a bandwidth 'matched' to the frequency spectrum of the PCM signal. Estimate the standard deviation of the noise present at the filter output.

18.2 A PCM transmitter sends out a 12-bit serial digital signal with 5V corresponding to a 1 and 0 V corresponding to a 0. The signal passes over a transmission link, affected by random noise, to a PCM receiver consisting of a low-pass filter and a comparator. The comparator input signal is the PCM signal with noise superimposed on it; the probability density $p(y)$ of this noise is the triangular function shown in Fig. Prob. 2. The comparator decides that

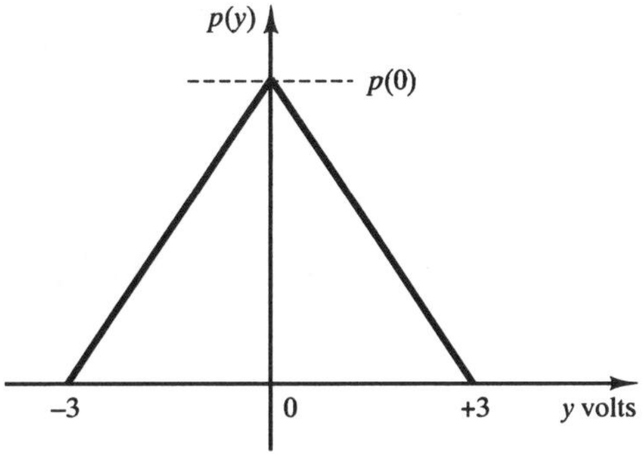

Fig. Prob. 2

a 1 has been transmitted if the comparator input signal is greater than 2.5 V, and that a 0 has been transmitted if the comparator input signal is less than or equal to 2.5 V.

(a) Calculate $p(0)$ such that $p(y)$ is normalised.
(b) Find the probability that y is greater than $+1.0$ V.
(c) Find the probability that a 0 is received as a 1.
(d) Find the probability that a 1 is received as a 0.
(e) What is the probability that a single error occurs in a complete code word?

18.3 Ten measurement signals are input to a multiplexer so that each one is sampled twice per second. The multiplexed signal is input to a serial digital transmitter incorporating a 10-bit ADC. The resulting PCM signal is converted into FSK such that 720 Hz corresponds to a 1 and 480 Hz corresponds to a 0. Estimate the bandwidth required by the FSK signal.

18.4 (a) Random noise is characterised by a Gaussian probability density function of standard deviation $\sigma = 1.0$ V and mean value $\bar{y} = 0$ V. Using the probability values given below, calculate the probability that the noise voltage:
(i) exceeds $+0.5$ V
(ii) lies between -1.0 and $+1.0$ V
(iii) is less than -1.5 V.

(b) Information in binary form is sent with a '0' represented by 0 V and a '1' represented by 5 V. During transmission the above random noise is added to it. The receiver decides that a '0' has been transmitted if the total input voltage is less than 2.5 V and that a '1' has been transmitted if the total input voltage is greater than 2.5 V. Use the probability values below to estimate the probable number of errors if 1600 bits of information are transmitted.

$$P_{y>\bar{y}+0.5\sigma} = 0.3085, \quad P_{y>\bar{y}+\sigma} = 0.1587$$

$$P_{y>\bar{y}+1.5\sigma} = 0.0668 \quad P_{y>\bar{y}+2.5\sigma} = 0.0062$$

Answers to numerical problems

Chapter 2

1 $6.06\,\mu V\,°C^{-1}$, $3.61 \times 10^{-3}\,\mu V\,°C^{-2}$, $2.59 \times 10^{-6}\,\mu V\,°C^{-3}$
2 $\beta = 2946\,K$, $\alpha = 1.86 \times 10^{-4}\,k\Omega$, $3.64\,k\Omega$
3 (a) +25.9% (b) 0,53.3 mV^{-1} (c) 19.3 $mV\,cm^{-1}$
4 13.2%
5 (b) 208.6 Hz, 0.6 Hz
6 $R_T = 100(1 + 3.908 \times 10^{-3}T - 5.82 \times 10^{-7}T^2)$
7 (a) $a = 4.0\,mA$, $K = 1.6\,mA\,bar^{-1}$, $K_I = +0.4\,mA\,°C^{-1}$, $K_M = 0.4\,mA\,bar^{-1}\,V^{-1}$
(b) 18.0 mA

Chapter 3

1 $\bar{E} = -0.425\,°C$, $\sigma_E = 1.93\,°C$
2 (a) 120.7 Pa (b) −500 Pa
3 a (i) 4.95 V a (ii) 4.97 V
4 (b) Increase of $10^{-6}\,rad\,V^{-1}$
5 (a) $\bar{E} = +5.0\,°C$, $\sigma_E = 2.6\,°C$
6 (a) $\bar{E} = +0.08\,m/s$, $\sigma_E = 0.35\,m/s$

Chapter 4

1 −29.4, −10.8, −0.5, +24.1, +0.7 °C
2 (a) $5 \times 10^{-3}\,m\,N^{-1}$, $20\,rad\,s^{-1}$, 0.3 (b) 1 cm
(c) $1.0 + 0.5\,[1 - e^{-6t}(\cos 19t + 0.32 \sin 19t)]$ cm
3 $50[1.07 \sin(10t - 3°) - 1.00 \sin 10t]$
$+ \frac{50}{3}[2.16 \sin(30t - 19°) - 1.00 \sin 30t)$
$+ \frac{50}{5}[1.62 \sin(50t - 156°) - 1.00 \sin 50t)$ N
4 (a) 0 to 0.1 $rad\,s^{-1}$ (b) 0 to 0.033 $rad\,s^{-1}$ (c) 0 to 0.33 $rad\,s^{-1}$
5 (a) 10 rad/s, 7.0, $\dfrac{0.1}{10^{-2}s^2 + 1.4s + 1}$
(b) approx. $\dfrac{1}{10^{-4}s^2 + 1.4 \times 10^{-2}s + 1}$
6 (a) $\omega_n = 10$ rad/s, $\xi = 0.1$, $V(t) = 0.49[1 - e^{-t}(\cos 10t + 0.1 \sin 10t)]$
7 (a) $50[0.734 \sin(10t + 40°) - 1.00 \sin 10t]$
$+ \frac{50}{3}[1.39 \sin(30t - 2°) - 1.00 \sin 30t)$
$+ \frac{50}{5}[2.44 \sin(50t - 79°) - 1.00 \sin 50t)$

Chapter 5

1 (a) $2.92 \times 10^{8}\,\Omega$, $0.15\,\text{pH mV}^{-1}$ (b) -20.5%
2 (a) $2.0\,\text{V cm}^{-1}$ (b) $500\,\Omega$, 50 V
3 -56 Pa
4 $$\frac{0.1\,s^2 + 10s + 1000}{5.1\,s^2 + 30s + 1100}$$

Chapter 6

1 (a) 0.15 V, 1.0 V
2 +1.0, +0.6, +0.2, −0.2, −0.6, −1.0, −0.6, −0.2, +0.2, +0.6, +1.0
3 (a) 10^{-4} W, 10^{-2} V, 10^{-2} V (b) −20 dB
(d) Increased to +10 dB (e) Increased to +30 dB
4 (a) 3.14 mV, 100 V (b) 4.15 mV, 15.85 mV
5 (a) 1.5 mW (b) 8.5 mW (c) −7.5 dB (d) 55 mV (e) 300 Hz (f) 92 mV

Chapter 7

1 (a) MDT = 6.2 hrs (b) MTBF = 10 074 hrs (c) $\lambda = 0.87\,\text{yr}^{-1}$ (d) $A = 0.99940$
2 (a) 0.62 (b) 0.24 (c) 0.31
3 TLOC = £19 000 for system (1), TLOC = £15 350 for system (2)

Chapter 8

1 (a) $3.91 \times 10^{-3}\,°\text{C}^{-1}$, $-5.85 \times 10^{-7}\,°\text{C}^{-2}$ (b) +0.76%
2 $R_1 = R_3 = 120.0025\,\Omega$, $R_2 = R_4 = 119.9975\,\Omega$
3 88.5, 55.3, 22.1 pF
4 7.6, 3.4 mH
5 521 mH, 5.6 mH
6 3.46 V, 367 Hz; 34.6 V, 3670 Hz
7 (a) −1.07%, −0.65% (b) $51.8\,\mu\text{V}\,°\text{C}^{-1}$, $8.68 \times 10^{-3}\,\mu\text{V}\,°\text{C}^{-2}$ (c) 248 °C
8 (a) $20\,\text{N m}^{-1}$, $0.51\,\text{Ns m}^{-1}$ (b) 0 to 1.24 cm (c) $1333\,\Omega$
9 −0.25 to +0.25 rad
10 (a) 1.2 mm (b) 0 to 0.17 mm
11 (a) $$G(s) = 0.02\,\frac{10^{-3}s}{(1 + 10^{-3}s)}\,\frac{5.4 \times 10^{10}}{(s^2 + 4.65 \times 10^3 s + 5.4 \times 10^{10})}$$
(c) $$G(s) = 0.002\,\frac{0.1s}{(1 + 0.1s)}\,\frac{5.4 \times 10^{10}}{(s^2 + 4.65 \times 10^3 s + 5.4 \times 10^{10})}$$
12 (a) 100 g (b) (ii) 1 V

Chapter 9

1 (a) $R_2 = 100\,\Omega$, $R_3 = 5770\,\Omega$, $R_4 = 5770\,\Omega$ (b) $R_2 = 100\,\Omega$, $R_3 = 6000\,\Omega$, $R_4 = 6000\,\Omega$
(c) $R_2 = 100\,\Omega$, $R_3 = 5000\,\Omega$, $R_4 = 6000\,\Omega$
2 $R_2 = 10\,\Omega$, $R_3 = 1650\,\Omega$, $R_4 = 1650\,\Omega$
3 (a) (i) 0 to 1.0 V (approx.) (a) (ii) − 1.5% (b) 0 to 0.6 V (approx.)
4 $R_2 = 1000\,\Omega$, $R_3 = 264\,\Omega$, $R_4 = 2370\,\Omega$, $V_s = 2.40$ V
5 (b) 9.1 mV
6 2584
7 0 to 44.2 mV
8 (a) 64.2 nF (b) 0.178 V (c) 0.3%
9 (a) Since $f_n = 32$ Hz, $\xi = 0.7$, $|G(j\omega)| = 1$ up to 10 Hz
(c) $+0.2 \sin 2000\pi t$ and $-0.2 \sin 2000\pi t$
11 $R_{IN} = 10\,\text{k}\Omega$, $R_F = 1\,\text{M}\Omega$, $C_{IN} = 0.159\,\mu F$, $C_F = 0.159$ nF
12 (a) 4.8×10^3 N/A, 19.3 N (b) 1.0, 0.2

13 81.6 kHz at 1 mm to 122.4 kHz at 3 mm
14 $|G(j\omega_n)| = 20$ and $\arg G(j\omega_n) = -90°$, between 2.60 and 4.56 kHz
15 (b) $L = 4.7$ mH (c) $|G(j\omega_n)| = 0.0295$ and $\arg G(j\omega_n) = -90°$ between 100 and 120 kHz
17 (b) 2.24 to 7.07 kHz (c) $|G(j\omega_n)| = 10^{-3}$ A/V, $\arg G(j\omega_n) = -180°$ over above range

Chapter 10

1 (a) (i) ±0.0122% (a) (ii) 000111000010, 100001101010 (b) 1C2, 86A
(c) (i) ±0.05005% (c) (ii) 0001 0001 0000, 0101 0010 0101
2 (a) 85.3 Ω (b) 1.97 V
3 (a) 0 to $1/\tau$ rad s^{-1} (d) 0 to $\pi/4\tau$ rad s^{-1}

Chapter 11

1 (a) 7.5 rad V^{-1}, 1.0 Hz, 35.6 (b) approx. 10 kΩ in series, 0.15 rad V^{-1}
2 (a) (a) 150 Ω in series (a) (ii) 333 Ω in parallel (b) 3.3 V
3 (a) 20 mV^{-1} (b) $\dfrac{0.8}{10^{-5}s^3 + 1.1 \times 10^{-3}s^2 + 10^{-2}s + 4 \times 10^{-2}}$ (c) 13.3 mV^{-1}

Chapter 12

1 (a) 10 ms^{-1} (b) 0.883 kg s^{-1} (c) 1.5×10^5 (d) 5.84 cm
2 (a) Re = 1.2×10^5 at max. flow (b) 7.68 cm
3 (a) 0.14 m (b) 2.86×10^5 Pa
4 (a) Re = 2.7×10^6 at max. flow (b) 0.146 m (c) 0.135 m
5 23.8 mV and 4.3 Hz at min. flow, 499 mV and 90.2 Hz at max. flow
6 5.4 cm, 423 Hz
7 432 pulses m^{-3}
8 $\tau = 150$ ms, $\dfrac{1}{fc} = 10$ ms
9 (a) 7200 kg hr^{-1} (b) 9.4×10^5 (d) 7610 kg hr^{-1}

Chapter 13

1 (b) (i) 0.200 to 1.00 bar (b) (ii) 0.202 to 1.007 bar
2 $E_{MAX} = 1.9\ \mu J$ safe to use with hydrogen–air

Chapter 14

1 $E_{OUT} = (3.93 + 6.55\sqrt{v})^{1/2}$
2 (a) $4.27 + 0.33\sqrt{v}$ (b) $\tau_V = 4$ m s, therefore unsuitable
3 $\tau_V = 2.7$ sec, i.e. bandwidth 0 to 0.06 Kz, therefore unsuitable
4 (a) $G(s) = \dfrac{1}{6.4 \times 10^4 s^2 + 1068s + 1}$
$|G(j\omega)| = 0.144$ at $\omega = 2\pi \times 10^{-3}$, therefore cannot follow variations
(b) $G(s) = \dfrac{1}{1600s^2 + 93s + 1}$
$|G(j\omega)| = 0.91$ at $\omega = 2\pi \times 10^{-3}$, can follow variations more closely
5 (a) 0 to 10.6 mV

Chapter 15

1 See Fig. Soln Prob. 15.1
2 Period $T_P = 10$ ms, maximum $T_D - T_s = 37.5$ °C, thermocouple constant $a_1 = 5 \times 10^{-2}$ mV °C^{-1}, output voltage range = 0 to 46.9 mV
3 0 to 4.5 mV

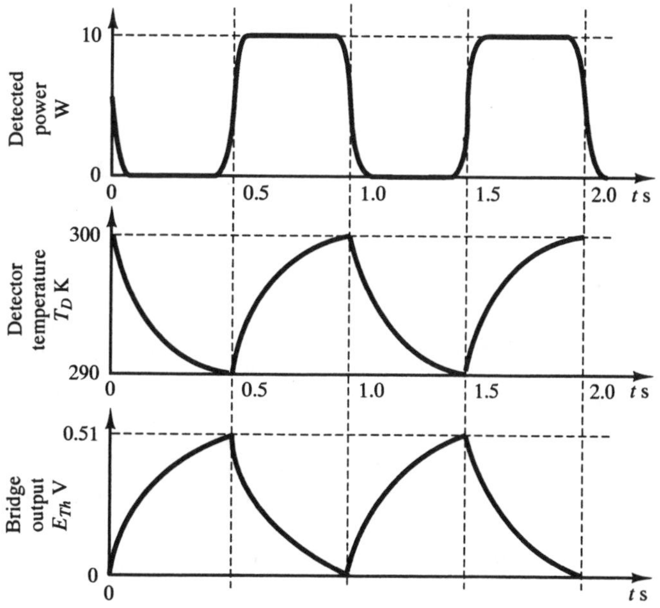

Fig. Soln Prob. 15.1

4 1.45×10^4

5 (a) 10 mW (b) 0.26, 15.1° (c) 0.47, 27.8° (d) 170 μW, 0.998, 169.7 μW
(e) 2.2 mW, 0.794, 1.75 mW (f) GLASS 169.7 μW, 93.3 μA
POLYMER 1.75 mW, 961 μA

6 500 to 1000 nm, 5.5

Chapter 16

1 (a) $\omega_n = 1.00 \times 10^6$ rad s^{-1}, $\omega_1 = 1.18 \times 10^6$ rad s^{-1}
(b) 50 Ω, $-2°\ 52'$ at ω_n, 14.1 kΩ, $-3°\ 27'$ at ω_1 (c) 0.02 AV^{-1}, $-177°$

2 (a) 7.7×10^{-3}, 7.7×10^{-3}, 7.3×10^{-10}, 7.3×10^{-10}

3 (a) 1 ms, 0.103 (b) $T_w = 30$ μs, $T_R = 10$ ms for example

4 (a) 6.7 kHz (b) 6×10^{-4} W

5 (a) See Fig. Soln Prob. 16.5(a)

6 See Fig. Soln Prob. 16.6

7 (a) 6.76×10^{-4} (b) 115 μs at 24 m s^{-1}, 1.5 μs at 0.3 m s^{-1}

8 (a) 5.8 mW (b) 4.96 mW (c) $5.8\ e^{-0.01n}(0.93)^{2n}$ mW

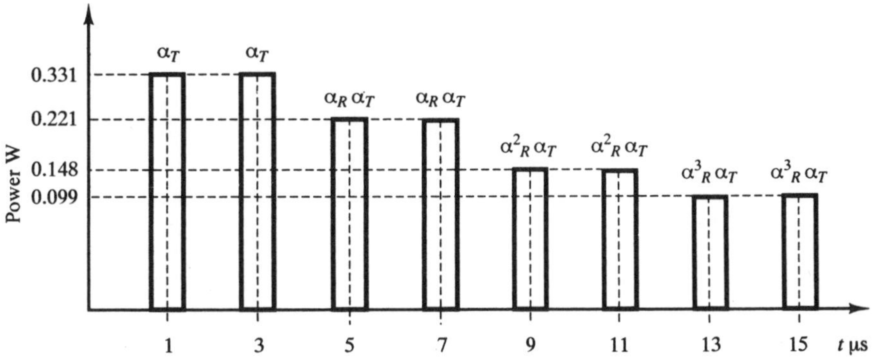

Fig. Soln Prob. 16.5

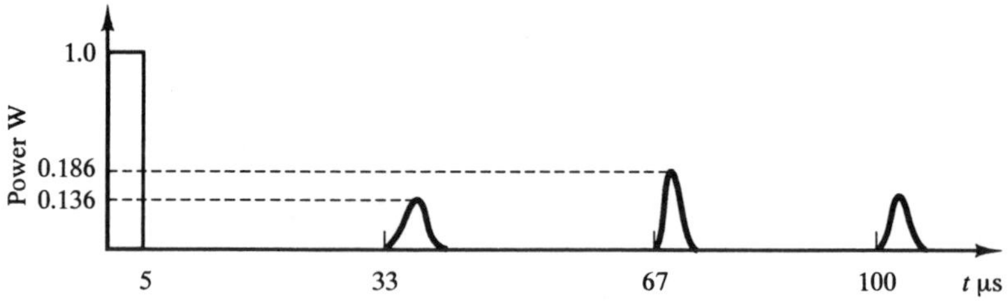

Fig. Soln Prob. 16.6

Chapter 17

1 (a) $2.63 \times 10^{-2}\,\text{ms}^{-1}$, 2.26 (b) 6 sec, 7 sec, 1.54
(c) average $N \approx 5400$, $1.85 \times 10^{-4}\,\text{m}$ (d) 18% O_2, 82% N_2

Chapter 18

1 (a) 15 samples s^{-1} for example (b) 240 samples s^{-1} (c) 4.17 ms (d) ±0.0122%
(e) 2880 bauds, 0 to 1440 Hz (f) 0.54 V

2 (a) 0.333 (b) 0.222 (c) 1.39×10^{-2} (d) 1.39×10^{-2} (e) 0.167

3 280 to 920 Hz

4 (a) 0.309, 0.683, 0.067 (b) 10

Index